Lecture Notes in Computer Science 2964

Edited by G. Goos, J. Hartmanis, and J. van Leeuwen

T0255312

Springer

Berlin
Heidelberg
New York
Hong Kong
London
Milan
Paris
Tokyo

Tatsuaki Okamoto (Ed.)

Topics in Cryptology – CT-RSA 2004

The Cryptographers' Track at the RSA Conference 2004
San Francisco, CA, USA, February 23-27, 2004
Proceedings

 Springer

Series Editors

Gerhard Goos, Karlsruhe University, Germany
Juris Hartmanis, Cornell University, NY, USA
Jan van Leeuwen, Utrecht University, The Netherlands

Volume Editor

Tatsuaki Okamoto
NTT Labs
Room 612A, 1-1 Hikarinooka, Yokosuka-shi, 239-0847 Japan
E-mail: okamoto@isl.ntt.co.jp

Cataloging-in-Publication Data applied for

A catalog record for this book is available from the Library of Congress.

Bibliographic information published by Die Deutsche Bibliothek
Die Deutsche Bibliothek lists this publication in the Deutsche Nationalbibliografie;
detailed bibliographic data is available in the Internet at <http://dnb.ddb.de>.

CR Subject Classification (1998): E.3, G.2.1, D.4.6, K.6.5, K.4.4, F.2.1-2, C.2, J.1

ISSN 0302-9743
ISBN 3-540-20996-4 Springer-Verlag Berlin Heidelberg New York

Springer-Verlag is a part of Springer Science+Business Media

springeronline.com

© Springer-Verlag Berlin Heidelberg 2004
Printed in Germany

Typesetting: Camera-ready by author, data conversion by PTP-Berlin, Protago-TeX-Production GmbH
Printed on acid-free paper   SPIN: 10986998   06/3142   5 4 3 2 1 0

Preface

The Cryptographers' Track (CT-RSA) is a research conference within the RSA conference, the largest, regularly staged computer security event. CT-RSA 2004 was the fourth year of the Cryptographers' Track, and it is now an established venue for presenting practical research results related to cryptography and data security.

The conference received 77 submissions, and the program committee selected 28 of these for presentation. The program committee worked very hard to evaluate the papers with respect to quality, originality, and relevance to cryptography. Each paper was reviewed by at least three program committee members. Extended abstracts of the revised versions of these papers are in these proceedings. The program also included two invited lectures by Dan Boneh and Silvio Micali.

I am extremely grateful to the program committee members for their enormous investment of time and effort in the difficult and delicate process of review and selection. Many of them attended the program committee meeting during the Crypto 2003 conference at the University of California, Santa Barbara.

I gratefully acknowledge the help of a large number of colleagues who reviewed submissions in their area of expertise: Masayuki Abe, Toru Akishita, Kazumaro Aoki, Gildas Avoine, Joonsang Baek, Harald Baier, Alex Biryukov, Dario Catalano, Xiaofeng Chen, Benoit Chevallier-Mames, J.S. Coron, Christophe De Cannière, Alex Dent, J.-F. Dhem, Matthias Fitzi, Marc Fossorier, Steven Galbraith, Pierrick Gaudry, Craig Gentry, Shai Halevi, Helena Handschuh, Javier Herranz Sotoca, Doi Hiroshi, Thomas Holenstein, Tetsu Iwata, Tetsuya Izu, Miodrag J. Mihaljevic, Jacques J.A. Fournier, Markus Jakobsson, Dominic Jost, Pascal Junod, Naoki Kanayama, Hiroki Koga, Yuichi Komano, Hugo Krawczyk, Dennis Kuegler, Noboru Kunihiro, Eyal Kushilevitz, Yi Lu, Christoph Ludwig, Philip MacKenzie, Keith Martin, Kazuto Matsuo, Jean Monnerat, Shiho Moriai, Christophe Mourtel, Sean Murphy, David Naccache, Koh-Ichi Nagao, Anderson Nascimento, Wakaha Ogata, Kenji Ohkuma, Satomi Okazaki, Elisabeth Oswald, Daniel Page, Kenny Paterson, Krzysztof Pietrzak, Zulfikar Ramzan, Renato Renner, Taiichi Saito, Ryuichi Sakai, Kouichi Sakurai, Arthur Schmidt, Katja Schmidt-Samoa, Junji Shikata, Atsushi Shimbo, Johan Sjödin, Ron Steinfeld, Makoto Sugita, Masahiko Takenaka, Jin Tamura, Bogdan Warinschi, Kai Wirt, Xun Yi, and Rui Zhang.

Electronic submissions were made possible by the Web Review system of K.U. Leuven. I would like to thank Bart Preneel for his kind support. Special thanks to Thomas Herlea, who greatly supported us by operating the Web Review system customized for CT-RSA 2004.

In addition, I would like to thank Mami Yamaguchi for her support in the review process and in editing these proceedings.

I am specially grateful to Burt Kaliski and Ari Juels of RSA Laboratories for interfacing with the RSA conference.

I wish to thank all the authors, who by submitting papers made this conference possible, and the authors of accepted papers for their cooperation.

December 2003   Tatsuaki Okamoto
                                                          Program Chair
                                                          CT-RSA 2004

RSA Cryptographers' Track 2004

February 23–27, 2004, San Francisco, CA, USA

The RSA Conference 2004 was organized by RSA Security Inc. and its partner organizations around the world. The Cryptographers' Track was organized by RSA Laboratories.

Program Chair

Tatsuaki Okamoto, NTT Labs, Japan

Program Committee

Junhui Chao .... Chuo U., Japan
Ronald Cramer .... Aarhus U., Denmark
Alex Dent .... Royal Holloway, UK
Anand Desai .... NTT MCL, USA
Rosario Gennaro .... IBM Research, USA
Goichiro Hanaoka .... U. of Tokyo, Japan
Martin Hirt .... ETH Zurich, Switzerland
Kwangjo Kim .... ICU, Korea
Mitsuru Matsui .... Mitsubishi Electric, Japan
Phong Nguyen .... ENS, France
Kazuo Ohta .... UEC, Japan
Pascal Paillier .... Gemplus, France
David Pointcheval .... ENS, France
Bart Preneel .... K.U. Leuven, Belgium
Jean-Jacques Quisquater .... UCL, Belgium
Tsuyoshi Takagi .... TU Darmstadt, Germany
Serge Vaudenay .... EPF Lausanne, Switzerland
Chung-Huang Yang .... NKNU, Taiwan
Moti Yung .... Columbia U., USA
Yuliang Zheng .... UNC Charlotte, USA

Steering Committee

Marc Joye .... Gemplus, France
Burt Kaliski .... RSA Lab, USA
Bart Preneel .... K.U. Leuven, Belgium
Ron Rivest .... MIT, USA
Moti Yung .... Columbia U., USA

Table of Contents

Side-Channel Attacks

Hardwares

Mode of Operations

Hash and Hash Chains

Visual Cryptography

Ellictic Curve Cryptosystems

Author Index

Online Encryption Schemes: New Security Notions and Constructions

Alexandra Boldyreva and Nut Taesombut

Dept. of Computer Science & Engineering,
University of California at San Diego, 9500 Gilman Drive,
La Jolla, California 92093, USA.
{aboldyre,ntaesomb}@cs.ucsd.edu
http://www-cse.ucsd.edu/users/{aboldyre,ntaesomb}

Abstract. We investigate new strong security notions for *on-line* symmetric encryption schemes, which are the schemes whose encryption and decryption algorithms operate "on-the-fly" and in one pass, namely can compute and return an output block given only the key, the current input block and the previous input and output blocks. We define the strongest achievable notion of privacy which takes into account both chosen-ciphertext attacks and the recently introduced blockwise-adaptive [15, 12] attacks. We show that all the schemes shown to be secure against blockwise-adaptive chosen-plaintext attacks are subject to blockwise-adaptive chosen-ciphertext attacks. We present an on-line encryption scheme which is provably secure under our notion. It uses any strong on-line cipher, the primitive introduced in [1]. We finally discuss the notion of authenticated on-line schemes and provide a secure construction.

1 Introduction

ONLINE ENCRYPTION SCHEMES. In this work we investigate strong security notions for on-line symmetric encryption schemes and analyze various constructions.

The on-line property of a function requires that each output block depend only on the key, the previous input and output blocks and the current input block. We say that an encryption scheme is on-line if for each key its encryption and decryption algorithms are *both* on-line[1]. We provide a more formal definition of on-line encryption schemes in Section 3. The on-line property allows schemes to be used in applications where encryption and decryption should perform "on-the-fly", in one pass, i.e. an output block should be returned as soon as the next input block is received. Example situations include the use of encryption modes in SSH protocol or when implemented in tamper-proof devices with low memory.

---

[1] Note that in the literature on-line encryption schemes require *only* the encryption algorithm to be on-line.

T. Okamoto (Ed.): CT-RSA 2004, LNCS 2964, pp. 1–14, 2004.

Most of the existing symmetric encryption schemes, e.g. CBC, CTR modes, are on-line [2].

PRIVACY OF ON-LINE SCHEMES. In the model for the standard notion of privacy (indistinguishability) against chosen-plaintext attacks (IND-CPA) [2] an adversary runs in two stages. In both stages it is given a symmetric encryption oracle. During the first stage the adversary has to output two equal-length messages. At the last stage it gets a challenge ciphertext, which is an encryption of one of the two messages. It wins if it can distinguish which message was encrypted.

As has been recently observed by Joux, Martinet and Valette in [15], the standard notion for IND-CPA security treats encryption and decryption as "atomic" operations and, therefore, is not strong enough for on-line encryption schemes, whose algorithms can be executed in the on-line ("non-atomic") manner[3]. For the same reason the existing proofs of security done in the standard model cannot guarantee security of an on-line scheme with on-line execution. To model the privacy notion of schemes with on-line encryption, [15] takes into account so-called "blockwise-adaptive attacks", when an adversary is allowed to generate two equal-length messages block-by-block, each time seeing the corresponding portion of the challenge ciphertext and being able to "adapt" the next blocks of the messages as a function of currently received ciphertext blocks.

The authors noted that blockwise-adaptive attacks do not seem to apply for fully parallelizable "not chained" encryption schemes such as CTR mode but showed that some popular chained encryption schemes, including CBC, are in fact insecure against such attacks. The notion of security of on-line encryption schemes against blockwise-adaptive chosen-plaintext attacks (IND-BLK-CPA) has been defined more formally by Fouque, Martinet and Poupard in [12]. They prove that CFB encryption mode [18] and DCBC, the modification to CBC scheme, are IND-BLK-CPA.

But nowadays preserving privacy only against chosen-plaintext attacks is not considered sufficient; it is desirable to achieve stronger notions of privacy, such as security against chosen-ciphertext attacks. The standard notion of privacy against chosen-ciphertext attacks (IND-CCA) is similar to the one against chosen-plaintext attacks (IND-CPA), except that the adversary is also given a decryption oracle with a restriction of not querying it on a challenge ciphertext. IND-CCA notion seems insufficient for on-line schemes since it does not allow blockwise-adaptive attacks. Security of on-line schemes against blockwise-adaptive chosen-ciphertext attacks has not been previously investigated. This is the focus of our work.

---

[2] These modes can also be executed in non on-line manner, outputting the whole plaintext (resp. ciphertext) only after the encryption (resp. decryption) algorithm completes.

[3] If an on-line scheme is implemented only for "atomic" (not on-line execution), then the standard notions of security are appropriate. In this paper when we refer to an on-line scheme we always assume the possibility of "non-atomic" execution of both encryption and decryption algorithms of the scheme (when the output blocks are returned on-the-fly.)

RELATED WORK. Very recently in their independent work Fouque et al. [11] investigated the notions of secure authenticated encryption for applications using low memory highly-secure devices such as smart cards together with insecure but powerful host devices. The applications have been topics of previous research known as "remotely-keyed encryption" (RKE) and "remotely-keyed authenticated encryption" (RKAE). The works in these areas include [6,17,7,14, 9]. The neat scheme proposed in [11] allows a smart card to partially decrypt a ciphertext produced by any secure authenticated encryption scheme (most of which have on-line encryption algorithms) in on-line manner while preserving privacy and authenticity. However, the host needs to perform a second pass over the data to complete the decryption (this phase is keyless). Hence, their scheme (as well as the schemes of [6,17,7,14,9]) does not fall under our definition of an on-line scheme, since their decryption algorithm is not done in one-pass.

The authors of [11] attempt to define the notion of privacy against blockwise-adaptive chosen-ciphertext attacks. While their notion is appropriate for schemes with on-line encryption, it is unachievable for schemes with on-line decryption, as we show below. Therefore, a new security notion is needed for analysis of on-line encryption schemes beyond chosen-plaintext attacks.

TOWARDS A STRONGER NOTION OF PRIVACY FOR ON-LINE SCHEMES. As we mentioned above we seek for an appropriate notion of privacy under blockwise-adaptive chosen ciphertext attacks for on-line encryption schemes and secure constructions. It turns out that finding the answers for these interesting theoretical and practically important questions requires some extra care.

For example, the approach taken by Fouque et al. [11] was to strengthen the IND-BLK-CPA notion defined by [15,12] by giving the adversary the particular decryption oracle which operates block-by-block, namely, takes ciphertext blocks and returns the corresponding plaintext blocks "on-the-fly". Indeed, this is the first definition that comes to mind. However, we point out that giving such a decryption oracle is equivalent to giving the adversary the standard decryption oracle. This can be realized by noting that unlike encryption, the decryption algorithm is always deterministic and any "blockwise-adaptive" queries $C[1], C[2], \ldots$ to the particular decryption oracle can be replaced by queries $C[1], C[1]\|C[2], \ldots$ to the standard decryption oracle.

Moreover, we claim that extending the IND-BLK-CPA notion with chosen-ciphertext attacks by giving the adversary the decryption oracle (which is equivalent to the notion proposed in [11]) is too strong for analyzing on-line encryption schemes. We show that by noting that no on-line encryption scheme can be secure even under the standard IND-CCA notion. We justify this claim in Section 3 by presenting a simple adversary which does not use any blockwise-adaptive attacks. Therefore, the standard IND-CCA notion and, moreover, its straightforward extension to blockwise-adaptive attacks are unachievable for on-line schemes.

We seek a strongest *achievable* privacy notion for on-line encryption schemes. In Section 3 we provide such a notion. The model takes into account both blockwise-adaptive chosen-plaintext attacks using the ideas from [15,12] and a class of chosen-ciphertext attacks by giving the adversary the *special* decryption oracle. The queries to this special decryption oracle are answered in such a way that the notion is achievable. Please refer to Definition 1 for the details of this security notion, which we call IND-BLK-CCA. One may argue that our notion does not exactly capture the whole class of chosen-ciphertext attacks. This is true, but one should realize that any scheme with on-line encryption and decryption execution is *always* subject to some chosen-ciphertext attacks as we claimed above and therefore the designers and implementors of on-line encryption schemes should not target for IND-CCA security. This does not mean, however, that IND-BLK-CPA security will be sufficient. Our notion captures the best possible security for schemes with on-line execution.

ATTACKS. In Section 4 we show that the schemes proved to resist blockwise-adaptive chosen-plaintext attacks (shown in [12] to be IND-BLK-CPA secure) are subject to blockwise-adaptive chosen-ciphertext attacks. Namely, we show that CFB and DCBC encryption modes are not IND-BLK-CCA secure. These attacks are very simple and are similar to the well-known IND-CCA attacks on these schemes. The result is not surprising since these schemes were not designed to resist such powerful attacks.

IND-BLK-CCA SECURE CONSTRUCTION. If one does not have a goal of having an on-line symmetric encryption scheme, it is easy to achieve the IND-CCA security. Having an IND-CPA secure symmetric encryption scheme $\mathcal{SE}$, e.g. CBC, and a message authentication code F strongly unforgeable against chosen-message attack (SUF-CMA secure), e.g. XCBC [5] or OMAC [13][4], one can obtain an IND-CCA secure scheme via "encrypt-then-MAC" paradigm [4]. Namely, in order to encrypt a message, first use the encryption algorithm of $\mathcal{SE}$ to get a ciphertext C and then apply the tag generation algorithm of F to C to get the tag τ. The resulting ciphertext is $C\|\tau$. In order to decrypt a ciphertext C', parse it as $C''\|\tau''$, verify the tag τ'', using the verification algorithm of F, and if it valid, decrypt C'' using the decryption algorithm of $\mathcal{SE}$ and return the result, otherwise, return a special symbol $\perp$. However, this "encrypt-then-MAC" paradigm does not allow to preserve the on-line property of the given encryption scheme. More precisely, the on-line property of decryption becomes violated: the resulting decryption algorithm needs the whole ciphertext to verify the tag. The same problem holds for specific authenticated encryption schemes such as, for example, OCB authenticated scheme [19], IACBC, IAPM [16] and XCBC [10], which are not constructed using the generic "encrypt-then-MAC" paradigm. They provide both privacy (IND-CPA security) and authenticity (INT-CTXT security) and, hence IND-CCA security [4] [5]. However, their decryption algorithm is not on-line, before being able to output the message, it first needs to verify the checksum, which depends on all the blocks of the message.

---

[4] These MACs are shown to be weakly unforgeable (WUF-CMA secure), however, for stateless deterministic MACs these two notions are equivalent.

[5] It is shown in [4] that if an encryption scheme is secure against chosen-plaintext attacks (IND-CPA secure) and provides integrity of a ciphertext (INT-CTXT secure) then it is also secure against chosen-ciphertext attacks (IND-CCA secure).

We look for an on-line encryption scheme provably secure against blockwise-adaptive chosen-ciphertext attacks. As a building block we use on-line ciphers, the notion introduced and analyzed by Bellare, Boldyreva, Namprempre and Knudsen in [1]. On-line ciphers are on-line, length-preserving permutations. Secure on-line ciphers are the ones which closely approximate truly random on-line length-preserving permutations. The authors [1] formally define the security notion for on-line ciphers against chosen-ciphertext attacks and present the construction called HPCBC, which is secure against chosen-ciphertext attacks assuming the underlying block cipher and the hash function are secure (the construction uses one block cipher and one keyed hash function invocation per input block). On-line ciphers are deterministic permutations and therefore are not even IND-CPA secure. The authors discuss how on-line ciphers can be used to obtain IND-CPA and INT-CTXT secure schemes. As we discussed above and as independently noted in [11], this does not immediately guarantee IND-BLK-CCA security.

In Section 5 we present our second main result. We give an on-line encryption scheme which is constructed using any on-line cipher. We prove it IND-BLK-CCA secure assuming the underlying cipher is a strong pseudorandom on-line permutation. Theorem 1 states the concrete security result. We next present a specific scheme based on HPCBC on-line cipher.

AUTHENTICATED ON-LINE ENCRYPTION SCHEMES. As we discussed above, all known schemes that provide authenticity (INT-CTXT security) are not on-line. Moreover, the standard notion of authenticity INT-CTXT is not appropriate to analyze on-line encryption schemes because it assumes that decryption is an atomic operation. In the full version of this paper [8] we define an appropriate notion of authenticity for on-line encryption schemes and provide an encryption scheme which simultaneously provides authenticity and IND-BLK-CCA security.

2   Preliminaries

NOTATION. A *string* is a member of $\{0,1\}^*$. Let $|X|$ denote the length of a string X. For strings X, Y let $X\|Y$ denote their concatenation. The empty string is denoted ε. If $d \geq 1, n \geq 1$, are integers, then $D_{d,n}$ denotes the set of all strings whose length is a positive multiple of n bits and at most dn bits (we borrow some notation from [1] for convenience). If $X \in D_{d,n}$, then $X[i]$ denotes its ith block, meaning $X = X[1]\| \ldots \|X[l]$ where $l = |X|/n$ and $|X[i]| = n$ for all $i = 1, \ldots, l$. We will mostly consider functions with inputs and outputs in $D_{d,n}$, hence both can be viewed as sequences of n-bit-long blocks.

CIPHERS AND ON-LINE CIPHERS. Informally, a cipher is a keyed deterministic length-preserving permutation. We recall the formal definitions presented in [1]. Let F: $Keys(F) \times Dom(F) \rightarrow Rng(F)$ be a function family, where $Keys(F)$ is the *key space* of F; $Dom(F)$ is the *domain* of F; and $Rng(F)$ is the *range* of F. F is a *cipher* if $Dom(F) = Rng(F)$ and each instance $F(K, \cdot)$ of F is a length-preserving permutation. If F is a cipher, then F^{-1} is the *inverse cipher*, defined by $F^{-1}(K, x) = F(K, \cdot)^{-1}(x)$ for all $K \in Keys(F)$ and $x \in Dom(F)$.

A function $f\colon D_{d,n} \to D_{d,n}$ is n-on-line if the first block of the output depends only on the first block of the input and the i-th block of the output (for $i \geq 2$) is determined completely by the $(i-1)$-th block of the input, $(i-1)$-th block of the output and the i-th block of the input[6]. A cipher is n-on-line if each instance is an on-line function. It is shown in [1] that the inverse of an on-line cipher is itself n-on-line. For convenience we will often omit n in the "n-on-line" term.

SECURITY OF ON-LINE CIPHERS. Let $\mathsf{OPerm}_{d,n}$ denote the family of all n-on-line, length-preserving permutations on $D_{d,n}$. A "secure" on-line cipher is one that closely approximates $\mathsf{OPerm}_{d,n}$. Let $F\colon Keys(F) \times D_{d,n} \to D_{d,n}$ be a family of functions with domain and range $D_{d,n}$. Let A be a distinguisher with access to two oracles.

We define the advantage of A attacking an on-line, pseudorandom permutation (OPRP) F chosen-ciphertext attacks as

$$\mathbf{Adv}_{F,A}^{\text{oprp-cca}} = \Pr\left[g \xleftarrow{R} F \ : \ A^{g,g^{-1}} = 1 \right] - \Pr\left[g \xleftarrow{R} \mathsf{OPerm}_{d,n} \ : \ A^{g,g^{-1}} = 1 \right].$$

This captures the *advantage* of the distinguisher in the task of distinguishing a random instance of F from a random, length-preserving, n-on-line permutation on $D_{d,n}$. The distinguisher can query the challenge instance and its inverse. For any $t, q_e, \mu_e, q_d, \mu_d$ we define the *advantage* of F as OPRP against chosen-ciphertext attacks as

$$\mathbf{Adv}_F^{\text{oprp-cca}}(t, q_e, \mu_e, q_d, \mu_d) = \max_A \left\{ \mathbf{Adv}_{F,A}^{\text{oprp-cca}} \right\},$$

where the maximum is over all A, having the time complexity t, making at most q_e queries to the challenge instance oracle, whose total length is at most μ_e, and making at most q_d queries to the challenge inverse oracle, whose total length is at most μ_d. An on-line cipher F is secure against chosen-ciphertext attacks (is a *strong* cipher) if the function $\mathbf{Adv}_F^{\text{oprp-cca}}()$ grows "slowly".

SYMMETRIC ENCRYPTION SCHEMES AND THEIR SECURITY. A symmetric encryption scheme $\mathcal{SE} = (\mathcal{K}, \mathcal{E}, \mathcal{D})$ consists of three algorithms. The randomized key generation algorithm $\mathcal{K}$ returns a key K; we write $K \xleftarrow{R} \mathcal{K}$. Associated with each encryption scheme there is a *message space* MsgSp, from which messages are drawn. In our context it will be important to make explicit the random coins underlying the randomized encryption algorithm $\mathcal{E}$. On input a key K, a plaintext $M \in$ MsgSp, and coin tosses R the randomized *encryption* algorithm $\mathcal{E}$ returns the ciphertext C; we write $C \leftarrow \mathcal{E}_K(M; R)$. The notation $C \xleftarrow{R} \mathcal{E}_K(M)$ is a shorthand for $r \xleftarrow{R} \mathsf{Coins}_{\mathcal{E}}\, ;\, C \leftarrow \mathcal{E}_K(M; R)$, where $\mathsf{Coins}_{\mathcal{E}}$ is the set from which randomness is drawn. The deterministic decryption algorithm $\mathcal{D}$ takes K and a ciphertext C and returns the message M or a symbol $\perp$; we write $M \leftarrow \mathcal{D}_K(C)$ or $\perp \leftarrow \mathcal{D}_K(C)$. We require that $\mathcal{D}_K(\mathcal{E}_K(M)) = M$ for all $M \in$ MsgSp.

An encryption scheme is said to provide privacy against chosen-plaintext attacks (be IND-CPA secure) if no adversary with reasonable resources can win

---

[6] For convenience we use a slightly stronger definition, than the one proposed in [1]. All the results and constructions of [1] satisfy our definition as well.

with non-negligible probability the game defined in the following experiment. An adversary runs in two stages. In both stages it is given a symmetric encryption oracle. During the first stage the adversary has to output two equal-length messages. One of the messages is chosen at random. At the last stage the adversary gets a challenge ciphertext, which is an encryption of this message. The adversary wins if it can distinguish which message was encrypted.

Seurity against chosen-plaintext attacks (IND-CCA security) is defined similarly, except the adversary in both stages is given a decryption oracle with a restriction of not querying it during the last stage on the challenge ciphertext.

3   Online Encryption Schemes and Their Security

Let d, n be integers. Let $\mathcal{SE} = (\mathcal{K}, \mathcal{E}, \mathcal{D})$ be a symmetric encryption scheme with $\mathrm{MsgSp} = D_{d,n}$. We will refer to n as the *block-length* of a message. We say that $\mathcal{SE}$ is a scheme with *n-on-line encryption* if for every key K, coins R and message M, $C[1]$ (where $C = \mathcal{E}_K(M; R)$) depends only on K and R; and $C[i]$ (for $i \geq 2$) depends only on $M[i-2], M[i-1], C[i-1], K$ and possibly, R[7]. Similarly, we say that $\mathcal{SE}$ is a scheme with *n-on-line decryption* if for every key K and ciphertext C, $M[i]$ (for $i \geq 1$) depends only on $C[i], C[i+1], K$ and possibly R. In this work we are interested in symmetric encryption schemes with n-on-line encryption and decryption. We call such a subclass *n-on-line* symmetric encryption schemes. It is easy to see that most of the popular block-cipher based encryption schemes (which are also called encryption modes) are on-line encryption schemes, e.g. CBC and CTR modes.

PRIVACY OF ON-LINE ENCRYPTION SCHEMES. We investigate stronger notions of security for on-line encryption schemes. As we mentioned in Section 1 we look for a notion of privacy which takes into account both blockwise-adaptive attacks and chosen-ciphertext attacks.

But let us first claim that even the standard notion of privacy against chosen-ciphertext attacks is unachievable for on-line encryption schemes, meaning no on-line encryption scheme can be IND-CCA. We prove the claim by presenting a simple adversary attacking a given on-line encryption scheme. It does not make use of any blockwise adaptive attacks, but simply employs the on-line property of the decryption algorithm. The adversary is given the decryption oracle. First, the adversary generates any two two-block messages with distinct first blocks. Given the challenge ciphertext, which is an encryption of one of the two messages, it removes its last block and queries the result to its decryption oracle. The query is different from the challenge and thus is being legitimate. Since the first block of the corresponding plaintext does not depend on the removed last block of the ciphertext due to the on-line property of decryption, the adversary will get back the first block of one of the messages it generated at the previous stage and hence can always win its game.

---

[7] For an example of on-line encryption scheme, see Figure 1 in Section 5.2. The definition can be easily modified to include nonces, counters, etc. to satisfy any intuitively on-line scheme, e.g. OCB.

The same attack would obviously apply when we enhance the IND-CCA notion with blockwise-adaptive attacks. We want to define the strongest achievable notion of privacy for on-line schemes. Let us start with some intuition.

As we discussed in the Introduction and above, in the applications which require on-line encryption schemes the standard notion of privacy against chosen-ciphertext attacks cannot be achieved. This means that there exist a particular type of chosen-ciphertext attacks which an on-line encryption scheme cannot resist, namely the ones where the chosen ciphertext has the same prefix as the challenge ciphertext. This does not mean, however, that the practitioners should not expect an on-line encryption scheme to resist various other chosen-ciphertext attacks. For example, assume an adversary wants to decrypt a ciphertext of some important message. Suppose that it flips a bit in each block of the ciphertext, makes the receiver decrypt the result and manages to see the corresponding message. It is highly undesirable that the adversary be able to infer some information on the original important message. This kind of attack is not captured by either of IND-CPA nor IND-BLK-CPA notions and as we show in the next section, even the schemes known to be IND-BLK-CPA secure are subject to this type of attacks. But there exist on-line encryption schemes which resist such attacks as well as blockwise-adaptive attacks. Therefore, such schemes would be desirable in on-line applications with stronger security requirements. Accordingly we devise a notion of security which captures all blockwise-adaptive chosen-plaintext attacks and in addition captures these type of chosen-ciphertext attacks.

The intuition behind the notion is as follows. As in IND-BLK-CPA notion the adversary gets an encryption oracle and outputs two equal-length messages block by block, each time receiving the new portion of the challenge ciphertext, which is an encryption of one of the messages. This captures blockwise-adaptive attacks since the adversary can adapt each next message-block pair depending on the previously received portions of the ciphertext. To capture the aforementioned class of chosen-ciphertext attacks we allow the adversary to obtain decryptions of ciphertexts of its choice unless the ciphertext has a common prefix with a challenge ciphertext. This captures the attack discussed above. If, however, there is a common prefix, say l first blocks, the adversary can only see the decrypted message starting from the $l + 1$ block. We now define the notion in detail.

We use the ideas for the model of security against blockwise-adaptive chosen-plaintext attacks [15,12] and will extend this model to capture on-line chosen-ciphertext attacks. An adversary A runs in several stages. In all stages A is given the encryption oracle and the *on-line decryption oracle*, which we define below. In the first stage find$_1$ the adversary outputs two messages $M_0[1], M_1[1]$ of a block-length (we will refer to such messages as to message-blocks), a state information s it wants to preserve and a flag *done* indicating whether these are the last message-blocks it wants to output (in which case *done* = 1). A challenge bit b and coin tosses R are chosen at random. In the next stage find$_2$ an encryption of $M_b[1]$ is computed using R and given to A along with s. A can continue outputting more message-block pairs $M_0[i], M_1[i]$ for $i = 2, \ldots$ which can be a function of previously seen information, and updating s. Each time A receives a current challenge ciphertext C which is an encryption of $M_b[1]\|\ldots\|M_b[i]$ using coins R. This process continues until A sets $done = 1$. At the last guess stage A is given the whole challenge ciphertext and s and has to guess the challenge bit b. Here is the formal definition.

Definition 1. *Fix $d, n \in N$. Let $\mathcal{OSE} = (\mathcal{K}, \mathcal{E}, \mathcal{D})$ be an on-line symmetric encryption scheme with $\mathrm{MsgSp} = D_{d,n}$. Let A be an adversary that has access to two oracles. For $b \in \{0, 1\}$ define the following experiments:*

Experiment $\mathbf{Exp}^{\text{ind-blk-cca-}b}_{\mathcal{OSE},A}$
    $K \xleftarrow{R} \mathcal{K}$; $R \xleftarrow{R} \mathsf{Coins}_{\mathcal{E}}$
    $i \leftarrow 1$; $C \leftarrow \varepsilon$; $s \leftarrow \varepsilon$
    Repeat
       $(M_0[i], M_1[i], s, done) \leftarrow A^{\mathcal{E}_K(\cdot),\mathcal{OD}_K(\cdot)}(\mathsf{find}_i, C, s)$
       $C \leftarrow \mathcal{E}_K(M_b[1]\|M_b[2]\|\ldots\|M_b[i]; R)$
       $i \leftarrow i + 1$
    Until done = 1
    $\bar{b} \leftarrow A^{\mathcal{E}_K(\cdot),\mathcal{OD}_K(\cdot)}(\mathsf{guess}, C, s)$
    Return $\bar{b}$

It is mandated that A outputs two message blocks of block-length n in each find stage and the total number of find stages is less or equal to d. The on-line decryption oracle $\mathcal{OD}_K(\cdot)$ takes inputs of length ln for some $l \in N$. In stage find_1 on input C' it returns $M = \mathcal{D}_K(C')$. At next stages it returns $M[k]\|M[k+1]\|\ldots\|M[l]$, where $M = \mathcal{D}_K(C')$ and $1 \le k \le l$ is the smallest index such that $C'[k] \ne C[k]$ (where C is the challenge ciphertext given to A at this stage).

We define the *advantage* of the adversary A as follows:

$$\mathbf{Adv}^{\text{ind-blk-cca}}_{\mathcal{OSE},A} = \Pr[\mathbf{Exp}^{\text{ind-blk-cca-0}}_{\mathcal{OSE},A} = 0] - \Pr[\mathbf{Exp}^{\text{ind-blk-cca-1}}_{\mathcal{OSE},A} = 0].$$

For any $t, \mu_t, q_e, \mu_e, q_d, \mu_d$ we define the *advantage* of $\mathcal{OSE}$ via

$$\mathbf{Adv}^{\text{ind-blk-cca}}_{\mathcal{OSE}}(t, \mu_t, q_e, \mu_e, q_d, \mu_d) = \max_A \left\{ \mathbf{Adv}^{\text{ind-blk-cca}}_{\mathcal{OSE},A} \right\},$$

where the maximum is over all A having time-complexity t, outputting message blocks during all find stages of total length at most μ_t in all find stages, making at most q_e queries to the encryption oracle, whose total length is at most μ_e, and making at most q_d queries to the on-line decryption oracle, whose total length is at most μ_d.

We say that on-line symmetric encryption scheme $\mathcal{OSE}$ is *secure against blockwise-adaptive chosen-ciphertext attacks* (or simply IND-BLK-CCA secure) if the function $\mathbf{Adv}^{\text{ind-blk-cca}}_{\mathcal{OSE}}()$ grows slowly. □

Let us briefly recall why we do not explicitly allow the adversary to query its on-line decryption oracle in "blockwise-adaptive" manner. The reason is that we could do it, but this would not add any power to the adversary. This is because unlike encryption, the decryption algorithm is deterministic. Therefore, to see the decryption of a ciphertext $C = C[1]\|C[2]\|\ldots\|C[l]$ block-by-block, our adversary can simply query its decryption oracle on $C[1]$, $C[1]\|C[2]$, ..., $C[1]\|C[2]\|\ldots\|C[l]$.

4   Analyses of Several Online Encryption Schemes

It was suggested in [15] that the fully parallelizable encryption modes such as counter mode CTR are not subject to blockwise-adaptive chosen-plaintext attacks. It is not hard to see that CTR is subject to a straightforward blockwise-adaptive chosen-ciphertext attack. An adversary can just choose two distinct one-block messages, flip a bit in the challenge ciphertext, query the result to the on-line decryption oracle and it will get back the one of the two challenge messages with a flipped bit in it. Thus the adversary can always win its game.

In [12] the authors prove that $\mathcal{CFB}$ encryption scheme [18] and $\mathcal{DCBC}$, the modification of $\mathcal{CBC}$ encryption scheme, are on-line encryption schemes secure against chosen-plaintext blockwise-adaptive attacks. We show that these schemes are unfortunately insecure against blockwise-adaptive chosen-ciphertext attacks. This is consistent with the design goal of these schemes which did not include chosen-ciphertext attack resistance. The attacks are simple and are similar to the one on CTR mode. We define the schemes and present the attacks in [8].

In the next section we propose schemes which are IND-BLK-CCA secure.

5   Proposed Schemes and Their Security

5.1   Online-Cipher-Based Online Encryption Schemes

It is suggested (without a proof) in [15] that the encryption scheme based on HPCBC on-line cipher [1] does not seem to be vulnerable to blockwise-adaptive chosen-plaintext attacks. We generalize, strengthen and formally justify this suggestion. We propose an encryption scheme based on any strong on-line cipher and formally prove that it resists not only blockwise-adaptive chosen-plaintext attacks but also blockwise-adaptive chosen-ciphertext ones.

It is shown in [1] that the use of on-line ciphers provide security against standard chosen-plaintext attacks, if the plaintext space has appropriate properties, namely, if the first block of a plaintext is a random string. We show that any secure strong on-line cipher applied to messages with a random first block (or to any message with prepended random block) is also an on-line symmetric encryption scheme secure against blockwise-adaptive chosen-ciphertext attacks.

Construction 1. Let n, d be integers, and let $F\colon Keys(F) \times D_{d,n} \to D_{d,n}$ be an on-line cipher. We associate to them the following symmetric encryption scheme $\mathcal{OCSE} = (\mathcal{K}, \mathcal{E}, \mathcal{D})$:

Algorithm $\mathcal{K}$	Algorithm $\mathcal{E}_K(M)$	Algorithm $\mathcal{D}_K(C)$
$K \xleftarrow{R} Keys(F)$	$R \xleftarrow{R} \{0,1\}^n$	$X \leftarrow F^{-1}(K, C)$
Return K	$X \leftarrow R\|M$	Parse X as $R\|M$ with $\|R\| = n$
	$C \leftarrow F(K, X)$	Return M
	Return C	

Given the fact that F is an on-line cipher, it is easy to see that $\mathcal{OCSE}$ is on-line encryption scheme. We want to show that it provides IND-BLK-CCA security when F is an on-line cipher secure against chosen-ciphertext attacks (a strong on-line cipher). The following theorem states the result.

Theorem 1. *Let d, n be integers, and let F: $\mathrm{Keys}(F) \times D_{d,n} \to D_{d,n}$ be an n-on-line cipher. Let $\mathcal{OCSE} = (\mathcal{K}, \mathcal{E}, \mathcal{D})$ be the on-line symmetric encryption scheme associated to it as per Construction 1. Then*

$$\mathbf{Adv}_{\mathcal{OCSE}}^{\mathrm{ind\text{-}blk\text{-}cca}}(t, \mu_t, q_e, \mu_e, q_d, \mu_d) \leq 2 \cdot \mathbf{Adv}_F^{\mathrm{oprp\text{-}cca}}(t', q_e', \mu_e', q_d, \mu_d) + \frac{q_e + q_d}{2^{n-1}},$$

where

$$t' = t + O(q_e, q_d, \mu_t), q_e' = q_e + \frac{\mu_t}{2n}, \mu_e' = \mu_e + nq_e + \frac{\mu_t^2 + 6\mu_t}{8n} .$$

Proof. Let A be an adversary attacking the $\mathcal{OCSE}$ scheme against blockwise-adaptive chosen-ciphertext attacks. We construct a distinguisher B attacking the pseudorandomness of on-line cipher F against chosen-ciphertext attacks. B has access to two oracles: g, which is either a random instance of the on-line cipher F, or a random instance of $\mathsf{OPerm}_{d,n}$, and g^{-1}, the inverse of g. B runs A as a subroutine and has to answer its oracle queries and provide the necessary inputs. We present pseudocode for B followed by some explanations and the analysis.

Adversary $B^{g,g^{-1}}$
  $R \xleftarrow{R} \mathsf{Coins}_{\mathcal{E}}$; $b \xleftarrow{R} \{0,1\}$
  $i \leftarrow 1$; $C \leftarrow \varepsilon$; $s \leftarrow \varepsilon$
  Repeat
    Run A on input (find_i, C, s)
      When A outputs $(M_0[i], M_1[i], s, done)$, do
        $C \leftarrow g(R \| M_b[1] \| M_b[2] \| \dots \| M_b[i])$
        $i \leftarrow i + 1$
        Return (find_i, C, s) to A
      On A's encryption query M, do
        $R' \xleftarrow{R} \mathsf{Coins}_{\mathcal{E}}$
        $Y \leftarrow g(R' \| M)$
        Return Y to A
      On A's on-line decryption query C', do
        $X \leftarrow g^{-1}(C')$, parse X as $R'' \| M'$ where $|R''| = n$
        $l \leftarrow |M'|/n$
        Return $(M'[k] \| M'[k+1] \| \cdots \| M'[l])$ to A,
          where k is the smallest index such that $C'[k] \neq C[k]$
    Until $done = 1$
  Run A on input (guess, C, s) replying to its oracle queries as before
    until it halts and returns $\bar{b}$
  If $(\bar{b} = b)$ then return 1 else return 0 EndIf

B picks a random bit b and a random n-bit string R and runs the adversary A answering its oracle queries and providing all the necessary inputs for it. When A outputs two messages block-by-block during find stages, each time B returns the current challenge ciphertext. The challenge ciphertext is computed by prepending R to the message determined by the bit b, and querying it to B's own oracle g. Similarly, B uses its oracles g, g^{-1} to answer A's encryption and on-line decryption oracle queries. This is possible since encryption and decryption algorithms use only oracle access to $F(K, \cdot)$. A little extra care is required when answering A's on-linedecryption oracle queries. According to the experiment $\mathbf{Exp}_{\mathcal{OSE},A}^{\text{ind-blk-cca-}b}$ defined in Definition 1, B returns to A only the message blocks starting from position at which the query block is different from the latest challenge ciphertext given to A. Finally, when A outputs its guess at the end of guess stage, B compares it with the bit b, and if they are equal, B concludes that g is a random instance of on-line cipher and outputs 1, otherwise, it outputs 0 to indicate its guess that g is a random on-line permutation.

We now proceed to the analysis. First consider the case when g is a random instance of the on-line cipher F. We claim that A's view in the simulated experiment is exactly as in experiment $\mathbf{Exp}_{\mathcal{OSE},A}^{\text{ind-blk-cca-}b}$. By the construction B will return 1 if and only if the adversary A can guess the random bit b correctly. Thus we have:

$$\Pr\left[g \xleftarrow{R} F : B^{g,g^{-1}} = 1\right] = \frac{1}{2} \cdot \Pr[\mathbf{Exp}_{\mathcal{OCSE},A}^{\text{ind-blk-cca-}0} = 0]$$
$$+ \frac{1}{2} \cdot (1 - \Pr[\mathbf{Exp}_{\mathcal{OCSE},A}^{\text{ind-blk-cca-}1} = 0])$$
$$= \frac{1}{2} + \frac{1}{2} \cdot \mathbf{Adv}_{\mathcal{OCSE}}^{\text{ind-blk-cca}}(A). \tag{1}$$

Next consider the case when g is a random instance of $\mathsf{OPerm}_{d,n}$. We claim that

$$\Pr\left[g \xleftarrow{R} \mathsf{OPerm}_{d,n} : B^{g,g^{-1}} = 1\right] \leq \frac{1}{2} + \frac{q_e + q_d}{2^n}. \tag{2}$$

After subtracting Equation (1) and Equation (2) and taking maximum we get the statement of the theorem. Due to lack of space we provide the proof of Equation (2) and check the resources used by the adversary B in [8].   □

5.2   HPCBC Encryption Scheme

We present the $\mathcal{HPCBC}$ encryption scheme, based on the HPCBC on-line cipher proposed in [1]. According to Construction 1, the encryption algorithm applies the HPCBC on-line cipher to an input message prepended with a random block. Here are the details of the scheme.

Let $d, n \in \mathbb{N}$, let $E: Keys(E) \times \{0,1\}^n \to \{0,1\}^n$ be a block cipher family and let $H: Keys(H) \times \{0,1\}^{2n} \to \{0,1\}^n$ be a family of functions. The message space of $\mathcal{HPCBC}$ is $D_{d,n}$ and the scheme consists of the following three algorithms:

Algorithm $\mathcal{K}$	Algorithm $\mathcal{E}_K(M)$	Algorithm $\mathcal{D}_K(C)$

$K_1 \xleftarrow{R} Keys(E)$
 $K_2 \xleftarrow{R} Keys(H)$
 $K \leftarrow K_1 \| K_2$
 Return K
	Parse K into $K_1 \| K_2$
	 Parse M into n-bit blocks
	 $\quad M[1]\|\ldots\|M[l]$
	 $R \xleftarrow{R} \{0,1\}^n$; $M[0] \leftarrow R$
	 $T \leftarrow 0^{2n}$; $Y \leftarrow H(K_2, T)$
	 $P \leftarrow Y \oplus M[0]$
	 $C[1] \leftarrow E(K_1, P) \oplus Y$
	 For $i = 2, \ldots, l+1$ do
	 $\quad T \leftarrow M[i-2]\|C[i-1]$
	 $\quad Y \leftarrow H(K_2, T)$
	 $\quad P \leftarrow Y \oplus M[i-1]$
	 $\quad C[i] \leftarrow E(K_1, P) \oplus Y$
	 EndFor
	 $C \leftarrow C[1]\|\ldots\|C[l+1]$
	 Return C
		Parse K into $K_1 \| K_2$
		 Parse C into n-bit blocks
		 $\quad C[1]\|\ldots\|C[l+1]$
		 $T \leftarrow 0^{2n}$; $Y \leftarrow H(K_2, T)$
		 $P \leftarrow E^{-1}(K_1, (C[1] \oplus Y))$
		 $M[0] \leftarrow Y \oplus P$
		 For $i = 1, \ldots l$ do
		 $\quad T \leftarrow M[i-1]\|C[i]$
		 $\quad Y \leftarrow H(K_2, T)$
		 $\quad P \leftarrow E^{-1}(K_1, C[i+1] \oplus Y)$
		 $\quad M[i] \leftarrow Y \oplus P$
		 EndFor
		 $M \leftarrow M[1]\|\ldots\|M[l]$
		 Return M

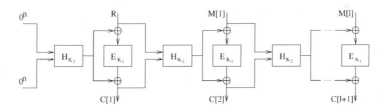

Fig. 1. The HPCBC encryption scheme

The $\mathcal{HPCBC}$ encryption scheme is depicted in Figure 1, from which it is easy to see that the scheme is on-line. The scheme makes one use of a keyed hash function in addition to that of a block cipher per each input block. HPCBC cipher has been proved to be a strong on-line cipher (secure against chosen-ciphertext attacks) [1] assuming E is a strong PRP family and H is almost-xor-universal function family. Hence, it follows from Theorem 1 that $\mathcal{HPCBC}$ is IND-BLK-CCA secure. The concrete security result can be easily obtained from the result of [1] and from Theorem 1.

Acknowledgements. We thank Mihir Bellare and Bogdan Warinschi for helpful discussions and the anonymous referees for their comments. Alexandra Boldyreva is supported in part by NSF Grant CCR-0098123 and NSF Grant ANR-0129617.

References

1. M. BELLARE, A. BOLDYREVA, L. KNUDSEN, C. NAMPREMPRE, "On-line Ciphers and the Hash-CBC Constructions," Available at www.cse.ucsd.edu/users/mihir/papers/olc.html. The abstract appeared in Crypto '01, LNCS Vol. 2139, J. Kilian ed., Springer-Verlag, 2001.
2. M. BELLARE, A. DESAI, E. JOKIPII AND P. ROGAWAY, "A Concrete Security Treatment of Symmetric Encryption: Analysis of the DES Modes of Operation," Proc. of the 38th Symposium on Foundations of Computer Science, IEEE, 1997.
3. M. BELLARE, J. KILIAN AND P. ROGAWAY, "The Security of Cipher Block Chaining," Crypto '94, LNCS Vol. 839, Y. Desmedt ed., Springer-Verlag, 1994.
4. M. BELLARE AND C. NAMPREMPRE, "Authenticated Encryption: Relations among Notions and Analysis of the Generic Composition Paradigm," ASIACRYPT '00, LNCS Vol. 1976, T. Okamoto ed., Springer-Verlag, 2000.
5. J. BLACK AND P. ROGAWAY, "CBC MACs for Arbitrary-Length Messages: The Three-Key Constructions," Crypto '00, LNCS Vol. 1880, M. Bellare ed., Springer-Verlag, 2000.
6. M. BLAZE, "High-bandwidth Encryption with Low-bandwidth Smart Cards," Fast Software Encryption '96, LNCS Vol. 1039, D. Gollmann ed., Springer-Verlag, 1996.
7. M. BLAZE, J. FEIGENBAUM AND M. NAOR, "A Formal Treatment of Remotely Keyed Encryption," Eurocrypt '98, LNCS Vol. 1403, K. Nyberg ed., Springer-Verlag, 1998.
8. A. BOLDYREVA AND N. TAESOMBUT, "Online Encryption Schemes: New Security Notions and Constructions," Full version of this paper. Available at http://www-cse.ucsd.edu/users/aboldyre.
9. Y. DODIS AND J.H. AN, "Concealment and Its Applications to Authenticated Encryption," Eurocrypt '03, LNCS Vol. 2656 , E. Biham ed., Springer-Verlag, 2003.
10. V. GLIGOR AND P. DONESCU, "Fast Encryption and Authentication: XCBC Encryption and XECB Authenticated Modes," Fast Software Encryption '01, LNCS Vol. 2355, M. Matsui ed., Springer-Verlag, 2001.
11. P.-A. FOUQUE, A. JOUX, G. MARTINET AND F. VALETTE, "Authenticated On-line Encryption", Selected Areas in Cryptography Workshop, SAC '03, 2003.
12. P.-A. FOUQUE, G. MARTINET, G. POUPARD, "Practical Symmetric On-line Encryption", Fast Software Encryption '03, LNCS, T. Johansson Ed., Springer Verlag, 2003.
13. T. IWATA AND K. KUROSAWA, "OMAC: One-Key CBC MAC," Fast Software Encryption '03, LNCS, T. Johansson Ed., Springer Verlag, 2003.
14. M. JAKOBSSON, J. STERN AND M. YUNG, "Scramble All, Encrypt Small", Fast Software Encryption '99, LNCS Vol. 1636, L. Knudsen ed., Springer-Verlag, 1999.
15. A. JOUX, G. MARTINET AND F. VALETTE, "Blockwise Adaptive Attackers - Revisiting the (In)security of Some Provably Secure Encryption Modes: CBC, GEM, IACBC", Crypto '02, LNCS Vol. 2442 , M. Yung ed., Springer-Verlag, 2002.
16. C. JUTLA, "Encryption Modes With Almost Free Message Integrity", Eurocrypt '01, LNCS Vol. 2045, B. Pfitzmann ed., Springer-Verlag, 2001.
17. S. LUCKS, "On the Security of Remotely Keyed Encryption," Fast Software Encryption '97, LNCS Vol. 1267, E. Biham ed., Springer-Verlag, 1997.
18. NIST. FIPS PUB81 - DES MODES OF OPERATION, December 1980.
19. P. ROGAWAY, M. BELLARE, J. BLACK, AND T. KROVETZ, "OCB: A Block-Cipher Mode of Operation for Efficient Authenticated Encryption," Eighth ACM Conference on Computer and Communications Security (CCS-8), ACM Press, 2001.

Related-Key Attacks on Triple-DES and DESX Variants

Raphael C.-W. Phan

Department of Engineering, Swinburne Sarawak Institute of Technology,
1st Floor, State Complex, 93576 Kuching, Malaysia
rphan@swinburne.edu.my

Abstract. In this paper, we present related-key slide attacks on 2-key and 3-key triple DES, and related-key differential and slide attacks on two variants of DESX. First, we show that 2-key and 3-key triple-DES are susceptible to related-key *slide* attacks. The only previously known such attacks are related-key *differential* attacks on 3-key triple-DES. Second, we present a related-key differential attack on DESX+, a variant of the DESX with its pre- and post-whitening XOR operations replaced with addition modulo 2^{64}. Our attack shows a counter-intuitive result, that DESX+ is weaker than DESX against a related-key attack. Third, we present the first known attacks on DES-EXE, another variant of DESX where the XOR operations and DES encryptions are interchanged. Further, our attacks show that DES-EXE is also weaker than DESX against a related-key attack. This work suggests that extreme care has to be taken when proposing variants of popular block ciphers, that it is not always newer variants that are more resistant to attacks.

1 Introduction

Due to the DES' small key size of 56 bits, variants of the DES under multiple encryption have been considered, including double-DES under one or two 56-bit key(s), and triple-DES under two or three 56-bit keys. Another variant based on the DES is the DESX [9].

In this paper, we consider the security of 2-key and 3-key triple-DES against *related-key slide attacks*, and the security of DESX variants against both *related-key slide and related-key differential attacks*. We point out that our results on the DESX variants do not invalidate the security proofs of [9,10], but serve to illustrate the limitations of their model. In particular, we argue that one should also consider a more flexible model that incorporates related-key queries [1,7,8].

1.1 Our Model

Related-key attacks are those where the cryptanalyst is able to obtain the encryptions of certain plaintexts under both the unknown secret key, K, as well as an unknown related key, K' whose relationship to K is known, or can even

T. Okamoto (Ed.): CT-RSA 2004, LNCS 2964, pp. 15–24, 2004.

be chosen [1,7,8]. Most researchers consider related-key attacks as strictly theoretical and which involves a strong and restricted attack model. However, as has been demonstrated by several researchers such as [7,8], some of the current real-world cryptographic implementations may allow for practical related-key attacks. Examples of such instances include key-exchange protocols and hash functions, details of which we refer the reader to [7,8].

1.2   Outline of the Paper

We briefly recall previous attacks on variants of triple-DES and DESX in Section 2. In Section 3, we present our related-key slide attacks on 2-key and 3-key triple-DES. We then present in Section 4 related-key attacks on DESX+ [9], a variant of DESX that replaces the pre- and post-whitening XOR operations with additions modulo 2^{64}; and DES-EXE [6], a DESX variant with its outer XOR operations interchanged with the inner DES encryption. We show that these variants are weaker than the original DESX against related-key attacks. We conclude in Section 5.

2   Previous Work

We review in this section, previous attacks on variants of triple-DES and of DESX.

Two-key triple-DES can be broken with a meet-in-the-middle (MITM) attack requiring 2^{56} *chosen-plaintexts* (*CP*s), 2^{56} memory and 2^{56} single DES encryptions [12]. There is also an attack by van Oorschot and Wiener [13] that requires m *known-plaintexts* (*KP*s), m words of memory and approximately $2^{120-log_2\ m}$ single DES encryptions. For $m = 2^{56}$, the number of encryptions is roughly 2^{114}.

Meanwhile, the most basic attack on three-key triple-DES is the MITM attack which requires 3 chosen plaintexts, 2^{56} memory and 2^{112} single DES encryptions. In [11], Lucks proposed an attack that requires 2^{32} known plaintexts, 2^{88} memory and roughly 2^{106} single DES encryptions. There is also a related-key differential attack by Kelsey et. al [7] that works with one known plaintext, one related-key chosen ciphertext (*RK-CC*), and 2^{56} single DES encryptions. [1]

As for DESX, Daemen proposed an attack [5] requiring 2^{32} chosen plaintexts and 2^{88} single DES encryptions, or 2 known plaintexts and 2^{120} single DES encryptions. Meanwhile, another attack by Kilian and Rogaway [9,10] requires m known plaintexts and $2^{118-log_2\ m}$ single DES encryptions. For $m = 2^{32}$, the number of encryptions is roughly 2^{113}. By making use of related-key queries, Kelsey et. al [8] demonstrated an attack that requires 2^6 related-key known plaintexts (*RK-KP*s) and 2^{120} single DES encryptions. Recently, Biryukov and

---

[1] As pointed out by an anonymous referee, our estimates are independent of the memory access time in contrast to the approach taken in [14], and hence we assume no difference between memory with slow access and memory with intensive access. Such a general approach has been adopted in this paper to maintain uniformity with other previous results.

Wagner [4] presented a more efficient attack requiring $2^{32.5}$ known plaintexts, $2^{32.5}$ memory and $2^{87.5}$ single DES encryptions.

3   Related-Key Slide Attacks on Triple-DES

3.1   Attacking 3-Key Triple-DES

We first consider the three-key triple-DES, which was attacked by a related-key differential attack in [7]. We denote such an encryption of P under key $K = (K_1, K_2, K_3)$ by:

$$C = E_{K_3}(E_{K_2}^{-1}(E_{K_1}(P))). \tag{1}$$

If we also obtain the three-key triple-DES decryption of another plaintext, $P' = E_{K_1}(P)$ under a related key $K' = (K_1, K_3, K_2)$, we will get the situation as shown in Fig. 1.

Fig.1. Sliding-with-a-twist on 3-key triple-DES of the form $E_{K_1} E_{K_2}^{-1} E_{K_3}$

We have in essence aligned the encryption, $E_{K_1} o E_{K_2}^{-1} o E_{K_3}$ under key K, with the decryption, $E_{K_2}^{-1} o E_{K_3} o E_{K_1}^{-1}$ under key K' in a *sliding with a twist* [4] style. The plaintexts, P and P', and the ciphertexts, C and C' are hence related by the following slid equations:

$$C' = E_{K_1}(P) \tag{2}$$

$$P' = E_{K_1}^{-1}(C) \tag{3}$$

Our related-key slide attack works as follows:
1. Obtain 2^{32} known plaintexts, P encrypted with three-key triple-DES under the key, $K = (K_1, K_2, K_3)$
2. Obtain another 2^{32} known ciphertexts, C' decrypted with three-key triple-DES under the key, $K' = (K_1, K_3, K_2)$. Store the values of (C', P') in a table, $T1$. By the birthday paradox, we would expect one slid pair (P, C) and (P', C') such that the slid equations (2) and (3) are satisfied.
3. Guess all 2^{56} values of K_1 and do:
(i) Partially encrypt all $2^{32} P$ under the key, K_1.
(ii) Search through $T1$ for a collision of the 1st element with the result of (i). Such a collision satisfies the slid equation in (2).
(iii) For such a collision, partially decrypt C under K_1 and check for a collision of this result with the 2nd element of $T1$. The latter collision satisfies the slid equation in (3).

The first step requires 2^{32} known plaintexts while Step 2 requires 2^{32} related-key known ciphertexts and $2^{32} \times 2 = 2^{33}$ words of memory. Step 3 requires $2^{56} \times 2^{32} = 2^{88}$ single DES encryptions, and no memory. To summarize, we have an attack on three-key triple-DES that requires 2^{32} known plaintexts, 2^{32} related-key known ciphertexts (RK-KCs), 2^{33} words of memory and 2^{88} single DES encryptions.

We note that a similar attack also applies to the case of three-key triple-DES of the form:

$$C = E_{K_3}(E_{K_2}(E_{K_1}(P))). \tag{4}$$

In this case, instead of sliding an encryption with a decryption, we slide two encryptions, one under the key $K = (K_1, K_2, K_3)$ and the other under $K' = (K_2, K_3, K_1)$, and obtain the situation as shown in Fig. 2.

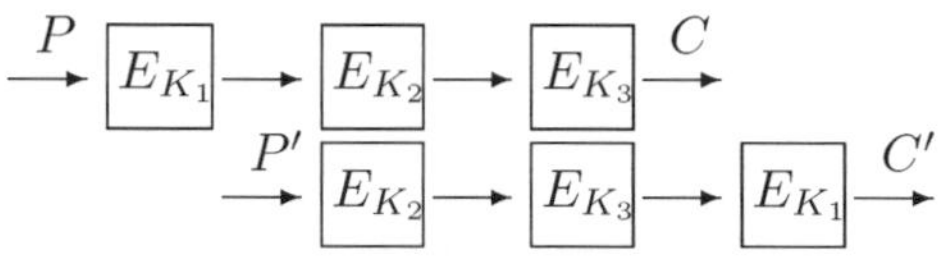

Fig.2. Sliding 3-key triple-DES of the form $E_{K_1} E_{K_2} E_{K_3}$

3.2   Attacking 2-Key Triple-DES

Two-key triple-DES is also vulnerable to a related-key slide attack. We slide an encryption under the key $K = (K_1, K_2)$, with a decryption under the key $K = (K_2, K_1)$. We then have the situation in Fig. 3.

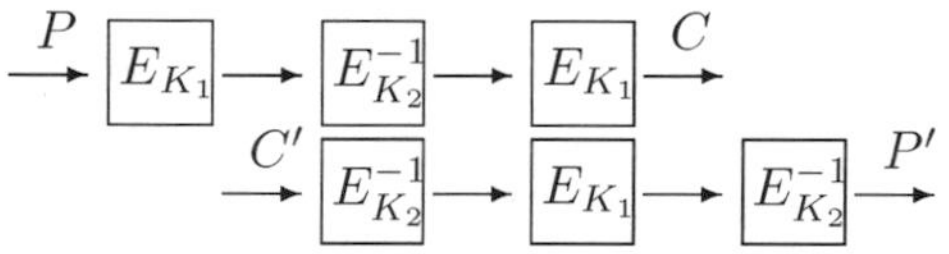

Fig.3. Sliding-with-a-twist on 2-key triple-DES of the form $E_{K_1} E_{K_2}^{-1} E_{K_1}$

We thus obtain the slid equations:

$$C' = E_{K_1}(P) \tag{5}$$

$$P' = E_{K_2}^{-1}(C) \tag{6}$$

The attack follows:

1. Obtain 2^{32} known plaintexts, P encrypted with two-key triple-DES under the key, $K = (K_1, K_2)$.

2. Obtain another 2^{32} known ciphertexts, C' decrypted with two-key triple-DES under the key, $K' = (K_2, K_1)$. Store the values of (C', P') in a table, $T1$. By the birthday paradox, we would expect one slid pair (P, C) and (P', C') such that the slid equations (5) and (6) are satisfied.

3. Guess all 2^{56} values of K_1 and do:

(i) Partially encrypt all $2^{32}P$ under the key, K_1.

(ii) Store $(E_{K_1}(P), C, K_1)$ in another table, $T2$.

4. Search through $T1$ and $T2$ for collisions in the first element, which immediately reveals the corresponding key, K_1. With $2^{56} \times 2^{32} = 2^{88}$ entries in $T2$, and a probability of 2^{-64} for a collision to occur, we expect $2^{88} \times 2^{-64} = 2^{24}$ values of K_1 to be suggested, and 2^{24} $(E_{K_1}(P), C, K_1)$ entries in $T2$ to survive this filtering.

5. For all 2^{24} remaining values of K_1, guess all 2^{56} values of K_2 and do:

(i) Partially decrypt $E_{K_1}(P)$ under the guessed key, K_2.

(ii) Further encrypt the result under K_1, and verify if the result is equal to C. The correct $K = (K_1, K_2)$ should satisfy this due to (5). Repeat with another plaintext-ciphertext pair if necessary.

The first step requires 2^{32} known plaintexts while Step 2 requires 2^{32} related-key known ciphertexts. Step 3 requires $2^{56} \times 2^{32} = 2^{88}$ single DES encryptions, and $2^{88} \times 3 \approx 2^{89.5}$ words of memory. Step 4 is negligible while Step 5 requires $2^{24} \times 2^{56} \times 2 = 2^{81}$ single DES encryptions, and no memory. To summarize, we have an attack on two-key triple-DES that requires 2^{32} known plaintexts, 2^{32} related-key known ciphertexts, $2^{89.5}$ words of memory and 2^{88} single DES encryptions.

4   Related-Key Attacks on DESX Variants

DESX encryption is denoted by:

$$C = K_b \oplus E_K(P \oplus K_a). \tag{7}$$

In this section, we will present related-key attacks on two DESX variants, namely the DESX+ [9] and the DES-EXE [6].

4.1   An Attack on DESX+

It was suggested in [9] to replace the XOR pre- and post-whitening steps in DESX by addition modulo 2^{64}, to obtain the DESX variant which we call DESX+, denoted by:

$$C = K_b + E_K(P + K_a) \tag{8}$$

where $+$ denotes addition modulo 2^{64}. We show here that this variant can be attacked by a related-key attack. The key observation is that if we obtain the DESX+ encryption of P under key, $K = (K_a, K, K_b)$, and also obtain the DESX+ encryption of P under key $K' = (K_a, K, K'_b) = (K_a, K, K_b \oplus \triangle)$, where $K_b \oplus K'_b = \triangle$ is any arbitrary known difference, then the two encryptions are related pictorially as in Fig. 4.

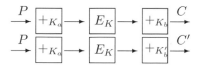

Fig.4. Related-key differential attack on DESX+

Here, $+_{K_a}$ denotes addition modulo 2^{64} with K_a. Notice that we started off with the same plaintext, P, and the similarity between the two encryptions remains until just before $+_{K_b}$.

Based on this observation, our related-key differential attack is given by:

1. Obtain the DESX+ encryption of P under key, $K = (K_a, K, K_b)$, and denote that as C.

2. Obtain the DESX+ encryption of P under key $K' = (K_a, K, K_b') = (K_a, K, K_b \oplus \triangle)$, and denote that as C'.

3. Guess all 2^{64} values of K_b, and do:

(i) Compute $X = C - K_b$.

(ii) Compute $X' = C' - K_b'$.

(iii) If $X = X'$, then the guessed K_b could be the right key value. The right key value would always satisfy this condition, whereas a wrong key value would satisfy this only with some probability, hence the number of possible values of K_b is reduced. Wrong key values can be easily checked against a trial encryption in the second analysis phase.

We have implemented this attack on a scaled-down generalization of DESX+, which we call FX32+, whose ciphertext, C is defined as:

$$C = K_b + F_K(P + K_a) \qquad (9)$$

Here, F denotes a random function, and P, C, K_a, K_b, and K are all 32 bits instead of 64. The execution takes just less than 1 minute on a Pentium 4, 1.8GHz machine with 256MB RAM, running on Windows XP. The correct K_b value is always suggested, while the number of wrong key values suggested ranges from $O(1)$ to $O(2^{31})$, depending on the hamming weight of the key difference, $\triangle$. The higher the hamming weight, the more efficient the filtering of wrong key values. An anonymous referee remarked that as the only difference between XOR and modulo addition lies in the carries, and that if the addition with K_b generates no carries, the attack on DESX+ would not work since in that case modulo addition would be the same as XOR. This possible complication can be overcome by using a $\triangle$ with a large hamming weight, or by repeating the attack with different plaintext-ciphertext pairs.

Once K_b is obtained in this way, the remaining keys K_a and K can be obtained from exhaustive search of 2^{120} single DES encryptions. But we can do better than that. We use K_b to peel off the $+_{K_b}$ operation, and apply a basic MITM attack on the remaining cipher that requires 2^{56} words of memory and 2^{56} DES encryptions [12]. Alternatively, we could reverse the roles of the plaintexts and ciphertexts, and repeat our attack to recover K_a in a similar way. What remains is then a single DES which can be attacked by exhaustive key search of 2^{56} values.

The main bulk of this attack is step 3, requiring $2^{64} \times 2 = 2^{65}$ modulo subtractions, which is negligible, so most of the work needed lies in the exhaustive key search of the remaining keys or an MITM attack on the remaining double-DES.

In summary, we have a related-key differential attack on DESX+ that requires 1 known plaintext, P encrypted under the secret key, K and related key, K', and 2^{120} single DES encryptions. The work complexity is similar to the attack on DESX in [8], but the text complexity is much less. Alternatively, our attack could work with the same text complexity but with 2^{56} words of memory and 2^{56} single DES encryptions. In this case, when memory is available, then both the text and work complexities are much less than those in [8].

Ironically, the original DESX with XOR for pre- and post-whitening is resistant to this attack. Therefore, this is the first attack for which the original DESX is stronger than the DESX+. This is counter-intuitive since the common belief is that the XOR operation is weaker than modulo addition. [2]

4.2   Attacks on DES-EXE

In [6], the authors posed the question of whether, DES-EXE, a DES variant of the form:

$$C = E_{K_b}(K \oplus E_{K_a}(P)))$$ (10)

would be stronger or weaker than DESX. Note that the DES-EXE is simply the DESX with its XOR operations in the pre- and post-whitening stage interchanged with the DES encryption in the middle.

Consider a key, $K = (K_a, K, K_b)$, and a related key, $K' = (K_b, K, K_a)$. Then, the encryptions under these two related keys could be slid as shown in Fig. 5.

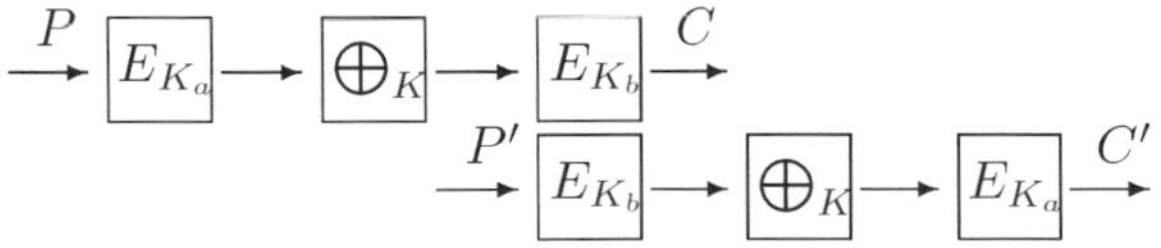

Fig.5. Sliding DESX-EXE

Therefore, we have the slid equations:

$$P' = E_{K_a}(P) \oplus K,$$ (11)

$$E_{K_a}^{-1}(C') = C \oplus K.$$ (12)

---

[2] Except in the work by Biham and Shamir [2,3] that showed how replacing XOR with addition in certain locations in the DES can significantly weaken the DES.

XORing (10) and (11), we obtain:

$$P' \oplus E_{K_a}^{-1}(C') = E_{K_a}(P) \oplus C. \tag{13}$$

A related-key slide attack proceeds as follows:

1. Obtain 2^{32} known plaintexts, P encrypted with DES-EXE under the key, $K = (K_a, K, K_b)$.

2. Obtain another 2^{32} known plaintexts, P' encrypted with DES-EXE under the key, $K = (K_b, K, K_a)$. Store the values of (P', C') in a table, $T1$. By the birthday paradox, we would expect one slid pair (P, C) and (P', C') such that the slid equations (10) and (11), and hence (12) are satisfied.

3. Guess all 2^{56} values of K_a, and do:

(i) Compute $E_{K_a}(P) \oplus C$ for all (P, C) and store $(E_{K_a}(P) \oplus C, K_a)$ in a table, $T1$.

(ii) Compute $P' \oplus E_{K_a}^{-1}(C')$ for all (P', C') and store $(P' \oplus E_{K_a}^{-1}(C'), K_a)$ in a table, $T2$.

4. Search through $T1$ and $T2$ for a collision in the first entry, which immediately reveals the key, K_a.

The remaining keys can be obtained via exhaustive search, or we could use K_a to peel off one layer and apply an MITM attack on the remaining two layers requiring 2^{56} words of memory and 2^{56} DES encryptions.

Step 1 requires 2^{32} known plaintexts while Step 2 requires 2^{32} related-key known plaintexts. Step 3 requires $2^{56} \times 2^{32} \times 2 = 2^{89}$ single DES encryptions, and $2^{88} \times 3 \times 2 \approx 2^{90.5}$ words of memory. Step 4 is negligible. Meanwhile, exhaustive search of the remaining keys requires $2^{56} \times 2^{64} = 2^{120}$ single DES encryptions, or an alternative MITM attack requires 2^{56} memory and 2^{56} DES encryptions. To summarize, we have an attack on DES-EXE that requires 2^{32} known plaintexts, 2^{32} related-key known plaintexts, $2^{90.5}$ words of memory and 2^{89} single DES encryptions.

A better attack works by observing that if we obtain the encryption, C of a plaintext, P under the key $K = (K_a, K, K_b)$, and subsequently obtain the decryption of C under the key $K' = (K_a', K, K_b) = (K_a \oplus \triangle, K, K_b)$ where $\triangle$ is any arbitrary known difference, then we get the situation as indicated in Fig. 6.

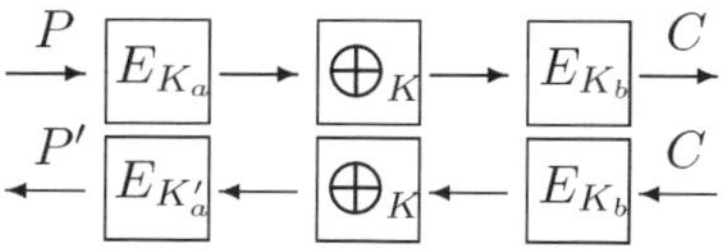

Fig.6. Related-key differential attack on DESX-EXE

Here, $\oplus_K$ denotes an XOR operation with the key, K. The following relation then applies:

$$P' = E_{K_{a'}}^{-1}(E_{K_a}(P)). \tag{14}$$

Table 1. Comparison of Attacks on Triple-DES Variants

Block Cipher	Texts	Memory	DES Encryptions	Source
2-key Triple-DES	$2^{56} CP$	2^{56}	2^{56}	[12]
2-key Triple-DES	$2^{56} KP$	2^{56}	2^{114}	[13,14]
2-key Triple-DES	$2^{32} KP, 2^{32} RK$-KC	$2^{89.5}$	2^{88}	This paper
3-key Triple-DES	$3\ CP$	2^{56}	2^{112}	[12]
3-key Triple-DES	$1\ KP,\ 1\ RK$-CC	2^{56}	2^{56}	[7]
3-key Triple-DES	$2^{32} KP$	2^{88}	2^{106}	[11]
3-key Triple-DES	$2^{32} KP, 2^{32} RK$-KC	2^{33}	2^{88}	This paper

Table 2. Comparison of Attacks on DESX Variants

Block Cipher	Texts	Memory	DES Encryptions	Source
DESX	$2^{32} CP$	-	2^{88}	[5]
DESX	$2\ KP$	-	2^{120}	[5]
DESX	$2^{32} KP$	-	2^{113}	[9,10]
DESX	$2^{6} RK$-KP	-	2^{120}	[8]
DESX	$2^{32.5} KP$	$2^{32.5}$	$2^{87.5}$	[4]
DESX+	$1\ RK, 1\ RK$-KP	-	2^{120}	This paper
DESX+	$1\ RK, 1\ RK$-KP	2^{56}	2^{56}	This paper
DES-EXE	$2^{32} KP, 2^{32} RK$-KP	$2^{90.5}$	2^{89}	This paper
DES-EXE	$1\ KP, 1\ RK$-CC	2^{56}	2^{56}	This paper

For all 2^{56} values of K_a, check that (13) satisfies, and K_a can be recovered with 2^{56} encryptions. Use this to peel of the first layer, and apply an MITM attack on the remaining two layers, requiring 2^{56} memory and 2^{56} encryptions [12]. In summary, we require one known plaintext, one related-key chosen ciphertext, 2^{56} words of memory and 2^{56} single DES encryptions. This shows that DES-EXE is much weaker than the original DESX against a related-key differential attack.

5 Conclusions

We have presented related-key slide attacks on 2-key and 3-key triple-DES. Our attacks are the first known related-key slide attacks on these triple-DES variants.

We have also presented attacks on DESX variants. In particular, we showed that contrary to popular belief, the DESX+, a DESX variant that uses addition modulo 2^{64} for its pre- and post-whitening, is weaker than DESX against a related-key differential attack. Our attacks on DES-EXE, another DESX variant with the outer XOR operations interchanged with the middle DES encryption, also show that DES-EXE is much weaker than the original DESX against related-key attacks. In Tables 1 and 2, we present a comparison of our attacks with previous attacks on variants of triple-DES and DESX.

Acknowledgements. We would like to thank David Naccache for his interest and comments on our attacks on triple-DES. We are grateful to the anonymous referees whose numerous suggestions helped to improve this paper. We also thank God for His many blessings (Ps. 33).

References

1. E. Biham, "New Types of Cryptanalytic Attacks Using Related Keys", Advances in Cryptology - Eurocrypt'93, Lecture Notes in Computer Science, Vol. 765, pp. 398–409, Springer-Verlag, 1994.
2. E. Biham and A. Shamir, "Differential Cryptanalysis of the Full 16-round DES", Advances in Cryptology - CRYPTO'92, Lecture Notes in Computer Science, Vol. 740, pp. 487–496, Springer-Verlag, 1993.
3. E. Biham and A. Shamir, "Differential Cryptanalysis of DES-like Cryptosystems", Journal of Cryptology, Vol.4, No.1, pp. 3–72, 1991.
4. A. Biryukov and D. Wagner, "Advanced Slide Attacks", Advances in Cryptology - Eurocrypt'00, Lecture Notes in Computer Science, Vol. 1807, pp. 589–606, Springer-Verlag, 2000.
5. J. Daemen, "Limitations of the Even-Mansour Construction", Advances in Cryptology - Asiacrypt'91, Lecture Notes in Computer Science, Vol. 739, pp. 495–498, Springer-Verlag, 1992.
6. B.S. Kaliski and M.J.B. Robshaw, "Multiple Encryption: Weighing Security and Performance", Dr. Dobb's Journal, 1996.
7. J. Kelsey, B. Schneier and D. Wagner, "Key-Schedule Cryptanalysis of IDEA, G-DES, GOST, SAFER and Triple-DES", Advances in Cryptology - Crypto'96, Lecture Notes in Computer Science, Vol. 1109, pp. 237–251, Springer-Verlag, 1996.
8. J. Kelsey, B. Schneier and D. Wagner, "Related-Key Cryptanalysis of 3-WAY, Biham-DES, CAST, DES-X, NewDES, RC2, and TEA", ICICS'97, Lecture Notes in Computer Science, Vol. 1334, pp. 233–246, Springer-Verlag, 1997.
9. J. Kilian and P. Rogaway, "How to Protect DES Against Exhaustive Key Search", Advances in Cryptology - Crypto'96, Lecture Notes in Computer Science, Vol. 1109, pp. 252–267, Springer-Verlag, 1996.
10. J. Kilian and P. Rogaway, "How to Protect DES Against Exhaustive Key Search (an Analysis of DESX)", Journal of Cryptology, Vol.14, No.1, pp. 17–35, 2001.
11. S. Lucks, "Attacking Triple Encryption", Advances in Cryptology - FSE'98, Lecture Notes in Computer Science, Vol. 1372, pp. 239–253, Springer-Verlag, 1998.
12. R.C. Merkle and M.E. Hellman, "On the Security of Multiple Encryption", Communications of the ACM, Vol. 24, No.7, 1981.
13. P.C. van Oorschot and M.J. Wiener, "A Known-plaintext Attack on Two-Key Triple Encryption", Advances in Cryptology - Eurocrypt'90, Lecture Notes in Computer Science, Vol. 473, pp. 318–325, Springer-Verlag, 1990.
14. P.C. van Oorschot and M.J. Wiener, "Parallel Collision Search with Cryptanalytic Applications", Journal of Cryptology, Vol.12, No.1, pp. 1–28, 1999.

Design of AES Based on Dual Cipher and Composite Field

Shee-Yau Wu, Shih-Chuan Lu, and Chi Sung Laih

Department of Electrical Engineering
National Cheng Kung University
No.1, Ta-Hsueh Road, Tainan 701, Taiwan
{henrywu,flamecloud}@crypto.ee.ncku.edu.tw
laihcs@eembox.ncku.edu.tw

Abstract. Recently, Barkan and Biham proposed the concept of dual ciphers and pointed out that there are 240 dual ciphers of AES (Dual AES). An interesting application of dual ciphers is to design a cipher which run faster than the original cipher. In this paper, we first generalize the dual AES and propose a complete setup procedure to determine all dual ciphers. Then, a hardware implementation of AES based on the combination of dual cipher and composite field is proposed. We demonstrate that our AES design not only offers better performance and smaller area requirement than the design proposed by Wolkerstorfer et al which uses a composite field only. Our results confirm Barkan et al.'s conjecture that it is possible to design an AES cipher more efficiency than ever.

1   Introduction

In October 2000, NIST (National Institute of Standards and Technology) selected the Rijndael algorithm as the AES (Advanced Encryption Standard) [1] to replace the original encryption standard DES. The AES is expected to be the standard for the next 30 years. Since then, both software and hardware implementations of AES are hot issues in the literature [3,4,5,6,7,8,9,10,15,16]. Some papers [16] mentioned about how to improve software efficiency and some [3,4, 5,6,7,8,9,10,15] mentioned about hardware implementation.

The AES is a block cipher. The four building blocks of the AES cipher are ShiftRows, SubBytes (S-Box), MixColumn and AddRoundKey. The inverse function in SubBytes offers AES the non-linear operation. The inverse function operates over Galois Field $GF(2^8)$. As for software implementation, however, it takes a lot of time to execute the inverse operation. Usually it is implemented with table-lookup to accelerate the execution, even though table-lookup is inefficient for hardware implementation. Because the table-lookup is built with ROM (read only memory) cells, it needs a large chip area for ASIC implementation. How to increase the efficiency of the inverse operation is a very important aspect for implementation. Using combinational logic cells to replace ROM cells can reduce the chip size, but unfortunately the circuitry for inversion over $GF(2^8)$ is complicated and it needs many logic cells to synthesize the inverse operation.

T. Okamoto (Ed.): CT-RSA 2004, LNCS 2964, pp. 25–38, 2004.

However, the inverse operation over a composite field is simpler than that of original [13,14]. Rudra et al [3] and Wolkerstorfer et al [4] mentioned to use a composite field $GF((2^4)^2)$ to implement the SubBytes operation of AES, demonstrating that it is practical to implement SubBytes with combinational logic cells. This procedure reduces the chip size very much.

Recently, Barkan and Biham [2] proposed the concept of dual ciphers and pointed out that there are 240 dual ciphers for the AES cipher. For example, let E denote the AES cipher, then the ciphers $E, E^2, E^4, E^8, E^{16}, E^{32}, E^{64}$ and E^{128} are the dual ciphers. The existence of these dual ciphers brings the investigation of the AES cipher to a wider view. However, the implementation and the speed of the dual ciphers is yet to be proposed.

In this paper, we generalize the dual AES and propose a realizable setup procedure to determine all dual ciphers. We also discuss the hardware implementation of the ciphers. Our results show that the combination of the dual cipher with a composite field can offer better speed with small area requirements than using a composite field only.

In section 2, we give a generalized description of the dual ciphers followed by proposing of a complete procedure to find out the dual cipher. In section 3, a model for the implementation of the original AES and the dual cipher is presented. In section 4, we present our design philosophy about the whole cipher. Section 5 is the conclusion.

2   The Dual AES Cipher

Barkan and Biham [2] first proposed the idea of dual ciphers of AES in 2002. The dual ciphers defined by Barkan and Biham are given as follows:

Definition 1. *[2] Two ciphers E and E' are called Dual Ciphers, if they are isomorphic, i.e., if there exists invertible transformations $f(\cdot), g(\cdot)$ and $h(\cdot)$ such that $\forall P, K f(E_k)(P)) = E'_{g(K)}(h(P))$, where P is the plaintext, and K is the secret key.*

If we modify the definition above as $E_{K(P)} = f^{-1}(E'_{g(K)}(h(P)))$, we can express the relation between dual AES and AES as illustrated in Figure 1. We use the transformation $T(P)$ to express the $h(P)$ function, and the transformation $T^{-1}(C')$ to express the $f^{-1}(C')$ function, as well as $T(K)$ to express $g(K)$.

2.1   Generalized Representation of AES

AES is a block cipher system based on algebraic operations over the algebraic finite field $GF(2^8)$. The dual cipher can be defined using the theory of finite field.

Definition 2. *[11] A mapping $f : G \to H$ of the group G into group H is called a homomorphism of G into H if f preserves the operation of G. That is, if $*$ and $\cdot$ are the operations of G and H, respectively, then f preserves the operation of G if for all $a, b \in G$, we have $f(a * b) = f(a) \cdot f(b)$. If f is a one-to-one homomorphism of G onto H, then f is called an isomorphism and we say that G and H are isomorphic. An isomorphism of G onto G is called an automorphism.*

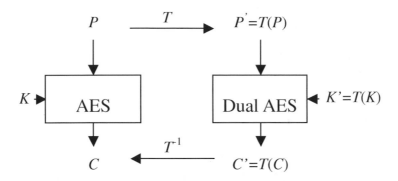

Fig. 1. The relation between AES and dual AES

This definition can be applied to mappings between rings. A mapping $\varphi :$ $R \to S$ from a ring R into a ring S is called a homomorphism if for any $a, b \in R$, we have

$$\varphi(a + b) - \varphi(a) + \varphi(b), \tag{1}$$

and

$$\varphi(ab) = \varphi(a)\varphi(b). \tag{2}$$

Theorem 1. *[11]The distinct automorphisms of $GF(q^m)$ over $GF(q)$ are exactly the mappings $\sigma_0, \sigma_1, \ldots, \sigma_{m-1}$, defined by $\sigma_j(\alpha) = \alpha^{q^j}$ for $\alpha \in GF(q^m)$ and $j = 0, 1, \ldots, m - 1$.*

The elements $\alpha, \alpha^q, \alpha^{q^2}, \ldots, \alpha^{q^{m-1}}$ are called the conjugates of α over $GF(q)$.

Over $GF(2^8)$, there are 8 elements in the set of conjugates of α, which are $\alpha, \alpha^2, \alpha^4, \alpha^8, \alpha^{16}, \alpha^{32}, \alpha^{64}$ and α^{128}. For each irreducible polynomial, there are 128 primitives over $GF(2^8)$ [11]. Each 8 primitives form a set of conjugates. There are 16 sets of conjugates over $GF(2^8)$.

As for AES, the irreducible polynomial representation is $R(x) = x^8 + x^4 + x^3 + x + 1$ (hexadecimal notation is $\{11B\}_x$). Sixteen sets of conjugates can be determined over $GF(2^8)$. However, only one set satisfies the equations 1 and 2 when $+$ and $\times$ operations are performed within AES. The set is $\{\{03\}_x, \{05\}_x, \{11\}_x, \{1A\}_x, \{4C\}_x, \{5F\}_x, \{E5\}_x, \{FB\}_x\}$. In finite field, a generator is defined as an element with its power terms can generate all the elements in the field. If we take the generator $\beta = \{03\}_x$, we denote this AES as $\{GF(2^8), \{11B\}_x, \{03\}_x\}$, or simply as $\{\{11B\}_x, \{03\}_x\}$. The set of 8-bits elements $\{01, 02, \ldots, FF\}$ is mapped onto the set of power representations of $\beta : \{0, 1, \beta^1, \beta^2, \ldots, \beta^{254}\}$. So, we can replace the original AES in terms of the power of the generator $\beta = \{03\}_x$. We call this as a **generalized representation of AES**.

We can choose any conjugate to generate a dual AES. For different generators we have different forms of dual ciphers. Over $GF(2^8)$ there are 30 irreducible polynomials found [11]. Only one set of conjugates is suitable to be a generator for each polynomial. Since there are 8 elements in the set, there are 240 different forms of dual ciphers. We can denote a dual cipher of AES simply as a dual AES.

In Appendix-A we list the 240 pairs of $\{R(x), b\}$ which are obtained by the setup procedure described in Section 2.2.

2.2   Mapping from AES to Dual AES over $GF(2^8)$

The original AES $\{\{11B\}_x, \{03\}_x\}$ can be mapped to a generalized AES. The generalized AES is the power form of the generator $\{03\}_x$. This generalized AES can be mapped onto another dual AES with a different generator or polynomial.Both the AES cipher and this dual cipher are isomorphic. Because the dual ciphers must follow the equations stated in [1] and [2], they can be ascertained easily.

Next, we propose a method to determine the dual AES $\{R(x), \beta\}$, where β is the generator and $R(x)$ is the polynomial. Assume $R(x)$ is known, and β is to be determined. The setup procedure is as follows:

1. Check to see if an element $a \in \{\{02\}_x, \ldots, \{FF\}_x\}$ is primitive or not. If it is not primitive, delete it from the set $\{\{02\}_x, \ldots, \{FF\}_x\}$. Repeat this procedure until the whole set is checked. The residual set will have 128 elements.
2. Pick one element from the residual set to be generator α. Choose any two elements α^k and α^l such that $\alpha^l = \alpha^k + \alpha^0$. Assume the power form elements α^k and α^l map to p and q in the dual AES, respectively. For $p \in \{\{02\}_x, \ldots, \{FF\}_x\}$, check to see if the equation $q = p + 1$ works or not. If it passes the test, a generator of dual AES has been found. Proceed directly to step 4.
3. Check the other conjugates , $\alpha^{2^j} = 1, 2, \ldots, 7$, to see if one is generator α. If any one of these works, a generator has found and you can proceed to step 4. If all of these fail, then return to Step 2 and test another one.
4. The primitive element α can be as a generator for building a dual AES, as can the other conjugates , $\alpha^{2^j} = 1, 2, \ldots, 7$. After choosing one of them as the generator β and mapping from the AES to the dual AES, we can form the matrix, $T = [\beta^0, \beta^{25}, \beta^{50}, \beta^{75}, \beta^{100}, \beta^{125}, \beta^{150}, \beta^{175}]$ is formed, where the β^i are binary vectors. The inverse matrix T^{-1}, mapping from the dual AES to the AES, is obtained by inverting the matrix T.
5. The overall operations of the dual AES$\{R(x), \beta\}$ are obtained as follows. **AddRoundKey.** This function keeps the same operation as the original AES. AddRoundKey is still a simple XOR operation of the intermediate state to the round key.
   ShiftRows. This function also keeps the same operation as the original AES. That is, ShiftRows is still cyclically shifted with the same offsets, as is the InvShiftRows.

MixColumns. The function MixColumns is the column-states multiplied by the 4 elements $[\beta^{25}, \beta^0, \beta^0, \beta]$, with the polynomial denoted as $\{\beta^{25}\}_x + \{1\}_x X + \{1\}_x X^2 + \{\beta\}_x X^3$. For the InvMixColumns, the four elements are $[\beta^{223}, \beta^{199}, \beta^{228}, \beta^{104}]$.

SubBytes. The SubBytes is composed of InverseMapping, Affine Transformation and InvAffine Transformation. The InverseMapping function of SubBytes and InvSubBytes keep the same operation as the original AES with the irreducible polynomial $R(x)$. The Affine Transformation of SubBytes is $G'(y) = T(Const) + (T \bullet Affine \bullet T^{-1}) \bullet y$, where $Affine$ is the same 8×8 matrix and $Const = \{63\}_x$ as the original AES.

Using this setup procedure, we can obtained any dual AES ciphers $\{R(x), \beta\}$.

3   Optimizing the SubBytes Implementation

Among the four functions in AES, SubBytes is the most important function because it provides a nonlinear transformation. From the aspect of implementation, the efficiency of AES hardware implementation is mainly determined by the implementation of SubBytes. In this section, we will only discuss the design of SubBytes.

3.1   SubBytes Implementation of the AES Proposed by Wolkerstorfer et al

The original AES applied with a composite field is discussed in [4]. The sub-fields and the polynomials are usually selected as follows [4].

$$\begin{cases} GF(2^4) : Q(y) = y^4 + y + 1, \\ GF((2^4)^2) : P(z) = z^2 + z + \{E\}_x \end{cases}$$

where $GF(2^4)$ is composed of $Q(y)$, and $GF((2^4)^2)$ is composed of $P(z)$.

Most of the functionality of SubBytes architecture can be implemented using two-input XOR gates. SubBytes implementation efficiency can be measured in terms of *space complexity* and *time complexity*.

In Table 1, the *space complexity* of SubBytes is measured by $\sharp XOR$(counts of XOR gates) and *time complexity* is measured by τ_{XOR} (total delay accumulated along the critical path). The total XOR gates count is 123. The critical path for encryption is composed of 18 XOR-gates. For decryption, it is only 17 because the InvAffine transformation has less complexity.

3.2   The Proposed SubBytes Implementation of the Dual AES

We can see in Figure 1 that if we want to implement the AES via the dual AES, then the cost of the transformation T and T^{-1} must be taken into consideration. How to lower this extra cost loss? by reconfiguring the hardware structure. In this section, we will only discuss the SubBytes implementation of the dual AES.

Table 1. Complexities of the SubBytes designed by Wolkerstorfer et al

block	τ_{XOR} of component	Amountin encrypt path	Amountin decrypt path	♯**XOR** of component	Amount in SubBytes	Total ♯**XOR**
Affine Tran	3	1	x	16	1	16
Affine $Tran^{-1}$	2	x	1	12	1	12
Mapping	2	1	1	11	1	11
$Mapping^{-1}$	3	1	1	15	1	15
Addition	1	2	2	4	3	12
Squaring	1	1	1	2	2	4
$ConstMul\{E\}_x$	2	1	1	5	1	5
Multiplication	2	1	1	12	3	36
Inversioon	3	1	1	12	1	12
Result	Max delay (Encryption)	18				
	Max delay (Decryption)	17		Sum of XOR gates	123

The Efficient Architecture of SubBytes. Because the cipher must contain both encryption and decryption, the SubBytes and the InvSubBytes must be designed within the same chip. The InverseMapping function can be shared between SubBytes and InvSubBytes. In the following discussion, we simply write "SubBytes" to represent both the SubBytes and the InvSubBytes. The design criterion of the SubBytes is to minimize the cost of the components used. The cost for the SubBytes is the sum of the components' cost of InverseMapping, Affine Transformation and InvAffine Transformation. For the design of the dual AES with its composite field, the design cost of SubBytes denoted by C_S is:

$$C_s = Cost(Affine) + Cost(InvAffine) + Cost(M) + Cost(M^{-1}) + Cost(ConstMul) \tag{3}$$

where the mapping matrix M denotes the mapping from $GF(2^8)$ to $GF((2^4)^2)$ and M^{-1} denotes the reverse mapping. ConstMul denotes a constant multiplication over $GF(2^4)$, which is required to compute the inversion. Cost analysis will be further discussed in Section 4.

Our Design of SubBytes. Proper selection of the dual AES can minimize the total delay along the critical path as well as reducing the gate counts of the SubBytes and InvSubBytes blocks. Because there are only a total of 240 dual AES ciphers, we were able to use exhaustive search to find the optimal dual AES and the composite field. In our design, the optimal fields and their polynomials have been selected as follows:

$$\begin{cases} GF(2^8) : R(x) = x^8 + x^4 + x^3 + x^2 + 1, \\ GF(2^4) : Q(y) = y^4 + y + 1, \\ GF((2^4)^2) : P(z) = z^2 + z + \{9\}_x. \end{cases}$$

The dual AES $\{\{11D\}_x, \{02\}_x\}$ is used. We selected $P(z) = z^2 + z + \{9\}_x$ as the irreducible polynomial over $\mathrm{GF}((2^4)^2)$. We describe the three functions of the SubBytes of our design below.

1. Affine Transformation. For the dual AES $\{\{11D\}_x, \{02\}_x\}$, the Affine Transformation is given as:

$$\begin{bmatrix} b_0 \\ b_1 \\ b_2 \\ b_3 \\ b_4 \\ b_5 \\ b_6 \\ b_7 \end{bmatrix} = \begin{bmatrix} 1 & 0 & 0 & 0 & 0 & 0 & 0 & 0 \\ 0 & 1 & 0 & 0 & 0 & 0 & 0 & 0 \\ 0 & 0 & 1 & 0 & 0 & 0 & 0 & 0 \\ 0 & 0 & 0 & 1 & 0 & 0 & 0 & 0 \\ 0 & 0 & 0 & 0 & 1 & 0 & 0 & 0 \\ 0 & 0 & 0 & 0 & 0 & 1 & 0 & 0 \\ 0 & 0 & 0 & 0 & 0 & 0 & 1 & 0 \\ 0 & 0 & 0 & 0 & 0 & 0 & 0 & 1 \end{bmatrix} \begin{bmatrix} c_0 \\ c_1 \\ c_2 \\ c_3 \\ c_4 \\ c_5 \\ c_6 \\ c_7 \end{bmatrix} + \begin{bmatrix} 0 \\ 0 \\ 1 \\ 0 \\ 0 \\ 1 \\ 1 \\ 0 \end{bmatrix}$$

Affine Transformation can be described as in the following (where the notation $\sim (\cdot)$ denotes the invert-operation).

$b = Affine\_Tran.(c)$	$b, c \in GF(2^8)$
b_0	$b_4 = c_0 \oplus c_1 \oplus c_4$
b_1	$b_5 = \sim (c_1 \oplus c_2 \oplus c_5)$
b_2	$b_6 = \sim (c_2 \oplus c_3 \oplus c_6)$
b_3	$b_7 = (c_3 \oplus c_4 \oplus c_7)$

2. InvAffine Transformation. For the dual AES $\{\{11D\}_x, \{02\}_x\}$, the InvAffine Transformation is given as:

$$\begin{bmatrix} b_0 \\ b_1 \\ b_2 \\ b_3 \\ b_4 \\ b_5 \\ b_6 \\ b_7 \end{bmatrix} = \begin{bmatrix} 1 & 0 & 0 & 0 & 0 & 0 & 0 & 0 \\ 0 & 1 & 0 & 0 & 0 & 0 & 0 & 0 \\ 0 & 0 & 1 & 0 & 0 & 0 & 0 & 0 \\ 1 & 0 & 0 & 1 & 0 & 0 & 0 & 0 \\ 1 & 1 & 0 & 0 & 1 & 0 & 0 & 0 \\ 0 & 1 & 1 & 0 & 0 & 1 & 0 & 0 \\ 1 & 0 & 1 & 1 & 0 & 0 & 1 & 0 \\ 0 & 1 & 0 & 1 & 1 & 0 & 0 & 1 \end{bmatrix} \begin{bmatrix} c_0 \\ c_1 \\ c_2 \\ c_3 \\ c_4 \\ c_5 \\ c_6 \\ c_7 \end{bmatrix} + \begin{bmatrix} 0 \\ 0 \\ 1 \\ 0 \\ 0 \\ 0 \\ 0 \\ 0 \end{bmatrix} \quad \text{InvAffine}$$

Transformation can also be described by the following equations:

$b = Affine\_Tran.^{-1}(c)$	$b, c \in GF(2^8)$
$b_0 = c_0$	$b_4 = c_0 \oplus (c_1 \oplus c_4)$
$b_1 = c_1$	$b_5 = c_1 \oplus c_2 \oplus c_5$
$b_2 = (c_2)$	$b_6 = (c_0 \oplus c_3) \oplus c_2 \oplus c_6$
$b_3 = (c_0 \oplus c_3)$	$b_7 = (c_1 \oplus c_4 \oplus c_3 \oplus c_7)$

3. InverseMapping. The inversion is computed over the composite field $\mathrm{GF}((2^4)^2)$. The InverseMapping consists of two main components: the computation over $\mathrm{GF}((2^4)^2)$ and the mapping matrices M and M^{-1}.

$$M = \begin{bmatrix} 1 & 1 & 0 & 1 & 0 & 0 & 0 & 0 \\ 0 & 1 & 0 & 0 & 0 & 1 & 0 & 0 \\ 0 & 1 & 0 & 1 & 1 & 0 & 0 & 1 \\ 0 & 1 & 1 & 0 & 0 & 1 & 0 & 0 \\ 0 & 0 & 0 & 0 & 1 & 0 & 0 & 1 \\ 0 & 1 & 0 & 1 & 1 & 0 & 0 & 0 \\ 0 & 0 & 1 & 0 & 0 & 1 & 1 & 0 \\ 0 & 0 & 0 & 0 & 0 & 1 & 0 & 0 \end{bmatrix}, \ M^{-1} = \begin{bmatrix} 1 & 0 & 1 & 0 & 1 & 0 & 0 & 0 \\ 0 & 1 & 0 & 0 & 0 & 0 & 0 & 1 \\ 0 & 1 & 0 & 1 & 0 & 0 & 0 & 0 \\ 0 & 1 & 1 & 0 & 1 & 0 & 0 & 1 \\ 0 & 0 & 1 & 0 & 1 & 1 & 0 & 0 \\ 0 & 0 & 0 & 0 & 0 & 0 & 0 & 1 \\ 0 & 1 & 0 & 1 & 0 & 0 & 1 & 1 \\ 0 & 0 & 1 & 0 & 0 & 1 & 0 & 0 \end{bmatrix}$$

The inversion over $GF(2^8)$ is computed in the composite field $GF((2^4)^2)$. An element $a \in GF(2^8)$ can be represented as a two-term polynomial:

$a = a_h z + a_l, where \, a \in GF(2^8), a_h, a_l \in GF(2^4)$.

For computing the inversion, the 1-element is given as: $(a_h z + a_l \bigotimes (a'_h z + a'_l)$ $= \{0\}_x z + \{1\}_x$, where $a_h, a_l, a'_h, a'_l \ GF(2^4)$. The inversion can then be derived as:

$(a_h z + a_l)^{-1} = a'_h z + a'_l = (a_h \otimes d)z + (a_l \oplus a_h) \otimes d$
$d = ((a_h^2 \otimes \{9\}_x) \oplus (a_h \otimes a_l) \oplus a_l^2)^{-1}$

For the element $a \in GF(2^8)$ and $a_h, a_l \in GF(2^4)$, the matrix M represents the transfer from a to $a_h z + a_l$. The mapping function of the matrix M is given by the function **Mapping**$(\cdot)$:

$a_h z + a_l = \mathbf{Mapping}(a)$	$a_h, a_l \in GF(2^4), a \in GF(2^8)$
$a_{l0} = a_0 \oplus (a_1 \oplus a_3)$	$a_{h0} = (a_4 \oplus a_7)$
$a_{l1} = a_1 \oplus a_5$	$a_{h1} = (a_1 \oplus a_3) \oplus a_4$
$a_{l2} = (a_1 \oplus a_3) \oplus (a_4 \oplus a_7)$	$a_{h2} = (a_0 \oplus a_5) \oplus a_6$
$a_{l3} = a_1 \oplus (a_2 \oplus a_5)$	$a_{h3} = a_5$

The inverse of $a \in GF(2^8)$ can be replaced by the inverse of $a_h z + a_l$, where $a_h, a_l \in GF(2^4)$. The computing of the inversion over $GF((2^4)^2)$ requires 1 inversion, 3 general multiplications, 2 squarings, 3 additions and 1 constant multiplication. All of these operations are expressed in the equations below.

A. Inversion. The inverse a^{-1} of an element $a \in GF(2^4)$ can be derived by solving the equation $a \cdot q \, mod \, Q(y) = 1$, where $q \in GF(2^4)$. The solution is given as follows:

$q = a^{-1} \, mod \, Q(y)$	$a, q \in GF(2^4)$
$q_o = a_0 \oplus a_1 \oplus a_2 \oplus a_0 a_2 \oplus a_1 a_2 \oplus a_0 a_1 a_2 \oplus a_3 \oplus a_1 a_2 a_3$	
$q_1 = a_0 a_1 \oplus a_0 a_2 \oplus a_1 a_2 \oplus a_3 \oplus a_1 a_3 \oplus a_0 a_1 a_3$	
$q_2 = a_0 a_1 \oplus a_2 \oplus a_0 a_2 \oplus a_3 \oplus a_0 a_3 \oplus a_0 a_2 a_3$	
$q_3 = a_1 \oplus a_2 \oplus a_3 \oplus a_0 a_3 \oplus a_1 a_3 \oplus a_2 a_3 \oplus a_1 a_2 a_3$	

B. Multiplication. Multiplication over $GF(2^4)$ is given by:

$q = a \otimes b = a\otimes b \, mod \, Q(y)$	a, b, q $\in GF(2^4)$
$q_0 = a_0 b_0 \oplus a_3 b_1 \oplus a_2 b_2 \oplus a_1 b_3$	
$q_1 = a_1 b_0 \oplus (a_0 \oplus a_3) b_1 \oplus (a_2 \oplus a_3) b_2 \oplus (a_1 \oplus a_2) b_3$	
$q_2 = a_2 b_0 \oplus a_1 b_1 \oplus (a_0 \oplus a_3) b_2 \oplus (a_2 \oplus a_3 b_3$	
$q_3 = a_3 b_0 \oplus a_2 b_1 \oplus a_1 b_2 \oplus (a_0 a_3) b_3$	

C. Squaring. The operation of squaring is given by:

$$\frac{q = a^2 \ mod \ Q(y) \quad \text{a, q} \in \text{GF}(2^4)}{\begin{array}{ll} q_0 = a_0 \oplus a_2 & q_1 = a_2 \\ a_2 = a_1 \oplus a_3 & q_3 = a_0 \end{array}}$$

D. Multiplication by constant $\{9\}_x$. Multiplication by $\{9\}_x$ is required in the inversion computation. The operation for this constant multiplication is given as:

$$\frac{q = a \otimes 9_x \ mod \ Q(y) \quad \text{a, q} \in \text{GF}(2^4)}{\begin{array}{ll} q_0 = a_0 \oplus a_1 & q_1 = a_2 \\ a_2 = a_3 & q_3 = a_0 \end{array}}$$

Table 2 is the implementation complexities of the our SubBytes design, with the dual AES $\{\{11D\}_x, \{02\}_x\}$. Comparing these results with those of Table 1, we can see that the design with dual AES is superior to that with standard AES.

Table 2. Complexities of our SubBytesdesign

block	τ_{XOR} of component	Amountin encrypt path	Amountin decrypt path	$\sharp$**XOR** of component	Amount in SubBytes	Total $\sharp$**XOR**
$AffineTran$	2	1		9	1	9
$AffineTran^{-1}$	2		1	9	1	9
$Mapping$	2	1	1	9	1	9
$Mapping^{-1}$	2	1	1	9	1	9
$Addition$	1	2	2	4	3	12
$Squaring$	1	1	1	2	2	4
$ConstMul\{9\}_x$	1	1	1	1	1	1
$Multiplication$	2	1	1	12	3	36
$Inversioon$	3	1	1	12	1	12
Result	Max delay (E/D)		15/15	$C_s(\tau_{XOR})$ (E/D)		7/7
	Sum of XOR (gates)		101	$C_s(\sharp XOR)$		37

The overall number of two-inputs XOR gates is 101. The critical paths for encryption and decryption are both composed of 15 XOR-gates in series. Compared with Wolkerstorfer et al.'s computations, the cost of XOR gates $C_S(\sharp XOR)$ is reduced by 17%, and the cost of path delay $C_S(\tau_{XOR})$ is reduced by 16% for encryption and 11% for decryption. In Table 3, we graphically compare our design with that of Wolkerstorfer et al.'s. The cost analysis of $C_S(\sharp XOR)$ and $C_S(\tau_{XOR})$ will be discussed in the next section.

In section 3.1, we discuss the implementation of SubBytes based on the original AES. In this section, we discuss the implementation of SubBytes based on the dual AES $\{\{11D\}x, \{02\}x\}$. Both of these use a composite field. Our findings indicate that implementation of SubBytes with dual AES is more efficient.

Table 3. Comparison between our design of SubBytes and Wolkerstorfer et al.'s

	$C_s(\sharp XOR)$	$C_s(\tau_{XOR})$ (E/D)	Max Delay (Encryption)	Max Delay (Decryption)	Sum of XOR gates
Wolkerstorfer et al.'s	59	10/9	18	17	123
Our Design	37	7/7	15	15	101

4   Design of AES via Dual Cipher

If we want to design the whole AES with ASIC, there are several different possible architectural styles [10]. In this paper, we focus on the two most commonly used: iterative circuits and pipeline circuits. Different circuitry or a different choice in building blocks can lead to differences in space or time complexity. For simplicity, we will limit our analysis of design complexity to the aspect of cost. Note that the cost analysis of the key expansion block is not included in our discussion.

4.1   Cost Analysis

The mapping from the AES to the dual AES is shown in Figure 1. Now, we want to implement an AES via a dual AES. We must take into consideration that the cost of transformation T and T^{-1} will increase when using the dual AES. The cost of T and T^{-1} must be as low as possible. Now, let $\text{Cost}(\phi)$ denote the cost of operation ϕ. The cost of SubBytes will be denoted by C_S, which is described in equation [3].

Compare the building blocks of the dual AES with those of AES. Because the function of the AddRoundKey, ShiftRows and InvShiftRows blocks are the same, we can ignore the cost difference between the design of the AES and the dual AES. In MixColumns and InvMixColumns, each column-state is multiplied by a specific polynomial. Those functions work like multiplication by a constant over $\text{GF}(2^8)$. Hence, the design criterion for these functions will be to minimize the cost C_P, where C_P is given as:

$C_P = C_{MC} + C_{IMC}$,
where $C(MC) = Cost(\times T(02)) + Cost(\times T(03))$,
and
$C(IMC) = Cost(\times T(0E)) + Cost(\times T(09)) + Cost(\times T(0D)) + Cost(\times T(0B))$

Here the notation $\times(\cdot)$ denotes constant multiplication and C_{MC} and C_{IMC} denote the costs of MixColumns and InvMixColumns , respectively.

We use the notation C_T to denote the sum of the cost of the transformation matrix T (denoted by $Cost(T)$) and its inverse T^{-1} (denoted by $Cost(T^{-1})$) as:
$C_T = (1 + k) \times Cost(T) + Cost(T^{-1})$,

where k denotes the cost weighing for the different length of the secret key. For the cost C_T for $\sharp$ XOR, the value of k is 1 for the 128-bits secret key, 1.5 for 192-bits, and 2 for 256-bits, respectively. For the cost C_T for τ_{XOR}, k is equal to 0 because the secret key can be transferred at the same time as the plaintext is being transferred.

The total cost for AES design is given by C. This value can be expressed by:
$C = t \times C_r + s \times C_s + p \times C_P$.
where t, s and p denote the cost weighting of the transformation matrices, architecture of SubBytes, and MixColumns operations, respectively. With different circuit architectures, the t, s and p values will change. Here we only discuss two frequently used architectures: the iterative circuit and the pipeline circuit.

Iterative circuit. In this architecture, one round of cipher is implemented.//designed for hardware implementation. For an n-round cipher, the hardware will be used n times. The advantage of using the iterative circuit is the hardware's area efficiency.On the other side, however, it must execute n times to encrypt or decrypt. Our design criterion is to minimize the cost $C(\sharp XOR)$ and $C(\tau_{XOR})$ which are stated as follows:
$C(\sharp XOR) = 16 \times C_r(\sharp XOR) + 16 \times n \times C_s(\sharp XOR) + 4 \times C_p(\sharp XOR)$.
$C(\tau_{XOR}) = n \times C_s(\tau_{XOR}) + (n-1) \times C_P(\tau_{XOR})$.
Pipelining circuit. In this architecture, each round of the cipher will be piped with hardware implementation. Registers between each round must be included. The advantage of using the pipeline circuit is the high data throughput, as encryption or decryption requires only an average of one clock cycle. However, the cost of area is also greatly increased. Our design criterion is to minimize the cost $C(\sharp XOR)$ and $C(\tau_{XOR})$, which are shown as follows.
$C(\sharp XOR) = 16 \times C_T(\sharp XOR) + 16 \times n \times C_s(\sharp XOR) + 4 \times (n-1) \times C_p(\sharp XOR))$.
$C(\tau_{XOR}) = C_s(\tau_{XOR}) + C_P(\tau_{XOR})$

Here, n means the number of rounds in AES. While the length of the secret key 128-, 192- or 256-bits, n is 10, 12 or 14, respectively.

4.2   The Feasibility of the Dual Ciphers' Application

The dual AES $\{\{11D\}x, \{02\}x\}$ discussed in the last section has special transformation matrices, T and T^{-1}, which are the same, and are given as follows:

$$T = \begin{bmatrix} 1 1 1 1 1 1 1 1 \\ 0 1 0 1 0 1 0 1 \\ 0 0 1 1 0 0 1 1 \\ 0 0 0 1 0 0 0 1 \\ 0 0 0 0 1 1 1 1 \\ 0 0 0 0 0 1 0 1 \\ 0 0 0 0 0 0 1 1 \\ 0 0 0 0 0 0 0 1 \end{bmatrix}, T^{-1} = \begin{bmatrix} 1 1 1 1 1 1 1 1 \\ 0 1 0 1 0 1 0 1 \\ 0 0 1 1 0 0 1 1 \\ 0 0 0 1 0 0 0 1 \\ 0 0 0 0 1 1 1 1 \\ 0 0 0 0 0 1 0 1 \\ 0 0 0 0 0 0 1 1 \\ 0 0 0 0 0 0 0 1 \end{bmatrix}$$

We can express T and T^{-1} with the following equations:

$b = \mathbf{Tran.}(a)(or\,\mathbf{Tran.}^{-1}(a))$	$b, c \in GF(2^8)$
$b_0 = a_0 \oplus a_1 \oplus (a_2) \oplus a_3 \oplus (a_4) \oplus (a_5 \oplus (a_6)) \oplus a_7)$	$b_4 = (a_4 \oplus a_5 \oplus (a_6 \oplus a_7))$
$b_1 = a_1 \oplus a_3 \oplus (a_5 \oplus a_7)$	$b_5 = (a_5 \oplus a_7)$
$b_2 = (a_2 \oplus a_3) \oplus (a_6 \oplus a_7)$	$b_6 = (a_6 \oplus a_7)$
$b_3 = a_3 \oplus a_7$	$b_7 = a_7$

We can calculate the cost $C_T(\sharp XOR)$ of the transformation matrices T and T^{-1} with $C_T(\sharp XOR) = (1 + k) \times Cost(T) + Cost(T^{-1})$, thus getting: $C_T(\sharp XOR) = (1 + 1) \times 12 + 12 = 36$. The cost $C_T(\sharp XOR)$ of the transformation matrices T and T^{-1} can be omitted because it is not affected by this evaluation.

For dual AES $\{\{11D\}x, \{02\}x\}$, the polynomial of the MixColumns is $\{03\}_x + \{01\}_x x + \{01\}_x x^2 + \{02\}_x x^3$. It is interesting to note that the coefficients of the polynomial are the same as those in the original AES, differing only in permutation. Therefore, the MixColumns' cost of this dual AES is the same as that of the original AES, as is the InvMixColumns'. For AES and this dual AES, the cost C_P is the same.

According to section 3 the cost $C_S(\sharp XOR)$ and $C_S(\tau_{XOR})$ of SubBytes are: $C_S(\sharp XOR) = 7$ for encryption, $C_S(\sharp XOR) = 7$ for decryption, and $C_S(\sharp XOR) = 37$. The cost $C(\sharp XOR)$ and $C(\tau_{XOR})$ for the whole dual AES design, then, is as follows:

For the iterative circuit: For the pipelining circuit:

$C(\sharp XOR) = 16 \times 36 + 16 \times 37 = 1168$ $C(\sharp XOR) = 16 \times 36 + 16 \times 10 \times 37 = 6496$

$C(\tau_{XOR}) = 10 \times 7 = 70.(for encryption)$ $C(\tau_{XOR}) = 7.(for encryption)$

$C(\tau_{XOR}) = 10 \times 7 = 70.(for decryption)$ $C(\tau_{XOR}) = 7.(for decryption)$

The cost formula derived above can still be used for the implementation stated by Wolkerstorfer et al [4]. The cost of transformation is equal to 0, because the implementation is done with the original AES. The cost $C(\sharp XOR)$ and $C(\tau_{XOR})$ for Wolkerstorfer et al.'s design are given as:

For the iterative circuit: For the pipelining circuit:

$C(\sharp XOR) = 16 \times 0 + 16 \times 59 = 944$ $C(\sharp XOR) = 16 \times 0 + 16 \times 10 \times 59 = 9440$

$C(\tau_{XOR}) = 10 \times 10 = 100.(for encryption)$ $C(\tau_{XOR}) = 10.(for encryption)$

$C(\tau_{XOR}) = 10 \times 9 = 90.(for decryption)$ $C(\tau_{XOR}) = 9.(for decryption)$

Table 4 compares the results produced with our design with Wolkerstorfer et al.'s. Using a pipelining circuit, our design is much better than Wolkerstorfer et al's in both time and space complexity. Using this architecture, we reduce the cost of area by 1/6 and the cost of delay by 1/4. Using the iterative circuit, our design has an even greater improvement in time complexity, but only slight increase in space complexity.

5   Conclusion

There are two major directions in ASIC design for AES. One is table-lookup and the other is with a composite field, however the table-lookup method is not efficient. Is it more efficient with composite field? We propose an approach using dual AES in combination with a composite field. In this article, we also present a generalization form for dual AES, and propose a method for ascertaining a dual AES. Based on the theory of the finite field, the dual AES may

Table 4. Cost comparison of AES with different circuitry

		Our Design	Wolkerstorefer et al.'s
Iterative Circuit	$C(\sharp XOR)$	1168	944
	$C(\tau(E))$	70	100
	$C(\tau(D))$	70	90
Pipelining Circuit	$C(\sharp XOR)$	6496	9440
	$C(\tau(E))$	7	10
	$C(\tau(D))$	7	9

have more attractive characteristics. Daemen and Rijmen called this special finite field RIJNDAEL-GF [10]. We can see that research and application on the dual AES is just beginning.

References

1. National Institute of Standards and Technology (NIST). Advanced Encryption Standard (AES). FIPS Publication 197, Nov. 2001. Available at http://csrc.nist.gov/encryption/aes/index.html.
2. E. Barkan, E. Biham. In How Many Ways Can You Write Rijndael. Asiacrypt 2002, pp.160–175, 2002.
3. A. Rudra, P. Dubey, C. Jutla, V.Kumar, J. Rao, P. Rohatgi. Efficient Rijndael Encryption Implementation with Composite Field Arithmetic. *CHES 2001, LNCS 2162,* pp.171–184 2001.
4. J. Wolkerstorfer, E. Oswald, M, Lamberger. An ASIC Implementation of the AES Sboxes. *CT-RSA 2002, LNCS 2271,* pp.67–78, 2002.
5. T. Ichikawa, T. Kasuya, and M. Matsui. Hardware Evaluation of the AES Finalists. The Third Advanced Encryption Standard Candidate Conference, pp.279–285, 2000, Available at http://csrc.nist.gov/encryption/aes/round2/conf3/papers/15-tichikawa.pdf
6. H. Kua, I. Verbauwhede. Architectural Optimization for a 1.82Gbits/sec VLSI Implementation of the AES Rijndael Algorithm. *CHES 2001, LNCS 2162,* pp.51–64, 2001.
7. A. Satoh, S. Morioka, K. Takano, and S. Munetoh. A Compact Rijndael Hardware Architecture with S-Box Optimization. *Asiacrypt 2001, LNCS 2248,* pp.239–254, 2001.
8. M. McLoone et al. High performance single-chip FPGA Rijndael algorithm implementations. *CHES 2001, LNCS 2162,* pp.65–76, 2001.
9. S. Morioka and A. Satoh. An Optimized S-Box Circuit Architecture for Low Power AES Design. *CHES 2002, LNCS 2523,* pp.172–186, 2003.
10. Joan Daemen, Vincent Rijmen. The Design of Rijndael. Springer printed in Germany, 2002.
11. R. Lidl, H. Niederreiter. *Introduction to finite fields and their applications.* Cambridge University Press, 1986.
12. F.J. MacWilliams, N.J.A. Sloane. *The Theory of Error-Correcting Codes.* North-Holland Publishing Company, 1978.

13. C. Paar. Efficient VLSI Architectures for Bit Parallel Computation in Galois Fields. PhD Thesis, Institute for Experimental Mathematics, University of Essen, Germany, 1994.
14. V. Rijmen. Efficient Implementation of the Rijndael S-box. Available at http://www.esat.kuleuven.ac.be/ rijmen/rijndael
15. Francois-Xavier Standaert, Gael Rouvroy, Jean-Jacques Quisquater, Jean-Didier Legat. Efficient Implementation of Rijndael Encryption in Reconfigurable Hardware: Improvements and Design Tradeoffs. Accepted at Workshop on Cryptographic Hardware and Embedded Systems (CHES2003), September 2003.
16. K.Y. Chen, P.D. Chen and C.S. Laih. Speed up AES with the modification of shift row table. Public Comments on the Draft Federal Information Processing Standard (FIPS), 2001.

Appendix A:

The dual cipher of AES is denoted by the pair $\{R(x), \beta\}$. All the 240 dual ciphers are listed as follows:

$R(x)$	Generator β of the dual AES $\{R(x), \beta\}$							
11B	03	05	11	1A	4C	5F	E5	FB
11D	02	04	10	1D	4C	5F	85	9D
12B	49	5C	8B	9B	9D	9F	A0	A7
12D	2A	3F	61	66	86	A8	CC	F0
139	2A	3F	60	67	88	A0	C3	F9
13F	4E	5B	CB	CC	E3	E5	E6	F2
14D	0D	18	1F	51	B7	E5	F6	FF
15F	0B	19	1E	45	80	8E	9D	DA
163	2F	3F	5B	5F	AC	BA	D9	DB
165	2A	3B	49	4C	B5	B6	C6	D1
169	21	23	31	32	A0	A5	C8	CC
171	17	3D	5C	64	93	95	AC	B8
177	16	29	4E	63	C6	D2	EA	EC
17B	26	35	49	5B	83	94	EB	FD
187	07	15	37	73	96	CA	E3	E9
18B	32	3E	6E	7D	85	BC	D8	FE
18D	21	2A	32	76	A2	C1	CB	E7
19F	06	14	24	3A	8F	AC	CD	E2
1A3	32	4F	52	75	A0	CE	DC	E8
1A9	37	47	76	7F	89	C0	D3	E3
1B1	1A	2A	53	6D	CA	D9	E8	F5
1BD	1C	68	91	A5	BF	E4	ED	F6
1C3	25	52	71	7F	9B	AA	B9	F1
1CF	21	47	5D	73	96	A3	B1	CC
1D7	1B	69	8E	92	9C	C8	D5	EF
1DD	1D	3F	46	6D	8C	96	A6	B5
1E7	06	14	2E	36	9A	A1	C6	F7
1F3	21	2B	6F	7C	81	B4	D7	FB
1F5	21	32	3F	71	9E	A7	AB	CF
1F9	07	15	25	75	97	9B	A6	E8

Periodic Properties of Counter Assisted Stream Ciphers

Ove Scavenius[1], Martin Boesgaard[1], Thomas Pedersen[1],
Jesper Christiansen[1], and Vincent Rijmen[2]

[1]Cryptico, Fruebjergvej 3, 2100 Copenhagen, Denmark
os@cryptico.com
[2]Cryptomathic, Lei 8A, 3000 Leuven, Belgium
vincent.rijmen@cryptomathic.com

Abstract. This paper analyses periodic properties of counter assisted stream ciphers. In particular, we analyze constructions where the counter system also has the purpose of providing additional complexity. We then apply the results to the recently proposed stream cipher Rabbit, and increase the lower bound on the internal state period length from 2^{158} to 2^{215}. With reasonable assumptions we illustrate that the period length of Rabbit is at least the period of the counter system, i.e. at least $2^{256} - 1$. The investigations are related to a "mod 3" characteristic of Rabbit. Attacks based on this characteristic are discussed and found infeasible.

Keywords. Stream cipher, period, counter, diversity, degeneracy, Rabbit

1 Introduction

An important problem in the construction of stream ciphers based on iterative number generators of the form $x_{i+1} = f(x_i) \bmod m$ is to secure that the state variable, x_i, does not enter into short periods. Most stream ciphers are based on Linear Feedback Shift Registers (LFSRs) which have provable period properties (see e.g. [1]). LFSRs are simple and linear and additional measures must be taken to ensure the security of such ciphers (see e.g. SNOW [2]). Another approach is to iterate a non-linear function f. However, the periodic properties then generally become uncertain. In [3] Shamir and Tsaban propose to use counter assisted generators with the iteration scheme $x_{i+1} = f(x_i) + c_i \bmod m$, where c_i is the independent counter state[1]. If $c_i \neq c_j$ in a counter period, N_c, for all $j - i \bmod N_c \neq 0$ then the generator will have the same period length as the counter system. In this construction the counter value is added before the pseudo-random data is extracted and the only secret part is located in the x_i variable [3]. For instance, by a linear combination of the output it might be possible to make the

---

[1] In this context '+' denotes the usual addition modulo m. However, any Latin square operation can be used, see [3] for details.

T. Okamoto (Ed.): CT-RSA 2004, LNCS 2964, pp. 39–53, 2004.
© Springer-Verlag Berlin Heidelberg 2004

counter system vanish [4]. Therefore, in some situations it might be beneficial to extract before the counter state is added, i.e. $x_{i+1} = f(x_i + c_i \bmod m) \bmod m$. In this way a potential additional complexity is included provided by the counter system, but this construction makes the period properties depend on the specific function f.

The aim of this paper is to extent parts of the analysis performed by Shamir and Tsaban [3] on the periodic properties of counter assisted stream ciphers. In particular, we aim at providing lower bounds for periods and diversities of generators where the complexity of the counter system is included as described above. As an example of such a construction we analyze the periodic properties of the recently proposed Rabbit stream cipher [5]. Strict bounds on the diversities and periods are determined and corresponding expectation values are calculated. The investigations are related to a "mod 3" characteristic of Rabbit. We analyze this characteristic in the last part of the paper.

The rest of the paper is organized as follows. In section two we perform an analysis of counter assisted generators and provide strict lower bounds on the internal state periods and diversities. In sections three and four we analytically and statistically, respectively, analyze the internal state period of Rabbit. We then analyze a "mod 3" characteristic of Rabbit in section five. We conclude and summarize in section six.

2   Period and Diversity Properties

In this section we provide basic definitions and strict lower bounds on the periods and diversities for simple systems.

2.1   Basic Definitions

We will need the following definitions.

Definition 1: Define the diversity, D_s, of a sequence, s_i, with period length N_s to be the number of distinct elements in the sequence. Note that $D_s \leqq N_s$.

Definition 2: Define the degeneracy, $d_f(u)$, of an output point, $u = f(s)$, to be the number of distinct input points, s, resulting in the same output point u. Furthermore, define the degeneracy, d_f, of the map $u = f(s)$ to be the maximal $d_f(u)$ of all $u \in \text{Im}(f)$.

Next, we provide diversity and period relations for simple systems to be used for more complicated constructions.

Consider a function $u = f(s) \bmod m$, as illustrated in the left part of Fig. 1, where $s \in \{0, \dots, m-1\}$, $u \in \text{Im}(f) \subseteq \{0, \dots, m-1\}$. Then we have

$$N_u \leqq N_s, \tag{1}$$

Fig. 1. The left figure illustrates the system, $u = f(s) \bmod m$, and the right figure illustrates the system, $u = s + t \bmod m$.

and

$$D_u \geqq \frac{D_s}{d_f}. \tag{2}$$

Note, that if f is bijective, i.e. $d_f = 1$, we have equalities in the above relations.

Also, consider the system defined by $u = s + t \bmod m$, where $s \in \{0, \ldots, m-1\}$, $t \in \{0, \ldots, m-1\}$ and $u \in \{0, \ldots, m-1\}$, as illustrated in the right part of Fig. 1. The following properties for this system hold true:

$$N_u \leqq \frac{N_s N_t}{\gcd(N_s, N_t)}, \tag{3}$$

and

$$D_u \geqq \begin{cases} D_s/D_t & \text{if } D_s \geqq D_t \\ D_t/D_s & \text{if } D_t \geqq D_s. \end{cases} \tag{4}$$

Definition 3: In the present context a counter assisted vector valued next-state function is defined in the following way

$$\vec{x}_{i+1} = \overrightarrow{F(\vec{y}_i)} \bmod m, \tag{5}$$

where

$$\vec{y}_i = (\vec{x}_i + \vec{c}_i) \bmod m, \tag{6}$$

such that $\vec{x}_i$ is the internal state variable. The counter state, $\vec{c}_i$, is generated by any of $\vec{F}$ independent generator and $\vec{c}_i \neq \vec{c}_j$ for $i - j \bmod N_c \neq 0$ where N_c is the (known) counter period, $\vec{y}_i$ is the counter modified internal state and m is the size of the state space for each vector component. The system is illustrated in Fig. 2.

2.2   Results for Counter Assisted Generators

In the following we analyze the general construction given in Definition 3 above[2].

---

[2] Note that eqs. (7) and (9) with proofs already appeared in Appendix C of [5], but they are included for completeness of the description.

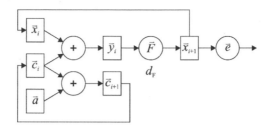

Fig. 2. Illustrates the counter assisted system described in Definition 3. The vector, $\vec{a}$, is the increment value for the counter system. Furthermore, $\vec{e}$ denotes the extraction function.

According to Lemma 4.2 in [3], $\vec{y}_i$ will have at least the period of the counter system, N_c:

Proof: Assume that there exists $\vec{y}_i = \vec{y}_j$ for $i - j \mod N_c \neq 0$, then $\vec{y}_{i+1} = \overrightarrow{F(\vec{y}_i)} + \vec{c}_{i+1}$ and $\vec{y}_{j+1} = \overrightarrow{F(\vec{y}_j)} + \vec{c}_{j+1}$. Moreover, we have: $\vec{c}_{i+1} \neq \vec{c}_{j+1}$, therefore, $\vec{y}_{i+1} \neq \vec{y}_{j+1}$. Finally, if $\vec{y}_{i-1} = \vec{y}_{j-1}$ this would imply that $\vec{y}_i \neq \vec{y}_j$ which is a contradiction. Thus, also $\vec{y}_{i-1} \neq \vec{y}_{j-1}$ ∎

The diversity, D_y, of the periodic sequence, $\vec{y}_i$, will at least be the square root of the diversity, D_c, of the counter system or at least the period, $N_c = D_c$, of the counter system, divided by the size of the image of $\vec{F}$, $|\text{Im}(\vec{F})|$, all according to Theorem 4.3 in [3].

It is not guarantied that $\vec{x}_i$ will have at least the period of the counter system (see Definition 3 or Fig. 2), but lower bounds for the period, N_x, and diversity, D_x, of the periodic sequence, $\vec{x}_i$, can still be obtained.

First, we note that there are relations between the counter period, N_c, the internal state period, N_x, and the period of the $\vec{y}$ variables, N_y:

$$N_y = aN_x = bN_c, \tag{7}$$

where a and b are integers greater than zero with $\gcd(a, b) = 1$.

Proof: According to eq. (1), we have $N_x \leq N_y$. In particular, N_x divides N_y, because, if we assume that this is not the case, then there would exist an i such that $\overrightarrow{F(\vec{y}_i)} = \vec{x}_{i+1} \neq \vec{x}_{i+1+\frac{N_y}{N_x}N_x} = \overrightarrow{F(\vec{y}_{i+\frac{N_y}{N_x}N_x})}$ which contradicts the N_y periodicity. Thus, there exists an integer, $a > 0$, such that $N_y = aN_x$. We also have that N_c divides N_y because if this was not the case then $\vec{c}_i \neq \vec{c}_{i+\frac{N_y}{N_c}N_c}$. We just showed that $\vec{x}_i = \vec{x}_{i+N_y}$ for all i, but $\vec{y}_i = \vec{x}_i + \vec{c}_i \neq \vec{x}_{i+\frac{N_y}{N_c}N_c} + \vec{c}_{i+\frac{N_y}{N_c}N_c} = \vec{y}_{i+N_y}$ which again contradicts the N_y periodicity. Therefore, there exists an integer, $b > 0$ such that $N_y = bN_c$ and consequently, $N_y = aN_x = bN_c$ ∎

Let $d_{\mathrm{F}}(\vec{x}_j)$ be the degeneracy of a point $\vec{x}_j$ in a periodic solution generated by the next-state function, $\vec{F}$. Furthermore, let $d_{\min}$ be the smallest $d_{\mathrm{F}}(\vec{x}_j)$ of all $\vec{x}_j$ belonging to the periodic solution. We then obtain the following bound:

$$N_{\mathrm{x}} \geq b \frac{N_{\mathrm{c}}}{d_{\min}}. \tag{8}$$

Proof: Since $N_{\mathrm{x}} = bN_{\mathrm{c}}/a$, we want to show that $d_{\min} \geq a$. For $a = 1$ this is trivially fulfilled. For $a > 1$ the periodicity gives: $\vec{x}_i = \vec{x}_{i+N_{\mathrm{x}}} = \vec{x}_{i+2N_{\mathrm{x}}} = \dots = \vec{x}_{i+(a-1)N_{\mathrm{x}}}$. On the other hand, the corresponding counter values are non-equal: $\vec{c}_i \neq \vec{c}_{i+N_{\mathrm{x}}} \neq \vec{c}_{i+2N_{\mathrm{x}}} \neq \dots \neq \vec{c}_{i+(a-1)N_{\mathrm{x}}}$, which is true since $\gcd(a,b) = 1$ and $\vec{c}_i \neq \vec{c}_j$ for $i - j \bmod N_{\mathrm{c}} \neq 0$. Therefore, it follows: $\vec{x}_i + \vec{c}_i \neq \vec{x}_{i+N_{\mathrm{x}}} + \vec{c}_{i+N_{\mathrm{x}}} \neq \vec{x}_{i+2N_{\mathrm{x}}} + \vec{c}_{i+2N_{\mathrm{x}}} \neq \dots \neq \vec{x}_{i+(a-1)N_{\mathrm{x}}} + \vec{c}_{i+(a-1)N_{\mathrm{x}}}$ or equivalently: $\vec{y}_i \neq \vec{y}_{i+N_{\mathrm{x}}} \neq \vec{y}_{i+2N_{\mathrm{x}}} \neq \dots \neq \vec{y}_{i+(a-1)N_{\mathrm{x}}}$. Because of the periodicity we have $\overrightarrow{F(\vec{y}_i)} = \overrightarrow{F(\vec{y}_{i+N_{\mathrm{x}}})} = \overrightarrow{F(\vec{y}_{i+2N_{\mathrm{x}}})} = \dots = \overrightarrow{F(\vec{y}_{i+(a-1)N_{\mathrm{x}}})}$. Therefore, the existence of such a periodic solution requires that the degeneracy of each point in the period must be at least a. Thus, we have $d_{\min} \geq a$ and eq. (8) follows ∎

Clearly, it is difficult to use the bound given eq. (8) in practice as b and $d_{\min}$ depend on the specific periodic solution. However, a more general bound follows trivially by noting that $b \geq 1$ and that $d_{\mathrm{F}} \geq d_{\min}$ where d_{F} is the degeneracy of the next-state function:

$$N_{\mathrm{x}} \geq \frac{N_{\mathrm{c}}}{d_{\mathrm{F}}}. \tag{9}$$

A lower bound for the diversity of the x_i sequence can also be specified. According to eq. (2), D_{x} is bounded by $D_{\mathrm{x}} \geq D_{\mathrm{y}}/d_{\mathrm{F}}$. For the present purpose, a part of Theorem 4.3 in [3] is useful: $D_{\mathrm{y}} \geq N_{\mathrm{c}}/D_{\mathrm{x}}$ which follows from eq. (4) and that $N_{\mathrm{c}} = D_{\mathrm{c}}$. Therefore, we can write

$$\frac{N_{\mathrm{c}}}{D_{\mathrm{x}}} \leq D_{\mathrm{y}} \leq D_{\mathrm{x}} \cdot d_{\mathrm{F}}, \tag{10}$$

to obtain the following bound:

$$D_{\mathrm{x}} \geq \sqrt{\frac{N_{\mathrm{c}}}{d_{\mathrm{F}}}}. \tag{11}$$

The above bounds given in eqs. (9) and (11) can be improved further. Clearly, using only the degeneracy of the next-state function might result in a rather pessimistic lower bound on both the period length, N_{x}, and the diversity, D_{x}, if not most of the points in $\mathrm{Im}(\vec{F})$ have d_{F} pre-images. In general, there will be a distribution of degeneracies, i.e. each output point will have a certain degeneracy, $d(x_j)$. We now define a sorted list, $\{d_{\mathrm{F}}(1), d_{\mathrm{F}}(2), \dots, d_{\mathrm{F}}(|\mathrm{Im}(\vec{F})|-1), d_{\mathrm{F}}(|\mathrm{Im}(\vec{F})|)\}$ of degeneracies of points in $\mathrm{Im}(\vec{F})$ such that $d_{\mathrm{F}}(j) \geq d_{\mathrm{F}}(i)$ if $j < i$. Thus, the bound $d_{\mathrm{F}} \cdot D_{\mathrm{x}} \geq D_{\mathrm{y}}$ generalizes to

$$\sum_{j=1}^{D_{\mathrm{x}}} d_{\mathrm{F}}(j) \geq D_{\mathrm{y}}, \tag{12}$$

Consequently D_x and N_x are bounded by

$$\sum_{j=1}^{D_x} d_F(j) \geqq \frac{N_c}{D_x}, \tag{13}$$

and

$$N_x \geqq \frac{N_c}{d_F(D_x)}, \tag{14}$$

where eqs. (8), (10) and (12) were used.

3   Strict Bound on the Internal State Period of Rabbit

In order to illustrate the use of the above results, we analyze Rabbit [5] in the following.

The core of the Rabbit algorithm is the iteration of the next-state function defined by

$$x_{j,i+1} = \begin{cases} g_{j,i} + (g_{j-1 \bmod 8,i} \lll 16) + (g_{j-2 \bmod 8,i} \lll 16) & \text{for } j \text{ even} \\ g_{j,i} + (g_{j-1 \bmod 8,i} \lll 8) + g_{j-2 \bmod 8,i} & \text{for } j \text{ odd} \end{cases} \tag{15}$$

$$g_{j,i} = \left((x_{j,i} + c_{j,i})^2 \oplus ((x_{j,i} + c_{j,i})^2 \gg 32)\right) \bmod 2^{32}, \tag{16}$$

where $j \in \{0, ..., 7\}$ and all additions are modulo 2^{32}. For convenience, we write the next-state function in the following way

$$\vec{x}_{i+1} = \overrightarrow{R(\vec{y_i})} = \overrightarrow{F(g(\vec{y_i}))} \bmod 2^{32}, \tag{17}$$

where

$$\vec{y_i} = (\vec{x}_i + \vec{c}_i) \bmod 2^{32}, \tag{18}$$

such that $\vec{x}_i$ is the internal state variable, $\vec{c}_i$ is the counter state, $\vec{g}$ is the vector of g-functions and $\vec{F}$ is the combining function containing the rotations and additions of the eight g-functions. The system is illustrated in Fig. 3.

According to eqs. (9) and (11) we need to know the counter period, N_c, as well as the maximal degeneracy, d_R, of the next-state function in order to obtain a lower bound on the period length, N_x, and the diversity, D_x of the internal state. If we use eq. (13) together with eq. (14) we also need to know the degeneracy distribution of the next-state function. The counter period is shown in [5] to be $N_c = 2^{256} - 1$. In the following we calculate bounds on the degeneracy as well as the degeneracy distribution of the next-state function of Rabbit.

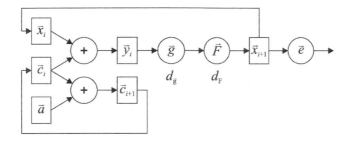

Fig. 3. Illustrates the Rabbit algorithm.

3.1   A Bound on the Degeneracy d_R

The degeneracy, d_R, is bounded by

$$d_R \leq d_F \cdot d_{\vec{g}} = d_F \cdot d_g{}^8, \tag{19}$$

where $d_{\vec{g}}$ is the degeneracy of the vector of g-functions, d_g is the degeneracy for each component of the $\vec{g}$-function and d_F is the degeneracy of the combining function, $\vec{F}$.

The degeneracy for the g-function, d_g, can easily be obtained by running through all its 2^{32} possible inputs. It turns out that there is one output with 18 inputs and all other images have smaller degeneracies, i.e. $d_g = 18$. The exact distribution is shown in Fig. 4. However, the degeneracy for the combining function, $\vec{F}$, cannot be obtained exactly by a measurement, but in the following we provide an upper bound.

Consider the three equation systems given in eqs. (20), (21) and (22) below

$$\vec{x} = \overrightarrow{F(\vec{y})} \bmod 2^{32}, \tag{20}$$

arising by replacing all the g-functions by identity functions but keeping the rotations and $\vec{y} \in \{0, \dots, 2^{32} - 1\}^8$,

$$\vec{w} = \overrightarrow{F(\vec{y})}, \tag{21}$$

same as eq. (20) but no addition modulus is performed, and finally

$$\vec{z} = \overrightarrow{F(\vec{y})} \bmod 2^{32} - 1, \tag{22}$$

where $\vec{y} \in \{0, \dots, 2^{32} - 2\}^8$. Furthermore the rotation operation entering in the $\vec{F}$-function can be written as

$$y \lll j = \begin{cases} 2^j y \bmod (2^{32} - 1) & \text{if } y < 2^{32} - 1 \\ 2^{32} - 1 & \text{if } y = 2^{32} - 1. \end{cases} \tag{23}$$

The system given in eq. (20) is non-linear. To circumvent this difficulty we first analyze the system defined in eq. (22) which is linear and can be written as

$$\vec{z} = B\vec{y} \bmod (2^{32} - 1). \tag{24}$$

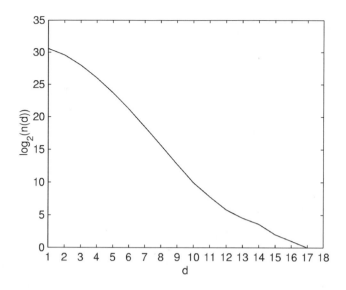

Fig. 4. The degeneracy distribution for the g-function, where $n(d)$ denotes the number of points having degeneracy d.

The matrix B is then given by

$$
B = \begin{pmatrix}
1 & 0 & 0 & 0 & 0 & 0 & 2^{16} & 2^{16} \\
2^8 & 1 & 0 & 0 & 0 & 0 & 0 & 1 \\
2^{16} & 2^{16} & 1 & 0 & 0 & 0 & 0 & 0 \\
0 & 1 & 2^8 & 1 & 0 & 0 & 0 & 0 \\
0 & 0 & 2^{16} & 2^{16} & 1 & 0 & 0 & 0 \\
0 & 0 & 0 & 1 & 2^8 & 1 & 0 & 0 \\
0 & 0 & 0 & 0 & 2^{16} & 2^{16} & 1 & 0 \\
0 & 0 & 0 & 0 & 0 & 1 & 2^8 & 1
\end{pmatrix} .
\tag{25}
$$

B is not invertible modulo $2^{32} - 1$, since 3 divides both the determinant and the modulus. B is therefore invertible modulo $(2^{32} - 1)/3$ but not invertible modulo 3. Furthermore, it can be verified that B modulo 3 is 3-to-1. Consequently, according to the Chinese Remainder Theorem (CRT), B is also 3-to-1 modulo $2^{32} - 1$. The kernel of the map consists of the three vectors: $\vec{\beta}_0 = (0, ..., 0)$, $\vec{\beta}_1 = ((2^{32} - 1)/3, ..., (2^{32} - 1)/3)$ and $\vec{\beta}_2 = (2(2^{32} - 1)/3, ..., 2(2^{32} - 1)/3)$.

Next we consider the case when the addition modulus is omitted, i.e. the system defined in eq. (21). Clearly, the images can be at most 3-to-1 when restricting the input vectors to $\vec{y} \in \{0, \dots, 2^{32} - 2\}^8$. Including the cases where one or more of the components of the input vector are $2^{32} - 1$ does not change the 3-to-1 property of eq. (21):

Proof: The particular case where $y_j = 2^{32} - 1$ for a $j \in \{0, \dots, 7\}$ and $y_i < 2^{32} - 1$ for $i \neq j$ corresponds to just adding $\overrightarrow{F(\hat{y})}$, where $\hat{y}_j = 2^{32} - 1$ and $\hat{y}_i = 0$ for

$i \neq j$, to a $\tilde{w}_j$ resulting from an $\tilde{y}_j = 0$ and $\tilde{y}_i = y_i$ for $j \neq i$. Of course, we could have that $\overrightarrow{F(\tilde{y})} = \overrightarrow{F(\vec{y} + \vec{\beta}_1 \bmod 2^{32} - 1)} = \overrightarrow{F(\vec{y} + \vec{\beta}_2 \bmod 2^{32} - 1)}$ but then none of those can be equal to $\overrightarrow{F(\vec{y})}$. As eq. (21) is 1-to-1 if $\vec{y} \in \{0, 2^{32} - 1\}^8$, we conclude that eq. (21) is at most 3-to-1 for all $\vec{y} \in \{0, \dots, 2^{32} - 1\}^8$ ∎

Furthermore, for every output vector of the systems eqs. (21) and (22) the sum of its elements is zero modulo 3, i.e.

$$\left(\sum_{j=0}^{7} w_j \right) \bmod 3 = \left(\sum_{j=0}^{7} z_j \right) \bmod 3 = 0. \tag{26}$$

This follows because the sums of the column elements of the matrix, B, are divisible by three, i.e. $1 + 2^8 + 2^{16} \bmod 3 = 0$ and $1 + 2^{16} + 1 \bmod 3 = 0$.

Finally, we investigate the system defined in eq. (20). In order for two vectors, $\vec{w}_1$ and $\vec{w}_2$, to be equal modulo 2^{32}, i.e. $\vec{x}_1 = \vec{x}_2$, we must have

$$\vec{w}_1 = \vec{w}_2 + 2^{32} \vec{k}, \tag{27}$$

where $\vec{k} \in \{-2, -1, 0, 1, 2\}^8$. Since $0 \leq w_j \leq 3(2^{32} - 1)$ then for each vector $\vec{w}$ there are 3^8 relevant vectors $\vec{k}$. Consequently, since the map is at most 3-to-1 without the addition modulus, an upper bound of how much the total system, eq. (20), is to one, is then given by $3 \cdot 3^8$-to-1. This can be lowered using eq. (26) since we must also have

$$\left(\sum_{j=0}^{7} 2^{32} k_j \right) \bmod 3 = 0. \tag{28}$$

This limits the numbers of $\vec{k}$-vectors, such that $d_F = 3 \cdot 3^8 / 3$. Consequently, the upper bound for d_R is then

$$d_R \leq 3^8 \cdot 18^8 \approx 2^{46}, \tag{29}$$

where eq. (19) was used. Eqs. (9) and (11) then provide the lower bound on the internal state period:

$$N_x \geq \frac{2^{256} - 1}{2^{46}} \approx 2^{210}, \tag{30}$$

and the diversity:

$$D_x \geq \sqrt{N_x} \approx 2^{105}. \tag{31}$$

3.2   An Improved Bound on d_R

The above bounds given in eqs. (30) and (31) do not take eqs. (13) (or equivalently eq. (12)) and (14) into account. The degeneracy distribution of the next-state function is not known, but we can obtain the degeneracy distribution of

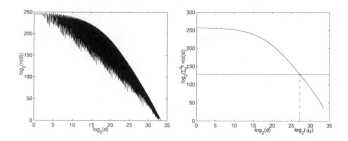

Fig. 5. The left figure shows the degeneracy distribution of the $\vec{g}$-function, and the right figure shows the accumulated distribution together with the horizontal line illustrating the lower bound, $D_y \geq 2^{128}$. In both figures $n(d)$ denotes the number of points having degeneracy d.

the $\vec{g}$-function. To accomplish that, we use the degeneracy distribution of the g-function and combine its degeneracies into the eight component vector in all possible ways. The result is shown in left part of Fig. 5. Using eq. (12) and that $D_y \geq 2^{128}$, we obtain $d_{\vec{g}}(D_x) \leq 2^{28}$, i.e. in any periodic solution of the $\vec{g}$-function at least one point will exist with degeneracy less than or equal to 2^{28}. The result is illustrated by the right part of Fig. 5. Thus, using eq. (14) the period is

$$N_x \geq \frac{2^{256} - 1}{3^8 \cdot 2^{28}} \approx 2^{215}, \tag{32}$$

and the diversity is $D_x > 2^{107}$.

4   Statistical Analysis of Periods and Diversities

Using the above analysis, we sample the degeneracy distribution of the system in eq. (20). Furthermore, this measurement can be used to argue that the internal state period will be larger than or equal to the counter period with a probability practically equal to one.

4.1   Measuring the Degeneracy Distribution of the $\vec{F}$-Function

The key observation is that for a given input vector, $\vec{y}$, we know all other input vectors which might result in the same output vector, $\vec{x} = \vec{w} \mod 2^{32}$. In order for $\vec{w}_1$ and $\vec{w}_2$ to be equal modulo 2^{32} we must have

$$\vec{w}_1 = \vec{w}_2 + 2^{32}\vec{k} = \vec{w}_2 + \vec{k} + (2^{32} - 1)\vec{k}, \tag{33}$$

where $\vec{k} \in \{-2, -1, 0, 1, 2\}^8$. Because of the linearity of the system given in eq. (22), we only need to find all pre-images of the vectors, $\vec{k}$, and add each one of them to the input vector, $\vec{y}$. More precisely, for all $\vec{k}$ calculate all $\vec{y}' = \vec{y} + B^{-1}\vec{k} \mod 2^{32} - 1$, then calculate all $\vec{x}' = \overrightarrow{F(\vec{y}')} \mod 2^{32}$ and look for matches[3]. However, by this procedure we only obtain the specific matches where all vector components of $\vec{y}'$ are different from $2^{32} - 1$. Therefore, we must also check the possibilities when one or more vector components are $2^{32} - 1$. This is done as follows: Whenever a vector component of $\vec{y}'$ is zero, we also try the vector where the zero component is replaced by $2^{32} - 1$.

In order to find all pre-images of the $\vec{k}$ vectors we do as follows. As described in section 3.1, the matrix B is 3-to-1. However, we can for a given output vector $\vec{k}$ find each one of the three pre-images by using the CRT:

$$B^{-1}\vec{k} = \left(a\frac{2^{32} - 1}{3}\vec{\kappa}_i + 3b\tilde{B}^{-1} \left(\vec{k} \bmod \frac{2^{32} - 1}{3} \right) \right) \bmod (2^{32} - 1), \qquad (34)$$

where a and b are determined by: $a(2^{32}-1)/3 \bmod 3 = 1$ and $3b \bmod (2^{32}-1)/3 = 1$, respectively. The matrix $\tilde{B}^{-1}$ denotes the inverse of B modulo $(2^{32} - 1)/3$. Finally, $\vec{\kappa}$ denotes one of the three pre-images of $\vec{k}$ modulo 3, which can easily be found by searching through the possible 3^8 input vectors to the matrix $B \bmod 3$. Now it is straightforward to sample the degeneracy distribution. It was done as follows:

1. Pick a random input vector $\vec{y} \in \{0, \ldots, 2^{32} - 2\}^8$.
2. Calculate $\vec{w} = \overrightarrow{F(\vec{y})}$ where the addition modulus is not performed.
3. Check each component, w_j, and select vectors $\vec{k}$ where k_j is in the corresponding interval, i.e.:
    - if $w_j \in \{0, \ldots, 2^{32} - 2\}$ select $k_j \in \{0, 1, 2\}$.
    - if $w_j \in \{2^{32} - 1, \ldots, 2 \cdot 2^{32} - 3\}$ select $k_j \in \{-1, 0, 1\}$.
    - if $w_j \in \{2 \cdot 2^{32} - 2, \ldots, 3 \cdot 2^{32} - 6\}$ select $k_j \in \{-2, -1, 0\}$.
4. This provides the candidates each of the form $\vec{y}' = \vec{y} + B^{-1}\vec{k} \bmod 2^{32} - 1$ for pre-images which can result in the same image as $\vec{y}$ under addition modulo 2^{32}. Furthermore, if $y_j' = 0$ then also try $y_j' = 2^{32} - 1$.
5. Repeat 10^8 times.

Fig. 6 shows the normalized result. It is normalized by dividing each count by the corresponding degeneracy.

4.2 Expectation Values of the Period and Diversity of the $\vec{g}$-Function

It seems reasonable to assume that the internal state variables, $\vec{x}_i$, and the counter variables, $\vec{c}_i$, are independent. Therefore, the expectation value of the diversity, D_y, of their sum should be much higher than the bound given above:

$$\langle D_y \rangle \approx \left(1 - e^{-1} \right) \cdot 2^{256} \approx 2^{255.34}, \qquad (35)$$

where the pre-factor, $1 - e^{-1}$, originates from the fact that if $n = 2^{256} - 1$ (minimal N_y) balls are thrown into $m = 2^{256}$ urns, a proportion of about $e^{-n/m}$

---

[3] Note, that we symbolically write B^{-1} even though B is not invertible modulo $2^{32} - 1$. In other words $B^{-1}\vec{k}$ is just a shorthand notation for the resulting pre-images.

50   O. Scavenius et al.

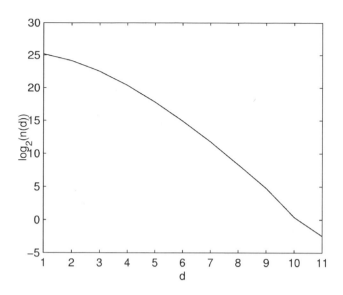

Fig. 6. The normalized measured degeneracy distribution for the combining function, $\vec{F}$, where $n(d)$ denotes the number of points having degeneracy, d.

of the urns will remain empty (see e.g. [6] for more details). According to eq. (8) then if a point with degeneracy one belongs to a periodic solution the period will be at least that of the counter period. The probability that no such vector belongs to the periodic solution can be calculated as follows. In the g-function 2721872779 input values are many-to-1 and 1573094517 input values are 1-to-1. Consequently, $1573094517^8 \approx 2^{244.41}$ input vectors to the $\vec{g}$-function are 1-to-1. The probability for an input vector to be many-to-1 is thus given by

$$1 - \frac{1573094517^8}{2^{256}} \approx 0.999676, \tag{36}$$

but the probability that $2^{255.34}$ different input vectors all to be many-to-1 is $0.999676^{2^{255.34}}$ which is by any measure zero. Consequently, it is reasonable to assume that the period of the $\vec{g}$ function, i.e. without the combining function $\vec{F}$, is N_y. The diversity, $D_{\vec{g}}$, can be calculated by the same method used in eq. (35) above: Let $n(d)$ be number of points having degeneracy d in the $\vec{g}$-function, then there are $|\mathrm{Im}(\vec{g})| = \sum_d n(d) \approx 2^{250.68}$ possible $\vec{g}$ images. Therefore, the expected diversity is

$$\langle D_{\vec{g}} \rangle \approx \left(1 - e^{-\frac{2^{255.34}}{2^{250.68}}}\right) \cdot 2^{250.68} \approx 2^{250.68}. \tag{37}$$

4.3   Expectation Values of the Internal State Period and Diversity

In order to calculate the expectation values of the internal state period, N_x, and diversity, D_x, we repeat the same analysis as in section 4.2. The probability for an input vector to be many-to-1 is

$$1 - \frac{37820209}{10^8} \approx 0.621798, \tag{38}$$

i.e number of samples minus the number of counted 1-to-1 points divided by number of samples (see Fig. 6). Consequently, the probability that all $2^{250.68}$ input vectors are many-to-1 is $0.621798^{2^{250.68}}$ which is again neglectible. Thus, according to eq. (8) we claim that the internal state period length is at least N_c. The expectation value of the diversity is obtained as in eq. (37): Let $n(d)$ be the measured number of points having degeneracy d in the $\vec{F}$-function, then there are $|\text{Im}(\vec{F})| \approx 2^{256}/10^8 \cdot \sum_d n(d) \approx 2^{255.36}$ possible $\vec{F}$ images. Thus,

$$\langle D_x \rangle \approx \left(1 - e^{-\frac{2^{250.68}}{2^{255.36}}}\right) \cdot 2^{255.36} \approx 2^{250.65}. \tag{39}$$

We conclude that the expectation value for the period of the counter assisted next-state function of Rabbit is at least that of the counter system and, moreover, a very large internal state diversity is expected. Finally, note that even if the strict bound for the diversities $D_y \geqq 2^{128}$ and $D_{\vec{g}} \geqq D_x \geqq \sqrt{2^{215}} = 2^{107}$ was used in the above calculations, this conclusion would remain the same.

5   Analysis of the Mod 3 Characteristic

As explained in section 3, eq. (26) there is a "mod 3" characteristic of the next-state function of Rabbit (see [7] for a general discussion on "mod n" attacks and see [8] for a "mod n" analysis of Rabbit). If the results of the additions are not reduced by modulus 2^{32}, then the sum of the eight output vector components is always zero modulo three. Taking the addition modulo 2^{32}, we still measure a distribution of the sum modulo three that is not uniform:

$$\left(\sum_{j=0}^{7} x_j \bmod 2^{32}\right) \bmod 3 = \begin{cases} 0 & \text{with probability } 0.3337 \\ 1 & \text{with probability } 0.3337 \\ 2 & \text{with probability } 0.3326. \end{cases} \tag{40}$$

This bias can to a high degree be reproduced by (tedious) analytical calculations (see [8] and [9] for details). If this bias was detectable in the extracted output, it would allow an attacker to mount a distinguishing attack. Since the property

Table 1. Numerical entropies. The figures for the joint distribution were obtained by computing the full distribution. The figures are therefore exact, except for rounding errors.

	$H(A)$	$H(B)$	$H(A,B)$
3	1.5850	12	13.5805
4	1.5850	16	17.5850

of $(\sum x_j)$ mod 3 does not depend in any way on the value of the key, it is not possible to recover the key from it.

Since the extracted output consists of the XOR values of the upper and lower halves of two x_j-values, there is no way we can reconstruct the sum of the vector components from the output (see [5] for details about the extraction function). Consequently, a necessary condition for detecting a bias in the output is that the following two distributions are dependent:

$$A(\vec{x}) = \sum_{j=0}^{7} (2^{16} x_{j,\mathrm{h}} + x_{j,\mathrm{l}}) \bmod 2^{32} \bmod 3, \tag{41}$$

$$B(\vec{x}) = (x_{0,\mathrm{h}} \oplus x_{3,\mathrm{l}}, x_{1,\mathrm{h}} \oplus x_{4,\mathrm{l}}, \dots, x_{7,\mathrm{h}} \oplus x_{2,\mathrm{l}}), \tag{42}$$

where the subscripts h and l denote the upper and lower 16 bits of x_j, respectively. We have done several measurements in order to detect any dependence of the two distributions, but have not found any. For instance, we computed the entropy, H, of the joint distribution (A, B) on down-scaled versions using three respectively four 8-bit registers instead of eight 32-bit registers. We used uniform distributions for the x_j. The results are presented in Table 1. It can be seen that the entropy of the joint distribution converges quickly to the sum of the entropies of the distributions of A and B. We conclude that when eight registers are used, the distributions will be almost completely independent. A fortiori, a cryptanalyst won't be able to notice the "mod 3" bias of the internal state in the output.

6   Conclusions

In this paper we analyzed the periodic properties of counter assisted generators where the counter system also provides additional complexity. As an example, we extended the lower bound on the internal state period of the recently proposed stream cipher Rabbit from 2^{158} to 2^{215}. By assuming that the internal state and the counter state are independent and applying the general analysis, we found that the internal state period of Rabbit is at least the counter period, i.e. $2^{256} - 1$. Furthermore, we discussed the "mod 3" characteristic of the next-state function. Attacks based on this characteristic were discussed and found infeasible.

Acknowledgements. The authors would like to thank Ivan Damgaard, Hans Anton Salomonsen and the reviewers for their helpful input.

References

1. S. Shelah and B. Tsaban, *Efficient linear feedback shift registers with maximal period*, Finite Fields and their Applications, 8, pp. 256–267 (2002)
2. P. Ekdahl and T. Johansson: *A New Version of the Stream Cipher SNOW*, Proceedings of Selected Areas in Cryptography 2002, Springer, pp. 49–61 (2002)

3. A. Shamir and B. Tsaban: *Guaranteeing the Diversity of Number Generators*, Information and Computation Vol.171, No.2, pp. 350–363 (2001) and http://xxx.lanl.gov/abs/cs.CR/0112014
4. D. Coppersmith, S. Halevi and C. Jutla, *Cryptanalysis of Stream Ciphers with Linear Masking*, Proceedings of Crypto 2002, Springer, (2002)
5. M. Boesgaard, M. Vesterager, T. Pedersen, J. Christiansen and O. Scavenius: *Rabbit: A New High-Performance Stream Cipher*, Proceedings of Fast Software Encryption (FSE) 2003, Springer, (2003)
6. P. Flajolet and A.M. Odlyzko: *Random Mapping statistics*, Advances in Cryptology - EUROCRYPT '89, Springer, pp. 329–354 (1990)
7. J. Kelsey, B. Schneier and D. Wagner: *Mod n Cryptanalysis, with Applications against RC5P and M6*, Fast Software Encryption, Sixth International Workshop Proceedings (March 1999), Springer, pp. 139–155 (1999)
8. *"mod n" Cryptanalysis of Rabbit*, white paper, version 1.0, www.cryptico.com (2003)
9. V. Rijmen, *Analysis of Rabbit*, unpublished report, www.cryptico.com (2003)

A Fast Correlation Attack via Unequal Error Correcting LDPC Codes

Maneli Noorkami and Faramarz Fekri

Center for Signal and Image Processing
School of Electrical & Computer Engineering
Georgia Institute of Technology
Atlanta, GA 30332, USA
{maneli,fekri}@ece.gatech.edu

Abstract. In this paper, an improved fast correlation attack on stream ciphers is presented. The proposed technique is based on the construction of an unequal error protecting LDPC code from the LFSR output sequence. The unequal error protection allows to achieve lower bit-error probability for initial bits of the LFSR in compared to the rest of the output bits. We show that constructing the unequal error protecting code has also the advantage of reducing the number of output bits involved in decoding to less than the available keystream output bits. Our decoding approach is based on combination of exhaustive search over a subset of information bits and a soft-decision iterative message passing decoding algorithm. We compare the performance of the proposed algorithm with the recent fast correlation attacks. Our results show that we can reduce the number of bits obtained by exhaustive search in half and still get better performance comparing to recent fast correlation attacks based on iterative decoding algorithm. Using the expected number of parity-check equations of certain weights, we find the lower bound on the number of information bits that needs to be obtained by the exhaustive search without compromising the performance.

Keywords. Stream ciphers, fast correlation attacks, linear feedback shift registers, cryptanalysis, LDPC codes.

1   Introduction

One of the most remarkable of all ciphers is the one-time-pad where the plaintext message is added bit by bit (or in general, character by character) to a random sequence of the same length. The remarkable fact about the one-time-pad is its perfect security. Assuming a ciphertext only attack, Shannon proved that even with infinite computing resources, the cryptanalyst could never separate the true plaintext from all other meaningful plaintexts. The disadvantage is the unlimited amount of key [1]. The appealing feature of the one-time-pad suggested building synchronous stream ciphers which encipher the plaintext by use of a pseudo-random sequence. This removes the requirement of an unlimited key. The

T. Okamoto (Ed.): CT-RSA 2004, LNCS 2964, pp. 54–66, 2004.
© Springer-Verlag Berlin Heidelberg 2004

pseudo-random sequence is under control of a secret key that is generated by a deterministic algorithm called the keystream generator. To ensure the security, it must not be possible to predict any portion of the keystream better than just random guessing, regardless of the number of keystream bits already observed.

Linear feedback shift registers (LFSR) are the basic components of most keystream generators. As shown in Figure 1, one method of generating the keystream is to combine a fixed number of LFSRs' outputs by means of a nonlinear function f. To resist cryptographic attacks using the Berlekamp-Massey algorithm, the function f is chosen so that a sequence with high linear complexity is obtained. The secret key is the initial state of each LFSR. The characteristic polynomial of each LFSR of length k_i is assumed to be known by the cryptanalyst. The total key bits required to specify the initial state of the stream cipher generator is $\sum_{i=1}^{R} k_i$, where R is the number of the LFSRs. In a brute force attack, $\prod_{i=1}^{R} 2^{k_i}$ possible states of the LFSRs are examined, which is not feasible in practical systems.

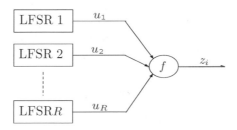

Fig. 1. Keystream generator.

In [2], Siegenthaler showed that if there exists a correlation between the keystream sequence and the outputs of LFSRs, it is possible to determine the initial state of each LFSR independently thereby reducing the cryptanalytic attack to a divide-and-conquer attack with approximate complexity of $\sum_{i=1}^{R} 2^{k_i}$. Siegenthaler's attack amounts to an exhaustive search through the state space of each individual LFSR.

Later, it was shown by Meier and Staffelbach that if the number of taps t of the characteristic polynomial is small, it is possible to determine the initial state of the LFSR by means of an iterative algorithm that has a complexity much less than the exhaustive search [3]. This work was followed by several papers, providing minor improvements to the initial results [4,5,6,7].

In [8], Johansson and Jonsson proposed a novel algorithm based on identifying an embedded low-rate convolutional code from the LFSR output sequence. This embedded low-rate convolutional code can then be decoded with low complexity using the Viterbi algorithm. Their approach considers a decoding algorithm that requires memory. This algorithm provides a remarkable improvement over previous methods. In [9], a new method for the fast correlation attack was proposed based on constructing and decoding turbo codes. For a fixed memory size, this technique provides better performance than the method in [8]. The price for the improved performance is an increased computational complexity. One of the advantages of the attacks based on the convolutional and turbo codes is that they do not require a low weight feedback polynomial. In [10], a new simple algorithm for fast correlation attacks on stream ciphers was presented. Although this algorithm is influenced by [8,9], it has the advantage that it reduces the memory requirement significantly. A detailed comparative study of the algorithms are covered in [11].

Canteaut and Trabbia showed in [12] that the Gallager iterative decoding algorithm using parity-check equations of weights four or five is more efficient than the attacks proposed in [8,9]. The large memory usage makes the complexity of the Viterbi decoding high. One advantage of the attack based on convolutional codes is the lower complexity of the preprocessing step, but this part of the attack is performed once while the decoding step is repeated for each new initialization of the system. An algorithm based on decoding of the binary block code whose parity-checks are of low weights was proposed in [13]. The authors employed combination of restricted exhaustive search over a set of hypotheses and a one-step or an iterative decoding technique. The exhaustive search is employed to provide a possibility for the construction of suitable parity-check equations needed for a high performance decoding. In [15], Chose and et al. presented some major algorithmic improvements. This improvement is achieved at the cost of some loss of parallelism over the method presented in [12]. They focused on the search for efficient algorithms to find and evaluate parity-check equations. The main idea is combined with the partial exhaustive search to yield an efficient cryptanalysis.

In this paper, we propose a new technique for fast correlation attack that is based on constructing a low-density parity-check codes (LDPC) from the LFSR output sequence. The key idea in our approach is to construct an LDPC code that provides a lower decoding error rate for the information bits (the initial state of the LFSR) compared to the rest of the LFSR output bits. We show that the structure of the underlying LDPC code constructed from the LFSR output sequence has a crucial role in the decoding performance. Motivated by the approach of [13,15], we also develop a novel algorithm for the construction of parity-check equations. Our decoding approach is based on combination of exhaustive search over a set of information bits and a soft-decision iterative message passing decoding algorithm [16].

The paper is organized as follows. Section 2 reviews the decoding model for the fast correlation attack. Section 3 points out the underlying idea for the proposed fast correlation attack, the procedure for the construction of the parity-check equations, and the decoding steps of the iterative message passing algorithm. The expected cardinalities of the parity-check equations are given in Section 4. Using the expected number of parity-check equations of certain weight, a lower bound on B is derived in Section 5. Section 6 presents the performance of the proposed algorithm. Finally, a comparison of the proposed algorithm with the recent fast correlation attacks is provided in Section 7.

2   Decoding Model for the Fast Correlation

Most authors view the problem of finding the initial state of the LFSR as a decoding problem. Assume that the target LFSR has length L and let γ denote the set containing all the distinct LFSR sequences. Clearly, $\gamma = 2^L$ and for a fixed length N, the truncated sequences from γ form a linear (N, L) block code. Therefore, the observed keystream sequence $\mathbf{z} = z_1, z_2, \ldots, z_N$ may be regarded as the received channel output and the LFSR sequence $\mathbf{x} = x_1, x_2, \ldots, x_N$ is regarded as a codeword from an (N, L) linear block code. Due to the correlation between x_i and z_i, we can view each z_i as the output of the binary symmetric channel (BSC) when x_i is transmitted. The correlation probability defined by

$$Pr(x_i = z_i) = 1 - p = 1/2 + \varepsilon$$

determines the crossover probability p of the BSC. This is shown in Figure 2.

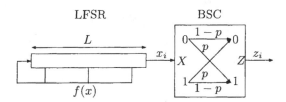

Fig. 2. Model of the correlation attack.

Therefore, the cryptanalyst's problem is to restore the LFSR's initial state $(x_1, x_2, \ldots, x_L)$ from the observed output sequence $\mathbf{z} = (z_1, z_2, \ldots, z_N)$, given the feedback polynomial of the LFSR of degree L. Let $H(p)$ be the binary entropy function $H(p) = -p \log_2 p - (1 - p) \log_2(1 - p)$. Then, for a code rate $R = L/N$ that is less than the capacity $C(p) = 1 - H(p)$ of the BSC, by Shannon's theory, there exists a code for which the decoding error probability approaches zeros as N goes to infinity.

3   Underlying Idea for the Fast Correlation Attack

The proposed algorithm construct a low-density parity-check codes (LDPC) from the LFSR output sequence that is defined by a sparse parity-check matrix. The resulting (N, L) LDPC code is an irregular LDPC code. An LDPC code is irregular if the number of 1's per column of H or the number of 1's per row of H is allowed to vary. An LDPC code can be represented by a bipartite graph consisting of check nodes and variable (bit) nodes. The $N - L$ rows of H specify the $N - L$ check node connections, and the N columns of H specify the N bit node connections. The ith bit node is connected to the jth check node if and only if the ij^{th} element of H is one. It is preferable to have a bit node with a high degree since it will receive more information from the check nodes, allowing for more accurate judgement of the correct bit value. On the other hand, it is more preferable to have a check node with a low degree since in this case the information that it sends to variable nodes is more valuable. Since they are connected in a bipartite graph, if the degree of the bit nodes is high (low), then so must the degree of the check nodes be high (low). But it is possible to put more weight of the parity-check equations on the information bits, raising the degrees of some variable nodes while keeping the degrees of check nodes constant.

In the fast correlation attack, we want to recover the initial state of the LFSR completely and we are not interested in the values of the rest of the output sequence. This motivated us to construct the underlying LDPC code such that we get lower bit error probability for the initial state of the LFSR than the rest of the output sequence. This can be achieved by increasing the degrees of the variable nodes that correspond to the initial state of the LFSR. Also raising the degrees of the output bits that are directly connected to the initial bits protects these bits more than the rest of the output sequence. Since they are directly connected to the initial bits, they help to lower the bit error probability of the initial bits.

A high performance decoding of LDPC codes requires low weight parity-check equations. Therefore, we must employ low weight parity-check equations that involve as many information bits as possible. Then, we use the parity-check equations that raise the degree of the output bits directly connected to the initial state of the LFSR. Constructing the H matrix in this way, some of the output nodes would have degree zero. Therefore, they do not get involve in decoding phase. This has the additional advantage of reducing the decoding complexity. In other words, instead of decoding a code of length N, a shorter sequence is decoded.

The bipartite structure of the proposed LDPC code is shown in Figure 3. Here, the bit nodes and the check nodes are represented by circles and squares, respectively. We have three sets of variable nodes: the initial bits of degree d_i, parity bits of degree d_j and parity bits of degree d_k. Note that we choose $d_i > d_j > d_k$ for unequal protection. There are two sets of parity check equations. Parity checks c_1 of degree dc_1 involve both information bits and the rest of the LFSR's output bits. Parity checks c_2 of degree dc_2 do not involve the initial bits.

3.1   Generating Parity Check Equations

The preprocessing step of our attack consists of generating low-weight parity-check equations that provide better performance and stronger error protection for the LFSR initial bits. The performance of a set of parity-check equations depends on their cardinality as well as on their weight distribution. Therefore,

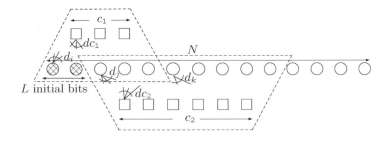

Fig. 3. Bipartite graph of the proposed LDPC code.

we are looking for sufficiently low weight parity-check equations. If the feedback polynomial of the LFSR is long and high weight, it is not possible to find sufficiently powerful parity-check equations when the length of the available keystream output sequence is short. Using the same idea as in [13,15], B bits of the initial state are found through exhaustive search and $L - B$ bits remain to be obtained using parity-check equations. This relaxes the constraint on parity-check equations. We require that the parity-check equations have low weight when the first B initial bits have arbitrary values.

We choose to construct the parity-check matrix H with parity equations of weights three and four. We first fill the matrix H with those parity equations that have the largest involvement of the information bits. Since the number of initial bits of the LFSR is much smaller than the output bits, this makes the degrees of the initial bits larger than the rest of the output sequence. Let the length of the available output sequence be N, and the degree of the feedback polynomial $f(x)$ be L. First, we generate parity-check equations that involve exactly one output bit and two or three information bits. These parity-check equations can be obtained as follows:

- Compute all the residues $q_i(x) = x^i \mod f(x)$ for $L < i < N$ and store them in a table $\mathbf{T}$.
- Search the table for parity-check equations that have weight four or less when the first B information bits take arbitrary values.

We use all the parity-check equations that are constructed by the above criteria. Then we employ parity-check equation that involve exactly two output bits and one or two information bits. These parities can be constructed as follows:

- Take all the choices of two out of $N - L$ entries of table $\mathbf{T}$ and generate a new table $\mathbf{R}$ by xoring those pairs.
- Search the table $\mathbf{R}$ for parity-check equations that have weight four or less when the first B information bits have arbitrary values.

Finding parity-check equations that involve three or four output bits by taking all the three or four choices from table $\mathbf{T}$ has a high complexity. We use the same idea as in [15] to find these parities with a much less complexity. To find parities that involve three output bits when the B information bits have arbitrary values, we proceed as follows:

- Find the indices of table **T** and **R** such that $(L - B)/2$ of the initial bits have a certain pattern.
- Among those entries, take an element from table **T** and an element from table **R** such that they have the same values in the remaining $(L - B)/2$ of initial bits.
- Repeat the above procedure for every possible pattern of the first $(L - B)/2$ of the initial bits of the LFSR.

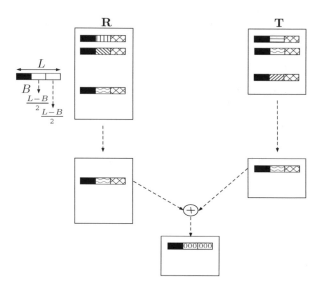

Fig. 4. Illustration of constructing parity-check equations of weight three.

The algorithm is illustrated by Figure 4. To find parities that involve three output bits with one information bit, we follow the same procedure as finding parities of three output bits, but each time we do not put any restriction on a certain initial bit (one of the $L - B$ bits). Among the parities found, we choose the ones that the selected bit has value one.

Same method is applied to find parities that only involve four output bits. But instead of searching both tables **T** and **R** for a certain pattern of the first $(L - B)/2$ initial bits, only table **R** needs to be searched. Among entries whose first $(L - B)/2$ initial bits are the same, we look for any two elements that have the same value in the next $(L - B)/2$ bits.

3.2 Iterative Decoding Algorithm

After generating the parity-check matrix H, we decode the keystream output sequence using the message passing algorithm discussed in [16] over the binary symmetric channel to recover the initial bits of the LFSR.

4 Expected Cardinalities of the Parity-Check Sets

Suppose we have N available output samples and B bits of the information bits are found through exhaustive search. Then by the following lemmas we give the expected cardinalities of the parity-check sets.

Lemma 1. *The expected number of parity-check equations that involve exactly one information bit and two output bits is*

$$(L - B)(N - L)(N - L - 1)2^{B-L-1}. \tag{1}$$

Lemma 2. *The expected number of parity-check equations that involve exactly two output bits and two information bits is*

$$(L - B)(L - B - 1)(N - L)(N - L - 1)2^{B-L-2} \tag{2}$$

Lemma 3. *The expected number of parity-check equations that involve only three output bits is*

$$(N - L)^2(N - L - 1)2^{B-L-1} \tag{3}$$

Lemma 4. *The expected number of parity-check equations that involve exactly three output bits and one information bit from the first half of the $(L - B)/2$ initial bits is equal to*

$$(N - L)^2(N - L - 1)2^{B-L} \tag{4}$$

Lemma 5. *The expected number of parity-check equations that involve exactly three output bits and one information bit from the remaining half of the $(L-B)/2$ initial bits is equal to*

$$(N - L)^2(N - L - 1)2^{B-L-1} \tag{5}$$

Our simulation results show the accuracy of the above approximation when N and L are 4000 and 40, respectively.

5   Lower Bound on B

Since finding the information bits through the exhaustive search is computationally expensive, ideally, we would like to minimize the number of those bits. On
the other hand, reducing B increases the initial state recovery error rate due
to the higher degrees of the check nodes. Here, we derive a lower bound on the
number of information bits that needs to be found by the exhaustive search when
the matrix H consists of mostly parity-check equations of weight three and some
parity-check equations of weight four that involves at least one information bit.

Ideally, the matrix H should have N columns and $N - L$ rows. From (3),(2)
and (1), we choose B such that the total number of parity-check equations of
weight three and some parity-check equations of weight four is greater than
$N - L$. Therefore, the following inequality should hold:

$$2^{B-L-1}\Big[(N - L)^2(N - L - 1) + (L - B)(N - L)(N - L - 1) +$$

$$\frac{1}{2}(L - B)(L_B - 1)(N - L)(N - L - 1)\Big] > N - L \qquad (6)$$

we find that the lower bound for B is 17 when N and L are 4000 and 40,
respectively.

Note that it is possible to reduce B further. However, we will not be able to
obtain sufficient parity-check equations of weight three and most of the parity-
check equations will then be of weight four. The performance of this code is
inferior to the case that most of the parity-check equations have weight three.

If we reduce B below than 17, then the number of parity-check equations that
relate two output bits to two information bits would be insufficient. Therefore,
we should employ parity-check equations that involve three output bits and one
information bit instead.

We can reduce B to the lowest value for which we can still obtain a few
parity-check equations for each bit of the LFSR initial state. We employ all the
parity-check equations we find for each bit and we fill the rest of the matrix H
with the parity-check equations that involve four output bits. Since the number
of these parity-check equations is large, they do not put any restriction on B.
For $L = 40$ and $N = 4000$, B can be reduced to 10.

6   Simulation Results

Here we present simulation results of our attack on an LFSR of length $L = 40$
with the feedback polynomial $f(x) = 1 + x^2 + x^3 + x^5 + x^9 + x^{11} + x^{12} + x^{17} +$
$x^{19} + x^{21} + x^{25} + x^{27} + x^{29} + x^{32} + x^{33} + x^{38} + x^{40}$. This is the same polynomial
considered in all the previous fast correlation attacks. We show that when the
majority of the parity-check equations have weight three, the minimum number
of information bits that needs to be found by the exhaustive search is equal
to 17. Table 1 presents the error rate of the LFSR initial state reconstruction

as a function of the correlation noise p when $B = 17, 18$ and the length of the available keystream output sequence is 4000. The error rate in Table 1 is obtained by averaging over a randomly selected set of 5000 sequences. This error rate is computed over the information bits only (the initial state of the LFSR).

Table 1. Error rate for the full recovery of the LFSR initial state.

p	$B = 17$	$B = 18$
0.46	0.000	0.000
0.47	0.240	0.004
0.48	0.990	0.740
0.49	1.000	1.000

To compare the performance of our unequal protecting LDPC code to that of the equal protecting code, we also constructed an equal protecting LDPC code. Table 2 summarizes the results for $B = 18$. As expected, the unequal LDPC code performs better than the equal LDPC code. The improvement in the recovery of the initial state of the LFSR is achieved in expense of increasing the bit error rate of the output bits.

Table 2. Comparison of an unequal protecting LDPC code with an equal protecting LDPC code for $B = 18$.

p	Unequal protecting LDPC code	Equal protecting LDPC code
0.45	0.000	0.000
0.46	0.000	0.100
0.47	0.004	1.000
0.48	0.740	1.000
0.49	1.000	1.000

When we reduce B to 16, we do not find sufficient parity-check equations of weight three. We construct the matrix H as before and fill the rest of the matrix H with weight-four parity-check equations that involve three output bits and one information bit. Since the number of these weight four parity-check equations is large, we can choose those parities for which all of the three output bits are already involved in the decoding. Therefore, the length of the output sequence involved in decoding is reduced to 1660 bits. Thus, the complexity of decoding is decreased. The results of decoding this code is given in Table 3. As expected the performance is degraded with respect to the results of Table 1.

As we discussed in the previous section, for the LFSR of length 40 with 16 feedback taps and an available output sequence of length 4000, B can be reduced to 10. The rest of the parity-check matrix is filled with parity-check equations that involve four output bits. The performance of this code is also given in Table

3. The results of the table suggests that $B = 10$ can be used if the noise level is less than 0.43.

Table 3. Error rate for the full recovery of the LFSR initial state for $B = 10$ and $B = 16$.

p	$B = 10$	$B = 16$
0.43	0.000	0.000
0.44	0.300	0.000
0.45	1.000	0.003
0.46	1.000	0.006
0.47	1.000	0.998
0.48	1.000	1.000

7   Comparison with the Previous Correlation Attacks

In this section, we compare our proposed attack with the previous fast correlation attacks when the LFSR is of length 40 and has 16 feedback taps. The comparison is presented in Table 4. Note that the noise limit of the proposed algorithm is the same as in [15] while the required output sample is only 4000 versus 80,000. The required output sample in [13] is very close to the proposed algorithm, but the noise limit is only 0.36. Furthermore, the value of B in [13] is more than twice as large as its value in our method. The results also suggest that by applying unequal error protection the length of the keystream output bits involved in decoding is smaller than the length of the available output bits.

Table 4. Comparison of noise limit and required samples of proposed attack with previous approaches.

Algorithm	Noise limit	Required samples	Specification
[12]	0.44	400,000	
[13] OSDA	0.34	40,000	$B = 22$
[13] IDA	0.36	4096	$B = 22$
[14]	0.469	400,000	
[15]	0.469	80,000	$B = 18$
Proposed algorithm	0.46	2558	$B = 17$
Proposed algorithm	0.44	1660	$B = 16$
Proposed algorithm	0.43	2735	$B = 10$

8   Conclusion

We proposed an improved fast correlation attack technique that is based on constructing an unequal error protecting LDPC code from the LFSR keystream output sequence. The unequal error protecting LDPC code is constructed such that the initial bits of the LFSR have higher degrees than the remainder of the output sequence. This provides a lower decoding error rate for the initial bits. The constructed code has the additional advantage of reducing the number of output bits involved in decoding to less than the available output sequence.

We employed parity-check equations of weights three and four only. These parities are found with a low complexity algorithm that is motivated by the approach of [15]. Since lower weight parity-check equations provide better performance, we constructed the parity-check matrix such that the majority of the parity-check equations have weight three. Our decoding approach is based on combination of exhaustive search over a set of information bits and a soft-decision iterative message passing decoding algorithm.

Our simulation results indicate that the proposed algorithm offers complete recovery of the LFSR initial state for correlation probability as high as 0.46 when the length of the available keystream output sequence is 4000 and 17 bits of information bits are found by the exhaustive search. The proposed algorithm is compared with the recent fast correlation attacks. Our simulation results indicate that we can reduce the size of exhaustive search in half compared to the previous works. Using the expected number of parity-check equations of certain weights, we also derived a lower bound on the number of information bits that needs to be obtained by the exhaustive search without compromising the performance.

Acknowledgement. The authors would like to thank Nazanin Rahnavard for providing input regarding unequal error protecting codes.

References

1. Rueppel, R.A.: Analysis and Design of Stream Ciphers. Spring-Verlag (1986)
2. Siegenthaler, T.: Correlation-immunity of nonlinear combining functions for cryptographic applications. IEEE Trans. on Information Theory **IT-30** (1984) 776–780
3. Meier, W., Staffelbach, O.: Fast correlation attacks on stream ciphers. In: Advances in Cryptology - EUROCRYPT '88. Workshop on the Theory and Application of Cryptographic Techniques. Proceedings. (1988) 301–314
4. Golic, J., Salmasizadeh, M., Clark, A., Khodkar, A., Dawson, E.: Discrete optimisation and fast correlation attacks. In: Cryptography: Policy and Algorithms. International Conference. Proceedings. (1996) 186–200
5. Penzhorn, W., Kuhn, G.: Computation of low-weight parity checks for correlation attacks on stream ciphers. In: Cryptography and Coding. 5th IMA Conference. Proceedings. (1995) 74–83
6. Chepyzhov, V., Smeets, B.: On a fast correlation attack on certain stream ciphers. In: Advances in Cryptology - EUROCRYPT '91. Workshop on the Theory and Application of Cryptographic Techniques Proceedings. (1991) 176–195

7. Mihaljevic, M., Golic, J.: A fast iterative algorithm for a shift register initial state reconstruction given the noisy output sequence. In: Advances in Cryptology-AUSCRYPT '90 International Conference on Cryptology. Proceeding. (1990) 165–185
8. Johansson, T., Jonsson, F.: Improved fast correlation attacks on stream ciphers via convolutional codes. In: Advances in Cryptology - EUROCRYPT '99. International Conference on the Theory and A pplications of Cryptographic Techniques. Proceedings. (1999) 347–362
9. Johansson, T., Jonsson, F.: Fast correlation attacks based on turbo code techniques. In: Advances in Cryptology - CRYPTO'99. 19th Annual International Cryptology Conference. Proceedings. (1999) 181–197
10. Chepyzhov, V., Johansson, T., Smeets, B.: A simple algorithm for fast correlation attacks on stream ciphers. In: Fast Software Encryption. 7th International Workshop, FSE 2000. Proceedings. (2000) 181–195
11. Jonsson, F.: Some results on fast correlation attacks. PhD thesis, Lund University (2002)
12. Canteaut, A., Trabbia, M.: Improved fast correlation attacks using parity-check equations of weight 4 and 5. In: Advances in Cryptology - EUROCRYPT 2000. International Conference on the Theory and Application of Cryptographic Techniques. Proceedings. (2000) 573–588
13. Mihaljevic, M., Fossorier, M., Imai, H.: A low-complexity and high performance algorithm for the fast correlation attack. Fast Software Encryption. FSE2000, Lecture Notes in Computer Science **1978** (2001) 196–212
14. Mihaljevic, M., Fossorier, M., Imai, H.: Fast correlation attack algorithm with list decoding and an application. Fast Software Encryption. 8th International Workshop, FSE 2002 Revised Papers **2355** (2002) 196–210
15. Chose, P., Joux, A., Mitton, M.: Fast correlation attacks: an algorithmic point of view. In: Advances in Cryptology - EUROCRYPT 2002. International Conference on the Theory and Applications of Cryptographic Techniques. Proceedings. (2002) 209–221
16. Kschischang, F., Frey, B., Loeliger, H.: Factor graphs and the sum-product algorithm. IEEE Trans. on Information Theory **47** (2001) 498–519

k-Resilient Identity-Based Encryption in the Standard Model

Swee-Huay Heng[1] and Kaoru Kurosawa[2]

[1] Department of Communications and Integrated Systems,
Tokyo Institute of Technology,
2-12-1 O-okayama, Meguro-ku, Tokyo 152-8552, Japan
shheng@crypt.ss.titech.ac.jp
[2] Department of Computer and Information Sciences,
Ibaraki University,
4-12-1 Nakanarusawa, Hitachi, Ibaraki 316-8511, Japan
kurosawa@cis.ibaraki.ac.jp

Abstract. We present and analyze an adaptive chosen ciphertext secure (IND-CCA) identity-based encryption scheme (IBE) based on the well studied Decisional Diffie-Hellman (DDH) assumption. The scheme is provably secure in the *standard model* assuming the adversary can corrupt up to a maximum of k users adaptively. This is contrary to the Boneh-Franklin scheme which holds in the *random-oracle model*.

Keywords: identity-based encryption, standard model

1 Introduction

The idea of identity-based encryption scheme (IBE) was formulated by Shamir [18] in 1984. Shamir's original motivation was to simplify certificate management in email systems. Some additional applications of IBE schemes include key escrow/recovery, revocation of public keys and delegation of decryption keys [2, 3].

An IBE scheme is an asymmetric system wherein the public key is effectively replaced by a user's publicly available identity information or any arbitrary string which derived from the user's identity. It enables any pair of users to communicate securely without exchanging public or private keys and without keeping any key directories. The service of a third party which we called Private Key Generator (PKG) is needed whose sole purpose is to generate private key for the user. The private key is computed using the PKG's master-key and the identity of the user. Key escrow is inherent in an IBE scheme since the PKG knows the private keys of all the users.

Since the presentation of the idea in 1984, several IBE schemes have emerged in the literature, based on various hard problems, for example [8,20,21,15,12,5, 19]. Unfortunately, most of the proposed schemes are impractical.

Recently, a practical and functional IBE scheme was proposed by Boneh and Franklin [2,3]. Their scheme is adaptive chosen ciphertext secure (IND-CCA) in

T. Okamoto (Ed.): CT-RSA 2004, LNCS 2964, pp. 67–80, 2004.

the random oracle model based on the Bilinear Diffie-Hellman (BDH) assumption, a natural analogue of the computational Diffie-Hellman problem. Specifically, the system is based on bilinear maps between groups realized through the Weil pairing or Tate pairing. The computational cost of the pairing is high compared to the computation of the power operation over finite fields.

However this scheme is still dissatisfactory due to two main issues: (1) its security was not proved under the standard model; (2) there is no evidence that BDH problem is indeed hard. Indeed, a security proof in the random oracle model is only a heuristic proof. These types of proofs have some limitations. In particular, they do not rule out the possibility of breaking the scheme without breaking the underlying intractablity assumption. There exist digital signature schemes and public key encryption schemes which are secure in the random oracle model, but for which any implementation yields insecure schemes, as shown by Canetti *et al.* [4]. It is mentioned in [22] that BDH is reducible to most of the older believed-to-be-hard discrete logarithm problems and Diffie-Hellman (DH) problems, but there is no known reduction from any of those problems to BDH. As a result, we have no evidence that BDH problem is indeed hard.

In this paper, we somewhat manage to answer part of the open problem posed by Boneh and Franklin [2,3], that is the possibility of building a chosen ciphertext secure IBE scheme under the standard computation model (rather than the random oracle model). We present and analyze an IND-CCA secure IBE scheme based on the DDH assumption. Our scheme is k-resilient, which means that the malicious adversary can corrupt up to a maximum of k users adaptively and thus possesses the k corresponding private keys; however she cannot obtain any information pertinent to ciphertexts that are encrypted with public identities not belong to the corrupt users.

We adopt the techniques of Cramer-Shoup [6,7] in our construction. More precisely, we use a polynomial-based approach as in [13,14,9], but their ultimate goal is different from us in that their concern is more on traitor tracing and revocation. For completeness, we also provide an IND-CPA secure IBE scheme for the non-adaptive setting and the adaptive setting respectively. The non-adaptive IND-CPA scheme is adapted from the El-Gamal scheme [11] and we incorporate the Pedersen commitments [16] in order to handle adaptive adversaries.

For the security proof, we adopt a simple variant of the chosen ciphertext security definition for IBE system in [2], which is slightly stronger than the standard definition for chosen ciphertext security [17]. First, the adversary is allowed to obtain from the PKG the private keys for at most k public identities of her choice adaptively. This models an adversary who obtains at most k private keys corresponding to some public identities of her choice and tries to attack some other public identity ID of her choice. Second, the adversary is challenged on an arbitrary public identity ID of her choice rather than a random public key.

A comparison of Boneh-Franklin (BF) scheme and our proposed scheme is given in the following table.

	Model	Assumption	# of malicious users
BF scheme	Random Oracle	BDH assumption	No limit
Proposed scheme	Standard	DDH assumption	At most k

The above table shows a trade-off between (model, assumption) and the number of malicious users. BF scheme requires a stronger (model, assumption), but there is no limit on the number of malicious users. Our scheme requires a weaker (model, assumption), but the number of malicious users is limited to at most k. This limitation arises at the sacrifice of the use of random oracles and thus it seems to be unavoidable.

We argue, however, that the limit on the number of malicious users is not a serious problem in the real world. Indeed, it is not easy to corrupt a large number of users normally, meaning that the size of a malicious coalition cannot be unreasonably large. In another paper [1], Boneh and Franklin mentioned that it may suffice for k to be a fairly small integer, e.g. on the order of 20, but this is applicable to the traitor tracing scheme. Since our goal is different from theirs, we may use larger k if necessary, depending on the application, for example in some cases $k = 100$ may be sufficient.

Further, our scheme may be even practical in some circumstances. For instance, in a company or organization whereby the total number of users is small and higher security level is of paramount importance, our scheme is preferred to the BF scheme since our scheme is more reliable in that it is provably secure in the standard model under the well-known DDH assumption (as compared to the BF scheme which is proven secure in the random oracle model based on the much less analyzed BDH assumption). Specifically, our scheme can provide optimum security in the particular scenario wherein the total number of users n is less than or equal to k.

The efficiency of our scheme is linear in k and it is independent of the total number of users n. In other words, there exists trade-off between the efficiency of our scheme and the resilience k, hence the security level.

Related Work. Independently, Dodis *et al.* showed key-insulated encryption schemes [10], where their schemes coincide with our schemes. However, [10] does not present any formal definition nor security proof on ID-based encryption.

The rest of the paper is organized as follows. Some preliminaries such as basic facts, definitions and security models are given in Section 2. In Section 3, we present our proposed k-resilient IND-CPA schemes in the non-adaptive and adaptive settings. In Section 4, a k-resilient adaptive IND-CCA scheme is presented. Finally, some concluding remarks are made in Section 5.

2 Preliminaries

Lagrange Interpolation. Let q be a prime and $f(x)$ a polynomial of degree k in Z_q; let $j_0, \ldots, j_k$ be distinct elements in Z_q, and let $f_0 = f(j_0), \ldots, f_k = f(j_k)$. Using Lagrange Interpolation, we can express the polynomial as

$f(x) \overset{def}{=} \sum_{t=0}^{k}(f_t \cdot \lambda_t(x))$, where $\lambda_t(x) \overset{def}{=} \prod_{0 \leq i \neq t \leq k} \frac{j_i - x}{j_i - j_t}, t = 0, \ldots, k$, are the Lagrange coefficients.

DDH Assumption. The security of our schemes will rely on the Decisional Diffie-Hellman (DDH) assumption in a group G: namely, it is computationally hard to distinguish a quadruplet $R = \langle g_1, g_2, u_1, u_2 \rangle$ of four independent elements in G from a quadruplet $D = \langle g_1, g_2, u_1, u_2 \rangle$ satisfying $\log_{g_1} u_1 = \log_{g_2} u_2$.

Collision-Resistant Hash Function. A family of hash functions is said to be collision resistance if given a randomly chosen hash function H from the family, it is infeasible for an adversary to find two distinct messages m and m' such that $H(m) = H(m')$.

IBE Scheme. An identity-based encryption scheme IBE is specified by four polynomially bounded algorithms: Setup, Extract, Encrypt, Decrypt where:

Setup: a probabilistic algorithm used by the PKG to set up all the parameters of the scheme. The Setup algorithm takes as input a security parameter 1^λ and a number k (i.e. the maximum number of users that can be corrupted) and generates the global system parameters params and master-key. The system parameters will be publicly known while the master-key will be known to the PKG only.

Extract: a probabilistic algorithm used by the PKG to extract a private key corresponding to a given public identity. The Extract algorithm receives as input the master-key and a public identity ID associated with the user; it returns the user's private key SK_{ID}.

Encrypt: a probabilistic algorithm used to encrypt a message m using a public identity ID. The Encrypt algorithm takes as input the system parameters params, a public identity ID and a message m and returns the ciphertext C.

Decrypt: a deterministic algorithm that takes as input the system parameters params, the private key SK_{ID} and the ciphertext C and returns the message m. We require that for all messages m, Decrypt(params, $SK_{ID}, C) = m$ where $C =$ Encrypt(params, ID, m).

Security. Chosen ciphertext security (IND-CCA) is the strongest notion of security for a public key encryption scheme. Hence, it is desirable to devise an IND-CCA secure IBE scheme. However, the definition of chosen ciphertext security in an identity-based system must be strengthened a bit for the following reason. When an adversary attacks a public identity ID, she might already possess the private keys of users $ID_1, ID_2, \ldots, ID_k$ of her choice. We refer to these users as corrupt users. Hence, the definition of IND-CCA must allow the adversary to issue a maximum of k private key extraction queries adaptively. That is, the adversary is permitted to obtain the private keys associated with a maximum of k public identities of her choice adaptively (other than the public identity ID being attacked). Another difference is that the adversary is challenged on a public identity ID of her choice as opposed to a random public key. The two amendments apply to adaptive IND-CPA definition as well based on the same reasoning. We give the attack scenarios for IND-CPA and IND-CCA as follows:

IND-CPA: First, Setup is run and the adversary A is given the system parameters params. Then, A enters the private key extraction query stage, where she is given oracle access to the extraction oracle. This oracle receives as input the public identity ID_i and returns the corresponding private key SK_i. This oracle can be called adaptively for at most k times.

In the second stage, A can query the encryption oracle (also known as left-or-right oracle) on any pair of messages m_0, m_1 and an identity ID on which it wishes to be challenged. [1] Then, σ is chosen at random from $\{0, 1\}$ and the encryption oracle returns the challenge ciphertext $C^* = \mathsf{Encrypt}(\mathsf{params}, \mathsf{ID}, m_\sigma)$. Without loss of generality, we can assume that the encryption oracle is called exactly once. At the end of this stage, A outputs a bit σ^* which she thinks is equal to σ. Define the *advantage* of A as $\mathsf{Adv}_{\mathsf{IBE},A}^{\mathsf{IND-CPA}}(\lambda) := |\Pr[\sigma^* = \sigma] - \frac{1}{2}|$.

Note: For the non-adaptive IND-CPA security, we must assume that the adversary has successfully corrupted the maximum of k users and thus obtained the k corresponding private keys before Setup takes place i.e. before the adversary learns the system parameters params. This is a weaker notion of security.

IND-CCA: The attack scenario is almost the same as that in the adaptive IND-CPA, except that now A has also access to the decryption oracle, which she can query on any pair $\langle \mathsf{ID}_i, C_i \rangle$ of her choice. A can call this oracle at any point during the execution, both in the first and in the second stage, arbitrarily interleaved with her other oracle calls. To prevent the adversary from directly decrypting her challenge ciphertext C^*, the adversary is disallowed to query the decryption oracle on the pair $\langle \mathsf{ID}, C^* \rangle$ which is the output from the encryption oracle (i.e. $\langle \mathsf{ID}_i, C_i \rangle \neq \langle \mathsf{ID}, C^* \rangle$). As before, we define the advantage as $\mathsf{Adv}_{\mathsf{IBE},A}^{\mathsf{IND-CCA}}(\lambda) := |\Pr[\sigma^* = \sigma] - \frac{1}{2}|$.

Definition 1. *(k-resilience of an IBE Scheme)*
Let $\mu \in \{IND\text{-}CPA, IND\text{-}CCA\}$. We say that an IBE scheme is k-resilient against a μ-type attack if the advantage, $\mathsf{Adv}_{\mathsf{IBE},A}^{\mu}(\lambda)$, of any probabilistic polynomial-time (PPT) algorithm A is a negligible function of λ.

Before continuing, we state the following useful lemma which would be referred later.

Lemma 1. *Let U_1, U_2 and F be events defined on some probability space. Suppose that $(U_1 \wedge \neg F)$ and $(U_2 \wedge \neg F)$ are equivalent events, then $|\Pr[U_1] - \Pr[U_2]| \leq \Pr[F]$.*

---

[1] For the sake of generality, we could have allowed A to interleave the calls to the extraction oracle and the encryption oracle. However, this definition is equivalent to the one we present.

3 k-Resilient IND-CPA Scheme

In this section, we present two IBE schemes, a basic scheme which is k-resilience in a non-adaptive setting of an IND-CPA attack and an adaptive IND-CPA scheme. Subsequent schemes can be build on the previous one, in an incremental way, so that it is possible to obtain increasing security at the cost of slight efficiency loss.

3.1   Non-adaptive IND-CPA Scheme (Basic Scheme)

First we describe the basic scheme achieving semantically secure against chosen plaintext attack, assuming DDH problem is hard in the group G. This scheme is k-resilience in a non-adaptive setting.

Setup: Given a security parameter 1^λ and k, the algorithm works as follows. The first step is to define a multiplicative group G of prime order q such that $p = 2q + 1$ is also prime, in which DDH is believed to hold. This is accomplished selecting a random prime q with the above two properties and a random element g of order q modulo p. The group G is then set to be the subgroup of Z_p^* generated by g, i.e. $G = \{g^i \bmod p : i \in Z_q\} \subset Z_p^*$. Then, a random k-degree polynomial $f(x) \overset{def}{=} \sum_{t=0}^{k} d_t x^t$ is chosen over Z_q. Finally, the algorithm publicizes the system parameters $\mathsf{params} = \langle g, g^{d_0}, \ldots, g^{d_k} \rangle$. The master-key is $f(x)$ which is known to the PKG only.

Extract: For a given public identity $\mathsf{ID} \in Z_q$, the algorithm computes $f_{\mathsf{ID}} = f(\mathsf{ID})$.

Encrypt: To encrypt a message $m \in G$ under the public identity ID, the algorithm computes $\mathcal{D}_{\mathsf{ID}} = \prod_{t=0}^{k} g^{d_t \mathsf{ID}^t} (= g^{f_{\mathsf{ID}}})$. Next, it selects $r \in Z_q$ randomly and set the ciphertext as $C = \langle g^r, m\mathcal{D}_{\mathsf{ID}}^r \rangle$.

Decrypt: Let $C = \langle c_1, c_2 \rangle$ be a ciphertext encrypted using the public identity ID. To decrypt C using the private key f_{ID}, the algorithm computes $m = c_2 / c_1^{f_{\mathsf{ID}}}$.

Recall that $D = \langle g_1, g_2, g_1^r, g_2^r \rangle$ and $R = \langle g_1, g_2, g_1^a, g_2^b \rangle$ where g_1, g_2 are generators and a, b and r are randomly chosen over Z_q. In the proof of the following theorem, by Lagrange Interpolation, $f(x)$ can be expressed as $f(x) \overset{def}{=} \sum_{t=0}^{k} (f_t . \lambda_t(x))$, where $f_t = f(\mathsf{ID}_t)$ and $\lambda_t(x) = \prod_{0 \leq i \neq t \leq k} \frac{\mathsf{ID}_i - x}{\mathsf{ID}_i - \mathsf{ID}_t}, t = 0, \ldots, k$.

Theorem 1. *The above basic scheme is k-resilient against the non-adaptive chosen plaintext attacks (IND-CPA) under the DDH assumption.*

Proof. Suppose that the adversary A attacks our encryption algorithm successfully in terms of non-adaptive IND-CPA security, we show that there is a PPT algorithm A_1 that distinguishes D from R with a non-negligible advantage ϵ_1.

Assume that A successfully corrupts up to k users of her choice and hence obtains the k corresponding private keys. Given the k private keys and the system parameters $\mathsf{params} = \langle g, g^{d_0}, \ldots, g^{d_k} \rangle$, A finds two messages m_0 and m_1

in G and outputs an identity ID on which it wishes to be challenged such that she can distinguish them by observing the ciphertext.

Let $\langle g_1, g_2, u_1, u_2 \rangle$ be the input of the DDH problem. The following algorithm A_1 shall decide whether $\langle g_1, g_2, u_1, u_2 \rangle$ is from D or R.

1. Choose k private keys $f_1, \ldots, f_k$ at random corresponding to the k corrupt users' public identities $\mathsf{ID}_1, \ldots, \mathsf{ID}_k$. Let $g = g_1, g^{d_0} = g_2$. Compute $g^{d_1}, \ldots, g^{d_k}$ as follows. Notice that, $f_1, \ldots, f_k$ can be written in the matrix form as follows:

$$
\begin{pmatrix} f_1 \\ f_2 \\ \vdots \\ f_k \end{pmatrix} = \begin{pmatrix} d_0 \\ d_0 \\ \vdots \\ d_0 \end{pmatrix} + \underbrace{\begin{pmatrix} \mathsf{ID}_1 & \mathsf{ID}_1^2 & \cdots & \mathsf{ID}_1^k \\ \mathsf{ID}_2 & \mathsf{ID}_2^2 & \cdots & \mathsf{ID}_2^k \\ \vdots & \vdots & \vdots & \vdots \\ \mathsf{ID}_k & \mathsf{ID}_k^2 & \cdots & \mathsf{ID}_k^k \end{pmatrix}}_{M} \cdot \begin{pmatrix} d_1 \\ d_2 \\ \vdots \\ d_k \end{pmatrix}
$$

where matrix M is a Vandermonde matrix. It is clear that M is non-singular since $\mathsf{ID}_1, \ldots, \mathsf{ID}_k$ are all distinct. Therefore, we have

$$
(d_1, \ldots, d_k)^T = M^{-1}(f_1 - d_0, \ldots, f_k - d_0)^T.
$$

Let $(b_{t1}, \ldots, b_{tk})$ be the tth row of M^{-1}. Then

$$
\begin{aligned}
d_t &= b_{t1}(f_1 - d_0) + \cdots + b_{tk}(f_k - d_0) \\
&= b_{t1}f_1 + \cdots + b_{tk}f_k - (b_{t1} + \cdots + b_{tk})d_0.
\end{aligned}
$$

Hence, $g^{d_t} = g_1^{b_{t1}f_1 + \cdots + b_{tk}f_k} / g_2^{b_{t1} + \cdots + b_{tk}}$, $t = 1, 2, \ldots, k$.
Let $f'(x) = \sum_{t=1}^{k} f_t \lambda_t(x)$ and $f(x) = f'(x) + d_0 \lambda_0(x)$ where $\lambda_t(x)$ is computed from $\mathsf{ID}_0 = 0$ and $\mathsf{ID}_1, \ldots, \mathsf{ID}_k$. Note that we do not know $d_0 = f_0$.

2. Feed the private keys $f_1, \ldots, f_k$ and the system parameters params $= \langle g_1, g_2, g^{d_1}, \ldots, g^{d_k} \rangle$ to A. A returns $m_0, m_1 \in G$ and an identity ID such that $\mathsf{ID} \notin \{\mathsf{ID}_1, \ldots, \mathsf{ID}_k\}$.

3. Randomly select $\sigma \in \{0, 1\}$ and encrypt m_σ as $C^* = \langle u_1, m_\sigma u_1^{f'(\mathsf{ID})} u_2^{\lambda_0(\mathsf{ID})} \rangle$.

4. Feed C^* to A and get a return σ^*. The algorithm outputs 1 if and only if $\sigma = \sigma^*$.

If $\langle g_1, g_2, u_1, u_2 \rangle$ is from D, $g = g_1, g_2 = g^{d_0}, u_1 = g^r, u_2 = g_2^r = g^{rd_0}$ and $u_1^{f'(\mathsf{ID})} u_2^{\lambda_0(\mathsf{ID})} = g^{r \sum_{t=1}^{k} f_t \lambda_t(\mathsf{ID})} g^{rd_0 \lambda_0(\mathsf{ID})} = g^{r \sum_{t=0}^{k} f_t \lambda_t(\mathsf{ID})} = g^{rf(\mathsf{ID})} = g^{rf_{\mathsf{ID}}} = \mathcal{D}_{\mathsf{ID}}^r$. Thus, C^* is the encryption of m_σ and $\Pr[A_1(g_1, g_2, u_1, u_2) = 1] = \Pr[A(C^*) = \sigma] = \frac{1}{2} + \epsilon_1$. Otherwise, since $u_1 = g_1^a, u_2 = g_2^b$, the distribution of C^* is the same for both $\sigma = 0$ and $\sigma = 1$. Thus, $\Pr[A_1(g_1, g_2, u_1, u_2) = 1] = \Pr[A(C^*) = \sigma] = \frac{1}{2}$. Therefore, A_1 distinguishes D from R with a non-negligible advantage ϵ_1. $\square$

3.2   Adaptive IND-CPA Scheme

In this section, we present an adaptive IND-CPA secure IBE scheme with the condition that the adversary can corrupt up to a maximum of k users adaptively.

For this scheme and the subsequent scheme, our proofs follow the structural approach advocated in [7] defining a sequence of attack games $\mathbf{G}_0, \mathbf{G}_1, \ldots, \mathbf{G}_l$, all operating under the same underlying probability space. Starting from $\mathbf{G}_0$, we make slight modifications to the behavior of the oracles, thus changing the way the adversary's view is computed, while maintaining the view's distributions indistinguishable among the games. We emphasize that the different games do not change the encryption algorithm (and decryption algorithm) but the encryption oracle (and decryption oracle) i.e. the method in which the challenge ciphertext is generated (and the ciphertext is decrypted) only. The actual encryption algorithm (and decryption algorithm) that the scheme (and hence the attacker) uses remains the same.

For any $1 \leq i \leq l$, let T_i be the event that $\sigma = \sigma^*$ in game $\mathbf{G}_i$. Our strategy is to show that for $1 \leq i \leq l$, the quantity $|\Pr[T_i] - \Pr[T_{i-1}]|$ is negligible. Also, it will be evident from the definition of game $\mathbf{G}_l$ that $\Pr[T_l] = \frac{1}{2}$, which will imply that $|\Pr[T_0] - \frac{1}{2}|$ is negligible.

Setup: It is almost the same as in the basic scheme except that in this scheme we use two generators. This is accomplished selecting a random element g_1 of order q modulo p. The group G is set to be the subgroup of Z_p^* generated by g_1, i.e. $G = \{g_1^i \bmod p : i \in Z_q\} \subset Z_p^*$. A random $w \leftarrow_R Z_q$ is then chosen and used to compute $g_2 = g_1^w$. Then, two random k-degree polynomials $p_1(x) \stackrel{def}{=} \sum_{t=0}^k d_t x^t$ and $p_2(x) \stackrel{def}{=} \sum_{t=0}^k d_t' x^t$ are chosen over Z_q. Next, the algorithm computes $D_0 = g_1^{d_0} g_2^{d_0'}, \ldots, D_k = g_1^{d_k} g_2^{d_k'}$. Finally, it publicizes the system parameters as params $= \langle g_1, g_2, D_0, \ldots, D_k \rangle$. The master-key is $\langle p_1, p_2 \rangle$ which is known to the PKG only.

Extract: For a given public identity ID $\in Z_q$, the algorithm computes $p_{1,\mathsf{ID}} = p_1(\mathsf{ID})$ and $p_{2,\mathsf{ID}} = p_2(\mathsf{ID})$ [2] and returns $SK_{\mathsf{ID}} = \langle p_{1,\mathsf{ID}}, p_{2,\mathsf{ID}} \rangle$.

Encrypt: To encrypt a message $m \in G$ under the public identity ID, the working steps of the encryption algorithm are given in Fig. 1.

Decrypt: To decrypt C using the private key $SK_{\mathsf{ID}} = \langle p_{1,\mathsf{ID}}, p_{2,\mathsf{ID}} \rangle$, the decryption algorithm is depicted in Fig. 1.

Theorem 2. *The above IBE scheme is k-resilient against adaptive chosen plaintext attacks (IND-CPA) under the DDH assumption.*

Proof. We shall define a sequence of "indistinguishable" modified attack games $\mathbf{G}_0$, $\mathbf{G}_1$, $\mathbf{G}_2$ and $\mathbf{G}_3$ where $\mathbf{G}_0$ is the original game and the last game clearly gives no advantage to the adversary.

Game $\mathbf{G}_0$. In game $\mathbf{G}_0$, the adversary A receives the system parameters params $= \langle g_1, g_2, D_0, \ldots, D_k \rangle$ and adaptively queries the extraction oracle for

---

[2] For conciseness, we will follow this notation throughout the paper.

Encryption algorithm	Decryption algorithm
$E1.\ r_1 \leftarrow_R Z_q$	$D1.\ s \leftarrow u_1^{p_{1,\mathsf{ID}}} \cdot u_2^{p_{2,\mathsf{ID}}}$
$E2.\ u_1 \leftarrow g_1^{r_1}$	$D2.\ m \leftarrow c \cdot s^{-1}$
$E3.\ u_2 \leftarrow g_2^{r_1}$	
$E4.\ \mathcal{D}_{\mathsf{ID}} \leftarrow \prod_{t=0}^{k} D_t^{\mathsf{ID}^t}$	
$E5.\ s \leftarrow \mathcal{D}_{\mathsf{ID}}^{r_1}$	
$E6.\ c \leftarrow m \cdot s$	
$E7.\ C \leftarrow \langle u_1, u_2, c \rangle$	

Fig. 1. Encryption and decryption algorithms for the IND-CPA scheme

a maximum of k public identities of her choice. Then, she outputs a challenge identity ID and queries the encryption oracle on (m_0, m_1). A receives the ciphertext C^* as the answer. At this point, A outputs her guess $\sigma^* \in \{0, 1\}$. Let T_0 be the event that $\sigma = \sigma^*$ in game $\mathbf{G}_0$.

Game $\mathbf{G}_1$. Game $\mathbf{G}_1$ is identical to game $\mathbf{G}_0$, except for a small modification to the encryption oracle. In game $\mathbf{G}_1$, steps $E4$ and $E5$ of the encryption algorithm in Fig. 1 are replaced with the following step:

$$E5'.\quad s \leftarrow u_1^{p_{1,\mathsf{ID}}} \cdot u_2^{p_{2,\mathsf{ID}}}$$

It is clear that step $E5'$ computes the same value as step $E5$. The point of this change is to make explicit any functional dependency of the above quantity on u_1 and u_2. Let T_1 be the event that $\sigma = \sigma^*$ in game $\mathbf{G}_1$. Clearly, it holds that $\Pr[T_0] = \Pr[T_1]$.

Game $\mathbf{G}_2$. To turn game $\mathbf{G}_1$ into game $\mathbf{G}_2$, we make another change to the encryption oracle. We replace $E1$ and $E3$ with the following:

$$E1'.\ r_1 \leftarrow_R Z_q,\quad r_2 \leftarrow_R Z_q \backslash \{r_1\}$$
$$E3'.\ u_2 \leftarrow g_2^{r_2}$$

Let T_2 be the event that $\sigma = \sigma^*$ in game $\mathbf{G}_2$. Notice that while in game $\mathbf{G}_1$ the values u_1 and u_2 are obtained using the same value r_1, in game $\mathbf{G}_2$ they are independent subject to $r_1 \neq r_2$. Therefore, using a standard reduction argument, any non-negligible difference in behaviour between $\mathbf{G}_1$ and $\mathbf{G}_2$ can be used to construct a PPT algorithm A_1 that is able to distinguish Diffie-Hellman tuples from totally random tuples with non-negligible advantage. Hence $|\Pr[T_2] - \Pr[T_1]| \leq \epsilon_1$ for some negligible ϵ_1.

Game $\mathbf{G}_3$. In this game, we again modify the encryption oracle as follows:

$$E6'.\quad e \leftarrow_R Z_q,\quad c \leftarrow g_1^e$$

Let T_3 be the event that $\sigma = \sigma^*$ in game $\mathbf{G}_3$. Due to this last change, the challenge no longer contains σ, nor does any other information in the adversary's view; therefore, we have that $\Pr[T_3] = \frac{1}{2}$. Moreover, we can prove that the adversary has the same chances to guess σ in both game $\mathbf{G}_2$ and $\mathbf{G}_3$, i.e. $\Pr[T_3] = \Pr[T_2]$. (The proof will be given in the full version of the paper.)

Finally, combining all the intermediate results, we can conclude that adversary A's advantage is negligible; more precisely: $\mathsf{Adv}_{\mathsf{IBE},A}^{\mathsf{IND-CPA}}(\lambda) \leq \epsilon_1$. $\qquad\square$

4 k-Resilient IND-CCA Scheme

We present an identity-based encryption scheme achieving adaptive chosen ciphertext security in this section. This scheme makes use of a hash function chosen randomly from a family of collision-resistant hash functions. Again, this adaptive IND-CCA IBE scheme is secure against the adversary which can corrupt a maximum of k users adaptively. We give the description of the scheme as follows:

Setup: As in the previous scheme, the first task is to select a random multiplicative group $G \subset Z_p^*$ of prime order q and two generators $g_1, g_2 \in G$. Then, six random k-degree polynomials are chosen over Z_q. That is, $f_1(x) \overset{def}{=} \sum_{t=0}^{k} a_t x^t, f_2(x) \overset{def}{=} \sum_{t=0}^{k} a'_t x^t, h_1(x) \overset{def}{=} \sum_{t=0}^{k} b_t x^t, h_2(x) \overset{def}{=} \sum_{t=0}^{k} b'_t x^t, p_1(x) \overset{def}{=} \sum_{t=0}^{k} d_t x^t$ and $p_2(x) \overset{def}{=} \sum_{t=0}^{k} d'_t x^t$. Next, the algorithm computes $A_t = g_1^{a_t} g_2^{a'_t}, B_t = g_1^{b_t} g_2^{b'_t}$ and $D_t = g_1^{d_t} g_2^{d'_t}$, for $t = 0, \ldots, k$ and chooses at random a hash function H from a family of collision-resistant hash functions. Finally, it publicizes the system parameters

$$\mathsf{params} = \langle g_1, g_2, A_0, \ldots, A_k, B_0, \ldots, B_k, D_0, \ldots, D_k, H \rangle.$$

The master-key is $\langle f_1, f_2, h_1, h_2, p_1, p_2 \rangle$ which is known to the PKG only.
Extract: For a given public identity $\mathsf{ID} \in Z_q$ the algorithm returns $SK_{\mathsf{ID}} = \langle f_{1,\mathsf{ID}}, f_{2,\mathsf{ID}}, h_{1,\mathsf{ID}}, h_{2,\mathsf{ID}}, p_{1,\mathsf{ID}}, p_{2,\mathsf{ID}} \rangle$.
Encrypt: To encrypt a message $m \in G$ under the public identity ID, the encryption algorithm works as depicted in Fig. 2.
Decrypt: To decrypt C using the private key $SK_{\mathsf{ID}} = \langle f_{1,\mathsf{ID}}, f_{2,\mathsf{ID}}, h_{1,\mathsf{ID}} h_{2,\mathsf{ID}}, p_{1,\mathsf{ID}}, p_{2,\mathsf{ID}} \rangle$, the decryption algorithm is given in Fig. 2.

Theorem 3. *The above IBE scheme is k-resilient against adaptive chosen ciphertext attacks (IND-CCA) suppose the DDH assumption holds for G and H is chosen from a family of collision-resistant hash functions.*

Proof. As in the proof of Theorem 2, we shall define a sequence of modified games $\mathbf{G}_i$, for $0 \leq i \leq 5$. Let T_i be the event that $\sigma = \sigma^*$ in game $\mathbf{G}_i$.
Game $\mathbf{G}_0$. In game $\mathbf{G}_0$, the adversary A receives the system parameters $\mathsf{params} = \langle g_1, g_2, A_0, \ldots, A_k, B_0, \ldots, B_k, D_0, \ldots, D_k, H \rangle$ and adaptively interleaves queries to the extraction oracle and the decryption oracle. For private key extraction query, A inputs the public identity ID_i of her choice while for the decryption query, A provides the oracle with the public identity and ciphertext pair $\langle \mathsf{ID}_i, C_i \rangle$ of her choice. A can query the extraction oracle for a maximum of k times adaptively. Then, A outputs a challenge identity ID and queries the encryption oracle on (m_0, m_1). She receives the ciphertext C^* as the

Encryption algorithm	Decryption algorithm
$E1.\ r_1 \leftarrow_R Z_q$	$D1.\ \alpha \leftarrow H(u_1, u_2, c)$
$E2.\ u_1 \leftarrow g_1^{r_1}$	$D2.$ Test if $v_{\mathsf{ID}} \leftarrow u_1^{f_{1,\mathsf{ID}}+h_{1,\mathsf{ID}}\alpha} \cdot u_2^{f_{2,\mathsf{ID}}+h_{2,\mathsf{ID}}\alpha}$;
$E3.\ u_2 \leftarrow g_2^{r_1}$	halt if this is not the case
$E4.\ \mathcal{A}_{\mathsf{ID}} \leftarrow \prod_{t=0}^{k} A_t^{\mathsf{ID}^t}$	$D3.\ s \leftarrow u_1^{p_{1,\mathsf{ID}}} \cdot u_2^{p_{2,\mathsf{ID}}}$
$\quad\ \mathcal{B}_{\mathsf{ID}} \leftarrow \prod_{t=0}^{k} B_t^{\mathsf{ID}^t}$	$D4.\ m \leftarrow c \cdot s^{-1}$
$\quad\ \mathcal{D}_{\mathsf{ID}} \leftarrow \prod_{t=0}^{k} D_t^{\mathsf{ID}^t}$	
$E5.\ s \leftarrow \mathcal{D}_{\mathsf{ID}}^{r_1}$	
$E6.\ c \leftarrow m \cdot s$	
$E7.\ \alpha \leftarrow H(u_1, u_2, c)$	
$E8.\ v_{\mathsf{ID}} \leftarrow \mathcal{A}_{\mathsf{ID}}^{r_1} \cdot \mathcal{B}_{\mathsf{ID}}^{r_1 \alpha}$	
$E9.\ C \leftarrow \langle u_1, u_2, c, v_{\mathsf{ID}} \rangle$	

Fig. 2. Encryption and decryption algorithms for the IND-CCA scheme

answer. Next, A can again query the decryption oracle, restricted only in that $\langle \mathsf{ID}_i, C_i \rangle \neq \langle \mathsf{ID}, C^* \rangle$. Finally, A outputs her guess $\sigma^* \in \{0, 1\}$.

Game $\mathbf{G}_1$. Game $\mathbf{G}_1$ is identical to game $\mathbf{G}_0$, except for a small modification to the encryption oracle. In game $\mathbf{G}_1$, steps $E4$ and $E5$ of the encryption algorithm in Fig. 2 are replaced with step $E5'$ and step $E8$ of the encryption algorithm in Fig. 2 is replaced with step $E8'$ as follows:

$$E5'.\ s \leftarrow u_1^{p_{1,\mathsf{ID}}} \cdot u_2^{p_{2,\mathsf{ID}}}$$
$$E8'.\ v_{\mathsf{ID}} \leftarrow u_1^{f_{1,\mathsf{ID}}+h_{1,\mathsf{ID}}\alpha} \cdot u_2^{f_{2,\mathsf{ID}}+h_{2,\mathsf{ID}}\alpha}$$

It is clear that steps $E5'$ and $E8'$ compute the same values as steps $E5$ and $E8$ respectively. The point of these changes is just to make explicit any functional dependency of the above quantities on u_1 and u_2. Clearly, it holds that $\Pr[T_0] = \Pr[T_1]$.

Game $\mathbf{G}_2$. To turn game $\mathbf{G}_1$ into game $\mathbf{G}_2$, we make another change to the encryption oracle. We replace $E1$ and $E3$ with the following:

$$E1'.\ r_1 \leftarrow_R Z_q, \quad r_2 \leftarrow_R Z_q \backslash \{r_1\}$$
$$E3'.\ u_2 \leftarrow g_2^{r_2}$$

Notice that while in game $\mathbf{G}_1$ the values u_1 and u_2 are obtained using the same value r_1, in game $\mathbf{G}_2$ they are independent subject to $r_1 \neq r_2$. Therefore, using a standard reduction argument, any non-negligible difference in behaviour between $\mathbf{G}_1$ and $\mathbf{G}_2$ can be used to construct a PPT algorithm A_1 that is able to distinguish Diffie-Hellman tuples from totally random tuples with non-negligible advantage. Hence $|\Pr[T_2] - \Pr[T_1]| \leq \epsilon_1$ for some negligible ϵ_1.

Game $\mathbf{G}_3$. To define game $\mathbf{G}_3$, we slightly modify the decryption oracle, replacing steps $D2$ and $D3$ with:

$$D2'.\ \text{Test if } u_2 = u_1^w \text{ and } v_{\mathsf{ID}} = u_1^{(f_{1,\mathsf{ID}}+h_{1,\mathsf{ID}}\alpha)+(f_{2,\mathsf{ID}}+h_{2,\mathsf{ID}}\alpha)w};$$
$$\text{halt if this is not the case}$$
$$D3'.\ s \leftarrow u_1^{p_{1,\mathsf{ID}}+wp_{2,\mathsf{ID}}}$$

Let R_3 be the event that, some decryption query that would have passed the test in step $D2$ used in game $\mathbf{G}_2$ fails the test in step $D2'$ in game $\mathbf{G}_3$. Obviously, $\mathbf{G}_2$ and $\mathbf{G}_3$ are identical until event R_3 occurs. In particular, the events $(T_2 \wedge \neg R_3)$ and $(T_3 \wedge \neg R_3)$ are identical. By Lemma 1, we have $|\Pr[T_3] - \Pr[T_2]| \le \Pr[R_3]$.

We introduce two more games, $\mathbf{G}_4$ and $\mathbf{G}_5$ in order to bound $\Pr[R_3]$.

Game $\mathbf{G}_4$. In this game, we again modify the encryption oracle as follows:

$$E6'. \quad e \leftarrow_{\mathrm{R}} Z_q, \quad c \leftarrow g_1^e$$

Due to this change, the challenge no longer contains σ, nor does any other information in the adversary's view; therefore, we have that $\Pr[T_4] = \frac{1}{2}$.

Let R_4 be the event that some decryption query that would have passed the test in step $D2$ used in game $\mathbf{G}_2$ fails the test in step $D2'$ in game $\mathbf{G}_4$. We show that those events happen with the same probability as the corresponding events of game $\mathbf{G}_3$. More precisely, we prove that $\Pr[T_4] = \Pr[T_3]$ and $\Pr[R_4] = \Pr[R_3]$. (The proof will be given in the full version of the paper.)

Game $\mathbf{G}_5$. This game is the same as game $\mathbf{G}_4$, except for the following modification. We modify the decryption oracle so that it applies the following *special rejection rule*, whose goal is to prevent the adversary from submitting illegal ciphertexts to the decryption oracle after she has received the challenge C^*.

Special rejection rule: After A receives her challenge $C^* = \langle u_1^*, u_2^*, c^*, v_{\mathsf{ID}}^* \rangle$, the decryption oracle rejects any query $\langle \mathsf{ID}_i, C_i \rangle$, with $C_i = \langle u_1, u_2, c, v_i \rangle$ such that $\langle u_1, u_2, c \rangle \neq \langle u_1^*, u_2^*, c^* \rangle$ but $\alpha = \alpha^*$. It does so before executing step $D2'$.

Let C_5 be the event that the adversary submits a decryption query that is rejected using the *special rejection rule*. Let R_5 be the event that A submits some decryption query that would have passed the test in step $D2$ used in game $\mathbf{G}_2$, but fails the test in step $D2'$ used in game $\mathbf{G}_5$. Clearly, $\mathbf{G}_4$ and $\mathbf{G}_5$ are identical until event C_5 occurs. In particular, the events $(R_4 \wedge \neg C_5)$ and $(R_5 \wedge \neg C_5)$ are identical. By Lemma 1, we have $|\Pr[R_5] - \Pr[R_4]| \le \Pr[C_5]$.

We need to show that events C_5 and R_5 occur with negligible probability. The argument to bound event C_5 is based on the collision-resistant assumption. Using a standard reduction argument, we can construct a PPT algorithm A_2 that breaks the collision-resistant assumption with non-negligible advantage. Hence, we have $\Pr[C_5] \le \epsilon_2$ for some negligible ϵ_2. Subsequently, we show that event R_5 occurs with negligible probability purely based on some information-theoretic considerations. That is, $\Pr[R_5] \le Q_A(\lambda)/q$, where $Q_A(\lambda)$ is an upper bound on the number of decryption queries made by the adversary. (The detailed proof will be given in the full version of the paper.)

Finally, combining all the intermediate results, we can conclude that adversary A's advantage is negligible; more precisely: $\mathsf{Adv}_{\mathsf{IBE},A}^{\mathsf{IND-CCA}}(\lambda) \le \epsilon_1 + \epsilon_2 + Q_A(\lambda)/q$. $\qquad \square$

5 Conclusion

We proposed an adaptive IND-CCA secure IBE scheme based on the DDH assumption. The scheme is provably secure in the standard model assuming the adversary can corrupt up to a maximum of k users adaptively. We also presented a non-adaptive IND-CPA secure IBE scheme and an adaptive IND-CPA secure IBE scheme based on the same assumption.

References

1. D. Boneh and M. Franklin. An efficient public key traitor tracing scheme. *Advances in Cryptology — CRYPTO '99, LNCS 1666*, pp. 338–353, Springer-Verlag, 1999.
2. D. Boneh and M. Franklin. Identity-based encryption from the weil pairing. *Advances in Cryptology — CRYPTO '01, LNCS 2139*, pp. 213–229, Springer-Verlag, 2001.
3. D. Boneh and M. Franklin. Identity-based encryption from the weil pairing. *Siam Journal of Computing, Vol. 32*, pp. 586–615, 2003. Updated version of [2].
4. R. Canetti, O. Goldreich and S. Halevi. The random oracle model, revisited. *30th Annual ACM Symposium on Theory of Computing — STOC '98*, pp. 209–218, 1998.
5. C. Cocks. An identity based encryption scheme based on quadratic residues. *Cryptography and Coding, LNCS 2260*, pp. 360–363, Springer-Verlag, 2001.
6. R. Cramer and V. Shoup. A practical public key cryptosystem provably secure against adaptive chosen ciphertext attack. *Advances in Cryptology — CRYPTO '98, LNCS 1462*, pp. 13–25, Springer-Verlag, 1998.
7. R. Cramer and V. Shoup. Design and analysis of practical public-key encryption scheme secure against adaptive chosen ciphertext attack. Manuscript, 2001. To appear in *Siam Journal of Computing*.
8. Y. Desmedt and J. Quisquater. Public-key systems based on the difficulty of tampering. *Advances in Cryptology — CRYPTO '86, LNCS 0263*, pp. 111–117, Springer-Verlag, 1986.
9. Y. Dodis and N. Fazio. Public key trace and revoke scheme secure against adaptive chosen ciphertext attack. *Public Key Cryptography — PKC '03, LNCS 2567*, pp. 100–115, Springer-Verlag, 2003. Full version available at http://eprint.iacr.org/.
10. Y. Dodis, J. Katz, S. Xu and M. Yung. Key-insulated public key cryptosystems. *Advances in Cryptology — EUROCRYPT '02, LNCS 2332*, pp. 65–82, Springer-Verlag, 2002.
11. T. El Gamal. A public-key cryptosystem and a signature scheme based on the discrete logarithm. *IEEE Transactions on Information Theory, vol. 31 No. 4*, pp. 469–472, 1985.
12. D. Hühnlein, M. J. Jacobson and D. Weber. Towards practical non-interactive public key cryptosystems using non-maximal imaginary quadratic orders. *Selected Areas in Cryptography — SAC '00, LNCS 2012*, pp. 275–287, Springer-Verlag, 2001.
13. K. Kurosawa and Y. Desmedt. Optimum traitor tracing and asymmetric schemes. *Advances in Cryptology — EUROCRYPT '98, LNCS 1403*, pp. 145–157, Springer-Verlag, 1998.

14. K. Kurosawa and T. Yoshida. Linear code implies public-key traitor tracing. *Public Key Cryptography — PKC '02, LNCS 2274*, pp. 172–187, Springer-Verlag, 2002.
15. U. Maurer and Y. Yacobi. Non-interactive public-key cryptography. *Advances in Cryptology — EUROCRYPT '91, LNCS 0547*, pp. 498–507, Springer-Verlag, 1991.
16. T. Pedersen. Non-interactive and information-theoretic secure verifiable secret sharing. *Advances in Cryptology — CRYPTO '91, LNCS 0576*, pp. 129–140, Springer-Verlag, 1992.
17. C. Rackoff and D. Simon. Non-interactive zero-knowledge proof of knowledge and chosen ciphertext attack. *Advances in Cryptology — CRYPTO '91, LNCS 0576*, pp. 433–444, Springer-Verlag, 1992.
18. A. Shamir. Identity-based cryptosystems and signature schemes. *Advances in Cryptology — CRYPTO '84, LNCS 0196*, pp. 47–53, Springer-Verlag, 1985.
19. R. Sakai, K. Ohgishi and M. Kasahara. Cryptosystems based on pairing over elliptic curve. *Symposium on Cryptography and Information Security — SCIS '01*, pp. 369–372, 2001 (In Japanese).
20. H. Tanaka. A realization scheme for the identity-based cryptosystem. *Advances in Cryptology — CRYPTO '87, LNCS 0293*, pp. 341–349, Springer-Verlag, 1987.
21. S. Tsuji and T. Itoh. An ID-based cryptosystem based on the discrete logarithm problem. *IEEE Journal on Selected Areas in Communication, vol. 7, no. 4*, pp. 467–473, 1989.
22. Y. Yacobi. A note on the bilinear Diffie-Hellman assumption. *IACR Cryptology ePrint Archive, Report 2002/113*. Available from http://eprint.iacr.org/2002/113/.

A Generic Construction for Intrusion-Resilient Public-Key Encryption

Yevgeniy Dodis[1], Matt Franklin[2], Jonathan Katz[3],
Atsuko Miyaji[4], and Moti Yung[5]

[1] Department of Computer Science, New York University.
dodis@cs.nyu.edu
[2] University of California, Davis.
franklin@cs.ucdavis.edu
[3] Department of Computer Science, University of Maryland.
jkatz@cs.umd.edu
[4] Japan Advanced Institute of Science and Technology.
miyaji@jaist.ac.jp
[5] Department of Computer Science, Columbia University.
moti@cs.columbia.edu

Abstract. In an *intrusion-resilient* cryptosystem [10], two entities (a user and a base) jointly evolve a secret decryption key; this provides very strong protection against an active attacker who can break into the user and base repeatedly and even simultaneously. Recently, a construction of an intrusion-resilient public-key encryption scheme based on specific algebraic assumptions has been shown [6]. We generalize this previous work and present a more generic construction for intrusion-resilient public-key encryption from any forward-secure public-key encryption scheme satisfying a certain homomorphic property.

1 Introduction

The exposure of secret keys can be a devastating attack against a cryptosystem. Especially when "standard" cryptanalytic techniques are infeasible, a determined attacker might find it much easier to obtain secret keys by hardware tampering, or via theft, bribery, or similar means. The problem of key exposure becomes more severe as cryptographic algorithms are increasingly used on inexpensive, lightweight, and portable consumer devices.

Key evolution is a powerful defense against the threat of key exposure. As an example, in a *forward-secure* scheme one's secret key is updated at each time period in such a way that key exposure during any time period compromises only future time periods (but not past time periods). Forward security was first formalized (in the context of signature and identification schemes) by Bellare and Miner [2], building on earlier ideas of Anderson [1]; numerous constructions of forward-secure signature schemes have been proposed (beginning with [2]). A forward-secure public-key encryption scheme has been constructed recently by Canetti, Halevi, and Katz [5].

T. Okamoto (Ed.): CT-RSA 2004, LNCS 2964, pp. 81–98, 2004.
© Springer-Verlag Berlin Heidelberg 2004

Key-insulated cryptosystems [7,3,8] extend the key evolution paradigm to further limit the damage from key exposure. As with forward security, the user (e.g., a mobile device) can perform all cryptographic operations during any particular time period on his own. However, to update the user's secret keys for the next time period, the user needs the help of a "base" (e.g., a desktop PC in the user's home). Using this model, one may guarantee that exposure of the user's keys during multiple time periods only compromises security for those specific time periods, and not for any other time periods either in the past or in the future. A key-insulated scheme is additionally termed "strong" if there is no security compromise when the adversary exposes the secrets stored on the base.

Intrusion-resilience (first proposed in the context of signature schemes by Itkis and Reyzin [10]) is a synthesis of forward security and key-insulated security. The system model is as in the key-insulated case: the user performs cryptographic operations on its own during each time period, and updates its key for the next time period with the help of the base. Here, however, a stronger security guarantee is provided. If the base and the user are exposed during the *same* time period, then all *prior* time periods remain secure (as in the case of forward security). Otherwise, repeated exposure of both the user and the base only compromise those specific time periods during which the user's secret keys were exposed (as in the case of key-insulated security).

The security provided by intrusion-resilient schemes may be further enhanced by allowing "refresh" operations between base and user in addition to "update" operations. Both of these are key-evolving functions. The difference is that an update operation is used only at the beginning of each time period, while any number of refresh operations can occur within a single time period. Someone who wants to interact with the user needs to know the current time period (i.e., number of update operations), but does not need to know how many refresh operations have occurred within each time period. Frequent refresh operations enhance security, since the attacker must expose user and base *between refreshes* in order to compromise future security.

Itkis and Reyzin [10] gave a construction of intrusion-resilient signatures based on the strong RSA assumption. Subsequently, Itkis [9] showed a *generic* construction of intrusion-resilient signatures from any one-way function. The first construction for intrusion-resilient public-key encryption is given in [6]. That construction relies on a very specific assumption (the BDH assumption [4]), and is based on the forward-secure encryption scheme of [5]. This raises the natural question of what assumptions are sufficient to achieve intrusion-resilient encryption. In this paper, we make progress on this question by presenting a more generic construction for intrusion-resilient public key encryption based on any forward-secure encryption scheme satisfying certain properties. In this sense, our work generalizes the previous work [6] which constructs an intrusion-resilient encryption schemes from a *specific* forward-secure scheme (i.e., that of [5]). It is hoped that our more generic construction will highlight those properties that enable intrusion-resilience and thus shed additional light on this primitive.

Indeed, the scheme in [6] is somewhat complicated and hard to parse. In particular, one has to be extremely careful when defining the order of operations in that scheme, it is not immediately really clear what specific properties of the forward-secure scheme of [5] are critically used, and, overall, what is the high level intuition behind that construction. This paper tries to clarify this point by presenting a more generic construction of intrusion-resilient encryption which clearly explains which special properties of the scheme of [5] are used. Specifically, we isolate two such crucial properties: a homomorphic structure of the key updating operation, and, more importantly, "separability" between the user's key material used for updating from that used for the actual decryption. Indeed, we will argue that without such separability it seems impossible (or very hard) to build an intrusion-resilient encryption scheme from a forward-secure scheme. For that reason, we also give a new, refined definition of forward-secure encryption which explicitly models this key separability, and argue that the scheme in [5] meets our definition. Then, we give a clean and intuitive construction of intrusion-resilient encryption from any such refined forward-secure encryption with an extra homomorphic property for key updating. Of course, since presently there exists only one specific forward-secure encryption of [5], we can currently instantiate our scheme in only one way — the one given in [6] — but our exposition hopefully clarifies and explains the design criteria for constructing intrusion-resilient encryption. In particular, shall a new forward-secure scheme be found, our construction pin-points the two natural extra properties which are needed to turn it into an intrusion-resilient scheme (from the same assumption). And since we argue that such extra properties also seem to be necessary, our work motivates the design of future forward-secure schemes which satisfy them as well. Thus, we believe that our generally will clarify and simplify future designs of both forward-secure and intrusion-resilient schemes.

As an additional contribution, we explore a number of alternative models and definitions for both forward-secure and intrusion-resilient encryption in Sections 2 and 3. Our generic construction appears in Section 4, and Section 5 contains the proof of security.

2   Forward-Secure Encryption

Our intrusion-resilient scheme is built from a forward-secure encryption scheme. The notation and model borrows from that of [5], slightly adapted for our purposes. We let $\mathbb{N}$ be the set of positive integers and let $[T] = \{1, 2, \ldots, T\}$ for $T \in \mathbb{N}$.

2.1   Functional Description

We assume a key-evolving encryption scheme in which the user's secret key can be "divided" into two components: an *update key* and a *local key*. An update key is used only to generate the update key and local key of the next time period, but is not used to decrypt a ciphertext. On the other hand, a local key is used only to decrypt a ciphertext in the corresponding time period, but is not used to generate the update key or local key for the next time period. Note that the forward-secure encryption scheme of [5] may be viewed in this way.

More formally, we specify a key-evolving encryption scheme (with the above-mentioned property) by the following tuple of polynomial-time algorithms:

fsKeyGen: key generation algorithm
Input: security parameter k, number of time periods T
Output: initial user key $\overline{sk}_0$, public key $\overline{pk}$

fsKeyUpd: key-update algorithm
Input: current user update key $\overline{sk}_t$ and time period t
Output: next user update key $\overline{sk}_{t+1}$ and next user local key $\overline{lsk}_{t+1}$

fsEnc: randomized encryption algorithm
Input: user public key $\overline{pk}$, current time period t, message M
Output: ciphertext C

fsDec: decryption algorithm
Input: user local secret key $\overline{lsk}_t$, ciphertext $C = \text{fsEnc}(\overline{pk}, t, M)$
Output: message M

The initial user update key $\overline{sk}_0$ is not actually used or stored (instead, fsKeyUpd is applied immediately to generate $\overline{sk}_1$ and $\overline{lsk}_1$). Therefore, the sets of keys which an adversary can access are defined as follows:

$$\overline{sk}^* = \{\overline{sk}_t | 1 \le t \le T\} \text{ and } \overline{lsk}^* = \{\overline{lsk}_t | 1 \le t \le T\}.$$

Remark: Note that in the definition of [5], a single secret key is used both for updates and for decryption (instead of having separate keys for updates and decryption, as above). We call such a scheme a *primitive* key-evolving scheme. A primitive scheme which is forward-secure is called a **PFSE** scheme, to distinguish it from forward-secure schemes which can additionally be cast as per the above definition (these are called **FSE** schemes).

2.2   Definition of Security

We now provide a definition of forward security for a key-evolving encryption scheme as defined in the previous section. Our definition is stronger than than the definition given in [5] in that we allow the adversary to obtain the local key (but not the update key) for time periods prior to the challenge time period. Formally, we accomplish this by giving the adversary access to two separate oracles: one of which returns local keys, and one of which returns update keys. Although this is a stronger definition than that given previously, note that the scheme of [5] satisfies it.

Let A be a probabilistic polynomial-time oracle Turing machine, which gets input $\overline{pk}$ and T, and interacts with the following oracles:

• Decryption oracle $O_{FSDec}(\overline{sk}_0, \cdot, \cdot)$, which on input $t \in [T]$ and a ciphertext C outputs a message M decrypted by $\overline{lsk}_t$ (where this key is derived in the appropriate way from $\overline{sk}_0$).

- Update-key oracle $O_{FSukey}(\overline{sk}_0, \cdot)$, which on input $t \in [T]$ outputs $\overline{sk}_t$ (again, this key is derived in the appropriate way from $\overline{sk}_0$).
- Local-key oracle $O_{FSlkey}(\overline{sk}_0, \cdot)$, which on input $t \in [T]$ outputs $\overline{lsk}_t$ (again, this key is derived in the appropriate way from $\overline{sk}_0$).
- Left-or-right oracle $O_{FSLR}(\overline{pk}, \cdot, LR_b(\cdot, \cdot))$ which on inputs $t^* \in [T]$ and equal-length messages m_0, m_1 returns a challenge ciphertext $C^* \leftarrow \texttt{fsEnc}(\overline{pk}, t^*, m_b)$. The bit b is chosen randomly at the outset of the experiment.

The adversary A may query all oracles adaptively, in any order it wants, subject to the following restrictions: queries t to O_{FSukey} satisfy $t > t^*$; queries t' to O_{FSlkey} satisfy $t' \neq t^*$; only a single query is made to O_{FSLR}; and the ciphertext C^* received from O_{FSLR} may not be queried to O_{FSDec} for time period t^*. Eventually, the adversary guesses a bit b' and halts. The adversary succeeds if $b' = b$. We define the adversary's advantage as the absolute value of the difference between its success probability and $1/2$.

Definition 1. *We say that a key-evolving encryption scheme* FSE *is forward secure against chosen-ciphertext attacks (*FS-CCA*) if the advantage of any* PPT *adversary* A *in the above experiment is negligible.*

Remark: We stress that separating the two oracles O_{FSukey} and O_{FSlkey} strengthens the notion of forward security as compared to [5]. Specifically, our model allows an adversary to get the local key corresponding to any $t' \neq t^*$.

3   Intrusion-Resilient Encryption

As mentioned in the introduction, intrusion-resilient encryption schemes achieve a stronger level of security than forward-secure encryption schemes, at the cost of introducing a second entity (i.e., the base). Our definition of security follows [10,6]. An adversary is allowed an adaptive chosen-ciphertext attack, can additionally obtain the secrets from the base and/or the user, and can eavesdrop on the communication between the base and user. As long as the user, the base, and the communication between user and base are not compromised at the same time period, the scheme remains secure for all time periods at which the user's key was not exposed. Furthermore, the scheme achieves forward security in case the user, base, and communication between user and base *are* compromised at the same time period. We now provide formal definitions.

3.1   Functional Description

The encryption scheme is specified by the following tuple of polynomial-time algorithms:

KeyGen: key generation algorithm
Input: security parameter k, number of time periods T, number of refreshes R
Output: initial user key $sk_{0.0}$, initial base key $skb_{0.0}$, public key pk

BaseUpd: base key-update algorithm
Input: current base key $skb_{t.r}$
Output: next base key $skb_{t+1.0}$, key update message sku_t

UserUpd: user key-update algorithm
Input: current user key $sk_{t.r}$, key update message sku_t
Output: next user key $sk_{t+1.0}$

BaseRef: base key-refresh algorithm
Input: current base key $skb_{t.r}$
Output: next base key $skb_{t.r+1}$, corresponding key refresh message $skr_{t.r}$

UserRef: user key-refresh algorithm
Input: current user key $sk_{t.r}$, key refresh message $skr_{t.r}$
Output: next user key $sk_{t.r+1}$

Enc: randomized encryption algorithm
Input: user public key pk, current time interval t, message M
Output: ciphertext C

Dec: decryption algorithm
Input: user secret key $sk_{t.r}$, ciphertext $C = \mathtt{Enc}(pk, t, M)$
Output: message M

The encryption scheme is run as follows:

$\mathtt{Syntactic}(k, T, R)$
    Set $(sk_{0.0}, skb_{0.0}, pk) \leftarrow \mathtt{KeyGen}(k, T, R)$.
    For $t = 0$ to $T - 1$:
        Set $(skb_{t+1.0}, sku_t) \leftarrow \mathtt{BaseUpd}(skb_{t.r})$ and $sk_{t+1.0} \leftarrow \mathtt{UserUpd}(sk_{t.r}, sku_t)$.
        For $r = 0$ to $R - 1$
            Set $(skb_{t.r+1}, skr_{t.r}) \leftarrow \mathtt{BaseRef}(skb_{t.r})$ and $sk_{t.r+1} \leftarrow \mathtt{UserRef}(sk_{t.r}, skr_{t.r})$.

Here the keys $sk_{t,0}$ and $skb_{t,0}$ for $0 \leq t \leq T$ are not actually used or stored. Key generation is immediately followed by an update, and each update is immediately followed by a refresh. Therefore, the secret keys which an adversary can potentially access are defined as follows:

- $sk^* = \{sk_{t.r} | 1 \leq t \leq T, 1 \leq r \leq R\}$
- $skb^* = \{skb_{t.r} | 1 \leq t \leq T, 1 \leq r \leq R\}$
- $sku^* = \{sku_t | 1 \leq t \leq T - 1\}$
- $skr^* = \{skr_{t.r} | 1 \leq t \leq T - 1, 0 \leq r \leq R - 1\} \setminus \{skr_{1.0}\}$

3.2 Definition of Security

We now define intrusion-resilience. Let A be a probabilistic polynomial-time oracle Turing machine which gets input pk, T, and R, and which may query the following oracles (each oracle is technically indexed by an initial tuple of keys $(sk_{0.0}, skb_{0.0}, pk)$ which is omitted for readability):

- Decryption oracle O_{Dec}, which on input $t \in [T]$, $r \in [R]$, and a ciphertext C outputs a message M decrypted using $sk_{t.r}$

- User key oracle O_{sk}, which on input $t \in [T]$ and $r \in [R]$ outputs $sk_{t.r}$
- Base key oracle O_{bk}, which on input $t \in [T]$ and $r \in [R]$ outputs $skb_{t.r}$
- Key update oracle O_u, which on input $t \in [T]$ outputs sku_t
- Key refresh oracle O_r, which on input $t \in [T]$ and $r \in [R]$ outputs $skr_{t.r}$
- Left-or-right oracle O_{LR}, which on input $t^* \in [T]$ and equal-length messages m_0, m_1, outputs challenge ciphertext $C^* \leftarrow \text{Enc}(pk, t^*, m_b)$ (for a bit b which is chosen at random at the beginning of the experiment).

The oracles O_{sk}, O_{bk}, O_u and O_r are generically called "key exposure oracles", and are denoted by O_{sec}. Queries to a particular oracle are indicated by including the appropriate string; thus, $O_{sec}(\text{"sk"}, t.r)$ denotes the query $O_{sk}(t, r)$.

The only restrictions for the adversary's queries are that key exposures must *respect erasure*. That is, if a value corresponding to a particular instant in time t_1 has been obtained by the adversary (via an oracle query), then a value corresponding to a prior instant in time (which would have been erased prior to t_1) cannot be obtained. More formally,

⋄ ("sk", $t.r$) must be queried before ("sk", $t'.r'$) if $t' > t$ or $t' = t$ and $r' > r$;
⋄ ("bk", $t.r$) must be queried before ("bk", $t'.r'$) if $t' > t$ or $t' = t$ and $r' > r$;
⋄ ("bk", $t.r$) must be queried before ("r", $t'.r'$) if $t' > t$ or $t' = t$ and $r' \geq r$;
⋄ ("bk", $t.r$) must be queried before ("u", t') if $t' \geq t$.

For a set Q of key exposure queries, we say that $sk_{t.r}$ is Q-*exposed* if one of the following is true:

- ("sk", $t.r$) $\in Q$;
- $r > 1$, ("r", $t.(r-1)$) $\in Q$, and $sk_{t.(r-1)}$ is Q-exposed;
- $r = 1$, ("u", $t-1$) $\in Q$, and $sk_{(t-1).R}$ is Q-exposed;
- $r < R$, ("r", $t.r$) $\in Q$, and $sk_{t.r+1}$ is Q-exposed.

A completely analogous definition may be given for Q-exposure of a base key $skb_{t.r}$. We say the scheme is (t^*, Q)-*compromised* if $sk_{t^*.r}$ is Q-exposed (for some r), or if both $sk_{t'.r}$ and $skb_{t'.r}$ are Q-exposed (for some r and $t' < t^*$).

We say that an adversary *succeeds* if it correctly guesses the bit b used by the O_{LR} oracle, subject to the following restrictions: (1) The system was not (t^*, Q)-compromised where O_{LR} was queried at time period t^*; and (2) The ciphertext C^* returned by O_{LR} was not queried to O_{Dec} (for the same time period t^*). An adversary's advantage is defined as the absolute value of the difference between its success probability and $1/2$.

Definition 2. *We say that an encryption scheme is intrusion-resilient against chosen-ciphertext attacks (*IR-CCA*) if the advantage of any* PPT *adversary A in the above experiment is negligible.*

Remark: We sometimes refer to the notion defined above as "full" intrusion resilience. In Appendix A, we define a security notion called *quasi-intrusion resilience* which lies "in between" key-insulated security and full intrusion-resilience. This intermediate notion helps describe the security level which is achieved by using a *primitive* key-evolving encryption scheme.

4   A Generic Construction of Intrusion-Resilient Encryption

In this section, we present a generic construction of a fully intrusion-resilient encryption scheme.

4.1   Preparations

Our idea is to extend a forward secure encryption scheme whose key-update algorithm is homomorphic in the sense we now describe. Assume a map

$$\phi : G_1 \to G_2 \times G_3,$$

where G_1, G_2, and G_3 are groups represented additively. We say that the map ϕ is homomorphic if for all $x, y \in G_1$ we have:

$$\phi(x + y) = \phi(x) + \phi(y).$$

More precisely, ϕ satisfies

$$\phi(x + y) = (x_1 + y_1, x_2 + y_2)$$

where $\phi(x) = (x_1, x_2)$ and $\phi(y) = (y_1, y_2)$.

To give a generic construction of fully intrusion resilient scheme, we specify the key-evolving encryption scheme FSE as generally as possible. Let S_1 be a set, and let G_1, G_2, and G_3 be groups (written additively). Let FSE be as follows:
FSE = (fsKeyGen, fsKeyUpd, fsEnc, fsDec):

- fsKeyGen: $\{0,1\}^* \times \mathbb{N} \to G_1 \times S_1$; fsKeyGen$(k, T) = (\overline{sk}_0, \overline{pk})$
- fsKeyUpd : $G_1 \to G_1 \times G_2$; fsKeyUpd$(\overline{sk}_t) = (\overline{sk}_{t+1}, \overline{lsk}_{t+1})$
Additionally, fsKeyUpd should be homomorphic; that is:

$$\text{fsKeyUpd}(x + y) = \text{fsKeyUpd}(x) + \text{fsKeyUpd}(y).$$

In other words, it satisfies:

$$\text{fsKeyUpd}(x + y) = (x_1 + y_1, x_2 + y_2),$$

where fsKeyUpd$(x) = (x_1, x_2)$ and fsKeyUpd$(y) = (y_1, y_2)$.
- fsEnc : $S_1 \times \mathbb{N} \times \{0,1\}^n \to \{0,1\}^n$; fsEnc$(\overline{pk}, t, M) = C$
- fsDec : $G_2 \times \{0,1\}^n \to \{0,1\}^n$; fsDec$(\overline{lsk}_t, C) = M$

4.2   Scheme Intuition

As intuition for our construction, we may note that a secret key of encryption scheme FSE consists of $\overline{sk}_t$ and $\overline{lsk}_t$, where the local key $\overline{lsk}_t$ is used only for decryption. We may notice that a user update key $\overline{sk}_t$ of FSE enables derivation of all the user secret keys for periods t through N, but none of the secret keys for periods $t' < t$. This will allow us to achieve forward security, as in [5]. However, in our model we also need to divide the user update key between the user and the base, so that we can derive the sharing for period $t + 1$ from that of period t and achieve future security also. To achieve this, we let the user store $\overline{lsk}_t$ — to enable decryption within the current time period — but additively share the user update key $\overline{sk}_t$ between the user and the base. In summary, let the user store $\overline{lsk}_t$ and the evolved share of $\overline{sk}_t$, and the base store the other evolved share of $\overline{sk}_t$. Intuitively, $\overline{lsk}_t$ by itself only allows the user to decrypt at period t, and the fact that the user update key $\overline{sk}_t$ is split ensures that exposure of the user cannot compromise any of the future periods. Security against compromises of the base follows similarly. This gives us intrusion-resilience.

The only issue to resolve is how to update a local key by using the separated shares of an update key. Both shares of the user and the base are evolved in each time period, which are executed independently by the user and the base. When each share is evolved by using the key-update algorithm of FSE, the algorithm outputs two elements: the sharing of the next-time-period update key and the sharing of the next-time-period local key. The base sends only the sharing of a local key to the user as the update message, and the user combines it with his own sharing of the local key by using the homomorphic property of the key-update algorithm; thus, the user derives the the next-time-period local key. As a result, the user and the base generate their own update keys independently and compute the next-time-period local key jointly. This step is immediately followed by a random refresh.

4.3 FISER

We now describe the fully intrusion-resilient encryption scheme FISER = (KeyGen, BaseUpd, UserUpd, BaseRef, UserRef, Enc, Dec). Let us note that each parameter is defined on the following set or groups:

◇ set of user public keys $:S_1$
◇ group of user secret keys $:G_3 = G_1 \times G_2$
◇ group of base secret keys $:G_1$
◇ group of key update message $:G_3 = G_1 \times G_2$
◇ group of key refresh message $:G_1$

Using the above notation, each function is described as follows:

KeyGen: $\{0,1\}^* \times \mathbb{N} \to G_3 \times G_1 \times S_1$; KeyGen$(k, T) = (sk_{0.0}, skb_{0.0}, pk)$
   1. Compute $(\overline{sk}_0, \overline{pk}) \leftarrow$ fsKeyGen(k, T).
   2. Let $\overline{sk}_0$ be divided in $\overline{sk}_0 = \overline{sks}_{0.0} + \overline{skb}_{0.0}$ for randomly chosen $\overline{sks}_{0.0} \in G_1$
   3. Set $pk = \overline{pk}$, $sk_{0.0} = (\overline{sks}_{0.0}, \cdot)$, and $skb_{0.0} = \overline{skb}_{0.0}$,
   4. Output $sk_{0.0}$, $skb_{0.0}$, and pk.
BaseUpd: $G_1 \to G_1 \times G_2$; BaseUpd$(skb_{t.r}) = (skb_{t+1.0}, sku_t)$
   For an input of base secret key $skb_{t.r} = \overline{skb}_{t.r}$
   1. Compute $(\overline{skb}_{t+1.0}, sku_t) \leftarrow$ fsKeyUpd$(\overline{skb}_{t.r})$.
   2. Output $skb_{t+1.0} = \overline{skb}_{t+1.0}$ and sku_t.

UserUpd: $G_3 \times G_2 \to G_3$; $\texttt{UserUpd}(sk_{t.r}, sku_t) = sk_{t+1.0}$

For inputs of user secret key $sk_{t.r} = (\overline{sks}_{t.r}, \overline{lsk}_t)$ and update message sku_t

1. Compute $(\overline{sks}_{t+1.0}, lsk_{t+1}) \leftarrow \texttt{fsKeyUpd}(\overline{sks}_{t.r})$.
2. Compute $\overline{lsk}_{t+1} = lsk_{t+1} + sku_t$.
3. Output $sk_{t+1.0} = (\overline{sks}_{t+1.0}, \overline{lsk}_{t+1})$.

BaseRef: $G_1 \to G_1 \times G_1$; $\texttt{BaseRef}(skb_{t.r}) = (skb_{t.r+1}, skr_{t.r})$

For an input of base secret key $skb_{t.r} = \overline{skb}_{t.r}$,

1. Compute $\overline{skb}_{t.r+1} = \overline{skb}_{t.r} - \overline{R}_{t.r}$ for a random secret $\overline{R}_{t.r} \in G_1$.
2. Output $skb_{t.r+1} = \overline{skb}_{t.r+1}$ and $skr_{t.r} = \overline{R}_{t.r}$.

UserRef: $G_3 \times G_1 \to G_3$; $\texttt{UserRef}(sk_{t.r}, skr_{t.r}) = sk_{t.r+1}$

For inputs of user secret key $sk_{t.r} = (\overline{sks}_{t.r}, \overline{lsk}_t)$ and refresh message $skr_{t.r} = \overline{R}_{t.r}$,

1. Compute $\overline{sks}_{t.r+1} = \overline{sks}_{t.r} + \overline{R}_{t.r}$.
2. Output $sk_{t.r+1} = (\overline{sks}_{t.r+1}, \overline{lsk}_t)$.

Enc: $S_2 \times \mathbb{N} \times \{0,1\}^n \to \{0,1\}^n$; $\texttt{Enc}(pk, t, M) = C$

For inputs of a public key pk, time t, and a message M,

1. Compute $C \leftarrow \texttt{fsEnc}(pk, t, M)$.
2. Output C.

Dec: $G_3 \times \{0,1\}^n \to \{0,1\}^n$; $\texttt{Dec}(sks_t, C) = M$

For inputs of user secret key $sk_{t.r} = (\overline{sks}_{t.r}, \overline{lsk}_t)$ and ciphertext C,

1. Compute $M \leftarrow \texttt{fsDec}(\overline{lks}_t, C)$.
2. Output M.

5   Security Analysis

We now prove security of the FISER given above. For simplicity, the time complexity of an adversary A is defined as the execution time of the experiment used to define the advantage of A, including the time taken for key generation and initialization, as well as the time required for the various oracles to compute replies to the adversary's queries.

Theorem 1. *Let A be an adversary of time complexity τ with at most Q queries to oracles $O \in \{O_{Dec}, O_{sec}, O_{LR}\}$ against FISER. If A has advantage δ, then there exists an adversary B performing a chosen-ciphertext attack against the underlying FSE with at least the same advantage. The time complexity of B is at most $\tau + O(\log k)$, and the number of queries is at most Q.*

Proof. We construct an adversary B that uses A to perform a chosen-ciphertext-and-key attack against FSE. B is allowed to ask queries to: a decryption oracle $O_{FSDec}(\cdot, \overline{pk}, \overline{sk}_{t.r}, \cdot)$; a user update-key oracle $O_{FSukey}(\overline{pk}, \overline{sk}_0, \cdot)$; a user local-key oracle $O_{FSlkey}(\overline{pk}, \overline{sk}_0, \cdot)$; and a left-or-right oracle $O_{FSLR}(\overline{pk}, \cdot, LR(\cdot, \cdot, b))$. Adversary B receives challenge ciphertext $C^* = \texttt{fsEnc}(\overline{pk}, t^*, m_b)$, and outputs a guess b'. Adversary B succeeds if $b' = b$.

B simulates A's environment as follows: first, B runs A until A outputs T and $R \in \mathbb{N}$. B also returns T. B runs $\texttt{fsKeyGen}(k, T)$ to produce $(\overline{sk}_0, \overline{pk})$. B chooses

$\overline{skb}_{0.0} \in G_1$ randomly and maintains a list U_1^{list}, which consists of tuples of the following form:

$$(t, r; \overline{sks}_{t.r}, \overline{skb}_{t.r}, \overline{R}_{t.r}) \in \mathbb{N} \times \mathbb{N} \times G_1 \times G_1 \times G_1.$$

We use the notation $(t, r; \overline{sks}_{t.r}, -, *)$ as follows: "$-$" is used if there is no list on $\overline{skb}_{t.r}$, i.e. empty , and "*" is used if we don't care about $\overline{R}_{t.r}$ like empty or not, or if we maintain the data after some operating. For example, "change $(t, r; *, *, -)$ to $(t, r; *, *, \overline{R}_{t.r})$" means that: change the data "-" to $\overline{R}_{t.r}$ while maintaining the data of $\overline{sks}_{t.r}$ and $\overline{skb}_{t.r}$ as they are.

To begin, B sets $pk = \overline{pk}$ and $U_1^{list} = \{(0, 0; -, \overline{skb}_{0.0}, -)\}$ and continues the execution of A on input pk using its oracles to respond A's queries as follows:

Decryption oracle. Let a query to $O_{Dec}(\cdot, \cdot, pk, sk_{t.r}, \cdot)$ be (t, r, C). B forwards (t, C) to its decryption oracle $O_{FSDec}(\cdot, \overline{pk}, \overline{sk}_0, \cdot)$, and returns the answer M to A. From the definition of O_{Dec}, the answer is exactly what A's decryption oracle would have answered.

Base key oracle. Let a query to $O_{bk}(\overline{skb}_{0.0}, pk, \cdot, \cdot)$ be (t, r). B conducts the following steps.

1. If there is $(t, r; *, \overline{skb}_{t.r}, *)$ in U_1^{list}, then pick $\overline{skb}_{t.r}$ from U_1^{list}.
2. Else if there is $(t, r; \overline{sks}_{t.r}, -, *)$ in U_1^{list}, which means exactly "simultaneous attack", then forward t to its user update-key oracle O_{FSukey}, get the answer $\overline{sk}_t$, compute

$$\overline{skb}_{t.r} = \overline{sk}_t - \overline{sks}_{t.r},$$

   and renew U_1^{list} by using $(t, r; \overline{sks}_{t.r}, \overline{skb}_{t.r}, *)$ instead of $(t, r; \overline{sks}_{t.r}, -, *)$.
3. Else if $r > 1$ and there is $(t, r-1; *, \overline{skb}_{t.r-1}, \overline{R}_{t.r-1})$ in U_1^{list}, then compute

$$\overline{skb}_{t.r} = \overline{skb}_{t.r-1} - \overline{R}_{t.r-1}$$

   and renew U_1^{list} by using $(t, r; *, \overline{skb}_{t.r}, *)$.
4. Otherwise, choose $\overline{skb}_{t.r} \in G_1$ randomly and renew U_1^{list} using $(t, r; -, \overline{skb}_{t.r}, *)$ in U_1^{list}.
5. Finally B returns $\overline{skb}_{t.r}$ to A.

Since $\overline{skb}_{t.r}$ was exactly what A's base key oracle would have answered, A's view is identical to its view in the attack against FISER.

User key oracle. Let a query to $O_{sk}(sk_{0.0}, pk, \cdot, \cdot)$ be (t, r). B conducts the following steps.

1. If there is $(t, r; \overline{sks}_{t.r}, *, *)$ in U_1^{list}, then pick $\overline{sks}_{t.r}$ from U_1^{list}.
2. Else if there is $(t, r; -, \overline{skb}_{t.r}, *)$ in U_1^{list}, which means exactly "simultaneous attack", then forward t to its user update key oracle O_{FSukey}, get the answer $\overline{sk}_t$, compute

$$\overline{sks}_{t.r} = \overline{sk}_t - \overline{skb}_{t.r},$$

and renew U_1^{list} by using $(t, r; \overline{sks}_{t.r}, \overline{skb}_{t.r}, *)$.

3. Else if $r > 1$ and there is $(t, r; \overline{sks}_{t.r-1}, *, \overline{R}_{t.r-1})$ in U_1^{list}, then compute

$$\overline{sks}_{t.r} = \overline{sks}_{t.r-1} + \overline{R}_{t.r-1}$$

and renew U_1^{list} by using $(t, r; \overline{sks}_{t.r}, -, *)$.
4. Otherwise, choose $\overline{sks}_{t.r} \in G_1$ randomly and renew U_1^{list} using $(t, r; \overline{sks}_{t.r}, -, *)$.
5. Finally B returns $\overline{sks}_{t.r}$ to A.

Since $\overline{sks}_{t.r}$ was exactly what A's user key oracle would have answered, A's view is identical to its view in the attack against FISER.

Refresh oracle. Let a query to $O_r(skb_{0.0}, pk, \cdot, \cdot)$ be (t, r). B conducts the following steps.

1. If there is $(t, r; *, *, \overline{R}_{t.r})$ in U_1^{list}, then pick $\overline{R}_{t.r}$ from U_1^{list}.
2. Else if either of the following are in U_1^{list}:

$$\{(t, r; \overline{sks}_{t.r}, *, -), (t, r; \overline{sks}_{t.r+1}, *, -)\}$$

or

$$\{(t, r; *, \overline{skb}_{t.r}, *), (t, r; *, \overline{skb}_{t.r+1}, *)\},$$

then compute

$$\overline{R}_{t.r} = \overline{sks}_{t.r+1} - \overline{sks}_{t.r} \text{ or } \overline{R}_{t.r} = \overline{skb}_{t.r} - \overline{skb}_{t.r+1},$$

and renew U_1^{list} by using $(t, r; \overline{sks}_{t.r}, *, \overline{R}_{t.r})$ or $(t, r; *, \overline{skb}_{t.r}, \overline{R}_{t.r})$, respectively.
3. Otherwise, choose $\overline{R}_{t.r} \in G_1$ randomly and renew U_1^{list} using $(t, r; *, *, \overline{R}_{t.r})$.
4. Finally B returns $\overline{R}_{t.r}$ to A.

Since $\overline{R}_{t.r}$ was exactly what A's refresh oracle would have answered, A's view is identical to its view in the attack against FISER.

Update oracle. Let a query to $O_u(skb_{0,0}, pk, \cdot)$ be t. B does as follows:

1. If there is $(t, R; *, \overline{skb}_{t.R}, *)$ in U_1^{list}, then compute

$$(\overline{skb}_{t+1.0}, sku_t) \leftarrow \text{fsKeyUpd}(\overline{skb}_{t.R}).$$

2. Else if there is $(t, R; \overline{sks}_{t.R}, -, *)$ in U_1^{list}, forward $t+1$ to its local-key oracle $O_{FSlkey}(\overline{pk}, \overline{sk}_0, \cdot)$, obtain the answer $\overline{lsk}_{t+1}$, and compute

$$(\overline{sks}_{t+1.0}, lsk_{t+1}) \leftarrow \text{fsKeyUpd}(\overline{sks}_{t.R}) \text{ and}$$
$$sku_t = \overline{lsk}_{t+1} - lsk_{t+1}.$$

3. Otherwise, choose randomly $\overline{sks}_{t.R} \in G_1$, forward $t+1$ to its local-key oracle $O_{FSlkey}(\overline{pk}, \overline{sk}_0, \cdot)$, obtain the answer $\overline{lsk}_{t+1}$, compute

$$(\overline{sks}_{t+1.0}, lsk_{t+1}) \leftarrow \texttt{fsKeyUpd}(\overline{sks}_{t.R}) \text{ and}$$
$$sku_t = \overline{lsk}_{t+1} - lsk_{t+1},$$

and renew U_1^{list} by using $(t, r; \overline{sks}_{t.R}, -, *)$.
4. Finally B returns sku_t to A.

Since sku_t was exactly what A's update key oracle would have answered, A's view is identical to its view in the attack against FISER.

Left-or-right oracle. Let a query to $O_{LR}(pk, \cdot, LR(\cdot, \cdot, b))$ be (t^*, m_0, m_1). Then B forwards (t^*, m_0, m_1) to its left-or-right oracle $O_{FSLR}(\overline{pk}, \cdot, LR(\cdot, \cdot, b))$, obtains a ciphertext C^*, and returns C^* to A. From the definition of Enc, the answer is exactly what A's left-or-right oracle would have answered.

When A outputs its guess bit b' and halts, B returns b' and halts. Note that even if A makes queries to more than one oracle $O \in O_{sec}$ for the same time/refresh period (t, r), adversary A does not see any inconsistencies among the answers from these oracle queries unless the scheme becomes (t^*, Q)-compromised (where Q represents the queries of A to O_{sec} up to and including that point in time); this assumes that A respects erasure. Furthermore, B queries O_{FSukey} if and only if A queries both O_{sk} and O_{bk} for the same time/refresh period (t, r). That is, the earliest time period queried to both O_{sk} and O_{bk} simultaneously by A is coincident with that time period submitted to O_{FSukey} by B. Therefore B succeeds whenever A does.

From the above simulation by B, we see that the time complexity of B is at most $\tau + \log k$ and that B makes at most Q queries to its oracles.

Table 1. Abstraction of each security notion

	underlying notion	achieved security level
KIS[3]	IBE	key-insulated
QISER	PFSE + IBE	quasi-intrusion-resilient
FISER	FSE	intrusion-resilient

6   Further Discussion

There are several security notions of key-evolving or key-updating encryption schemes: forward-secure encryption as defined by [5] (called PFSE), forward-secure encryption (FSE) as defined here (recall, in our model the secret key is split into a key used for decryption and a key used for updates), key-insulated encryption [7], and intrusion-resilient encryption [6]. These notions and the notion of ID-based encryption (IBE) [4] are related; this has already been noted in [7,3,8]. We summarize the relation here.

Any secure ID-based encryption scheme IBE with a certain homomorphic property can be transformed to achieve key-insulated security, following [3].[1] We denote this construction by KIS. Unfortunately, this scheme is insecure in case both user and base are corrupted (indeed, the scheme was not designed with this security property in mind).

Our results shows that FSE with a certain homomorphic property is sufficient to achieve intrusion resilience. Then, we may raise the natural question as to whether a generic PFSE scheme can be transformed to achieve intrusion resilience. Unfortunately, the answer seems to be "no" in general (at least using a "simple" construction as shown here) even if we assume that the key-update algorithm is appropriately homomorphic. More formally, any PFSE scheme which can be converted in this way can actually be cast as an FSE scheme anyway.

We briefly discuss why. Intuitively, both the user and the base must share the secret key of the PFSE scheme in order to achieve intrusion resilience. This requires that no single entity can have enough control to cause any security concerns. On the other hand, the user needs to decrypt a ciphertext. This indicates some separation between keys used for decryption and keys used for key updates. It would be interesting to formalize and rigorously prove the above informal reasoning.

This may raise another question of what level of security is achieved by using PFSE. We show that any primitive forward secure encryption scheme together with any secure ID-based encryption scheme that satisfies a certain homomorphic property can be transformed to achieve quasi-intrusion-resilience in Appendix B. The construction is called QISER, and the definition of quasi-intrusion-resilience is given in Appendix A. These abstraction of each security notion is shown in Table 1.

Remark: The Boneh-Franklin ID-based encryption scheme satisfies the necessary homomorphic property. Therefore, a forward-secure encryption scheme (e.g., [5]) combined with this IBE scheme satisfies quasi-intrusion-resilience.

References

1. R. Anderson. "Two remarks on public-key cryptology." Invited Lecture, *ACM-CCCS'97*.
   Available at http://www.cl.cam.ac.uk/ftp/users/rja14/forwardsecure.pdf.
2. M. Bellare and S. K. Miner. "A forward-secure digital signature scheme." *Advances in Cryptology — Crypto '99*, LNCS vol. 1666, Springer-Verlag, 1999.
3. M. Bellare, and A. Palacio. "Protecting against key exposure: strongly key-insulated encryption with optimal threshold." Available at http://eprint.iacr.org.

---

[1] The construction in [3] is based on the Boneh-Franklin ID-based encryption scheme [4], but may be extended to use any ID-based encryption scheme with a certain homomorphic property.

4. D. Boneh and M. Franklin. "Identity based encryption from the Weil pairing." *Advances in Cryptology — Crypto 2001*, LNCS vol. 2139, Springer-Verlag, 2001. Full version to appear in *SIAM J. Computing* and available at `http://eprint.iacr.org/2001/090`.

5. R. Canetti, S. Halevi, and J. Katz. "A forward-secure public-key encryption scheme." *Advances in Cryptology — Eurocrypt 2003*, LNCS vol. 2656, Springer-Verlag, 2003.

6. Y. Dodis, M. Franklin, J. Katz, A. Miyaji and M. Yung. "Intrusion-resilient public-key encryption." *RSA — Cryptographers' Track 2003*, LNCS 2612, Springer-Verlag, 2003.

7. Y. Dodis, J. Katz, S. Xu, and M. Yung. "Key-insulated public-key cryptosystems." *Advances in Cryptology — Eurocrypt 2002*, LNCS vol. 2332, Springer-Verlag, 2002.

8. Y. Dodis, J. Katz, S. Xu, and M. Yung. "Strong key-insulated signature schemes." *Public Key Cryptography 2003*, LNCS vol. 2567, Springer-Verlag, 2003.

9. G. Itkis. "Intrusion-resilient signatures: generic constructions; or defeating a strong adversary with minimal assumptions." *Security in Communication Networks 2003*, LNCS vol. 2576, Springer-Verlag, 2002.

10. G. Itkis and L. Reyzin. "SiBIR: signer-base intrusion-resilient signatures." *Advances in Cryptology — Crypto 2002*, LNCS, vol. 2442, Springer-Verlag, 2002.

A   Definition of Quasi-intrusion Resilience

We introduce the notion of quasi intrusion resilience, which lies "in-between" key-insulated security and intrusion resilience. Informally, the security obtained is as follows: corrupting both the base and the user at the same time period means that any period before the first user corruption is secure; otherwise, repeated exposure of the user and the base only compromises those specific time periods during which the user's secret keys were exposed.

We generalize the notion of (t^*, Q)-compromise from Section 3.2 by considering two disjoint scenarios, *simultaneous* and *non-simultaneous* corruption. We call a corruption *simultaneous* if both user and base were compromised for the same time period and refresh period; otherwise we call the corruption *non-simultaneous*. More formally, we say the scheme is (t^*, Q)-simultaneous-compromised if one of the following is true:

- $sk_{t^*.r}$ is Q-exposed (for some r); or
- both $sk_{t'.r}$ and $skb_{t'.r}$ are Q-exposed (for some t' and r).

We say the scheme is (t^*, Q)-non-simultaneous-compromised if:

- $sk_{t^*.r}$ is Q-exposed (for some r); or
- both $sk_{t'.r}$ and $skb_{t'.r}$ are Q-exposed (for some r and $t' < t^*$); or
- both $sk_{t'.r}$ and $skb_{t'.r}$ are never both Q-exposed (for any r and $t' > t^*$).

One can consider definitions in which (t^*, Q)-simultaneous-compromise is disallowed, or in which (t^*, Q)-non-simultaneous-compromise is disallowed. A scheme secure against any adversary who is not (t^*, Q)-simultaneous-compromise is called non-simultaneous-compromise secure; the opposite case gives a system which is simultaneous-compromise secure. Obviously, an encryption scheme is (fully) intrusion resilient if and only if it is both simultaneous-compromise and non-simultaneous-compromise secure.

By slightly modifying the condition of (t^*, Q)-non-simultaneous-compromised, we may define a system as (t^*, Q)-quasi-non-simultaneous-compromised if:

- $sk_{t.r}$ is Q-exposed (for some r and $t \leq t^*$); or
- both $sk_{t'.r}$ and $skb_{t'.r}$ are Q-exposed (for some r and $t' < t^*$); or
- both $sk_{t'.r}$ and $skb_{t'.r}$ are never both Q-exposed (for any r and $t' > t^*$).

Let us define (t^*, Q)-quasi-non-simultaneous-compromised and CCA1 variation of intrusion-resilience as quasi-simultaneous-compromise secure; that is, the adversary does not query after receiving the challenge ciphertext c from O_{LR} and the scheme is not (t^*, Q)-quasi-non-simultaneous-compromised. The notion of quasi-intrusion-resilience is given as follows.

Definition 3. *We say that an encryption scheme is quasi-intrusion-resilient against chosen ciphertext attacks (QIR-CCA) if it is both quasi-simultaneous-compromise and non-simultaneous-compromise secure.*

We may note that the definition of quasi-intrusion-resilience is rather ad-hoc in the current version.

B   Generic Quasi-intrusion-Resilient Encryption

B.1   Preparations

Our idea is to combine a primitive forward-secure encryption scheme with a secure ID-based encryption scheme, where key-extract algorithm has a *homomorphic-like* property. Let us define a homomorphic-like property of map,

$$\phi : G_1 \times S \to G_2,$$

where both G_1 and G_2 are groups and S is a set. The operation of G_1 and G_2 is represented additively, and S does not have to be a group. Then, we say that the map ϕ has a homomorphic-like property if for all $s_1, s_2 \in G_1$ and all $t \in S$

$$\phi(s_1 + s_2, t) = \phi(s_1, t) + \phi(s_2, t).$$

Now we give a general construction of quasi-intrusion-resilient scheme. Let S_2 and S_3 be sets, which are used in PFSE. We do not require any group property for PFSE. Let G_4 and G_5 be groups, which are used in IBE. The operations are represented additively.

- PFSE $=$ (pfsKeyGen, pfsKeyUpd, pfsEnc, pfsDec):
  ◇ pfsKeyGen: $\{0,1\}^* \times \mathbb{N} \to S_3 \times S_2$; pfsKeyGen$(k, T) = (\overline{sk}_0, \overline{pk})$
  ◇ pfsKeyUpd $: S_3 \to S_3$; pfsKeyUpd$(\overline{sk}_t) = \overline{sk}_{t+1}$
  ◇ pfsEnc $: S_2 \times \mathbb{N} \times \{0,1\}^n \to \{0,1\}^n$; pfsEnc$(\overline{pk}, t, M) = C$
  ◇ pfsDec $: S_3 \times \{0,1\}^n \to \{0,1\}^n$; pfsDec$(\overline{sk}_t, C) = M$

ID-based encryption consists of key-generation, key-extraction, encryption, and decryption algorithms.

- IBE = (IBKeyGen, IBKeyExt, IBEnc, IBDec):
- ⋄ IBKeyGen: $\{0,1\}^* \to G_4 \times G_5$; IBKeyGen$(k)=(s_0, pk_1)$

Input: security parameter k

Output: master secret s_0, public key pk_1

- ⋄ IBKeyExt: $G_4 \times \mathbb{N} \to G_5$; IBKeyExt$(s_0, t) = isk_t$

Input: user ID t and secret s_0

Output: user secret key isk_t

IBKeyExt has to satisfy a homomorphic-like property for G_4 and G_5:

$$\text{IBKeyExt}(s_1 + s_2, t) = \text{IBKeyExt}(s_1, t) + \text{IBKeyExt}(s_2, t).$$

- ⋄ IBEnc: $G_5 \times \mathbb{N} \times \{0,1\}^n \to \{0,1\}^n$; IBEnc$(pk_1, t, M) = C$

Input: public key pk_1, user ID t, message M

Output: cipher text C

- ⋄ IBDec: $G_5 \times \{0,1\}^n \to \{0,1\}^n$; IBDec$(isk_t, C) = M$

Input: user secret key isk_t, ciphertext $C = \text{IBEnc}(pk_1, t, M)$

Output: message M

B.2   QISER

Let us describe the quasi-intrusion-secure encryption scheme QISER = (KeyGen, BaseUpd, UserUpd, BaseRef, UserRef, Enc, Dec). Here, user secret keys, user public keys, base secret keys, key update message, and key refresh message are defined on the following sets or groups:

- ⋄ set of user public keys   : $S_5 = S_2 \times G_5$
- ⋄ set of user secret keys   : $S_4 = S_3 \times G_4 \times G_5$
- ⋄ group of base secret keys   : G_4
- ⋄ group of key update message : $G_6 = G_4 \times G_5$
- ⋄ group of key refresh message : G_4

KeyGen: $\{0,1\}^* \times \mathbb{N} \to S_4 \times G_4 \times S_5$; KeyGen$(k, T)=(sk_{0.0}, skb_{0.0}, pk)$

For inputs of security parameter k and time T,

1. Set $(s_0, pk_1) \leftarrow$ IBKeyGen(k) and $isk_0 \leftarrow$ IBKeyExt$(s_0, 0)$.
2. Let s_0 be divided in $s_0 = sks_{0.0} + skb_{0.0}$ for randomly chosen $sks_{0.0} \in G_4$
3. Compute $(\overline{sk_0}, \overline{pk}) \leftarrow$ pfsKeyGen(k, T).
4. Set $pk = (\overline{pk}, pk_1)$ and $sk_{0.0} = (\overline{sk_0}, sks_{0.0}, isk_0)$.
5. Output $sk_{0.0}$, $skb_{0.0}$, and pk.

BaseUpd: $G_4 \to G_4 \times G_6$; BaseUpd$(skb_{t.r}) = (skb_{t+1.0}, sku_t)$

For an input of base secret key skb_t r,

1. Compute $skb_{t+1.0} = skb_{t.r} - l_t$ for a random secret $l_t \in G_4$
2. Compute $u_t \leftarrow$ IBKeyExt$(skb_{t+1.0}, t + 1)$.
3. Output $skb_{t+1.0}$ and $sku_t = (l_t, u_t)$.

UserUpd: $S_4 \times G_6 \to S_4$; UserUpd$(sk_{t.r}, sku_t) = sk_{t+1.0}$

For inputs of user secret key $sk_{t.r} = (\overline{sk_t}, sks_{t.r}, isk_t)$ and update message

$sku_t = (l_t, u_t)$,
1. Compute $sks_{t+1.0} = sks_{t.r} + l_t$.
2. Compute $isk_{t+1} = \texttt{IBKeyExt}(sks_{t+1.0}, t+1) + u_t$.
3. Compute $\overline{sk}_{t+1} \leftarrow \texttt{pfsKeyUpd}(\overline{sk}_t)$.
4. Output $sk_{t+1.0} = (\overline{sk}_{t+1}, sks_{t+1.0}, isk_{t+1})$.

$\texttt{BaseRef}: G_4 \to G_4 \times G_4$; $\texttt{BaseRef}(skb_{t.r}) = (skb_{t.r+1}, skr_{t.r})$
For an input of base secret key $skb_{t.r}$,
1. Compute $skb_{t.r+1} = skb_{t.r} - l_{t.r}$ for a random secret $l_{t.r} \in G_4$.
2. Output $skb_{t.r+1}$ and $skr_{t.r} = l_{t.r}$.

$\texttt{UserRef}: S_4 \times G_4 \to S_4$; $\texttt{UserRef}(sk_{t.r}, skr_{t.r}) = sk_{t.r+1}$
For inputs of user secret key $sk_{t.r} = (\overline{sk}_t, sks_{t.r}, isk_t)$ and refresh message $skr_{t.r}$,
1. Compute $sks_{t.r+1} = sks_{t.r} + skr_{t.r}$.
2. Output $sk_{t.r+1} = (\overline{sk}_t, sks_{t.r+1}, isk_t)$.

$\texttt{Enc}: S_5 \times \mathbb{N} \times \{0,1\}^n \to \{0,1\}^n$; $\texttt{Enc}(pk, t, M) = C$
For inputs of a public key pk, time t, and a message M,
1. Compute $C \leftarrow \texttt{IBEnc}(pk_1, t, \texttt{pfsEnc}(\overline{pk}, t, M))$.
2. Output C.

$\texttt{Dec}: S_4 \times \{0,1\}^n \to \{0,1\}^n$; $\texttt{Dec}(sk_{t.r}, C) = M$
For inputs of user secret key $sk_{t.r} = (\overline{sk}_t, sks_{t.r}, isk_{t.r})$ and a ciphertext C,
1. Compute $M \leftarrow \texttt{pfsDec}(\overline{sk}_t, \texttt{IBDec}(isk_{t.r}, C))$.
2. Output M.

B.3   Security Analysis

The following theorems will be proved in the final version.

Theorem 2. *Let A be an adversary of time complexity τ with at most Q queries to oracles $O \in \{O_{Dec}, O_{sec}, O_{LR}\}$ against* QISER. *If A has non-negligible advantage under non-simultaneous compromise, then there exists an adversary B performing a chosen ciphertext attack against the underlying* IBE *with at least the same advantage. The time complexity of B is at most $\tau + O(\log k)$, and the number of queries is at most Q.*

Theorem 3. *Let A be an adversary of time complexity τ with at most Q queries to oracles $O \in \{O_{Dec}, O_{sec}, O_{LR}\}$ against* QISER. *If A has non-negligible advantage under quasi-simultaneous-compromise, then there exists an adversary B performing a chosen ciphertext attack against the underlying* PFSE *with at least the same advantage. The time complexity of B is at most $\tau + O(\log k)$, and the number of queries is at most Q.*

A Certificate-Based Signature Scheme

Bo Gyeong Kang, Je Hong Park, and Sang Geun Hahn

Department of Mathematics,
Korea Advanced Institute of Science and Technology,
373-1 Guseong-dong, Yuseong-gu, Daejeon, 305-701, Korea
{snubogus,Jehong.Park,sghahn}@kaist.ac.kr

Abstract. In this paper, we propose the security notion of certificate-based signature that uses the same parameters and certificate revocation strategy as the encryption scheme presented at Eurocrypt 2003 by Gentry. Certificate-based signature preserves advantages of certificate-based encryption, such as implicit certification and no private key escrow. We present concrete certificate-based signature schemes derived from pairings on elliptic curves and prove their security in the random oracle model assuming that the underlying group is GDH. Additionally, we propose a concrete delegation-by-certificate proxy signature scheme which is derived from a certificate-based signature scheme after simple modifications. Our proxy scheme is provably secure in the random oracle model under the security notion defined by Boldyreva, Palacio and Warinschi.

1 Introduction

1.1 Certificate-Based Cryptosystem

In traditional public key signatures (PKS), the public key of the signer is essentially a random bit string picked from a given set. So, the signature does not provide the authorization of the signer by itself. This problem can be solved via a certificate which provides an unforgeable and trusted link between the public key and the identity of the signer by the CA's signature. And there is a hierarchical framework that is called by *public key infrastructure* (PKI) to issue and manage certificates. In general, the signer registers its own public key with its identity in certificate server and anyone wishing to obtain the signer's public key requests it by sending the server the identity of the signer and gets it. Before verifying a signature using the signer's public key, however, a verifier must obtain the signer's certification status, hence in general make a query on the signer's certificate status to the CA. It is called by *third-party query*. As mentioned in [11], even though the third party query has some problems in public-key encryptions, those problems can be surmounted in signature schemes simply by transmitting the certificate for its valid public key with its signature. Despite of this settlement, a verifier must verify the certificate first and if authorization of the CA about the signer's public key is valid then verifies the signed message with given public key from the signer. In the point of a verifier, two verification

T. Okamoto (Ed.): CT-RSA 2004, LNCS 2964, pp. 99–111, 2004.
© Springer-Verlag Berlin Heidelberg 2004

steps for independent signatures are needed. As a consequence, this system requires a large amount of computing time and storage when the number of users increases rapidly.

To simplify key management procedures of conventional PKIs, Shamir asked for ID-based cryptography (IBC) in 1984 [18], but recently Boneh and Franklin [2] proposed a practical ID-based encryption (IBE) scheme based on bilinear maps. Subsequently, several ID-based signature (IBS) schemes which share system parameters with the IBE scheme of Boneh and Franklin are proposed [17, 12,8]. The main practical benefit of IBC is in greatly reducing the need for, and reliance on, the public key certificates. But IBC uses a trusted third party called a *Private Key Generator* (PKG). The PKG generates the secret keys of all of its users, so a user can decrypt only if the PKG has given a secret key to it (so, certification is implicit), hence reduces the amount of storage and computation. On the other hand, private key *escrow* is inherent and secret keys must be sent over *secure channels*, making private key distribution difficult [11].

To import several merits of IBC into conventional PKIs, the concept of *certificate-based encryption* (CBE) was introduced by Gentry [11]. A CBE scheme which is created by combining a public key encryption (PKE) scheme and an IBE scheme consists of a certifier and users. Each user generates its own secret and public key pair and requests a certificate from the CA, then the CA uses the user private key generation algorithm in the Boneh-Franklin IBE scheme to generate certificates. That gives us implicit certification by virtue of the fact that a certificate can be used as a signing key, and so allows to eliminate third-party queries on certificate status. But this ordinary CBE scheme is inefficient when the CA has a large number of users and performs frequent certificate updates, so Gentry suggests to use subset covers to overcome inefficiency [11].

1.2   Our Contributions

We give the first construction of a certificate-based signature (CBS) scheme that can use the same parameters and certificate revocation strategy as the CBE scheme of [11]. In Section 2, we define a formal security model of CBS, and describe two similar pairing-based CBS schemes which are secure in the random oracle model in Section 3. It is obvious that our schemes maintain most of the advantages of CBE over PKE and IBE. Both of these schemes use an IBS scheme for signing phase and the BLS signature scheme [4] for certificate issuing phase, but one scheme uses multisignatures, while the other does aggregate signatures as a temporary signing key, which provide implicit certification. Since the CA does not know user's personal secret key, CBS does not suffer from the key escrow property which is inherent in IBC and since the CA's certificate need not be kept secret, there is no secret key distribution problem.

We show in Section 4 that a delegation-by-certificate proxy signature scheme immediately follows. A proxy signature permits an entity to delegate its signing rights to another entity. The basic model of proxy signature schemes is that the original signer creates a signature on delegation information and gives it to the proxy signer, and then the proxy signer uses it to generate a proxy key pair. That is analogous to the certificate issuing and temporary signing key generation phases in CBS. Based on this fact, we make slight modifications to our CBS scheme, and prove that the resulting proxy signature scheme is secure in the random oracle model, assuming that the underlying group is GDH.

1.3   Related Works

The general notion of self-certified signatures (SCS) proposed by Lee and Kim [16] is that a signer computes a temporary signing key with its secret key and certification information together, and generates a signature on a message and certificate information using the temporary signing key. Then a verifier verifies both signer's signature on the message and related certification information together. We can easily see that both SCS and CBS provide the authenticity of a digital signature and the authorization of a public key simultaneously. But there are some different aspects between SCS and CBS. The former does not concern the certificate revocation problem which is the main contribution of the latter. It only specifies how to sign a message and verify a signature using a long-lived key pair and the corresponding certificate together, and provides explicit authentication of a public key.

The notion of certificateless public key signature (CL-PKS) presented by Al-Riyami and Paterson [1] does not require the use of certificates. In CL-PKS, the Key Generation Center (KGC) supplies an user with a partial secret key which the KGC computes from the user's identity and a master key, and then the user combines its partial secret key and the KGC's public parameters with some secret information to generate its actual secret key and public key respectively. In this way, an user's secret key is not available to the KGC, whereas the KGC must send the partial secret keys over secure channels. In this case, it is assumed that the KGC is trusted not to replace users' public keys because a new public key could have been created by the KGC and it cannot be easily decided which is the case. This rather strong security assumption can be reduced by a slight modification that an user must first generate its public key and then bind it with its identity as the new identity of the user. The user sends it to the KGC to generate a partial secret key. This technique makes the trust level of the KGC apparent and equivalent to that of the CA in conventional PKIs. Independently, Chen, Zhang and Kim [9] apply the same idea of above modification to the IBS scheme of Cha and Cheon [8]. Although these schemes remove the key escrow property, they still require secure channels and are less efficient than our CBS scheme.

2   Preliminaries

In this section, we review some definitions and provide a formal security model necessary to build our signature scheme. We refer the reader to [2,4,10,13] for a discussion of how to build a concrete instance using supersingular curves and compute the bilinear map.

2.1   Cryptographic Assumptions

Let $\mathbb{G}_1$ and $\mathbb{G}_2$ be two cyclic groups of some large prime order q. We view $\mathbb{G}_1$ as an additive group and $\mathbb{G}_2$ as a multiplicative group. A bilinear map $\hat{e} : \mathbb{G}_1 \times \mathbb{G}_1 \to \mathbb{G}_2$ between these two groups which is called *admissible pairing* must satisfy the following properties [2,11]:

1. Bilinear: $\hat{e}(aQ, bR) = \hat{e}(Q, R)^{ab}$ for all Q, $R \in \mathbb{G}_1$ and all a, $b \in \mathbb{Z}$.
2. Non-degenerate: $\hat{e}(Q, R) \neq 1$ for some Q, $R \in \mathbb{G}_1$.
3. Computable: There is an efficient algorithm to compute $\hat{e}(Q, R)$ for any Q, $R \in \mathbb{G}_1$.

From an admissible pairing $\hat{e}$, decisional Diffie-Hellman (DDH) problem in $\mathbb{G}_1$ can be easily solved, since $\hat{e}(aP, bP) = \hat{e}(P, abP)$ implies that (P, aP, bP, cP) is a valid Diffie-Hellman tuple.

Definition 1. A prime order group $\mathbb{G}$ is a *GDH group* if there exists an efficient algorithm which solves the DDH problem in $\mathbb{G}$ and there is no polynomial time algorithm which solves the CDH problem.

A *GDH parameter generator* $\mathcal{IG}$ is a randomized algorithm that takes a security parameter $k \in \mathbb{N}$, runs in time polynomial in k, and outputs the description of two groups $\mathbb{G}_1$ and $\mathbb{G}_2$ of the same prime order q and the description of an admissible pairing $\hat{e} : \mathbb{G}_1 \times \mathbb{G}_1 \to \mathbb{G}_2$. We say that $\mathcal{IG}$ satisfies the *GDH assumption* if the following probability is negligible (in k) for all PPT algorithm $\mathcal{A}$:

$$\Pr[\mathcal{A}(\mathbb{G}_1, \mathbb{G}_2, \hat{e}, P, aP, bP) = abP \,|\, (\mathbb{G}_1, \mathbb{G}_2, \hat{e}) \leftarrow \mathcal{IG}(1^k), P \leftarrow \mathbb{G}_1^*, a, b \leftarrow \mathbb{Z}_q^*].$$

As noted in [8], a BDH parameter generator $\mathcal{IG}_{\mathsf{BDH}}$ [2] satisfying the BDH assumption can also be viewed as a GDH parameter generator $\mathcal{IG}_{\mathsf{GDH}}$ satisfying the GDH assumption because the BDH assumption is stronger than the GDH assumption.

2.2   ID-Based Signature Scheme of Cha and Cheon

Recently, Cha and Cheon [8] proposed an IBS scheme which is not only efficient but also provably secure in the random oracle model assuming that the underlying group is GDH. This scheme consists of the following algorithms:

IBS.Setup: Choose a generator P of $\mathbb{G}_1$, pick a random $s \in \mathbb{Z}/q\mathbb{Z}$ and set $P_{\mathrm{pub}} = sP$. Choose hash functions $H_1 : \{0,1\}^* \to \mathbb{G}_1$ and $H_3 : \{0,1\}^* \times \mathbb{G}_1 \to \mathbb{Z}/q\mathbb{Z}$. The system parameter is $(\mathbb{G}_1, \mathbb{G}_2, \hat{e}, P, P_{\mathrm{pub}}, H_1, H_3)$ and the master secret key is s.

IBS.Extr: Given an identity ID, compute the public key $Q_{\mathrm{ID}} = H_1(\mathrm{ID})$ and output the secret key $D_{\mathrm{ID}} = sQ_{\mathrm{ID}}$ associated to ID.

IBS.Sign: Given a secret key D_{ID} and a message m, pick a random number $r \in \mathbb{Z}/q\mathbb{Z}$ and output a signature $\sigma = (U, V)$ where $U = rQ_{\mathrm{ID}}$, $h = H_3(m, U)$ and $V = (r + h)D_{\mathrm{ID}}$.

IBS.Vrfy: To verify a signature $\sigma = (U, V)$ of a message m for an identity ID, check whether $(P, P_{\mathrm{pub}}, U + hQ_{\mathrm{ID}}, V)$, where $h = H_3(m, U)$ is a valid Diffie-Hellman tuple.

2.3   BLS Signature and Multisignature Schemes

Here, we introduce a pairing-based signature scheme of Boneh, Lynn and Shacham [4], and a multisignature scheme of Boldyreva [5]. Let $\mathbb{G}_1$ be a GDH group of prime order q and let P be a generator of $\mathbb{G}_1$. The global information PARAMS contains P, q and a description of a hash function H mapping arbitrary strings to the elements of $\mathbb{G}_1^*$. A *BLS signature scheme* consists of the following algorithms:

BLS.Key: Given a security parameter PARAMS, choose a random element $x \in \mathbb{Z}_q^*$ and return a pair $(SK, PK) = (x, xP)$.

BLS.Sign: Given a secret key SK and a message $m \in \{0,1\}^*$, compute $H(m)$ and return a signature $\sigma = xH(m)$.

BLS.Vrfy: To verify a signature σ of a message m, check whether $(P, PK, H(m), \sigma)$ is a valid Diffie-Hellman tuple. If valid then return 1, else return 0.

Suppose n users each has a secret and public key pair (SK_i, PK_i) via running BLS.Key algorithm. For simplicity we are assigning the members consecutive integer identities $1, 2, \ldots, n$. Suppose user i signs a message $m \in \{0,1\}^*$ to obtain the signature σ_i via BLS.Sign algorithm. The multisignature of an arbitrary subset of $L \subseteq [n]$ is computed simply as $\sigma = \prod_{i \in L} \sigma_i \in \mathbb{G}_1^*$. To verify the multisignature σ on condition that public keys of all users in L are given, compute $PK_L = \prod_{i \in L} PK_i$ and check whether $(P, PK_L, H(m), \sigma)$ is a valid Diffie-Hellman tuple. If valid, then return 1 else return 0.

Both the BLS and induced multisignature scheme are proven to be secure in the random oracle model assuming that the underlying group $\mathbb{G}_1$ is GDH.

2.4   The Model

We now provide a formal definition of certificate-based signature schemes and their security. Our definition parallels the definition of a CBE scheme of Gentry. As stated in [11], it does not necessarily have to be "certificate updating". Two main entities involved in CBS are a certifier and a user. This model does not require a secure channel between the two entities.

Definition 2. A *certificate-updating certificate-based signature scheme* consists of the following algorithms:

CBS.Gen$_{\mathsf{IBS}}$, the *IBS key generation* algorithm, takes as input a security parameter 1^{k_1} and (optionally) the total number of time periods t. It returns SK_C (the certifier's master secret) and public parameters PARAMS that include a public key PK_C, and the description of a string space $\mathcal{S}$.

CBS.Gen$_{\mathsf{PKS}}$, the *PKS key generation* algorithm, takes as input a security parameter 1^{k_2} and (optionally) the total number of time periods t. It returns a secret key SK_U and public key PK_U (the user's secret and public keys).

CBS.Upd1, the *certifier update* algorithm, takes as input SK_C, PARAMS, i, string $s \in \mathcal{S}$ and PK_U at the start of time period i. It returns CERT'_i, which is sent to the user.

CBS.Upd2, the *user update* algorithm, takes as input PARAMS, i, CERT'_i and (optionally) CERT_{i-1} at the start of time period i. It returns CERT_i.

CBS.Sign, the *signature generation* algorithm, takes $(m, \mathrm{PARAMS}, \mathrm{CERT}_i, SK_U)$ as input in time period i. It computes the temporary signing key $SK = f(SK_U, \mathrm{CERT}'_i)$ where f is public algorithm, and outputs a signature σ.

CBS.Vrfy, the *verification* algorithm, takes $(\sigma, m, i, PK_C, PK_U)$ as input and outputs a binary value 0 (invalid) or 1 (valid).

As the formal model for CBE, CBS is designed as a combination of PKS and IBS, where the signer need both its personal secret key and a certificate from the CA to sign. The string s includes a message that the certifier signs and may be changed depending on the scheme.

Security. Roughly speaking, we are concerned with two different types of attacks by an uncertified user and by the certifier, as considered in CBE. We want CBS to be secure against each of these entities, even though each basically has *half* of the secret information needed to sign. Accordingly, we define different two games and the adversary chooses one game to play. In Game 1, the adversary essentially assumes the role of an uncertified user. After proving knowledge of the secret key corresponding to its claimed public key, it can make CBS.Sign and CBS.Upd1 queries. In Game 2, the adversary essentially assumes the role of the certifier. After proving knowledge of the master secret corresponding to its claimed PARAMS, it can make CBS.Sign queries. Let PID $= (i, PK_C, PK_U, \mathrm{USINFO})$ be a match for a user U's ID in IBC and call it by *pseudo ID*.

Game 1: The challenger runs IBS.Gen$(1^{k_1}, t)$, and gives PARAMS to the adversary. The adversary then issues CBS.Cert and CBS.Sign queries. These queries are answered as follows:
- On certification query (PID, SK_U), the challenger checks that SK_U is the secret key corresponding to PK_U in PID. If so, it runs CBS.Upd1 and returns CERT'_i; else returns $\bot$.
- On sign query (PID, SK_U, m), the challenger checks that SK_U is the secret key corresponding to PK_U in PID. If so, it generates CERT_i and outputs a valid signature CBS.Sign$(m, \mathrm{PARAMS}, \mathrm{CERT}_i, SK_U)$; else it returns $\bot$.

The adversary outputs (PID, m, σ), where PID $= (i, PK_C, PK_A, \mathrm{ASINFO})$, m is a message and σ is a signature, such that PID and (PID, m) are not equal to the inputs of any query to CBS.Cert and CBS.Sign, respectively. The adversary wins the game if σ is a valid signature of m for i.

Game 2: The challenger runs CBS.Gen$_{\text{PKS}}(1^{k_2}, t)$, and gives PK_U to the adversary. The adversary then issues CBS.Sign query.

- On CBS.Sign query (PID, SK_C, PARAMS, m), the challenger checks that SK_C is the secret key corresponding to PK_C in PARAMS. If so, it generates CERT$_i$ and outputs a valid signature CBS.Sign(m, PARAMS, CERT$_i$, SK_U); else returns $\perp$.

The adversary outputs (PID, m, σ), such that (PID, m) is not equal to the inputs of any query to CBS.Sign. The adversary wins the game if σ is a valid signature of m for i.

Definition 3. A certificate-updating certificate-based signature scheme is *secure against existential forgery under adaptively chosen message and pseudo ID attacks* if no PPT adversary has non-negligible advantage in either Game1 or Game2.

3   Concrete Certificate-Based Signature Schemes

We describe two concrete certificate-based signature schemes called CBSm and CBSa. They use an IBS scheme in common, but as a temporary signing key, the former uses multisignatures and the latter does aggregate signatures[1]. Let k be the security parameter given to the setup algorithm, and let $\mathcal{IG}$ be a GDH parameter generator. Both of them have the same setup and certificate update algorithm.

CBS.Setup: The CA runs $\mathcal{IG}$ on input k to generate groups $\mathbb{G}_1$, $\mathbb{G}_2$ of some prime order q and an admissible pairing $\hat{e} : \mathbb{G}_1 \times \mathbb{G}_1 \rightarrow \mathbb{G}_2$. Then picks an arbitrary generator $P \in \mathbb{G}_1$ and a random secret $s_C \in \mathbb{Z}/q\mathbb{Z}$, and sets $PK_C = s_C P$. Chooses cryptographic hash functions $H_1 : \{0, 1\}^* \rightarrow \mathbb{G}_1$, and $H_3 : \{0, 1\}^* \times \mathbb{G}_1 \rightarrow \mathbb{Z}/q\mathbb{Z}$.

The system parameters are PARAMS $= (\mathbb{G}_1, \mathbb{G}_2, \hat{e}, P, PK_C, H_1, H_3)$ and the CA's master secret key is $SK_C = s_C \in \mathbb{Z}/q\mathbb{Z}$. The CA uses its parameters and its secret to issue certificates. And Alice computes a secret and public key pair as $(SK_A, PK_A) = (s_A, s_A P)$ according to the parameters issued by the CA.

CBS.Cert: Alice obtains a certificate from his CA as follows.
1. Alice sends ALICESINFO to the CA, which includes his public key $s_A P$ and any necessary additional identifying information, such as his name.
2. The CA verifies Alice's information;
3. If satisfied, the CA computes $P_A = H_1(i, PK_C, PK_A, \text{ALICESINFO}) \in \mathbb{G}_1$ in period i.
4. The CA then computes CERT$_A = s_C P_A$ and sends this certificate to Bob.

---

[1] We refer the reader to [3] for a discussion of multisignatures and aggregate signatures

In CBSm, before signing a message $m \in \{0,1\}^*$, Alice signs ALICESINFO, producing $s_A P_A$ and then computes $S_A = s_C P_A + s_A P_A = \text{CERT}_A + s_A P_A$, which is a two person multisignature. Alice will use this multisignature as his temporary signing key.

CBSm.Sign: To sign $m \in \{0,1\}^*$ using ALICESINFO, picks a random $r \in \mathbb{Z}/q\mathbb{Z}$
 and outputs a signature $\sigma = (U, V)$ where $U = rP_A$, $h = H_3(m, U)$ and
 $V = (r+h)S_A = (r+h)(s_C + s_A)P_A$.
CBSm.Vrfy: To verify a signature $\sigma = (U, V)$ of a message m, checks whether
 $e(s_C P + s_A P, U + hP_A) = e(P, V)$, where $h = H_3(m, U)$.

In CBSa, before signing a message $m \in \{0,1\}^*$, Alice also signs ALICESINFO, producing $s_A P'_A$ where $P'_A = H_1(\text{ALICESINFO})$. And she computes her temporary signing key $S_A = s_C P_A + s_A P'_A = \text{CERT}_A + s_A P'_A$, which is a two person aggregate signature.

CBSa.Sign: To sign $m \in \{0,1\}^*$ using ALICESINFO, Alice does the following:
 1. Computes $P_A = H_1(i, PK_C, s_A P, \text{ALICESINFO}) \in \mathbb{G}_1$.
 2. Picks a random $r \in \mathbb{Z}/q\mathbb{Z}$ and outputs a signature $\sigma = (U_1, U_2, V)$
    where $U_1 = rP_A$, $U_2 = rP'_A$, $h = H_3(m, U_1, U_2)$ and $V = (r+h)S_A = (r+h)(s_C P_A + s_A P'_A)$.
CBSa.Vrfy: To verify a signature $\sigma = (U_1, U_2, V)$ of a message m, checks
 $e(PK_C, U_1 + hP_A) \cdot e(PK_A, U_2 + hP'_A) = e(P, V)$, where $h = H_3(m, U_1, U_2)$.

As a note, CBSm can be vulnerable to the following "chosen-key" attack [3] induced by multisignatures. If a malicious signer $\mathcal{A}$ given a secret and public key pair $(s_{\mathcal{A}}, s_{\mathcal{A}}P)$ sets $PK'_{\mathcal{A}} = s_{\mathcal{A}}P - s_C P$ as its public key whose corresponding secret key it does not know, then $\mathcal{A}$ can generate a valid temporary signing key $S_{\mathcal{A}} = s_{\mathcal{A}}P_A(= \text{CERT}_A + (s_{\mathcal{A}} - s_C)P_A)$ by himself, without a valid certificate of period i. To prevent this attack, CBSm is required to have one assumption that the signer must provide a separate proof that it knows the secret key corresponding to its claimed public key and then the verifier must check this separate proof, but the verifier only has to do this once over the lifetime of the public key of the signer, so this extra verification cost may be amortized. For example, the notion of "long-lived certificate" in [11] can be used directly as a role of the separate proof. This auxiliary assumption prevents some malicious users from doing chosen-key attacks and, what is more, not a burden to the implementation of CBSm.

On the other hand, CBSa uses the temporary signing keys obtained by aggregate signatures instead of multisignatures. It prevents the chosen-key attack even if we do not consider above assumption and is provably secure under the security notion in subsection 2.4, naturally. But the verification time is slower than that of CBSm (3 pairing computations instead of 2 are needed).

Remark 1. As stated above, even though CBSm has an auxiliary assumption, it is quite acceptable since extra cost may be ignored. Furthermore, CBSm induces proxy signatures immediately and efficiently in conventional PKIs. So, we intend to focus on multisignature based CBS (i.e. CBSm) for the rest of this paper.

Remark 2. The IBS scheme of Hess [12] can be used in place of the scheme of Cha and Cheon, but the latter is rather efficient than the former in general case [7].

Security Proof. A pseudo ID is the input value of H_1 to derive a certificate from the CA in period i. We modify the notion of security in subsection 2.4, which is acceptable for CBSm schemes. An adversary in Game 2 is allowed to make BLS.Sign queries. Independently, we need to prohibit a signature forged by the chosen-key attack from being accepted as a valid one. Thus it is required that a forged signature (PID, m, σ) is accepted as a *valid* one only when it comes with the secret key corresponding to the public key in PID. Without loss of generality, we say that a certificate-based signature scheme is *secure against existential forgery under adaptively chosen message and pseudo ID attacks* if no PPT adversary has a non-negligible advantage in the following one of games:

Game 1: As the Game 1 in subsection 2.4 except the notion of forged signature validity.

Game 2: Addition to the queries of the Game 2 in subsection 2.4, the adversary is allowed to issue BLS.Sign query. This query is answered as follows:
 - On BLS sign query (PID, SK_C), the challenger checks that SK_C is the secret key corresponding to the public key PK_C in PARAMS. If so, it returns BLS.Sign(SK_U, PID); else returns $\perp$.

The adversary outputs (PID, m, σ) and SK_C. It is valid when PID and (PID, m) are not equal to the inputs of any query to BLS.Sign and CBS.Sign respectively, and SK_C is the secret key corresponding to the public key in PID. The adversary wins the game if σ is a valid signature of m for PID.

For notational purposes, the result of the CBS.Sign query will be denoted by (PID, m, U, h, V) where (U, V) is the output of the signing algorithm of our scheme and $h = H_1(m, U)$.

Theorem 1. *Our certificate-based signature scheme is secure against existential forgery under adaptively chosen message and pseudo ID attacks assuming the underlying group is GDH.*

The proof of the above theorem is in the full version of this paper [14].

Theorem 2. *An aggregate signature based CBS scheme (CBSa) is also secure against existential forgery under adaptively chosen message and pseudo ID attacks.*

As stated above, this theorem can be proved exactly under the security notion in subsection 2.4. The proof of the above theorem is in the full version of this paper [14].

Remark 3. As a certificate-based encryption scheme, adapting CBS to a hierarchy of CAs is fairly obvious. In this case, we need to use aggregate signatures instead of multisignatures because of certification of CAs.

4   Proxy Signature Schemes

Next, we show an application of CBS to proxy signatures. The concept of proxy signatures was first introduced by Mano, Usuda and Okamoto in 1996. A proxy signature scheme which consists of an original signer, a proxy signer and verifiers, allows the original signer to delegate its signing capability to the proxy signer, to sign messages on its behalf. From a proxy signature, anyone can check both the original signer's delegation and the proxy signer's digital signature.

Recently, Boldyreva, Palacio and Warinschi [6] formalize a notion of security for proxy signatures and show that secure proxy signature schemes can be derived from secure standard signature schemes. But they focus on the case that one digital signature scheme is used for standard signing, proxy designation and proxy signing, simultaneously.

4.1   Definition and Security Notion of Proxy Signature Schemes

The basic idea to implement a secure proxy signature scheme is that the original signer creates a signature on the delegation information (warrant[2]) and then the proxy signer uses it to generate a proxy secret key and signs on the delegated message. Since the proxy key pair is generated using the original signer's signature on delegation information, any verifier can check the original signer's agreement from the proxy signature. For simplicity, let users be identified by natural numbers, PK_i denote the public key of user $i \in \mathbb{N}$, and SK_i denote the corresponding secret key. Then a *proxy signature scheme* consists of eight algorithms. Three algorithms PS.Key, S.Sign and S.Vrfy are as in ordinary signature schemes. The other five algorithms provide the proxy signature capability.

(PS.Del, PS.Pro), a pair of interactive algorithms forming the *proxy designation protocol*, takes as input (PK_i, PK_j) for the original signer i and the proxy signer j in common. Each PS.Del and PS.Pro also takes as input SK_i and SK_j, respectively. As result of the interaction, PS.Pro outputs a proxy signing key SKP.

PS.Sign, the *proxy signature generation* algorithm, takes as input (m, SKP), and outputs a proxy signature $p\sigma$.

PS.Vrfy, the *proxy verification* algorithm, takes as input $(p\sigma, m, PK_i)$, and outputs 0 (invalid) or 1 (valid).

PS.Iden, the *proxy identification* algorithm, takes as input $p\sigma$, and outputs PK_j.

For all messages m and all users $i, j \in \mathbb{N}$, if SKP is a proxy signing key for user j on behalf of user i, then PS.Vrfy(PS.Sign(m, SKP), m, PK_i) = 1 and PS.Iden(PS.Sign(m, SKP)) = PK_j.

Chosen message attack capabilities are formed by providing the adversary access to two oracles: a standard signing oracle and a proxy signing oracle. The first oracle takes input a message m, and returns a standard signature for m

---

[2] A warrant is a message containing the public key of the designated proxy signer and possibly restrictions on the messages the proxy signer is allowed to sign.

by user 1. The second oracle takes input a tuple (i, l, m), and if user 1 was designated by user i at least l times, returns a proxy signature for m created by user 1 on behalf of user i, using the l-th proxy signing key. The goal of the adversary is to produce one of the following forgeries:

1. a standard signature by user 1 for a message that was not submitted to the standard signing oracle.
2. a proxy signature for a message m by user 1 on behalf of some user i such that either user i never designated user 1 or m was not in a query (i, l, m) made to the proxy signing oracle, or
3. a proxy signature for a message m by some user i on behalf of user 1, such that user i was never designated by user 1.

We refer the reader to [6] for the notion of security for delegation-by-certificate proxy signatures.

4.2   A Concrete Proxy Signature Scheme

We construct a delegation-by-certificate proxy signature scheme which is derived from CBS. Contrary to the examples in [6], we use the BLS signature scheme for standard signing different from the proxy signing. After all, our proxy signature scheme employs the BLS signature scheme for standard signing and for delegation, and allows an IBS scheme for proxy signing.

Assume that there are two participants, Charlie and Alice with secret and public key pairs $(s_C, s_C P)$ and $(s_A, s_A P)$ respectively, and that they have the common system parameters PARAMS $= (\mathbb{G}_1, \mathbb{G}_2, \hat{e}, P, H_1, H_3)$.

S.Sign: A standard signature for message m is obtained by signing the result using BLS.Sign.

S.Vrfy: The verification of a signature σ for a message m is done by computing BLS.Vrfy.

(PS.Del, PS.Pro): In order to designate Alice as a proxy signer, Charlie simply sends to Alice an appropriate warrant w together with a signature $\text{CERT}_A = s_C P_A$, where $P_A = H_1(PK_C, PK_A, w)$. The corresponding proxy signing key of Alice is $SKP_A = \text{CERT}_A + s_A P_A$.

PS.Sign: A proxy signature for message m produced by Alice on behalf of Chalie, contains a warrant w, the public key of the proxy signer PK_A, and signature $\sigma = (U, V)$ where $U = r P_A$, $h = H_3(m, U)$ and $V = (r + h) SKP_A = (r + h)(s_C + s_A) P_A$.

PS.Vrfy: To verify a signature $(PK_C, m, (PK_A, w, \sigma))$, checks whether $e(PK_C + PK_A, U + h P_A) = e(P, V)$, where $h = H_3(m, U)$.

PS.Iden: The identification algorithm is defined as PS.Iden$(PK_A, w, \sigma) = PK_A$.

In our proxy signature scheme, the role of the CA in CBS schemes is transformed to the original signer, so trust of certificate provider may be removed. And the signer's information to be signed by the CA for certification in CBS schemes is issued conversely from the original signer as a warrant for delegation.

Due to a merit of CBS, our proxy scheme does not need to include a signature for the warrant under the secret key of the original signer in the proxy signature. And it does not require a secure channel for proxy designation [15]. The following theorem shows that our proxy scheme is secure under the security notion of [6].

Theorem 3. *The scheme defined above is a secure proxy signature scheme in the random oracle model assuming that the underlying group is GDH.*

The proof of the above theorem is in the full version of this paper [14].

Remark 4. Recently, the proxy signature scheme using the same idea is proposed by Zhang, Safavi-Naini and Lin [19]. It is based on the IBS scheme of Hess [12], and uses the BLS signature scheme for standard signature and for certification of warrant. Though their scheme also holds desirable and implicit security conditions, it does not guarantee a provable security. We make sure that our work bridges this gap.

5   Conclusion

In this paper, we defined the security notion of certificate-based signature using the same parameters and certificate revocation strategy as the encryption scheme by Gentry. We presented and compared two concrete CBS schemes, and provided proofs of security in the random oracle model assuming that the underlying group is GDH. Our scheme may be useful to construct an efficient PKI combining the CBE scheme of Gentry. Additionally, we derived a concrete delegation-by-certificate proxy signature scheme from a certificate-based signature scheme through simple modifications and proved its security under the security notion defined by Boldyreva, Palacio and Warinschi.

Acknowledgements. The authors would like to thank Craig Gentry, Jung Hee Cheon, Adriana Palacio and the anonymous referees of CT-RSA 2004 for many helpful discussions and comments. This work was supported by grant no. R01-2002-000-00151-0 from the Basic Research Program of the KOSEF.

References

1. S.S. Al-Riyami and K.G. Paterson. Certificateless public key cryptography. *Cryptology ePrint Archive*, Report **2003/126**. An extended abstract will appear in *Advances in Cryptology - ASIACRYPT 2003*, C.S. Laih (Ed.), Lecture Notes in Comput. Sci. **2139**, Springer-Verlag, 2003.
2. D. Boneh and M. Franklin. Identity-based encryption from the Weil pairing. *SIAM J. Comput.*, **32**(3), pp. 586–615 (2003). A preliminary version appeared in *Advances in Cryptology - CRYPTO 2001*, J. Kilian (Ed.), Lecture Notes in Comput. Sci. **2139**, Springer-Verlag, pp. 231–229 (2001).

3. D. Boneh, C. Gentry, B. Lynn and H. Shacham. Aggregate and verifiably encrypted signatures from bilinear maps. *Advances in Cryptology - EUROCRYPT 2003*, E. Biham (Ed.), Lecture Notes in Comput. Sci. **2656**, Springer-Verlag, pp. 416–432 (2003).
4. D. Boneh, B. Lynn and H. Shacham. Short signatures from the Weil pairing. *Advances in Cryptology - ASIACRYPT 2001*, C. Boyd (Ed.), Lecture Notes in Comput. Sci. **2248**, Springer-Verlag, pp. 514–532 (2001).
5. A. Boldyreva. Threshold signatures, multisignatures and blind signatures based on the gap-Diffie-Hellman-group signature scheme. *Public Key Cryptography - PKC 2003*, Y.G. Desmedt (Ed.), Lecture Notes in Comput. Sci. **2567**, pp. 31–46 (2003).
6. A. Boldyreva, A. Palacio and B. Warinschi. Secure proxy signature schemes for delegation of signing rights. *Cryptology ePrint Archive*, Report **2003/096**.
7. X. Boyen. Multipurpose identity-based signcryption - A Swiss army knife for identity-based cryptography. *Advances in Cryptology - CRYPTO 2003*, D. Boneh (Ed.), Lecture Notes in Comput. Sci. **2729**, Springer-Verlag, pp. 382–398 (2003).
8. J.C. Cha and J.H. Cheon. An identity-based signature from gap Diffie-Hellman groups. *Public Key Cryptography - PKC 2003*. Lecture Notes in Comput. Sci. **2567**, Y.G. Desmedt (Ed.), Springer-Verlag, pp. 18–30 (2003).
9. X. Chen, F. Zhang and K. Kim. A new ID-based group signature scheme from bilinear pairings. *Cryptology ePrint Archive*, Report **2003/116**.
10. S.D. Galbraith. Supersingular curves in cryptography. *Advances in Cryptology - ASIACRYPT 2001*, Lecture Notes in Comput. Sci. **2248**, C. Boyd (Ed.), Springer-Verlag, pp. 495–513 (2001).
11. C. Gentry. Certificate-based encryption and the certificate revocation problem. *Advances in Cryptology - EUROCRYPT 2003*, Lecture Notes in Comput. Sci. **2656**, E. Biham (Ed.), Springer-Verlag, pp. 272–293 (2003).
12. F. Hess. Efficient identity based signature scheme based on pairings. *Selected Areas in Cryptography - SAC 2002*, K. Nyber and H. Heys (Eds.), Lecture Notes in Comput. Sci. **2595**, Springer-Verlag, pp. 310–324 (2003).
13. A. Joux. The Weil and Tate pairings as building blocks for public key cryptosystems. *Algorithmic Number Theory - ANTS V*, C. Fieker and D.R. Kohel (Eds.), Lecture Notes in Comput. Sci. **2369**, Springer-Verlag, pp. 20–32 (2002).
14. B.G. Kang, J.H. Park and S.G. Hahn. A certificate-based signature scheme. Full version of this paper. Available at http://crypt.kaist.ac.kr/.
15. J.-Y. Lee, J.H. Cheon and S. Kim. An analysis of proxy signatures: is a secure channel necessary? *Topics in Cryptology - CT-RSA 2003*, M. Joye (Ed.), Lecture Notes in Comput. Sci. **2612**, Springer-Verlag, pp. 68–79 (2003).
16. B. Lee and K. Kim. Self-certified signatures. *Progress in Cryptology - INDOCRYPT 2002*, A. Menezes and P. Sarkar (Eds.), Lecture Notes in Comput. Sci. **2551**, Springer-Verlag, pp. 199–214 (2002).
17. K.G. Paterson. ID-based signatures from pairings on elliptic curves. *Electron. Lett.*, **38**(18), 1025–1026 (2001).
18. A. Shamir. Identity-based cryptosystems and signature schemes. *Advances in Cryptology - CRYPTO'84*, Lecture Notes in Comput. Sci. **196**, G.R. Blakley and D. Chaum (Eds.), Springer-Verlag, pp. 47–53 (1985).
19. F. Zhang, R. Safavi-Naini and C.-Y. Lin. New proxy signature, proxy blind signature and proxy ring signature schemes from bilinear pairings. *Cryptology ePrint Archive*, Report **2003/104**.

Identity Based Undeniable Signatures

Benoît Libert and Jean-Jacques Quisquater

UCL Crypto Group
Place du Levant, 3. B-1348 Louvain-La-Neuve. Belgium
{libert,jjq}@dice.ucl.ac.be
http://www.uclcrypto.org/

Abstract. In this paper, we give a first example of identity based undeniable signature using pairings over elliptic curves. We extend to the identity based setting the security model for the notions of invisibility and anonymity given by Galbraith and Mao in 2003 and we prove that our scheme is existentially unforgeable under the Bilinear Diffie-Hellman assumption in the random oracle model. We also prove that it has the invisibility property under the Decisional Bilinear Diffie-Hellman assumption and we discuss about the efficiency of the scheme.

Keywords. ID-based cryptography, undeniable signatures, pairings, provable security.

1 Introduction

Identity based public key cryptography is a paradigm proposed by Shamir in 1984 ([37]) to simplify key management and remove the necessity of public key certificates. To achieve this, the trick is to let the user's public key be an information identifying him in a non ambiguous way (e-mail address, IP address, social security number...). The removal of certificates allows avoiding the trust problems encountered in today's public key infrastructures (PKIs). This kind of cryptosystem involves trusted authorities called private key generators (PKGs) that have to deliver private keys to users after having computed them from their identity information (users do not generate their key pairs themselves) and from a master secret key. End-users do not have to enquire for a certificate for their public key. Although certificates are not completely removed (the PKG's public key still has to be certified since it is involved in each encryption or signature verification operation), their use is drastically reduced since many users depend on the same authority. Several practical identity based signature schemes (IBS) have appeared since 1984 ([23],[17],[36]) but a practical identity based encryption scheme (IBE) was only found in 2001 ([4]) by Boneh and Franklin who took advantage of the properties of suitable bilinear maps (the Weil or Tate pairing) over supersingular elliptic curves. Many other identity based primitives based on pairings were proposed after 2001: digital signatures, authenticated key exchange, non-interactive key agreement, blind and ring signatures, signcryption, ... ([7],[9],[15],[25],[32],[38],[39], ...).

T. Okamoto (Ed.): CT-RSA 2004, LNCS 2964, pp. 112–125, 2004.

Undeniable signatures are a concept introduced by Chaum and van Antwerpen in 1989 ([10]). It is a kind of signatures that cannot be verified without interacting with the signer. They are useful in situations where the validity of a signature must not be universally verifiable. For example, a software vendor might want to embed signatures into his products and allow only paying customers to check the authenticity of these products. If the vendor actually signed a message, he must be able to convince the customer of this fact using a confirmation protocol and, if he did not, he must also be able to convince the customer that he is not the signer with a denial protocol. These proofs have to be non-transferable: once a verifier is convinced that the vendor did or did not sign a message, he should be unable to transmit this conviction to a third party.

In some applications, a signer needs to decide not only when but also by whom his signatures can be verified. For example a voting center can give a voter a proof that his vote was counted without letting him the opportunity to convince someone else of his vote. That is the motivation of designated verifier proofs for undeniable signatures. This kind of proof involves the verifier's public key in such a way that he is not able to convince a third party that a signer actually signed a message or not because he is able to produce such a valid proof himself using his private key. Several proof systems were proposed for undeniable signatures ([18],[26],[33],...). The use of non-transferable designated verifier proofs ([26]) can provide non-interactive confirmation and denial protocols.

Several examples of undeniable signature schemes based on discrete logarithm were proposed ([10],[11],[12]) and the original construction of Chaum and van Antwerpen ([10]) was proven secure in 2001 by Okamoto and Pointcheval ([31]) thanks to new kind computational assumptions. Several convertible [1] undeniable signatures were proposed ([6],[16],[29],...). RSA-based undeniable signatures were designed by Gennaro, Krawczyk and Rabin ([21]) and Galbraith, Mao and Paterson ([19]). However, no secure identity based undeniable signature has been proposed so far. A solution was proposed in [24] but it was shown in [40] to be insecure. In this paper, we show how bilinear maps over elliptic curves can provide such a provably secure scheme. It is known ([30]) that an undeniable signature can be built from any public key encryption scheme and a similar result is likely to hold in the ID-based setting. However, the scheme described here can offer a security that is more tightly related to some computational problem than a scheme derived from the Boneh-Franklin IBE ([4]).

Chaum, van Heijst and Pfitzmann introduced the notion of 'invisibility' for undeniable signatures. Intuitively, it corresponds to the inability for a distinguisher to decide whether a message-signature pair is valid for a given user or not. The RSA-based schemes described in [19] and [21] do not provide invisibility. In [20], Galbraith and Mao describe a new RSA-based undeniable signature that provides invisibility under the so-called composite decision Diffie-Hellman

---

[1] See [6]. Convertible undeniable signatures are undeniable signatures that can be converted by the signer into universally verifiable signatures.

assumption and they show that invisibility and anonymity [2] are essentially equivalent security notions for undeniable signature schemes satisfying some particular conditions. In this paper, we extend these two security notions to the identity based setting and we prove in the random oracle model that our scheme is both existentially unforgeable and invisible under some reasonable computational assumptions. Invisibility and anonymity can also be shown to be equivalent in the context of identity based cryptography and we will not do it here.

In section 2, we first recall the properties of pairings over elliptic curves before formally describing security notions related to identity based undeniable signatures. In section 3, we describe the different components of our scheme. We then show their correctness and we discuss about their efficiency. The rest of the paper is made of a security analysis of the scheme in the random oracle model.

2 Preliminaries

2.1 Overview of Pairings and Bilinear Problems

Let us consider groups $\mathbb{G}_1$ and $\mathbb{G}_2$ of the same prime order q. We need a bilinear map $\hat{e} : \mathbb{G}_1 \times \mathbb{G}_1 \to \mathbb{G}_2$ satisfying the following properties:

1. Bilinearity: $\forall\, P, Q \in \mathbb{G}_1$, $\forall\, a, b \in \mathbb{Z}_q$, we have $\hat{e}(aP, bQ) = \hat{e}(P, Q)^{ab}$.
2. Non-degeneracy: for any $P \in \mathbb{G}_1$, $\hat{e}(P, Q) = 1$ for all $Q \in \mathbb{G}_1$ iff $P = \mathcal{O}$.
3. Computability: some efficient algorithm can compute $\hat{e}(P, Q)\ \forall\, P, Q \in \mathbb{G}_1$.

Typical admissible bilinear maps are obtained from a modification of the Weil pairing (see [4]) or from the Tate pairing (the original Weil pairing is defined over a non-cyclic group, see [4] for details). The security of the schemes described in this paper relies on the hardness of the following problems.

Definition 1. *Given groups $\mathbb{G}_1$ and $\mathbb{G}_2$ of prime order q, a bilinear map $\hat{e} : \mathbb{G}_1 \times \mathbb{G}_1 \to \mathbb{G}_2$ and a generator P of $\mathbb{G}_1$,*

- *the **Bilinear Diffie-Hellman problem** (BDH) in $(\mathbb{G}_1, \mathbb{G}_2, \hat{e})$ is to compute $\hat{e}(P, P)^{abc}$ given (P, aP, bP, cP).*
- *The **Decisional Bilinear Diffie-Hellman problem** (DBDH) is, given (P, aP, bP, cP) and $z \in \mathbb{G}_2$, to decide whether $z = \hat{e}(P, P)^{abc}$ or not. The advantage of a distinguisher $\mathcal{D}$ for the DBDH problem is defined as*

$$Adv(\mathcal{D}) = \left| Pr_{a,b,c \in_R \mathbb{Z}_q, h \in_R \mathbb{G}_2}[1 \leftarrow \mathcal{D}(aP, bP, cP, h)] \right.$$
$$\left. - Pr_{a,b,c \in_R \mathbb{Z}_q}[1 \leftarrow \mathcal{D}(aP, bP, cP, \hat{e}(P, P)^{abc})] \right|.$$

- *The **Gap Bilinear Diffie Hellman problem** is to solve a given instance (P, aP, bP, cP) of the BDH problem with the help of a DBDH oracle that is able to decide whether a tuple $(P, a'P, b'P, c'P, z)$ is such that $z = \hat{e}(P, P)^{a'b'c'}$ or not. Such tuples will be called DBDH tuples.*

---

[2] This security notion is related to the inability for an adversary to decide which user generated a particular message-signature pair in a multi-user setting.

The DBDH problem was introduced in [13] where it is shown to be not harder than the decisional Diffie-Hellman problem in $\mathbb{G}_2$. It is not known whether the Bilinear Diffie-Hellman problem is strictly easier than the computational Diffie-Hellman problem in $\mathbb{G}_1$ or not. It is also an open question whether the DBDH problem is strictly easier than the BDH one (although it is obviously not harder). Nevertheless, no probabilistic polynomial time algorithm is known to solve any of them with a non-negligible advantage so far.

2.2   Identity Based Undeniable Signatures

An identity based undeniable signature is made of three algorithms and two possibly interactive protocols.

Setup: the PKG takes as input a security parameter k and produces a public/private key pair (s, P_{pub}) and the system's public parameters **params**. s is the system's master key and P_{pub} is the PKG's public key that must be certified.

Keygen: given a user's identity ID, the PKG uses its master secret key s to compute the corresponding private key d_{ID} and transmit it to the user through a secure channel.

Sign: given a message $M \in \{0,1\}^*$ and his private key d_{ID}, the user generates a signature σ associated to M for his identity ID.

Confirm: is a protocol between a signer and a (possibly designated) verifier that takes as input a message $M \in \{0,1\}^*$, an identity $ID \in \{0,1\}^*$, the associated private key d_{ID} and a valid signature σ for the pair (M, ID). The output of the protocol is a (possibly non-interactive) non-transferable proof that σ is actually a valid signature on M for the identity ID.

Deny: takes as input an invalid signature σ for a given pair (M, ID) and the private key d_{ID} corresponding to ID. Its output is a proof that σ is not a valid signature for a message M and an identity ID.

Confirm and Deny may be a single protocol. In our scheme, they are distinct.

2.3   Security Notions for Identity Based Undeniable Signatures

The first security notion for ID-based undeniable signature is close to the one for other existing identity based signatures: it is the notion of existential unforgeability under chosen-message attacks.

Definition 2. *An identity based undeniable signature scheme is said to be **existentially unforgeable** under chosen-message attacks if no probabilistic polynomial time (PPT) adversary has a non-negligible advantage in this game:*

1. *The challenger runs the setup algorithm to generate the system's parameters and sends them to the adversary.*
2. *The adversary $\mathcal{F}$ performs a series of queries:*
   - *Key extraction queries: $\mathcal{F}$ produces an identity ID and receives the private key d_{ID} corresponding to ID.*

- *Signature queries: $\mathcal{F}$ produces a message M and an identity ID and receives a signature on M that was generated by the signature oracle using the private key corresponding to the public key ID.*
- *Confirmation/denial queries: $\mathcal{F}$ produces a pair message-signature (M, σ) and an identity ID and gives them to the signature oracle that runs the confirmation/denial protocol to convince $\mathcal{F}$ that σ is actually related to M and ID or that it is not (in a non-transferable way) using the private key d_{ID} corresponding to ID.*

3. *After a polynomial number of queries, $\mathcal{F}$ produces a tuple (ID, M, σ) made of an identity ID, whose corresponding private key was not asked to the challenger during stage 2, and a message-signature pair (M, σ) that was not issued by the signature oracle during stage 2 for the identity ID.*

The forger $\mathcal{F}$ wins the game if it is able to provide a non-transferable proof of validity of the signature σ for message M and identity ID. Its advantage is defined to be its probability of success taken over the coin-flippings of the challenger and $\mathcal{F}$.

A second security notion, introduced by Chaum, van Heijst and Pfitzmann ([12]), is called 'invisibility'. Informally, this notion corresponds to the inability for a dishonest verifier to decide whether a given signature on a given message was issued by some signer even after having observed several executions of confirmation/denial protocols by the same signer for other signatures. Galbraith and Mao ([20]) proposed a general definition for this security notion. In the identity based setting, we need to strengthen it a little to consider the fact that a dishonest user might be in possession of private keys associated to other identities before trying to validate or invalidate an alleged signature on a message for an identity without the help of the alleged signer.

Definition 3. *An identity based undeniable signature scheme is said to satisfy the **invisibility** property if no PPT distinguisher $\mathcal{D}$ has a non-negligible advantage against a challenger in the following game:*

1. *The challenger performs the setup of the scheme and sends the system's public parameters to $\mathcal{D}$.*
2. *The distinguisher $\mathcal{D}$ performs a polynomially bounded number of queries: key extraction queries, signature queries and confirmation/denial queries of the same kind as those of the previous definition.*
3. *After a first series of queries, $\mathcal{D}$ asks for a challenge: it produces a pair (M, ID) made of a message and an identity for which the associated private key was not asked in step 2. The challenger then flips a coin $b \leftarrow_R \{0, 1\}$. If $b = 0$, the challenger sends $\mathcal{D}$ a valid signature σ on M for the identity ID. Otherwise, $\mathcal{D}$ receives from the challenger a random element $\sigma \leftarrow_R \mathcal{S}$ taken at random from the signature space $\mathcal{S}$.*
4. *The distinguisher $\mathcal{D}$ then performs a second series of queries. This time, it is not allowed to perform a confirmation/denial query for the challenge (σ, M, ID) nor to ask the private key associated to ID.*

5. *At the end of the game, $\mathcal{D}$ outputs a bit b' (that is 0 if $\mathcal{D}$ finds that (σ, M, ID) is a valid message-signature-identity tuple and 1 otherwise) and wins the game if $b = b'$.*

$\mathcal{D}$'s advantage in this game is defined to be $Adv^{inv}(\mathcal{D}) := 2Pr[b = b'] - 1$. where the probability is taken over the coin flippings of the distinguisher $\mathcal{D}$ and the challenger.

Similarly to what is done in [20], we also consider the notion of anonymity. This notion is slightly strengthened in the identity based setting

Definition 4. *We say that an identity based undeniable signature scheme satisfies the **anonymity** property if no probabilistic polynomial time distinguiser $\mathcal{D}$ has a non-negligible advantage in the following game:*

1. *The challenger performs the setup of the scheme and sends the system's public parameters to $\mathcal{D}$.*
2. *The distinguisher $\mathcal{D}$ performs a polynomially bounded number of queries: key extraction queries, signature queries and confirmation/denial queries of the same kind as those of definition 2.*
3. *After a first series of queries, $\mathcal{D}$ requests a challenge: it produces a message M and a pair of identities ID_0, ID_1 for which it did not ask the associated private keys in the first stage. The challenger then flips a coin $b \leftarrow_R \{0,1\}$ and computes the signature σ on M with the private key associated to ID_b. σ is sent as a challenge to $\mathcal{D}$.*
4. *$\mathcal{D}$ performs another series of queries. This time, it is not allowed to perform a confirmation or denial query for the challenge σ on identities ID_0, ID_1 nor to request the private key associated to these identities.*

At the end of the game, $\mathcal{D}$ outputs a bit b' for which it finds that σ is a valid signature on M for the identity $ID_{b'}$. It wins the game if $b' = b$. Its advantage is defined as the previous definition.

It is shown in [20] that the notions of invisibility and anonymity are essentially equivalent for undeniable and confirmer signature schemes satisfying some particular properties. It is straightforward (by using the techniques of [20]) to show that this equivalence also holds in the identity based setting. In the next section, we describe an example of identity based undeniable signature and we show its existential unforgeability and its invisibility in the random oracle model.

3   An Identity Based Undeniable Signature

Our ID-based undeniable signature scheme is made of the following algorithms.

Setup: given security parameters k and l, the PKG chooses groups $\mathbb{G}_1$ and $\mathbb{G}_2$ of prime order $q > 2^k$, a generator P of $\mathbb{G}_1$, a bilinear map $\hat{e} : \mathbb{G}_1 \times \mathbb{G}_1 \to \mathbb{G}_2$ and hash functions $H_1 : \{0,1\}^* \to \mathbb{G}_1, H_2 : \{0,1\}^* \times \{0,1\}^l \times \{0,1\}^* \to \mathbb{G}_1$,

$H_3 : \mathbb{G}_2{}^3 \to \mathbb{Z}_q$ and $H_4 : \mathbb{G}_2{}^4 \to \mathbb{Z}_q$. It chooses a master secret $s \leftarrow_R \mathbb{Z}_q$ and computes $P_{pub} = sP \in \mathbb{G}_1$ that is made public. The system's public parameters are

$$\texttt{params} := \{q, \mathbb{G}_1, \mathbb{G}_2, \hat{e}, P, P_{pub}, H_1, H_2, H_3, H_4\}.$$

Keygen: given an identity ID, the PKG computes $Q_{ID} = H_1(ID) \in \mathbb{G}_1$ and the associated private key $d_{ID} = sQ_{ID} \in \mathbb{G}_1$ that is transmitted to the user.

Sign: to sign a message $M \in \{0,1\}^*$, the signer Alice uses the private key d_{ID_A} associated to her identity ID_A.

1. She picks a random string $r \leftarrow_R \{0,1\}^l$ to compute $H_2(M, r, ID_A) \in \mathbb{G}_1$.
2. She then computes $\gamma = \hat{e}(H_2(M, r, ID_A), d_{ID_A}) \in \mathbb{G}_2$. The signature on M is given by

$$\sigma = (r, \gamma) = (r, \hat{e}(H_2(M, r, ID_A), d_{ID_A})) \in \{0,1\}^l \times \mathbb{G}_2.$$

Confirm: to verify a signature σ on a message M, a verifier of identity ID_B needs the help of the signer Alice. He sends her the pair (M, σ), where $\sigma = < r, \gamma > \in \{0,1\}^l \times \mathbb{G}_2$ is the alleged signature. Alice then runs the following confirmation protocol to produce a non-interactive designated-verifier proof that σ is a valid signature on M for her identity ID_A:

a. She first computes $Q_{ID_B} = H_1(ID_B)$.
b. She picks $U, R \leftarrow_R \mathbb{G}_1$ and $v \leftarrow_R \mathbb{Z}_q$ and computes

$$c = \hat{e}(P, U)\hat{e}(P_{pub}, Q_{ID_B})^v \in \mathbb{G}_2$$
$$g_1 = \hat{e}(P, R) \in \mathbb{G}_2 \text{ and } g_2 = \hat{e}(H_2(M, r, ID_A), R) \in \mathbb{G}_2.$$

c. She takes the hash value $h = H_3(c, g_1, g_2) \in \mathbb{Z}_q$.
d. She computes $S = R + (h + v)d_{ID_A}$.

The proof is made of (U, v, h, S). To check its validity, the verifier first computes $c' = \hat{e}(P, U)\hat{e}(P_{pub}, Q_{ID_B})^v$, $g_1' = \hat{e}(P, S)\hat{e}(P_{pub}, Q_{ID_A})^{h+v}$ and $g_2' = \hat{e}(H_2(M, r, ID_A), S)\gamma^{h+v}$ and accepts if and only if $h = H_3(c', g_1', g_2')$.

Deny: in order to convince a designated verifier of identity ID_B that a given signature $\sigma = < r, \gamma >$ on a message M is not valid for her identity ID_A,

a. Alice computes $Q_{ID_B} = H_1(ID_B) \in \mathbb{G}_1$ and picks random $U \leftarrow_R \mathbb{G}_1$, $v \leftarrow_R \mathbb{Z}_q$ to compute $c = \hat{e}(P, U)\hat{e}(P_{pub}, Q_{ID_B})^v$.
b. She computes a commitment $C = \left(\frac{\hat{e}(H_2(M, r, ID_A), d_{ID_A})}{\gamma}\right)^\omega$ for a randomly chosen $\omega \leftarrow_R \mathbb{Z}_q$.
c. She proves in a zero-knowledge way that she knows a pair $(R, \alpha) \in \mathbb{G}_1 \times \mathbb{Z}_q$ such that

$$C = \frac{\hat{e}(H_2(M, r, ID_A), R)}{\gamma^\alpha} \quad \text{and} \quad 1 = \frac{\hat{e}(P, R)}{\hat{e}(P_{pub}, Q_{ID_A})^\alpha} \tag{1}$$

To do this,

1. She picks $V \leftarrow_R \mathbb{G}_1$, $v \leftarrow_R \mathbb{Z}_q$ to compute

$$\rho_1 = \hat{e}(H_2(M, r, ID_A), V)\gamma^{-v} \in \mathbb{G}_2 \text{ and } \rho_2 = \hat{e}(P, V)y^{-v} \in \mathbb{G}_2$$

where $y = \hat{e}(P_{pub}, Q_{ID_A})$.
2. She computes $h = H_4(C, c, \rho_1, \rho_2) \in \mathbb{Z}_q$.
3. She computes $S = V + (h + v)R \in \mathbb{G}_1$ and $s = v + (h + v)\alpha \in \mathbb{Z}_q$.

The proof is made of (C, U, v, h, S, s). It can be verified by the verifier of identity ID_B who rejects the proof if $C = 1$ and otherwise computes $c' = \hat{e}(P, U)\hat{e}(P_{pub}, Q_{ID_B})^v$, $\rho_1' = \hat{e}(H_2(M, r, ID_A), S)\gamma^{-s}C^{-(h+v)}$ and $\rho_2' = \hat{e}(P, S)y^{-s}$ where $y = \hat{e}(P_{pub}, Q_{ID_A})$. The verifier accepts the proof if and only if $h = H_4(C, c', \rho_1', \rho_2')$.

The confirmation protocol is inspired from a designated verifier proof ([26]) proposed by Jakobsson, Sako and Impagliazzo that allows a prover to convince a designated verifier of the equality of two discrete logarithms. The denial protocol is an adaptation of a protocol proposed by Camenisch and Shoup ([8]) to prove the inequality of two discrete logarithms. Both adaptations are non-transferable proofs of respectively equality and inequality of two inverses of the group isomorphisms $f_Q : \mathbb{G}_1 \rightarrow \mathbb{G}_2, Q \rightarrow f_Q(U) = \hat{e}(Q, U)$ with $Q = P$ and $Q = H_2(M, r, ID_A)$. In an execution of the confirmation protocol, the verifier B takes the signature as valid if he is convinced that $f_P(d_{ID_A}) = \hat{e}(P_{pub}, Q_{ID_A})$ and γ have equal pre-images for isomorphisms $f_P(.) = \hat{e}(P, .)$ and $f_{H_2(M,r,ID_A)}(.) = \hat{e}(H_2(M, r, ID_A), .)$. In the denial protocol, he takes the signature as invalid on M for identity ID_A if he is convinced that these inverses differ.

Completeness and soundness of the confirmation protocol: it is easy to see that a correct proof is always accepted by the verifier B: if (U, v, h, S) is correctly computed by the prover, we have $\hat{e}(P, S) = \hat{e}(P, R)\hat{e}(P, d_{ID_A})^{h+v}$ and $\hat{e}(P, d_{ID_A}) = \hat{e}(P_{pub}, Q_{ID_A})$. We also have

$$\hat{e}(H_2(M, r, ID_A), S) = \hat{e}(H_2(M, r, ID_A), S)\hat{e}(H_2(M, r, ID_A), d_{ID_A})^{h+v}.$$

To show the soundness, we notice that if a prover is able to provide two correct answers S_1, S_2 for the same commitment (c, g_1, g_2) and two different challenges h_1 and h_2, we then have $\hat{e}(P, (h_1 - h_2)^{-1}(S_1 - S_2)) = \hat{e}(P_{pub}, Q_{ID_A})$ and $\hat{e}(H_2(M, r, ID_A), (h_1 - h_2)^{-1}(S_1 - S_2)) = \gamma$. This shows that both inverses of $f_P^{-1}(\hat{e}(P_{pub}, Q_{ID_A}))$ and $f_{H_2(M,r,ID_A)}^{-1}(\gamma)$ are equal.

Completeness and soundness of the denial protocol: one easily checks that a honest prover is always accepted by the designated verifier. To prove the soundness, one notices that if the prover is able to provide a proof of knowledge of a pair (R, α) satisfying equations (1), then the second of these equations implies $R = \alpha f_P^{-1}(y)$ with $y = \hat{e}(P_{pub}, Q_{ID_A})$ by the bilinearity of the map. If we substitute this relation into the first equation of (1), it comes that

$$C = \left(\frac{\hat{e}(H_2(M, r, ID_A), f_P^{-1}(y))}{\gamma} \right)^\alpha.$$

Since the verifier checks that $C \neq 1$, it comes that $\hat{e}(H_2(M, r, ID_A), f_P^{-1}(y)) \neq \gamma$ and the signature γ is actually invalid. The soundness of the proof of knowledge in step c is easy to verify.

Non-transferability: in order for the proofs to be non-transferable, both protocols need a trapdoor commitment $\mathtt{Commit}(U, v) = \hat{e}(P, U)\hat{e}(P_{pub}, Q_{ID_B})^v$ that allows the owner of the private key d_{ID_B} to compute commitment collisions: indeed, given a tuple $(U, v, \mathtt{Commit}(U, v))$, B can easily use d_{ID_B} to find a pair (U', v') such that $\mathtt{Commit}(U, v) = \mathtt{Commit}(U', v')$. This is essential for the proof to be non-transferable: the verifier B cannot convince a third party of the validity or of the invalidity of a signature since his knowledge of the private key d_{ID_B} allows him to produce such a proof himself. Indeed, given a message-signature pair (M, σ), with $\sigma = < r, \gamma > \in \{0, 1\}^l \times \mathbb{G}_2$, B can choose $S \leftarrow_R \mathbb{G}_1$, $x \leftarrow_R \mathbb{Z}_q$ and $U' \leftarrow_R \mathbb{G}_1$ to compute $c = \hat{e}(P, U')$, $g_1 = \hat{e}(P, S)\hat{e}(P_{pub}, Q_{ID_A})^x$, $g_2 = \hat{e}(H_2(M, r, ID_A), S)\gamma^x$ and $c = H_3(c, g_1, g_2)$. He can then compute $v = x - h \bmod q$ and $U = U' - v d_{ID_B} \in \mathbb{G}_1$ where d_{ID_B} is the verifier's private key. (U, v, h, S) is thus a valid proof built by the verifier with the trapdoor d_{ID_B}. This trapdoor also allows him to produce a false proof of a given signature's invalidity using the same technique with the denial protocol.

Efficiency considerations: From an efficiency point of view, the signature generation algorithm requires one pairing evaluation as a most expensive operation. The confirmation and denial protocols are more expensive: the first one requires 4 pairing evaluations (3 if $\hat{e}(P_{pub}, Q_{ID_B})$ is cached in memory: this can be done if the verifier often performs verification queries), one exponentiation in $\mathbb{G}_2$ and one computation of the type $\lambda_1 P + \lambda_2 Q$ in $\mathbb{G}_1$. The verifier needs to compute 3 pairings (2 if $\hat{e}(P_{pub}, Q_{ID_A})$ is cached), 3 exponentiations and 3 multiplications in $\mathbb{G}_2$. In the denial protocol, the prover must compute 5 pairings (4 if $\hat{e}(P_{pub}, Q_{ID_B})$ is cached), 4 exponentiations and 4 multiplications in $\mathbb{G}_2$, one computation of the type $\lambda_1 P + \lambda_2 Q$ and some extra arithmetic operations in $\mathbb{Z}_q$. The verifier must compute 4 pairings (3 if $\hat{e}(P_{pub}, Q_{ID_B})$ is cached), 2 exponentiations, 1 multi-exponentiation and 3 multiplications in $\mathbb{G}_2$. To improve the efficiency of the confirmation and denial algorithms, one can speed up the computation of commitments. Indeed, the prover can pre-compute $\hat{e}(P, P)$ once and for all. To generate a commitment in an execution of the confirmation protocol, he then picks $u, v, x \leftarrow_R \mathbb{Z}_q$ and computes $c = \hat{e}(P, P)^u \hat{e}(P_{pub}, Q_{ID_B})^v$, $R = xP$ $g_1 = \hat{e}(P, P)^x$, $g_2 = \hat{e}(H_2(M, r, ID_A), R)$. The answer to the challenge h must then be computed as $S = R + (h + v)d_{ID_A}$ and the proof is made of (u, v, h, S). This technique can also be applied to the denial protocol. It allows replacing 2 pairing evaluations by 2 scalar multiplications, a exponentiation and a multi-exponentiation in $\mathbb{G}_2$ (to compute c). A single pairing evaluation is then required for the prover at each run of the confirmation and denial protocols.

Globally, it turns out that a signature verification is more expensive than a signature generation even if a pre-computation is performed. Our ID-based undeniable signature solution is nevertheless reasonable.

If we consider the length of signatures, the binary representation of a pairing is about 1000 bits long (1024 if we use the same curve as in [4]) while the length l of the binary string can be of the order of 100 bits. This provides us with signatures of about 1100 bits. This is roughly one half of the size of the RSA-based undeniable signature proposed in [20] for 1024-bit moduli. If we compare our scheme with the original undeniable signature proposed by Chaum and van Heijst and proven secure by Okamoto and Pointcheval ([31]), both lengths are similar if the Chaum-van Heijst scheme is used over a group like $\mathbb{Z}_p$ with $|p| = 1000$ (this is no longer true if this scheme is used over a suitable elliptic curve). However, it remains an open problem to devise identity based undeniable signature schemes with shorter signatures than ours.

Convertible signatures: It is really easy to notice that issued signatures can be selectively turned into universally verifiable signatures by the signer. In order to convert a genuine signature $\sigma = < r, \hat{e}(H_2(M, r, ID_A), d_{ID_A}) >$, the signer Alice just has to take a random $x \leftarrow_R \mathbb{Z}_q$ and compute $R = xP$, $g_1 = \hat{e}(P, P)^x$, $g_2 = \hat{e}(H_2(M, r, ID_A), R)$, the hash value $h = H(g_1, g_2)$ and the answer $S = R + hd_{ID_A}$. The proof, given by $(h, S) \in \mathbb{Z}_q \times \mathbb{G}_1$, is easily universally verifiable by a method similar to the verification in the confirmation protocol. Alice can also give a universally verifiable proof that a given signature is invalid for her identity by using the non-designated verifier counterpart of the denial protocol.

Removing key escrow: In order to prevent a dishonest PKG from issuing a signature on behalf of a user and from compromising the invisibility and anonymity properties, one can easily use the generic transformation proposed by Al-Riyami and Paterson ([1]) to turn the scheme into a certificateless undeniable signature. Unfortunately, the advantage of easy key management is lost since the resulting scheme no longer supports human-memorizable public keys. On the other hand, key escrow, which is often an undesirable feature in signature schemes, is then removed as well as the need for public key certificates.

4   Security Proofs for the ID-Based Undeniable Signature

The following theorem claims the scheme's existential unforgeability under adaptive chosen-message attacks.

Theorem 1. *In the random oracle model, if there exists an adversary $\mathcal{F}$ that is able to succeed in an existential forgery against the identity based undeniable* *signature scheme described in the previous section with an advantage ϵ within a time t and when performing q_E key extraction queries, q_S signature queries, q_{CD} confirmation/denial queries and q_{H_i} queries on hash oracles H_i, for $i = 1, \ldots, 4$, then there exists an algorithm $\mathcal{B}$ that is able to solve the Bilinear Diffie-Hellman problem with an advantage*

$$\epsilon' \geq \frac{\epsilon - (2q_{H_3} + q_{CD} + 1)/2^k}{e^2(q_E + 1)(q_{CD} + 1)}$$

in a time $t' \leq t + 6\mathcal{T}_p + (q_E + q_{H_1} + q_{H_2})\mathcal{T}_m + (q_S + q_{CD})\mathcal{T}_e + 2q_{CD}\mathcal{T}_{me}$ where $\mathcal{T}_p$ denotes the time required for a pairing evaluation, $\mathcal{T}_m$ is the time to perform a multiplication in $\mathbb{G}_1$, $\mathcal{T}_e$ is the time to perform an exponentiation in $\mathbb{G}_2$, $\mathcal{T}_{me}$ the time for a multi-exponentiation in $\mathbb{G}_2$ and e is the base for the natural logarithm.

Proof. given in the full paper ([28]).   □

In order for the proof to hold, we must have $q_{H_2} \ll 2^l$, where l is the size of the random salt r. We then take $l = 100$. We note that the reduction is not really efficient: if we take $q_E \approx q_{CD} \leq 2^{30}$ and $q_{H_3} < 2^{80}$, we then have $(2q_{H_3} + q_{CD} + 1)2^{-k} \approx 1.65 \times 10^{-64}$. If we assume $\epsilon - (2q_{H_3} + q_{CD} + 1)2^{-k} > \epsilon/2$, we obtain the bound $\epsilon' > \epsilon/2^{64}$. However, we have a proof with a tighter bound if the underlying assumption is the hardness of the Gap Bilinear Diffie-Hellman problem. The use of a DBDH oracle allows algorithm $\mathcal{B}$ to perfectly simulate the confirmation/denial protocols. The advantage of algorithm $\mathcal{B}$ is then

$$\epsilon' \geq \frac{\epsilon - (2q_{H_3} + q_{CD} + 1)2^{-k}}{e(q_E + 1)} > \epsilon/2^{32}.$$

We note that using the techniques of Katz and Wang ([27]) easily allows replacing the random salt r by a single bit and then obtaining signatures that are about 100 bits shorter without losing security guarantees.

The theorem below claims the scheme's invisibility in the sense of Galbraith and Mao (see [20]) under the Decisional Bilinear Diffie-Hellman assumption.

Theorem 2. *In the random oracle model, the identity based undeniable signature presented in section 3 satisfies the invisibility property provided the Decisional Bilinear Diffie-Hellman problem is hard. More formally, if we assume that no algorithm is able to forge a signature in the game of definition 2 with a non-negligible probability and if a distinguisher $\mathcal{D}$ is able to distinguish valid signature from invalid ones for a messages and an identity of its choice with a non-negligible advantage ϵ after having asked q_E key extraction queries, then there exists a distinguisher $\mathcal{B}$ that has an advantage $\epsilon' \geq \frac{\epsilon}{e(q_E+1)}$ for the DBDH problem within a running time bounded as in theorem 1.*

Proof. given in the full paper ([28]).   □

It is possible to directly show that the scheme also satisfies the anonymity property in the random oracle model under the Decisional Bilinear Diffie-Hellman assumption. However, since anonymity and invisibility are essentially equivalent, the anonymity of our signature derives from its invisibility property.

5   Another Application

Our construction provides an application of independent interest which is the possibility to design an identity based signature with a 'tighter' security proof than all other existing provably secure identity based signatures ([9],[17],[23],[25]) for which the security proofs make use of the forking lemma ([34],[35]): indeed by concatenating a signature produced by the undeniable scheme with a non-designated verifier proof of its validity (using the non-designated verifier counterpart of the confirmation protocol), we obtain a universally verifiable identity based signature for which the security is the most tightly related to some hard computational problem. Recall that all existing identity based signatures have a security proof built on the forking lemma of Pointcheval and Stern that involves a degradation in security during the reduction as pointed out in [22]: if q_H denotes the number of message hash queries and t the forger's running time, then the upper bound on the average running time to solve the problem is $q_H t$ (if we assume $q_H < 2^{80}$, this makes a great degradation for the bound on the running time). This new ID-based signature may be viewed as an adaptation of the signature recently proposed by Goh and Jarecki ([22]). It is less efficient than all the other ones but is more tightly related to some computational assumption than those in ([9],[25]), which are only loosely related to the computational Diffie-Hellman problem, or the scheme in ([23]) that has a security loosely related to the RSA assumption. As for the proof of unforgeability of the undeniable signature under the Gap Bilinear Diffie-Hellman assumption, one can show that, if the forger's advantage is ϵ, then the average time to solve the BDH problem is smaller than $2^{32} t/\epsilon$ where t is the running time of the forger. The corresponding bound for the identity based signature described in [25] is roughly $2^{146} t/\epsilon$ if 2^{30} identity hash queries and 2^{80} message hash queries [3] are allowed to the attacker.

6   Conclusions

We showed in this paper a first construction for a provably secure identity based undeniable signature and we extended the panel of primitives for identity based cryptography ([1]). A proof of existential unforgeability under the Bilinear Diffie-Hellman assumption is given in the full paper ([28]). We pointed out that a 'tighter' proof can be made under the Gap Bilinear Diffie-Hellman assumption. We also extended the notions of invisibility and anonymity of Galbraith and Mao ([20]) to the identity based setting and we proved the invisibility of our scheme in the random oracle model under the Decisional Bilinear Diffie-Hellman assumption.

As a side effect, our construction allows the design of an identity based signature scheme with a security more tightly related to the hardness of some hard problem than any other existing provably secure identity based signature.

[3] Hashing onto a finite field may be viewed as an operation of unit cost while hashing onto an elliptic curve requires some extra computation as explained in [2].

References

1. S.-S. Al-Riyami , K.G. Paterson, *Certificateless Public Key Cryptography*, Advances in Cryptology - Asiacrypt'03, LNCS Series, 2003.
2. P.-S.-L.-M. Barreto, H.-Y. Kim, *Fast hashing onto elliptic curves over fields of characteristic 3*, eprint available at http://eprint.iacr.org/2001/098/.
3. M. Bellare, P. Rogaway, *Random oracles are practical: A paradigm for designing efficient protocols*, Proc. of the 1st ACM Conference on Computer and Communications Security, pp. 62–73, 1993.
4. D. Boneh, M. Franklin, *Identity Based Encryption From the Weil Pairing*, Advances in Cryptology - Crypto'01, LNCS 2139, Springer-Verlag, pp. 213–229, 2001.
5. D. Boneh, B. Lynn, H. Shacham, Short signatures from the Weil pairing, Advances in Cryptology - Asiacrypt'01, LNCS 2248, Springer-Verlag, pp.514–532, 2001.
6. J. Boyar, D. Chaum, I. Damgård, T. Pedersen, *Convertible undeniable signatures*, Advances in Cryptology - Crypto'90, LNCS 0537, Springer-Verlag, pp. 189–208, 1990.
7. X. Boyen, *Multipurpose Identity-Based Signcryption. A Swiss Army Knife for Identity-Based Cryptography*, Advances in Cryptology - Crypto'03, LNCS 2729, Springer-Verlag, pp. 383–399, 2003.
8. J. Camenisch, V. Shoup, *Practical Verifiable Encryption and Decryption of Discrete Logarithms*, Advances in Cryptology - Crypto'03, LNCS 2729, Springer-Verlag, pp. 126.-144, 2003.
9. J.C. Cha, J.H. Cheon, *An Identity-Based Signature from Gap Diffie-Hellman Groups*, proceedings of PKC 2003. Springer-Verlag, LNCS 2567, pp. 18–30, Springer-Verlag, 2003.
10. D. Chaum, H. van Antwerpen, *Undeniable signatures*, Advances in Cryptology - Crypto'89, LNCS 0435, Springer-Verlag, pp. 212–216, 1989.
11. D. Chaum, *Zero-knowledge undeniable signatures*, Advances in Cryptology - Crypto'90, LNCS 0473, Springer-Verlag, pp. 458–464, 1990.
12. D. Chaum, E. van Heijst, B. Pfitzmann, *Cryptographically strong undeniable signatures, unconditionally secure for the signer*, Advances in Cryptology - Crypto'91, LNCS 0576, Springer-Verlag, pp. 470-484, 1991.
13. J.H. Cheon, D. H. Lee, *Diffie-Hellman Problems and Bilinear Maps*, eprint available at http://eprint.iacr.org/2002/117/.
14. C. Cocks, *An Identity Based Encryption Scheme Based on Quadratic Residues*, Proc. of Cryptography and Coding, LNCS 2260, Springer-Verlag, pp. 360–363, 2001.
15. R. Dupont, A. Enge, *Practical Non-Interactive Key Distribution Based on Pairings*, available at http://eprint.iacr.org/2002/136.
16. I. Damgård, T. Pedersen, *New convertible undeniable signature schemes*, Advances in Cryptology - Eurocrypt'96, LNCS 1070, pp. 372–386, Springer-Verlag, 1996.
17. A. Fiat, A. Shamir, *How to Prove Yourself: Practical Solutions to Identification and Signature Problems*, Advances in Cryptology - Crypto'86, LNCS 0263, Springer-Verlag, pp. 186–194, 1986.
18. A. Fujioka, T. Okamoto, K. Ohta, *Interactive Bi-Proof Systems and undeniable signature schemes*, Advances in Cryptology - Eurocrypt'91, LNCS 0547, pp. 243–256, Springer-Verlag, 1991.
19. S. Galbraith, W. Mao, K.G. Paterson, *RSA-based undeniable signatures for general moduli*, Topics in Cryptology - CT-RSA 2002, LNCS 2271, pp. 200–217, Springer-Verlag, 2002.

20. S. Galbraith, W. Mao, *Invisibility and Anonymity of Undeniable and Confirmer Signatures.*, Topics in Cryptology - CT-RSA 2003, LNCS 2612, pp. 80–97, Springer-Verlag, 2003.
21. R. Gennaro, H. Krawczyk, T. Rabin, *RSA-based undeniable signatures*, Advances in Cryptology - Crypto'97, LNCS 1294, Springer-Verlag, pp. 132–149, 1997.
22. E.-J. Goh, S. Jarecki, *A Signature Scheme as Secure as the Diffie-Hellman Problem*, Advances in Cryptology - Eurocrypt'03, LNCS 2656, Springer-Verlag, pp. 401-415, 2003.
23. L. Guillou, J-J. Quisquater, *A "Paradoxical" Identity-Based Signature Scheme Resulting From Zero-Knowledge*, Advances in Cryptology - Crypto'88, LNCS 0403, Springer-Verlag, pp. 216–231, 1988.
24. S. Han, K.Y. Yeung, J. Wang, *Identity based confirmer signatures from pairings over elliptic curves*, proceedings of ACM conference on Electronic commerce, pp. 262–263, 2003.
25. F. Hess, *Efficient identity based signature schemes based on pairings*, proceedings of SAC'02. LNCS 2595, pp. 310-324, Springer-Verlag, 2002.
26. M. Jakobsson, K. Sako, R. Impagliazzo, *Designated Verifier Proofs and Their Applications*, Advances in Cryptology - Eurocrypt'96, LNCS 1070, Springer-Verlag, pp. 143–154, 1996.
27. J. Katz, N. Wang, *Efficiency Improvements for Signature Schemes with Tight Security Reductions*, to appear at ACM Conference on Computer and Communications Security 2003.
28. B. Libert, J-J. Quisquater, *Identity Based Undeniable Signatures*, extended version, eprint available at http://eprint.iacr.org/2003/206.
29. M. Michels, M. Stadler, *Efficient Convertible Undeniable Signature Schemes.* Proc. of 4th Annual Workshop on Selected Areas in Cryptography, SAC'97, August, 1997.
30. T. Okamoto, *Designated Confirmer Signatures and Public Key Encryption are Equivalent*, Advances in Cryptology - Crypto'94, LNCS 0839, Springer-Verlag, pp. 61–74, 1994.
31. T. Okamoto, D. Pointcheval *The Gap-Problems: A New Class of Problems for the Security of Cryptographic Schemes*, Proc. of of PKC'01, LNCS 1992, Springer-Verlag, pp.104–118, 2001.
32. K.G. Paterson, *ID-based signatures from pairings on elliptic curves*, eprint available at http://eprint.iacr.org/2002/004/.
33. D. Pointcheval, *Self-Scrambling Anonymizers*, Proceedings of Financial Cryptography 2002, LNCS 1962, Springer-Verlag, pp. 259–275, 2001.
34. D. Pointcheval, J. Stern, *Security proofs for signature schemes*, Advances in Cryptology - Eurocrypt'96, LNCS 1070, Springer-Verlag, pp. 387–398, 1996.
35. D. Pointcheval, J. Stern, *Security arguments for digital signatures and blind signatures*, Journal of Cryptology, vol. 13-Number 3, pp. 361–396, 2000.
36. R. Sakai, K. Ohgishi, M. Kasahara, *Cryptosystems based on pairing*, In The 2000 Sympoium on Cryptography and Information Security.
37. A. Shamir, *Identity Based Cryptosystems and Signature Schemes*, Advances in Cryptology - Crypto' 84, LNCS 0196, Springer-Verlag, 1984.
38. N.P. Smart, *An identity based authenticated key agreement protocol based on the Weil pairing*, Electronic Letters, 38(13): 630–632, 2002.
39. F. Zhang, K. Kim, *ID-Based Blind Signature and Ring Signature from Pairings.* Advances in Cryptology - Asiacrypt'02, LNCS 2501, Springer-Verlag, 2002.
40. F. Zhang, R. Safavi-Naini, W. Susilo, *Attack on Han et al.'s ID-based Confirmer (Undeniable) Signature at ACM-EC'03*, eprint available at http://eprint.iacr.org/2003/129/.

Compressing Rabin Signatures

Daniel Bleichenbacher

Bell Labs – Lucent Technologies

Abstract. This note presents a method to compress a Rabin signature [Rab78] to about half of its length.

Let $n > 1$ be a (square free) public key of the Rabin signature scheme and let (s, m) be a Rabin signature of a message m. I.e.,

$$s^2 \equiv h(m) \pmod{n},$$

where h is a message formatting function. The goal of this note is to replace s by a positive integer v smaller than $\sqrt{n}$ such that v, n and m are sufficient to recover the signature s, without knowledge of the secret key. The paper assumes that h is deterministic, i.e., the value $h(m)$ can be computed without the knowledge of s. For example PKCS #1 v.1.5 [RSA93] signatures use deterministic formatting, but RSA-PSS [BR96] signatures do not.

Previously, Coron and Naccache [CN03] and independently Bernstein [Ber] have shown that a Rabin signature can be reconstructed if e.g., more than half of the most significant bits of s are known. They use Coppersmith's LLL-based root finding method [Cop96]. This method leads to a slow decompression when the fraction of known bits is close to $1/2$. Bernstein notes however, that a fast decompression method can be found when at least $2/3$ of the bits are given.

Continued Fractions: Let α be a real positive number. Define $\alpha_0 = \alpha$, $q_i = \lfloor \alpha_i \rfloor$ and define recursively $\alpha_{i+1} = 1/\{\alpha_i\}$ for all $i \geq 0$ until $\{\alpha_i\} = 0$. Then the partial convergents u_i/v_i of s can be computed by $u_0 = q_0, v_0 = 1, u_1 = q_0 q_1, v_1 = q_1 + 1$ and $u_{i+2} = q_{i+2} u_{i+1} + u_i, v_{i+2} = q_{i+2} v_{i+1} + v_i$. The theory of continued fractions asserts that the principal convergents u_i/v_i are close rational approximations of α. In particular the following equation is satisfied (see e.g. [Knu81, §4.5.3, Eq. (12)],[Lan95, Ch. 1, Theorem 5]).

$$|v_i \alpha - u_i| \leq 1/v_{i+1} \tag{1}$$

If α is rational then there exists an integer k with $\{\alpha_k\} = 0$ and $u_k/v_k = \alpha$.

Compression: We compress a signature (s, m) as follows: If $\gcd(s, n) \neq 1$ then output an error and stop. Let $u_i/v_i, i = 1, \ldots, k$ be the principal convergents of the continued fraction expansion of s/n. Let ℓ be such that $v_\ell < \sqrt{n} \leq v_{\ell+1}$. Then (v_ℓ, m) is the compressed Rabin signature.

T. Okamoto (Ed.): CT-RSA 2004, LNCS 2964, pp. 126–128, 2004.

Verification and Decompression: Let (v, m) be a compressed signature. If $\gcd(v, n) \neq 1$ then output an error and stop. Otherwise, compute $0 \leq t < n$ such that

$$t \equiv h(m)v^2 \pmod{n}.$$

The compressed signature is valid if and only if t is a square in $\mathbb{Z}$. If it is valid then set $w = \sqrt{t}$ and $s \equiv w/v \pmod{n}$ and output (s, m).

Analysis: Neither compression nor decompression need to use the secret key. The following theorem shows that any valid Rabin signature can be converted into a valid compressed signature and vice versa. Thus Rabin signatures and compressed signatures are equally difficult to forge.

Theorem 1. *Let n be a Rabin public key that is square free.*

(I) If (s, m) is a valid Rabin signature then the compression algorithm generates a valid compressed signature for m or finds a nontrivial factor of n.

(II) If (v, m) is a valid compressed signature then the decompression algorithm generates a valid Rabin signature for m.

Proof. (I) By assumption (s, m) is a valid signature, if $s^2 \equiv h(m) \pmod{n}$ and $\gcd(s, n) = 1$. The later condition implies $v_k = n$ for the last principal convergent $u_k/v_k = s/n$. Since the denominators $v_1, \ldots, v_k$ are strictly increasing there exists $\ell : v_\ell < \sqrt{n} \leq v_{\ell+1}$ and therefore the compression is well defined. Let (v_ℓ, m) be the signature computed by the compression algorithm. Setting $\alpha = s/n$ in (1) implies

$$|v_\ell s - u_\ell n| < n/v_{\ell+1} \leq \sqrt{n}.$$

Hence there exists $r \in \mathbb{Z} : |r| \leq \sqrt{n}$ such that $r \equiv v_\ell s \pmod{n}$. By assumption n is not a square, hence $0 \leq r^2 < n$ and $r^2 \equiv (v_\ell)^2 h(m) \pmod{n}$. Thus the value t computed in the decompression is indeed a square in $\mathbb{Z}$. Hence (v_ℓ, m) is a valid compressed signature unless $\gcd(v_\ell, n) > 1$.

(II) The compressed signature (v, m) is by assumption valid if $0 \leq t \leq n$: $t \equiv v^2 h(m) \pmod{n}$ is a square in $\mathbb{Z}$. Then $w = \sqrt{t}$ is an integer and thus $s^2 \equiv w^2/v^2 \equiv h(m) \pmod{n}$. Hence (s, m) is a valid Rabin signature. $\square$

Time Complexity: Compression requires a continued fraction expansion and takes time $O(\log(n)^2)$. Decompression requires two multiplications and an inverse over $\mathbb{Z}/n\mathbb{Z}$ and a square root in $\mathbb{Z}$ and hence also takes time $O(\log(n)^2)$. Note, that these bounds are obtained by using school book methods. Asymptotically faster algorithms (e.g. FFT based gcd) are not optimal for typical key sizes.

A Variant: An alternative compressed signature is $(|r|, m)$, where $r \in \mathbb{Z}$ is such that $|r| \leq n$ and $r \equiv v_\ell s \pmod{n}$. In the proof of Theorem 1 it is shown that such an r exists when $v_\ell < \sqrt{n} < v_{\ell+1}$. A compressed signature is valid if $h(m)/r^2 \bmod n$ is a square in $\mathbb{Z}$. Decompression is done using the equality

$(v_\ell)^2 \equiv h(m)/r^2 \pmod{n}$. This variant is more expensive, because the verifier has to compute an additional modular inverse. But the variant has the advantage that the verification accepts both compressed and uncompressed signatures without modification.

RSA: The method can be extended to RSA signatures with small public exponent (i.e. $e = 3$), but the benefits are smaller. For $e = 3$ the signature can be compressed to $2/3$ of its size as follows.

Assume that

$$s^3 \equiv h(m) \pmod{n},$$

is an RSA signature, where h is again a deterministic formatting function. To compress a signature one computes the continued fraction expansion of s/n and selects the principal convergent u_ℓ/v_ℓ satisfying $v_\ell < n^{1/3} \leq v_{\ell+1}$. The compressed RSA signature is (v_ℓ, m).

Equation (1) implies

$$|v_\ell s - u_\ell n| \leq n^{1/3},$$

and thus there exists $r \in \mathbb{Z}$ with $|r| \leq n^{1/3}$ and $r^3 \equiv h(m)(v_\ell)^3 \pmod{n}$. Given $h(m)$ and v_ℓ this value r can be found by checking whether either of $h(m)(v_\ell)^3 \bmod n$ or $n - h(m)(v_\ell)^3 \bmod n$ is a cube in $\mathbb{Z}$. Finally, one can reconstruct the signature noting by setting $s \equiv r/v_\ell \pmod{n}$.

Acknowledgment. I'm thanking the program committee for their suggestions.

References

[Ber]   D. J. Bernstein. Squeezing Rabin signatures. unpublished.

[BR96]   Mihir Bellare and Phillip Rogaway. The exact security of digital signatures - how to sign with RSA and Rabin. In Ueli Maurer, editor, *Advances in Cryptology – EUROCRYPT '96*, volume 1070, pages 399–416. Springer Verlag, 1996.

[CN03]   Jean Sebastien Coron and David Nacacche. Procédé de réduction de la taille d'une signature RSA ou Rabin. Patent FR28293333, March 2003.

[Cop96]   Don Coppersmith. Finding a small root of a univariate modular equation. In *Advances in Cryptology – EUROCRYPT '96*, volume 1070 of *Lecture Notes in Computer Science*, pages 155–165, Berlin, 1996. Springer Verlag.

[Knu81]   Donald E. Knuth. *The art of computer programming, Seminumerical Algorithms*, volume 2. Addison Wesley, 2nd edition, 1981.

[Lan95]   Serge Lang. *Introduction to Diophantine Approximations*. Springer Verlag, new expanded edition, 1995.

[Rab78]   Michael O. Rabin. Digitalized signatures. *Foundation of Secure Computation*, pages 155–169, 1978.

[RSA93]   RSA Data Security, Inc. *PKCS #1: RSA Encryption Standard*. Redwood City, CA, November 1993. Version 1.5.

A Key Recovery System as Secure as Factoring

Adam Young[1] and Moti Yung[2]

[1] Cigital Labs
ayoung@cigital.com
[2] Dept. of Computer Science, Columbia University
moti@cs.columbia.edu

Abstract. There has been a lot of recent work in the area of proving in zero-knowledge that an RSA modulus N is in the correct form. For example, protocols have been given that prove that N is the product of: two safe primes, two primes nearly equal in size, etc. Such proof systems are rather remarkable in what they achieve, but may be regarded as being heavyweight protocols due to the computational and messaging overhead they impose. In this paper an efficient zero-knowledge protocol is given that simultaneously proves that N is a Blum integer and that its factorization is recoverable. The proof system requires that the RSA primes p and q be such that $p \equiv q \equiv 3 \bmod 4$ and another sematically secure encryption. The solution is therefore amenable for use with systems based on PKCS #1. A proof is given that shows that our algorithm is secure under the integer factorization problem (and can be turned into a non-interactive roof in the random oracle model).

Keywords: RSA, Rabin, Blum integer, quadratic residue, pseudosquare, zero-knowledge, public key cryptography, PKCS #1, semantic Security, chosen ciphertext security, standard compatibility, key recovery.

1 Introduction

The RSA [27] algorithm has gained widespread use in the security industry. However, many existing systems trust user's to generate RSA key pairs honestly without any form of compliance checking. Without knowing a given user's private key, the most simple compliance tests include verifying that: N is not divisible by small primes, that N is not prime, and that N is not a perfect power. However, using zero-knowledge techniques, much more can be proven about the form of N.

A zero-knowledge (ZK) protocol has been given for showing that N is the product of two safe primes [6]. Also, a zero-knowledge protocol has been given that proves that N is *square-free* (see Section 2 for definition, with a soundness error of $1/2^K$ where K is the number of rounds in the proof [1]. A protocol that does not leak too many bits has been given that proves that N is the product of two nearly equal primes, assuming that N has already been proven to be the product of two distinct primes [17]. Also, a proof system has been given for proving that N is a Blum integer [16]. Finally, a zero-knowledge proof of membership

T. Okamoto (Ed.): CT-RSA 2004, LNCS 2964, pp. 129–142, 2004.

in the set of pseudosquares mod N has been presented [10]. A PVSS for secret sharing of factoring based private keys was presented by Fujisaki and Okamoto [8]. This PVSS relies on a non-standard strong RSA assumption. Related work that encompasses verifiable encryptions is by Camenisch and Damgaard [5].

When key recovery is needed, a higher level of compliance verification is verifying that a given user's private key is recoverable by a designated recovery authority. Ideally such systems will insure that only the user knows his or her own private key until such time as it is needed by the key recovery authority or authorities. A key recovery system for RSA was presented by Poupard and Stern that has the advantage that the ciphertext of the escrowed private key is very small [24], in fact, considerably smaller than the non-interactive zero-knowledge proof transcript that is proposed here. The system is based on the Paillier cryptosystem that is semantically secure under the *composite residuosity class assumption* [23]. It is argued that the system can also be based on the Naccache-Stern [20] and Okamoto-Uchiyama [21] cryptosystems. The Naccache-Stern cryptosystem is semantically secure under the *Prime Residuosity Assumption*. The Okamoto-Uchiyama cryptosystem uses public keys of the form $n = p^2 q$ and is semantically secure under the *p-subgroup assumption*. We remark that Paillier's cryptosystem[1] uses public keys that are not compatible with PKCS #1 as currently defined. In this paper a factoring based key recovery system is presented and it is proven secure assuming that there exists a public key cryptosystem that is semantically secure against plaintext attacks for single messages in the uniform model. Since there exist such cryptosystems under the factoring assumption (e.g., Optimal Asymmetric Encryption [3,28] based on Rabin [25]), we prove that our solution is secure if and only if factoring a Blum integer is hard. We remark that OAEP is semantically secure against adaptive chosen ciphertext attacks in the random oracle model under the partial one-wayness of the underlying permutation [9].

Though the space reduction is remarkable and the combination of methods ingenious, a significant drawback to the Poupard-Stern approach is that it relies on cryptosystems that are not secure against chosen ciphertext attacks[2]. The reason that the Paillier cryptosystem was used as the primary example is that it is not clear how a chosen ciphertext attack can be used effectively against it in this scenario. So, to be on the safe side their solution imposes the restriction that the recovery authority decrypts messages upon request and does not reveal the decryptions themselves. This is not the case in the solution we propose here. The escrow authority can use any public key encryption algorithm including one that is secure against adaptive chosen ciphertext attacks. This implies that the range of application of our solution is larger (this scenario was advocated by Camenisch and Shoup [7]).

The efficiency of the Poupard-Stern solution is as follows [24]. The probability of a cheating strategy to succeed during a proof of fairness is smaller than $1/B^\ell$,

---

[1] Paillier's cryptosystem is remarkable since it constitutes an additive homomorphism under multiplication modulo $p^2 q^2$.

[2] e.g., the Paillier scheme is malleable.

so $\ell|B|$ must be large enough, e.g., $\ell|B| = 80$ in order to guarantee a high-level of security. Also, the workload for a third party to defeat the security is $O(\sqrt{B})$ in the worst case so B cannot be too large. They recommend setting $\ell = 2$ and $B = 2^{40}$. The value of ℓ is the number of iterations in the proof system. The system therefore has a recovery time vs. space trade-off. By increasing ℓ, the size of the non-interactive transcript goes up. By increasing B the size of the transcript goes down, but recovery takes longer. By taking $B = 2$, the transcript winds up being the size of a typical non-interactive ZK transcript. The solution therefore has obvious advantages in practice in terms of space efficiency, but does not constitute an asymptotic improvement over existing methods for the exponentially small advantage cases.

The solution we propose makes no assumptions regarding the form of the escrow authority public key. However, having the escrow authority use a composite public key N_1 is a good choice. It is a slightly optimized version of an ealier version of the algorithm [29]. Our solution requires that the modulus whose private key is being escrowed be a Blum integer that is not a perfect power, but otherwise it is an RSA number. This implies that the solution is highly compatible with PKCS #1. Consider a PKCS #1 key generator that outputs primes p and q, among other values. If p ends in the two bits 01 or if q ends in the two bits 01 then the key pair can be rejected and a new one generated. Only when p ends in 11 and q ends in 11 is the key pair accepted. This way, acceptance/rejection is used to guarantee that $p \equiv q \equiv 3 \mod 4$. Note that multiprecision library calls are not needed to verify this. For example, assuming that the prime p can be obtained as a binary number in the form of a byte array, the least significant byte u ends in the two bits 11 if and only if the bitwise logical AND of u and 0x03 equals 0x03 in hexadecimal. Compatibility with PKCS #1 is important since RSA is so widely used.

It is shown that in the random oracle model, $2m$ iterations of the non-interactive version of the protocol are needed to achieve a computational zero-knowledge error of at most $1/2^m$ (note that since PKCS #1 compatibility is a major goal and since PKCS #1 uses OAEP which is random oracle based, assuming the use of a random oracle for our non-interactive version makes sense). For concreteness, suppose that N is 768 bits and N_1 is 1024 bits. Also, suppose that $m = 40$. As shown in Appendix B, the non-interactive zero-knowledge transcript requires about 28 kilobytes in this case.

So, with respect to previous work on this problem we argue that our solution has the following novel features:

1. **High PKCS #1 Compatibility**: The users modulus N needs to be a Blum integer a check that is easy to implement inside the key generation as argued above.
2. **Weaker Theoretical Security Assumption**: Unlike previous proposals which utilize new assumptions such as the composite residuosity class assumption, we only assume that factoring is hard.
3. **Protection Against Adaptive Attacks**: The escrow authority can use a key and corresponding cryptosystem that is secure against adaptive chosen

ciphertext attacks. This allows the system to be used in applications where the system is exposed to such attacks.

2   Notation and Definitions

Let $J(x/N)$ denote the Jacobi symbol of x with respect to N. Let $\mathbb{Z}_N^*$ denote the set of integers contained in $\{0, 1, 2, ..., N-1\}$ that are relatively prime to N. Also, let $\mathbb{Z}_N^*(+1)$ denote those elements in $\mathbb{Z}_N^*$ with Jacobi symbol 1 with respect to N. Recall that a is a *pseudosquare* mod N provided that a is not a quadratic residue mod N and that $a \in \mathbb{Z}_N^*(+1)$. Two roots d_0 and d_1 of a quadratic residue a *mod* N are called *ambivalent roots* of a if $d_0 \neq d_1$ and $d_0 \neq -d_1$ *mod* N. A number N is *square-free* if m^2 does not divide N for any $m > 1$. If the prime powers appearing in the factorization of N all have odd exponents then we say that N is *free of squares* (not to be confused with being square-free).

The proof system of van de Graaf and Peralta utilizes the following definitions for Blum integers and Class II integers. A Blum integer is defined as the product of two prime powers p^r and q^s such that $p, q = 3$ *mod* 4 with r and s odd. The composite $N = p^{2r}q^s$ is a class II integer if p, q are prime, $p \equiv 3$ *mod* 4, $q \equiv 1$ *mod* 4, and s is odd. Their proof system is based on the following key observation: If N is a Blum integer then you are guaranteed that x or $-x$ is a quadratic residue mod N, whereas if N is not a Blum integer or of class II, then the probability that one or both of x and $-x$ are quadratic residues mod N is less than or equal to $1/2$. This observation allows the prover to convince the verifier that N is a Blum or a class II integer. A method based on computing Jacobians is used to convince the verifier that N is not a class II integer.

In the sequel, we will let $N = VW$ where W is the part of N that is free of squares and V is a square. Let w denote the number of distinct prime factors of W and let v denote the number of distinct prime factors of V. From van de Graaf and Peralta [16] (page 130) we have the following fact.

Fact 1: If $a \in_R \mathbb{Z}_N^*(+1)$ then $a \in QR_N$ with probability $(1/2)^{v+w-1}$.

The well-known definitions of ensembles, polynomial indistinguishability, and multiple-message indistinguishability of encryptions is given in Appendix A. In this paper we will only consider adversaries contained in BPP.

Let $E_e(\cdot, \cdot)$ denote a randomized single-message semantically secure encryption function in the uniform model that uses public key e (in fact, E_e can be OAEP which is secure against adaptive chosen ciphertext attacks [9]; it can be based on RSA). Let $D_e(\cdot)$ denote the corresponding decryption function. Let M be the message space and let S_1 denote the set from which the random string is drawn to form the probabilistic encryption. Let C denote the ciphertext space. To encrypt $m \in M$ to get the ciphertext $c \in C$, we choose $r \in_R S_1$ and compute $c = E_e(m, r)$. Thus, $m = D_e(E_e(m, r))$.

3   Background: ZK Proof for $N = p^r q^s$ Is a Blum Integer

All of the theorems and algorithms in this section were taken directly from van de Graaf and Peralta [16]; a reader familiar with the system can skip it. N is a Blum integer if $N = p^r q^s$, p and q are distinct primes, r and s are odd, and $p \equiv q \equiv 3 \; mod \; 4$. The following theorems provide the basis for proving the form of N.

Theorem 1. *Suppose N is a Blum integer or of class II. Let a be a random number in $\mathbb{Z}_N^*(+1)$. Then either a or $-a$ is a quadratic residue mod N.*

Theorem 2. *Suppose $N \equiv 1 \; mod \; 4$ and has more than two distinct prime factors. If a is a random element in $\mathbb{Z}_N^*(+1)$ then the probability that at least one of a or $-a$ is a quadratic residue modulo N is less than or equal to $1/2$.*

The following protocol proves that N is a Blum integer.

AtomicProtocolForN:
1. P and V choose $a \in_R \mathbb{Z}_N^*(+1)$ together randomly
2. P and V choose $\mu \in_R \{1, -1\}$ together randomly
3. P sends V a square root β of a or $-a$ with Jacobi symbol equal to μ
4. V accepts iff all of the following hold:
5. N is not a square and N is not a prime power
6. $N \equiv 1 \; mod \; 4$
7. $\beta^2 = a \; mod \; N$ or $\beta^2 = -a \; mod \; N$
8. $J(\beta/N) = \mu$

The value $N = p^{15}q^3$ where p and q are prime could form a valid Blum integer. Hence, in this example $N = (p^5 q)^3$ is a perfect cube. It is straightforward to add an additional check to verify that N is not a perfect power (as explained below).

Note that one need only show zero-knowledge with respect to an honest verifier since the verifier does not send any challenges to the prover (rather a random challenge is jointly chosen; typically using a simulatable coin flipping protocol). Honest verifier zero-knowledge can be shown using a standard simulation argument.

4   Interactive Atomic Proof of Recoverability

Recently it has been shown that key recovery systems for public key infrastructures can be implemented as efficiently in terms of protocol overhead as PKI systems without a key recovery mechanism [30,31,32]. This is accomplished by having the user who generates his or her own key pair provide the CA with a public key and certificate of recoverability prior to having the user's public key certified. A certificate of recoverability certifies that the private key is recoverable by a designated key recovery authority given the public key and certificate of recoverability. The Certificate Authority certifies the public key if and only if the certificate of recoverability is valid. This is the approach that is taken here.

Bellare and Rogaway formalized the Fiat-Shamir heuristic into what is known as the random oracle model [2]. In section 5.2 of that paper, they give a generic construction that shows how to turn any three round computational ZK atomic proof system achieving error probability $1/2$ into a non-interactive proof system using a random oracle. They prove that the resulting proof system is computational zero-knowledge in the random oracle model, and assume that the first and last messages are from the prover to the verifier, and that the middle message is from the verifier to the prover. More specifically they prove that to achieve an error of $2^{-k(n)}$ in the non-interactive proof, $2k(n)$ iterations are sufficient. Here n is a security parameter and $k(n) = \omega(log\ n)$ is given.

There are two important things to note about section 5.2. First, it is assumed in the simplifying assumptions section that the atomic protocol is only honest verifier zero-knowledge. This is evidenced by the fact that the ensemble for the simulator S' is constructed with a bit b chosen uniformly at random (not by a subroutine V^* representing a potentially cheating verifier). Thus, the proof of zero knowledge of the resulting non-interactive proof in section 5.2 simulates the view of an honest verifier since that is all that is necessary in the non- interactive random oracle case. Hence, we can assume that V can be trusted to generate truly random values (e.g., values uniformly at random from $\mathbb{Z}_N^*(+1)$) in the interactive atomic protocol.

Second, upon close inspection of their simplifying assumptions it is clear that they have in fact shown this not only for perfect ZK proof systems, but for computational ZK proof systems as well. This is because they assume that the ensemble corresponding to transcripts produced between the prover and the verifier and the ensemble corresponding to the simulated transcripts are only *computationally indistinguishable*. This is important since the use of the semantically secure encryption function E_e only guarantees computational indistinguishability of the two transcript ensembles.

4.1   The Atomic Protocol

The intuition behind the recovery aspect of the protocol is as follows. The prover is given the public key of the escrow authority. The prover has a Blum integer N and wishes to prove to a verifier that a factor of N is recoverable by the escrow authority. To do so, the prover generates two ambivalent roots of a quadratic residue modulo N. One root has Jacobi 1 and the other has Jacobi -1. These roots are encrypted under the public key of the escrow authority. The two ciphertexts along with the quadratic residue are given to the verifier. The verifier chooses a sign 1 or -1 at random and forces the prover to open a ciphertext containing a root that has a Jacobi that matches the randomly chosen sign.

Let N_1 be the public key of the escrow authority. Recall that this protocol requires that N be a Blum integer that is not a perfect power. The atomic protocol will now be described.

$SQRT AtomicProtocol$:

1. P chooses $\nu \in_R QR_N$
2. P computes d_0 to be a root of ν with Jacobi 1
3. P computes d_1 to be a root of ν with Jacobi -1
4. P chooses $r_0, r_1 \in_R S_1$
5. P computes $c_j = E_{N_1}(d_j, r_j)$ for $j = 0, 1$
6. P sends V the tuple (N, ν, c_0, c_1)
7. V sends (b, μ, a) to P where $b \in_R \{0,1\}$, $\mu \in_R \{1,-1\}$, $a \in_R \mathbb{Z}_N^*(+1)$
8. P computes a square root β of a or $-a$ with Jacobi symbol equal to μ
9. P sends V the tuple $(D, R, \beta) = (d_b, r_b, \beta)$
10. V accepts iff all of the following hold:
11. N is not a prime and N is not a perfect power
12. $N \equiv 1 \ mod \ 4$
13. $\beta^2 = a \ mod \ N$ or $\beta^2 = -a \ mod \ N$
14. $J(\beta/N) = \mu$
15. $D \in \mathbb{Z}_N^*$
16. $D^2 = \nu \ mod \ N$
17. $J(D/N) = -1^b$
18. $R \in S_1$ and $c_b = E_{N_1}(D, R)$

It is well known that using a binary search, N can be tested for being a perfect power in time $O((log^3 N) \ log \ log \ log \ N)$ [19]. Improvements to perfect power testing have been given [4] where the worst case running time is shown to be $O(log^3 N)$ using a modification of Newton's method.

Recall that a non-trivial factor of N is a factor that is less than N and greater than 1. The escrow authority attempts to recover a non-trivial factor ψ of N by computing $\psi_1 = gcd(d_0 + d_1, N)$ and $\psi_2 = gcd(d_0 - d_1, N)$. The reason this works is the following. Let $\omega, -\omega$ be the two non-trivial roots of unity. It follows that $d_1 = \pm \omega d_0 \Rightarrow d_1^2 - (\pm \omega d_0)^2 \equiv 0 \ mod \ N \Rightarrow d_1^2 - d_0^2 = kN$ for some integer k. Hence, $N \mid (d_0 - d_1)(d_0 + d_1)$. But, N does not divide $(d_0 - d_1)$ and does not divide $(d_0 + d_1)$. Thus, ψ_1 or ψ_2 must be a non-trivial factor of N. It is not hard to show that when N is a Blum integer with no integer roots (not a perfect power), a non-trivial factor of N can be used to efficiently recover the full factorization of N.

The following is the transcript of the protocol.

$$t_{P,V} = (N, \nu, c_0, c_1, b, \mu, a, D, R, \beta)$$

Theorem 3. *If $N = pq$ where p and q are distinct primes and -1 is a pseudosquare mod N then P can always compute a square root β in step 8.*

Proof. This follows immediately from Theorem 1. $\diamond$

From Theorem 3 it is not hard to see that the atomic protocol is complete.

4.2   Security of the Atomic Protocol

The intuition behind the proof of zero knowledge is as follows. First, a probabilistic poly-time simulator S is given. It is then argued that the transcripts it produces are polynomially indistinguishable from true transcripts. This is proven by assuming otherwise, thereby implying the existence of a transcript distinguisher D. A probabilistic poly-time algorithm D' is then described that uses D as an oracle to break the single-message indistinguishability of E in the uniform model, thereby reaching a contradiction. The key aspect behind the construction of D' is as follows. D' takes as input (X_n, Y_n, p, q, e, c_e) where X_n is a random variable corresponding to the unopened plaintext in a true transcript. Y_n is a random variable corresponding to the unopened plaintext in a simulated transcript. $Y_n = 1^{|N|}$ with probability unity. The primes p and q constitute a priori information concerning these plaintexts. The value e is the public encryption function and c_e is the ciphertext for which we are trying to distinguish as encrypting X_n or Y_n. The algorithm D' constructs exactly one transcript and has no idea whether it corresponds to a simulated or true transcript. It does this by using (p, q) to compute a root of $X_n^2 \bmod pq$ which is ambivalent with respect to X_n. This root is then encrypted under e to obtain a second ciphertext. If c_e encrypts X_n then the resulting transcript is a true transcript. If c_e encrypts $1^{|N|}$ then the resulting transcript is a simulated one. D' gives the resulting transcript to D to let D decide. The output of D is the final output.

Theorem 4. *(computational ZK - uniform model) If E is single-message semantically secure then $SQRTAtomicProtocol$ is honest verifier computational zero-knowledge.*

Proof. (Sketch). Assume that E is single-message semantically secure in the uniform model. Thus, by Theorem 5.2.15 of Goldreich it follows that E is single-message secure in the sense of ciphertext indistinguishability in the uniform model, since single-messages is a special case of Theorem 5.2.15. Let V be the honest verifier. Consider the following simulator.

$S(N)$:
1. choose $b \in_R \{0, 1\}$, $t \in_R \{1, -1\}$, and $r_0, r_1 \in_R S_1$
2. if $b = 0$ then set $u = 1$ else set $u = -1$
3. choose $D \in_R \mathbb{Z}_N^*$ s.t. $J(D/N) = u$
4. compute $\nu = D^2 \bmod N$
5. choose $\beta \in_R \mathbb{Z}_N^*$
6. compute $\mu = J(\beta/N)$ and $a = t\beta^2 \bmod N$
7. compute $c_0 = E_{N_1}((1^{|N|})^b D^{1-b}, r_0)$ and $c_1 = E_{N_1}((1^{|N|})^{1-b} D^b, r_1)$
8. output $t_S = (N, \nu, c_0, c_1, b, \mu, a, D, r_b, \beta)$ and halt

Let $U_{P,V}$ denote the probability ensemble over the transcripts $t_{P,V}$. Let U_S denote the probability ensemble over the transcripts t_S. These two ensembles are polynomially indistinguishable. To show this, assume for the sake of contradiction that they are not. Then it follows from Definition 2 that there exists a distinguisher $D \in BPP$, a polynomial p, and a sufficiently large n such that D distinguishes these two ensembles correctly with advantage at least $1/p(n)$. Hence,

$$Pr[D(U_{P,V}, 1^n) = 1] - Pr[D(U_S, 1^n) = 1] \geq 1/p(n)$$

It will be shown how to use D as an oracle in an algorithm D' that distinguishes single-message encryptions. Let c_e denote an encryption of X_n or Y_n.

$D'(Z_n, e, c_e)$:
1. if Z_n is not a 4-tuple then output a random bit and halt
2. set $(X_n, Y_n, p, q) = Z_n$ and compute $N = pq$
3. compute $\nu = X_n^2 \bmod N$
3. compute D randomly s.t. $D^2 = \nu \bmod N$ and s.t. $J(D/N) = -J(X_n/N)$
4. if $J(D/N) = 1$ then set $b = 0$ else set $b = 1$
5. set $c_{1-b} = c_e$
6. choose $r_b \in_R S_1$, $\beta \in_R \mathbb{Z}_N^*$, and $t \in_R \{1, -1\}$
7. compute $\mu = J(\beta/N)$ and $a = t\beta^2 \bmod N$
8. compute $c_b = E_{N_1}(D, r_b)$
9. set $t = (N, \nu, c_0, c_1, b, \mu, a, D, r_b, \beta)$
10. output $D(t, 1^n)$ and halt

Let X_n denote the probability distribution over the plaintext messages in c_e corresponding to the prover P interacting with V. Let Y_n denote the probability distribution over the plaintext messages in the unopened value c_e corresponding to the simulator S. In this case $Pr[Y_n = 1^{|N|}] = 1$. Define the random variable W_n such that $Pr[W_n = (p, q)] = 1$ where $N = pq$. Finally, take $Z_n = X_n Y_n W_n$. By taking $G_1(1^n) = N_1$ it is not hard to see that,

$$Pr[D'(Z_n, N_1, E_{N_1}(X_n)) = 1] = Pr[D(U_{P,V}, 1^n) = 1]$$

and

$$Pr[D'(Z_n, N_1, E_{N_1}(Y_n)) = 1] = Pr[D(U_S, 1^n) = 1]$$

Hence, D' distinguishes the encryptions with non-negligible advantage. But this implies that E is not secure in the sense of indistinguishability for single-messages in the uniform setting. Hence, a contradiction has been reached. So, the assumption the two ensembles are distinguishable is wrong. $\diamond$

Define L to be the set of Blum integers that are not perfect powers.

Theorem 5. *If $N \notin L$ or P does not provide a correct pair of roots (to the escrow authority), then P passes with probability at most $1/2$.*

Proof. Assume that $N \notin L$ or that P does not provide a correct pair of roots. From van de Graaf and Peralta's proof of soundness [16] it follows that if N is not a Blum integer then P passes with probability at most $1/2$. If N is a perfect power then P passes with probability 0. Hence, if $N \notin L$ then P passes with probability at most $1/2$. It remains to consider the case that $N \in L$ but that P does not provide a correct pair of roots. Since $N = p^r q^s \in L$ it follows that $J(-1/p^r) = J(-1/q^s) = -1$ due to the fact that r and s are odd.

Consider the case that neither c_0 nor c_1 encrypts a root of ν. In this case P fails the proof with probability unity due to step 16. Now consider the case that only one root is encrypted in the pair of encryptions (c_0, c_1). Let p_0 denote the probability that c_0 is the proper encryption of a root. Note that p_0 may depend on the transcript generated thus far. Since the verifier is honest, V asks for c_0 to be opened with probability $1/2$. With probability $1 - p_0$, c_1 is the proper encryption of a root. It follows that the probability that P fools V in step 16 is at most $p_0(1/2) + (1 - p_0)(1/2) = 1/2$.

Finally, consider the case that c_0 and c_1 encrypt square roots. Since it is assumed that P is cheating, the plaintexts in c_0 and c_1 cannot have differing Jacobi symbols. To see this note that if this were the case the roots would form an ambivalent pair of roots. Since the verifier chooses the challenge honestly it follows that the probability that P fools V in step 17 is at most $1/2$.

It has been shown that in all cases P passes with probability at most $1/2$. $\diamond$

Observe that the atomic protocol is a standard three round protocol in which a message is sent from P to V, then from V to P, and finally from P to V. It was shown to be honest verifier computational ZK, and was shown to be sound with error $1/2$. It follows that the transformation into NIZK of Bellare and Rogaway in Section 5.2 directly applies.

5   Conclusion

A key recovery system for factoring based private keys was presented. It was proven secure in the random oracle model based on the integer factorization problem. It was shown to be highly compatible with existing PKCS #1 implementations. The scheme is meant to be an alternative to existing schemes since it is based directly on thedifficulty of factoring.

References

1. J. Boyar, K. Friedl, C. Lund. Practical Zero-Knowledge Proofs: Giving Hints and Using Deficiencies. In *Journal of Cryptology*, 1991.
2. M. Bellare, P. Rogaway. Random oracles are practical: A paradigm for designing efficient protocols. In *Proc. First Annual Conference on Computer and Communications Security*, pages 62–73, ACM, 1993 (on-line version dated Oct. 20, 1995).
3. M. Bellare, P. Rogaway. Optimal Asymmetric Encryption- How to Encrypt with RSA. In *Advances in Cryptology—Eurocrypt '94*, LNCS 950, pages 92–111, 1994.
4. E. Bach, J. Sorenson. Sive Algorithms for Perfect Power Testing. In *Algorithmica*, volume 9, pages 313–328, 1993.
5. J. Camenisch, I. Damgaard. Verifiable encryption, group encryption, and their applications to separable group signatures and signature sharing schemes. In *Advances in Cryptology—Asiacrypt '00*,LNCS 1976, Springer-Verlag, 2000.

6. J. Camenisch, M. Michels. Proving in Zero-Knowledge that a Number is the Product of Two Safe Primes. In *Advances in Cryptology—Eurocrypt '99*.
7. J. Camenisch, V. Shoup. Practical Verifiable Encryption and Decryption of Discrete Logarithms. In *Advances in Cryptology—Crypto '03*.
8. E. Fujisaki, T. Okamoto. A Practical and Provably Secure Scheme for Publicly Verifiable Secret Sharing and Its Applications. In *Advances in Cryptology—Eurocrypt '98*, pages 32–46.
9. E. Fujisaki, T. Okamoto, D. Pointcheval, J. Stern. RSA-OAEP is Secure under the RSA Assumption In *Advances in Cryptology—Crypto '01*, pages 260–274, LNCS 2139, 2001.
10. S. Goldwasser, S. Micali, C. Rackoff. The Knowledge Complexity of Interactive Proof Systems. In *SIAM Journal on Computing*, v. 18, pages 186–208, 1989.
11. R. Gennaro, D. Micciancio, T. Rabin. An Efficient Non-Interactive Statistical Zero-Knowledge Proof System for Quasi-Safe Prime Products. In *The 5th ACM Conference on Computer and Communications Security*, 1998.
12. O. Goldreich. Introduction to Complexity Theory: Non-Uniform Polynomial Time - *P/poly*. Lecture number 8, (Goldreich's web page at http://www.wisdom.weizmann.ac.il/~oded/PS/CC/l8.ps).
13. O. Goldreich. Modern Cryptography, Probabilistic Proofs and Pseudo-randomness. Appendix A.2, page 113, Springer, 1999.
14. O. Goldreich. Foundations of Cryptography. Chapter 1 - Introduction, February 27, 1998.
15. O. Goldreich. Fragments of a chapter on Encryption Schemes. Chapter 5, section 2, February 10, 2002.
16. J. van de Graaf, R. Peralta. A simple and secure way to show the validity of your public key. In *Advances in Cryptology –Crypto '87*, v. 293, LNCS, pages 128–134.
17. M. Liskov, R. Silverman. A Statistical Limited-Knowledge Proof for Secure RSA Keys. Submitted to IEEE P1363 working group.
18. M. Luby. Pseudorandomness and Cryptographic Applications. Princeton University Press, page 4, 1996.
19. A. Menezes, P. Orschoot, S. Vanstone. Handbook of Applied Cryptography. CRC Press, page 89, 1997.
20. D. Naccache, J. Stern. A new candidate trapdoor function. In *5th ACM Symposium on Computer and Communications Security*, 1998.
21. T. Okamoto, S. Uchiyama. An efficient public-key cryptosystem. In *Advances in Cryptology—Eurocrypt '98*, pages 308–318, 1998.
22. PKCS #1-RSA Cryptography Standard, version 2.1, available from www.rsa.com/rsalabs/pkcs.
23. P. Paillier. Public Key Cryptosystems based on Composite Degree Residuosity Classes. In *Advances in Cryptology—Eurocrypt '99*, pages 223–238, 1999.
24. G. Poupard, J. Stern. Fair Encryption of RSA Keys. In *Advances in Cryptology—Eurocrypt '00*, LNCS 1807, 2000.
25. M. Rabin. Digitalized signatures and public-key functions as intractable as factorization. , TR-212, MIT Laboratory for Computer Science, January 1979.
26. S. Ross. A First Course in Probability Theory. 4th Edition, Prentice-Hall, page 126, 1994.
27. R. Rivest, A. Shamir, L. Adleman. A method for obtaining Digital Signatures and Public-Key Cryptosystems. In *Communications of the ACM*, volume 21, n. 2, pages 120–126, 1978.
28. V. Shoup. OAEP Reconsidered. In *Advances in Cryptology—Crypto '01*, 2001.

29. A. Young, M. Yung. Auto-Recoverable and Auto-certifiable cryptosystems with RSA or factoring based keys. United States Patent 6,389,136. Filed September 17, 1997. Issued May 14, 2002.
30. A. Young, M. Yung. Auto-Recoverable Auto-Certifiable Cryptosystems. In *Advances in Cryptology—Eurocrypt '98*, 1998.
31. A. Young, M. Yung. Auto-Recoverable Auto-Certifiable Cryptosystems (a survey). CQRE, Springer-Verlag, LNCS, 1999.
32. A. Young, M. Yung. RSA Based Auto-Recoverable Cryptosystems. In *Proceedings of Public Key Cryptography—PKC '00*, 2000.

A   Appendix: Cryptographic Definitions

Recall from probability theory that a *random variable* is formally defined as a real-valued *function* that is defined over a sample space [26].

Definition 1. *(ensembles): Let I be a countable index set. An ensemble indexed by I is a sequence of random variables indexed by I. Namely, $X = \{X_i\}_{i \in I}$, where the X_i's are random variables, is an ensemble indexed by I.*

If X is a random variable and f is a function then $f(X)$ is the random variable defined by evaluating f on an input chosen according to X [18]. Thus, $f(X, X)$ would normally represent a random variable defined by evaluating f on a pair where each value in the pair is chosen according to X. However, in this work we adopt the following convention from Section 1.2.1 of [14]. *All occurrences of the same symbol in a probabilistic statement refer to the same (unique) random variable.* Thus, $f(X, X)$ refers to the random variable found by evaluating f on an input pair (x, x) where x is chosen according to X.

The following is Definition 3.2.2 of O. Goldreich.

Definition 2. *(polynomial-time indistinguishability):*
Two ensembles $X \stackrel{\text{def}}{=} \{X_n\}_{n \in \mathbb{N}}$ and $Y \stackrel{\text{def}}{=} \{Y_n\}_{n \in \mathbb{N}}$ are indistinguishable in poly-time if for every probabilistic poly-time algorithm D, every polynomial $p(\cdot)$, and all sufficiently large n,

$$Pr[D(X_n, 1^n) = 1] - Pr[D(Y_n, 1^n) = 1] < 1/p(n).$$

Goldreich makes the distinction between uniform adversaries (those contained in BPP) and non-uniform adversaries (those contained in $P/poly$). For an explanation of these types of adversaries, see [12,13]. An ensemble $X \stackrel{\text{def}}{=} \{X_n\}_{n \in \mathbb{N}}$ is said to be poly-time constructible if there exists a probabilistic poly-time algorithm S so that for every n, the random variables $S(1^n)$ and X_n are identically distributed.

Let $G(1^n)$ be a key pair generator. Define $G(1^n) = (G_1(1^n), G_2(1^n))$ where $G_1(1^n)$ is the public key and $G_2(1^n)$ is the private key. We may assume that $|G_1(1^n)|$ and $|G_2(1^n)|$ are polynomially related to n. The following definition is from [15].

Definition 3. *(indistinguishability of encryptions - uniform-complexity):*
An encryption scheme, (G, E, D), has uniformly indistinguishable encryptions in the public-key model if for every polynomial t, every probabilistic polynomial time algorithm D', every polynomial-time constructible ensemble $\overline{T} \overset{\text{def}}{=} \{\overline{T}_n = \overline{X}_n \overline{Y}_n Z_n\}_{n \in \mathbb{N}}$, with $\overline{X}_n = \{X_n^{(1)}, ..., X_n^{(t(n))}\}$, $\overline{Y}_n = \{Y_n^{(1)}, ..., Y_n^{(t(n))}\}$, and $|X_n^{(i)}| = |Y_n^{(i)}| = poly(n)$,

$$Pr[D'(Z_n, G_1(1^n), \overline{E}_{G_1(1^n)}(\overline{X}_n)) = 1]-$$
$$Pr[D'(Z_n, G_1(1^n), \overline{E}_{G_1(1^n)}(\overline{Y}_n)) = 1] < 1/p(n)$$

for every positive polynomial p and all sufficiently large n's.

We stress that $\overline{X}_n$ is a sequence of random variables, which may depend on one another. The random variable Z_n captures a-priori information about the plaintexts for which encryptions should be distinguished.

B   Appendix: Storage Requirements

Let H be a random function such that,

$$H : \{0,1\}^* \to \{0,1\}^{2m} \times \{1, -1\}^{2m} \times (\mathbb{Z}_N^*(+1))^{2m}$$

The amount of information sent from P to V can be reduced due to redundancy. This is illustrated in the NIZK version of the protocol given below.

$SQRTNIZKProof(N)$:
1. for $i = 1$ to $2m$ do:
2. P chooses $\nu_i \in_R QR_N$
3. P computes $d_{i,0}$ to be a root of ν_i with Jacobi 1
4. P computes $d_{i,1}$ to be a root of ν_i with Jacobi -1
5. P chooses $r_{i,0}, r_{i,1} \in_R S_1$
6. P computes $c_{i,j} = E_{N_1}(d_{i,j}, r_{i,j})$ for $j = 0, 1$
7. P computes $(b_1, b_2, ..., b_{2m}, \mu_1, \mu_2, ..., \mu_{2m}, a_1, ..., a_{2m}) = H(N, (\nu_1, c_{1,0}, c_{1,1}), ..., (\nu_{2m}, c_{2m,0}, c_{2m,1}))$
8. for $i = 1$ to $2m$ do:
9. P computes a square root β_i of a_i or $-a_i$ with Jacobi symbol equal to μ_i
10. P sends to V the tuple $(N, (c_{1,1-b_1}, d_{1,b_1}, r_{1,b_1}, \beta_1), ..., (c_{2m,1-b_{2m}}, d_{2m,b_{2m}}, r_{2m,b_{2m}}, \beta_{2m})))$
11. for $i = 1$ to $2m$ do:
12. V computes $\nu_i = d_{i,b_i}^2 \bmod N$
13. V computes $c_{i,b_i} = E_{N_1}(d_{i,b_i}, r_{i,b_i})$
14. V computes $(b_1, b_2, ..., b_{2m}, \mu_1, \mu_2, ..., \mu_{2m}, a_1, ..., a_{2m}) = H(N, (\nu_1, c_{1,0}, c_{1,1}), ..., (\nu_{2m}, c_{2m,0}, c_{2m,1}))$
15. V accepts iff all of the following hold:

16. N is not a prime and N is not a perfect power
17. $N \equiv 1 \bmod 4$
18. $\beta_i^2 = a_i \bmod N$ or $\beta_i^2 = -a_i \bmod N$ for $i = 1, 2, ..., 2m$
19. $J(\beta_i/N) = \mu_i$ for $i = 1, 2, ..., 2m$
20. $d_{i,b_i} \in \mathbb{Z}_N^*$ for $i = 1, 2, ..., 2m$
21. $J(d_{i,b_i}/N) = -1^{b_i}$ for $i = 1, 2, ..., 2m$
22. $r_{i,b_i} \in S_1$ for $i = 1, 2, ..., 2m$

For concreteness suppose that N_1 is a 1024 bit RSA key, E_{N_1} is OAEP, and N is a 768 bit public key. Also, suppose that the random string used in OAEP is about 256 bits in length. It follows that $1024 + 768 + 256 + 768 = 2816$ bits of information is transmitted from P to V due to iteration i. This corresponds to $(c_{i,1-b_i}, d_{i,b_i}, r_{i,b_i}, \beta_i)$. Recall that there are $2m$ iterations. Taking $m = 40$ the $2m$ iterations occupy $28, 160$ bytes.

Server Assisted Signatures Revisited

Kemal Bicakci and Nazife Baykal

Middle East Technical University, Informatics Institute
06531 Ankara, Turkey
{bicakci,baykal}@ii.metu.edu.tr

Abstract. One of the main objectives of server-assisted computation is to reduce the cost of generating public key signatures for ordinary users with their constrained devices. On the other hand, based on nothing more than a one-way function, one-time signatures provide an attractive alternative to public key signatures. This paper revisits server assisted computation for digital signatures to show server assisted one-time signature (SAOTS) that combines the benefits of these two efficiency solutions. The proposed protocol turns out to be a more computational and round-efficient protocol than previous verifiable-server approaches. In addition, SAOTS offers other advantages like verification transparency, getting rid of public key operations for the ordinary user and proving the server's cheating without storing the signatures.

Keywords: server-assisted signature, one-time signature, digital signature, pervasive computing.

1 Introduction

Broadly speaking, the most important design criteria for any kind of security protocol should of course be the "security". At this point, note that absolute security may in practice be impossible to reach, thus the security quality could be relative. In a security protocol, the second criteria that has to be considered is "efficiency", skillfulness in avoiding wasted time, effort and other resources.

In the early days of computers where saving a few clock cycles has a meaning and available cryptographic tools are unoptimized and very slow, being inefficient might mean being unusable. Today, we see that in spite of high-performance computing and really fast cryptographic algorithms, efficiency still remains a remarkable concern in designing a security protocol. This is because with more efficient security protocols three important targets can be attained:

- Cost cuts are possible e.g., by buying a less expensive and less powerful server machine.
- "Pervasive computing" vision can be realized securely where computer applications are hosted on a wide range of platforms, including many that are small, mobile and regarded today as devices having only limited computational capabilities.
- Security level can be increased e.g., by using a longer key length.

T. Okamoto (Ed.): CT-RSA 2004, LNCS 2964, pp. 143–156, 2004.
© Springer-Verlag Berlin Heidelberg 2004

Digital signatures, quickly find applications in many spheres of computing, are among the most fundamental and valuable building blocks of security protocols. Basically, a digital signature provides three essential security services all in once:

Authentication: assurance of the identity of the signer.

Integrity: assurance that the message is not altered after it is signed.

Nonrepudiation: blocking a sender's false denial that he or she signed a particular message, thus enabling the receiver to easily prove that the sender actually did sign the message.

While there are other means like message authentication codes (MACs) to ensure data integrity and authentication, digital signatures are better in one important respect. They can be used to solve the nonrepudiation problem. Moreover, the MAC approach is inadequate in a multicast setting because it is based on a shared secret among participants.

Most current techniques for generating digital signatures are based on public key cryptography (based on complex mathematical problems such as factoring or discrete logarithms e.g., RSA [1] or DSS [2]), but it is well known that public key cryptography has efficiency problems. Moreover, some mobile devices may have 8-bit microcontrollers running at very low CPU speeds, so public key cryptography at any kind may not even be an option for them.

One-time signatures (OTS), on the other hand, provide an attractive alternative to public key based signatures. Unlike signatures based on public key cryptography, OTS is based on nothing more than a one-way function (OWF). Examples of conjectured OWFs are SHS [3] and MD5 [4]. OTSs are computationally more efficient since no complex arithmetic is involved.

We observe that; at one hand, despite the performance advantages provided by OTSs, they have gained attention in security world only in very recent years. On the other, server assisted computation was well-studied in the cryptography literature and more or less seems to reach its maturity. Hence the question to think about is: "Can we combine the efficiency of one-time signatures with server-assisted computation?"

This paper tries to answer this question and revisits server assisted computation for digital signatures to show server assisted one-time signature (SAOTS), that turns out to be more computational efficient and offers other advantages over previous alternatives.

The rest of this paper is organized as follows. In the next section some background material is provided. Previous studies on server-assisted digital signature protocols are summarized in section 3. The proposed SAOTS protocol is presented in section 4. Section 5 and 6 are for the security analysis and performance evaluation of SAOTS protocol, respectively. Finally, section 7 concludes the paper.

2   Background

2.1   One-Way Functions

"One-way functions" (OWFs) are functions that are relatively easy to compute but significantly harder to reverse. That is, given x it is easy to compute $f(x)$, but given $f(x)$, it is hard to compute x. OWFs are public functions; no secret keys are involved. The security is in their one-wayness. Having been used for computer science for a long time, "hash functions" take a variable-length input and convert it to a fixed-length generally smaller output. And finally, "one-way hash functions" are hash functions that work in one direction or in another view, they are like digital fingerprints: small pieces of data that can serve to identify much larger digital objects (e.g., SHS [3] and MD5 [4].)

SHS and MD5 were originally designed as one-way hash functions but they can be easily used as one-way functions when the input message length is set to be equal to the length of output.

2.2   Hash Chains

The idea of "hash chain" was first proposed by Lamport [5] in 1981 and suggested to be used for safeguarding against password eavesdropping. However being an elegant and versatile low-cost technique, the hash chain construction finds alot of other applications. A hash chain of length N is constructed by applying a one-way hash function $h()$ recursively to an initial seed value (s).

$$K^N = h^N(s) = \underbrace{h(h(h(...h(s)...)))}_{N\ times}$$

The last element K^N resembles the public key in public key cryptography i.e., by knowing K^N, K^{N-1} can not be generated by those who does not know the value s. This property of hash chains has been directly evolved from the property of one-way hash functions.

In most of the hash-chain applications, first K^N is securely distributed and then the elements of the hash chain is spent one by one by starting from K^{N-1} and continuing until the value of s is reached. At this point the hash chain has been exhausted and a new hash chain needs to be generated to proceed.

2.3   One-Time Signatures

The OTS concept was first proposed by Lamport, too. [6]. The idea behind OTS concept is again very easy to understand. A message sender prepares for a digital signature by generating a random number r, which is retained as the private value. He then securely distributes the hash of r, $h(r)$, where h is a one-way function; this represents the public value and is used by receivers as the signature certificate to verify the signature. The signature is sent by distributing the value r itself. Receivers verify that this message could only be sent by the sender by applying h to r to get $h(r)$. If this matches the value for $h(r)$ in the signature certificate, then the OTS is considered to be verified, since only the sender can know r. This, in effect, allows the signing of a predictable 1-bit value. In order to sign any 1-bit value, two random numbers $(r1, r2)$ are needed; this way, both $h(r1)$ and $h(r2)$ are pre-distributed but at most one of $(r1, r2)$ is revealed as a signature. In the original proposal [6], 160 random numbers out of 320 are revealed as the signature of 160-bit hash value of any given message.

Other than the computational efficiency mentioned previously, one-time signatures can also be regarded as a more secure solution since OTS is based only on OWF whereas public-key signatures are based on complex mathematical problem as well as OWF, using OTSs allow us to elliminate one more point of vulnerability.

Despite these advantages provided by OTSs, they could not find practical usage in security world. We believe this is due to two main reasons:

First of all, one-time signatures are longer than traditional signatures that might result in storage and bandwidth constraints. Recent studies succeeded in decreasing the length of one-time signatures in some extent. It was previously realized that p out of n random numbers are sufficient to sign a b-bit length message if the following inequality holds for a given n and p [7].

$$2^b \geq C(n, p) = \frac{n!}{p! * (n - p)!} \tag{1}$$

To sign an arbitrary length message by OTS, just like the public-key based signatures, we can reduce the length of the message m by computing the hash value of the message, $h(m)$ and then sign $h(m)$. This means for instance for $b = 160$ (e.g. SHS), n must be at least 165 with subsets of size 75 ($p = 75$).

The extra length of one-time signatures (which was a concern two decades ago) is negligible today owing to the high speed of modern networks hence we think that the second disadvantage is a more serious one, that is one-time signatures can be used to sign only one message per one public key in its simple form. Since the public key requires to be distributed in a secure fashion which is done most typically using a public key signature, the benefit of using quick and efficient hash function is apparently lost.

There is also a bunch of clever approaches to overcome this limitation. One of which is on-line/off-line signatures [8] where the public key of one-time signatures is signed by using public key techniques off-line before the message is known. When the message to be signed is in hand, there will not be any necessity to perform public key operation so that the response time (real-time efficiency) is improved.

Due to constraints of mobile devices, we might want to minimize or eliminate the number of public key operations no matter it is off-line or not. Then Merkle's proposal [9] can be preferred where one-time signatures can be embedded in a tree structure, allowing the cost of a single public key signature to be amortized over a multitude of OTS. The problem in this formulation is the longer lengths of signatures. Now we face a more severe storage and bandwidth requirement than one-time signatures in its simple form since the length of signatures increases as the number of signatures generated using the same tree structure increase.

3   Previous Studies

Another promising approach to use one-time signatures more than once by using a single pre-distributed public key is the Server-Assisted One-Time Signature (SAOTS) protocol we propose in this paper which is based on a third party (the server). But before introducing it, in a more general view we would like to give a very short summary of the previous studies on employing a powerful server to decrease the computation requirements to generate digital signatures.

Server assisted signatures can be explained in three subgroups depending on the trust relationship between the user and the server. More specifically the server employed may be either (1) fully trusted, (2) untrusted or (3) verifiable.

In the first category, after receiving an authentic message from a user (A MAC algorithm which can be implemented very efficiently may be used for authentication), a more powerful proxy server on behalf of the user generates a public key digital signature for the message [10]. Notice that the user himself does not need to perform any public key operation, he just computes a MAC using secret key cryptography [1]. The drawback here is that this simple design is only applicable when the user fully trusts the proxy server i.e. the server can generate forged signatures and that cheating cannot be proven by the user.

As the opposite, a totally untrusted server might be utilized i.e. the server only executes computations for the user. Now the goal of securely reducing computational costs on the user's machine becomes more difficult to accomplish and in fact most of the schemes proposed so far have been found not to be secure. For instance the protocol proposed by Bequin and Quisquater [12] was later broken by Nguyen and Stern [13]. Up to our knowledge, for RSA signatures, designing a secure server-assisted protocol that utilizes an untrusted server is still an open problem. But the situation for DSA is not the same. A secure (unbroken) example for DSA is the interesting approach of Jakobson and Wetzel [14]. However we see that in their approach to generate the signature public key operations although in reduced amount are still needed to be performed on the constrained device.

3.1   Verifiable-Server Assisted Signature Protocols

The last server-assisted signature alternative is to employ a verifiable server (VS). A VS is the one whose cheating can be proven. This approach can be considered in somewhere between the other two since the server in this case can cheat but subsequently the user would have the ability to prove this situation

---

[1] In the literature, this method is sometimes called "proxy protocol with full delegation". Proxy signatures have other variations mostly designed for the purpose of restricting the server's signing rights. These variations are less-efficient than traditional public key signatures hence are not of interest in our discussion [11].

to other parties (e.g. an arbiter). We see that in some papers, VS is named as semi-trusted server.

In traditional methods of digital signature generation, the signer usually obtains a public key certificate from a certification authority (CA). In order to trust the legitimacy of signatures, the receiver must trust the CA's certificate-issuance procedures. For instance the CA can issue a fake certificate for a particular user and then impersonate the user by generating a forged signature. However, if a contract between the CA and the user was signed in the certification process, in dispute the signer can prove the CA's cheating by asking this contract from the CA. Notice the similarities between the trust relationship between the signer and CA in traditional methods and the signer and the server in verifiable-server assisted signature protocols.

The first work that aims to reduce the computational costs to generate digital signatures for low-end devices by employing a powerful VS is SAS protocol [15]. In [16], the authors extend this work by providing implementation results as well as other details of the scheme. We now would like to provide a brief summary of SAS protocol. For a comprehensive treatment, please refer to the original papers [15,16].

There is an initialization phase in SAS where each user (originator) gets a certificate from an offline certification authority for K^N (the last element of a hash chain of length N). In addition, each user should register to a VS (which has the traditional public-key based signing capability) before operation. Then the SAS protocol works in three rounds:

1. The originator (O) sends m and K^i to VS where
   - m is the message
   - K^i is the ith element of the hash chain. The counter i is initially set to $N-1$ and decremented after each run.
2. Having received O's request, VS checks the followings:
   - Whether O's certificate is revoked or not [2].
   - Whether $h^{N-i}(K^i) = K^N$ or in a more efficient way $h(K^i) = K^{i+1}$ since K^{i+1} has already been received.

   If these checks are OK, VS signs m concatenated with K^i and sends it back to O.
3. After receiving the signed message from VS, O verifies the VS's signature, attaches K^{i-1} to this message and sends it to the receiver R.

Upon receipt of the signed message, the receiver verifies VS's signature and checks whether $h(K^{i-1}) = K^i$.

---

[2] Another important issue for server-assisted signature protocols is "revocation", which is explored in Appendix A.

3.2   SAS Protocol Weaknesses

We have observed that SAS protocol has several drawbacks. These are:

1. *Verifying VS's signature*: In step 3 of the SAS protocol, before sending the signed message to R, O should verify the VS's signature otherwise an attack can be performed as follows:
   An attacker modifies the message that O sends to VS and if VS signs this new message instead, O's revealing of K^{i-1} without verifying VS's signature results in a forged signature for the message the attacker has generated.
   Remember that for some constrained devices, public key cryptography is simply untenable no matter it is used for signing or verifying.
   Even when public key cryptography is acceptable for the user's device, the efficiency provided by this protocol is based on an assumption which is not always valid. More precisely, if the VS uses RSA [1] signature scheme, where verification is much more efficient with respect to signing, SAS brings some efficiency. On the other hand, if instead of RSA, other digital signature schemes like DSS [2] where verification is at least as costly as signing are used to sign the message, SAS protocol apparently becomes less-efficient than traditional signing methods.

2. *Incompatible Verification*: As stated in [16], unlike the proxy signatures explained in section 3, SAS signatures are not compatible with other primary signature types. Therefore, the receiver must utilize the custom-built verification method of SAS protocol.

3. *Storing VS's signatures*: In SAS protocol, the signer must store VS's signatures to prove its cheating [15]. For some devices having a limited storage capacity, this also might put a burden on the operation.

4. *Network Overhead*: One of the factors that affects the overall performance of the SAS protocol is the round-trip delay between O and VS. To decrease the network delay, one can try to decrease the number of rounds from three to two in SAS (you can not do better than two rounds in a server assisted protocol). However, to reduce number of rounds to two, if O attaches the hash element K^{i-1} to the first message he sends to VS, an attacker can forge a signed message easily by modifying the message while in transit.
   As a result, SAS protocol cannot be a two-round protocol like the SAOTS protocol that will be introduced in the next section this is basically because the signature is not binded with the message itself in a two-round case.

4   Server Assisted One-Time Signature (SAOTS) Protocol

In this section, we propose the server assisted one-time signature protocol (SAOTS) which is the first VS based approach where the user does not need to perform any public key operation at all. SAOTS is completely transparent to verifiers (the signatures are indistinguishable from standard signatures). Moreover, in our proposed protocol unlike other alternatives the server not the user is required to save the signatures for dispute resolution. Operating in two rounds as opposed to three, SAOTS eliminates all the four aforementioned drawbacks of SAS protocol.

4.1   The Basic Idea

Our protocol is built on top of one-time signature idea. Similar to proxy signatures where an efficient MAC algorithm is employed to establish an authentic channel between the user and the server, in SAOTS the user sends the message to the server after he signs it with a one-time signature. However this basic idea needs to be enhanced otherwise we face again with the inherent problem of OTS, signing only one message per one public key. We will explain how we have solved this problem in subsequent paragraphs. Our previous solution for this problem that gets benefit of the idea of hash chains is summarized in Appendix B [17].

4.2   Setup

As a setup, every user registers to a server and generates a one-time private key (random numbers) and a one-time public key (hash values of these random numbers) and in a secure fashion he distributes the public key to the server. This can be accomplished by a public key signature if he has already a capability of traditional signing or he can directly get a certificate from a CA for the one-time public key he has.

In addition, to produce public-key signatures, the server generates a private key on behalf of each registered user and obtains a certificate from a CA for the corresponding public key (in the certification process the user confirms that the public key belongs to himself).

4.3   Operation

The protocol works in two rounds as illustrated in Figure 1:

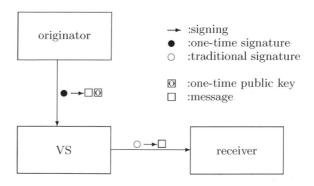

Fig. 1. Operation of the server assisted one-time signature (SAOTS) protocol

1. The user precomputes a second one-time private key - public key pair. When the message to be signed is ready, he concatenates the message with the new one-time public key and signs this by his previous one-time private key. He then sends the message and the new one-time public key as well as the one-time signature to the server.

2. Having already received securely the one-time public key of the user's signature on the message, the server verifies the one-time signature. He stores the new one-time public key the user has signed for the verification of next message. It is now ready to sign the message with the user's private key (if his certificate is not revoked). Finally, the signed message is transmitted to the intended receiver(s).

Since the signature is indistinguishable from a standard signature, receivers can transparently verify the signature by using the user's public key [3]. One can easily prove that this protocol provides all the three security services asked from a digital signature but only if the server does not cheat i.e. it does not sign any message on behalf of the user without user's approval. We will show in the next section how the user can prove the server's cheating. If he cannot prove, the other parties conclude that the user is the one who actually signs the message.

The user can sign any further messages easily by repeating the step 1. The server can always verify the one-time signature since it has securely received the one-time public key in the previous run of the protocol. The server must store all previous messages (as well as one-time signatures and one-time public keys) for secure operation but the user does not need to store anything to prove the server's cheating. This becomes more clear when we make the security analysis in the next section.

We would like to point that the "chaining" technique we use that attaches the one-time public key for the next message to the current message before signing is previously suggested by [19] in order to sign infinite length digital streams more efficiently.

5   Security Analysis of SAOTS

5.1   Security of Underlying Components

For secure operation, we need to prove the security of signatures of both the user and the server. Since the server's signature is a traditional one, we conclude that if the traditional signature algorithm used is a secure one, then the server's signature is also secure. Secondly, we note that the security of the chaining technique used in the one-time signature the user generates has been studied previously. For the security proofs please refer to [19].

---

[3] In some applications the server's public key can be initially embedded in the verification software and the receiver himself can not obtain securely the public key of all possible signers. Then it is better to have the server to sign the message with his own public key after appending a statement on the message saying that it has received it from the user [18]. Another advantage of this method is that the server avoids to obtain a new certificate for each registered user.

5.2   Dispute Resolution

Provided that the underlying signatures building the SAOTS protocol are secure, we now want to show how a dispute can be resolved. In case of a dispute, the receiver submit the message and its public-key signature received from the server to an arbiter. The arbiter will verify the followings:

- the public key of the user is certified by the CA.
- the public key signature is valid.

If these checks are successful, then the user is allowed to take the oppurtunity to repudiate the message. There will be two checks to decide whether the user's claim is true or not:

- CA is asked to prove that the user's one-time public-key was registered.
- The server is asked to prove that the message was signed by the user's one-time private key.

As a proof, the server shows all the signed messages received from the user starts from the first one and continues until the message in dispute is reached. The arbiter verifies all these one-time signatures. If both CA and server successfully shows that they did not cheat, the arbiter concludes that the user is dishonest and claims falsely that he has not sent the message.

5.3   Denial of Service Attacks

In the SAS protocol [16], unlike traditional signature schemes, denial of service (DoS) attacks aiming to deny the server's service to the users are of concern. The basic idea behind these attacks is simple, by sending legitimate (well-formed) requests, an adversary can force the server to perform alot of signing tasks so that it cannot response timely to real requests coming from registered users [16]. However if SAOTS is preferred, these attacks are eliminated because an adversary cannot forge users' one-time signatures and therefore cannot generate legitimate requests.

6   Performance Evaluation of SAOTS

Table 1 shows the comparison of SAOTS and SAS protocols with respect to computation requirements on the participating entities. Note that, an efficient encoding method was presented in [7] which costs less than one hash operation. Encoding refers to computation of which subset of random numbers should be revealed as the OTS of the message.

In SAOTS protocol, the server needs to perform one hash to get the hash of the message and 75 hash operations ($p = 75$ if SHS is used) to verify the OTS if all of 165 hash values constitute the public key. By a simple trick and with a cost of additional hash operation for the server we can reduce the length of

Table 1. Computational comparison of SAS and SAOTS protocols

	SAS	SAOTS
Originator	$1H + 1V$	$1H + 1M$
Server	$2H + 1S$	$(p+2)H + 1M + 1S$
Receiver	$1V + 2H$	$1V + 1H$

H: hash computation
S: traditional signing by a public key
V: verification of public key signature
M: encoding computation (costs less than one hash)
p: number of hash computations to verify OTS

public key to a single hash value. The idea is simple: as the public key, calculate the hash of concatenation of all the 165 hashes. Now to be able to verify the OTS the sender should send the chosen 75 random numbers and the other 90 random number's hash value. In each run of the protocol the user should send one signature and one public key so if the length of random number is equal to the length of hash value, in overall the signer should send $75 + 90 + 1 = 166$ hash values to the server.

Despite this reduction in size, there is a trade-off between SAS and SAOTS protocols with respect to communication efficiency because the round efficiency of SAOTS protocol as explained in subsection 3.2 comes with an increase in the length of the message transmitted from the user to the server (the messages in SAS protocol are significantly smaller in size). In Appendix B, we present a variant of SAOTS where the length of messages exchanged are shortened.

To have a more concrete comparison of SAS and SAOTS, we have implemented both of them using MIRACL library [20]. A PC running Windows 2000 with an 800 MHz Pentium III and a 128 MB memory was chosen as the VS and a PC running Windows 95 with a 200 MHz Pentium I and a 32 MB memory was chosen as the users' machine. Note that today's high-end PDA's and palmtops have a processor speed around 200 MHz. RSA with a 1024 bit key and SHS with a 160 bit output was used and $m = 165$, $p = 75$ were the OTS parameters.

Table 2. Performance measurements of cryptography primitives (ms)

	Pentium III 800 Mhz	Pentium I 200 Mhz
SHS	0.028	0.156
RSA(verifying)	2.220	13.893
RSA(signing)	9.454	59.162
Encoding	0.02	0.1

Table 2 gives the performance measurements of cryptography primitives on two platforms used and Table 3 summarizes our findings of the experiments. From these tables, we conclude that SAOTS is a more efficient protocol with respect to computational requirements for ordinary users. In our opinion, the increase in server's computation is not a big problem since in practice a much more powerful machine is usually employed as the server.

Table 3. Experimental comparison of SAS and SAOTS protocols (ms)

	SAS	SAOTS
Originator's computation	14.049	0.256
Server's computation	9.482	11.650
Receiver's computation	14.205	14.049

7 Conclusion

To generate signatures, getting help from a verifiable-server has an advantage over proxy-based solutions since as opposed to proxy-server, verifiable-server's cheating can be proven. Verifiable-server assisted signatures were proposed in the past but they could not eliminate public key operations for the signer. In this paper, we propose a new alternative called SAOTS (server assisted one-time signature) where just like proxy signatures generating a public key signature is possible without performing any public key operations at all.

Verification transparency, no necessity to store past signatures to prove server's cheating, reduced number of rounds are other advantages of SAOTS. The only drawback of the proposed protocol is the increased length of the message transmitted from the user to the server's machine.

References

1. R.L. Rivest, A. Shamir, and L.M. Adleman. A method for obtaining digital signatures and public-key cryptosystems. Communications of the ACM, 21(2), 1978.
2. National Institute for Standards and Technology. Digital Signature Standard (DSS). Federal Register, 56(169), August 30, 1991.
3. National Institute of Standards and Technology. FIPS Publication 180: Secure Hash Standard (SHS), May 11, 1993.
4. R.L. Rivest. The MD5 message-digest algorithm. Internet Request for Comments, April 1992. RFC 1321.
5. L. Lamport. Password authentication with insecure communication. Communications of the ACM, 24(11), 1981.
6. L. Lamport. Constructing digital signatures from a one-way function. Technical Report CSL-98, SRI International, October 1979.
7. K. Bicakci, G. Tsudik, and B. Tung. How to construct optimal one-time signatures. Computer Networks (Elsevier), Vol.43(3), October 2003.
8. S. Even, O. Goldreich, and S. Micali. On-line/off-line digital signatures. CRYPTO 1989, LNCS No. 435, Springer-Verlag, 1990.

9. R.C. Merkle. A digital signature based on a conventional encryption function. CRYPTO 87, LNCS No. 293, Springer-Verlag, 1988.
10. M. Burnside, D. Clarke, T. Mills, A. Maywah, S. Devadas, and R. Rivest. Proxy-Based Security Protocols in Networked Mobile Devices. Proceedings of the 17th ACM Symposium on Applied Computing (Security Track), March 2002.
11. A. Boldyreva, A. Palacio, and B. Warinschi. Secure Proxy Signature Schemes for Delegation of Signing Rights. Cryptology ePrint Archive, Report 2003/096, 2003, http://eprint.iacr.org.
12. P. Beguin and J. J. Quisquater. Fast server-aided RSA signatures secure against active attacks. CRYPTO 95, LNCS No. 963, Springer-Verlag, 1995.
13. P. Nguyen and J. Stern. The Beguin-Quisquater server-aided RSA protocol from Crypto '95 is not secure. ASIACRYPT 98, LNCS No. 1514, Springer-Verlag, 1998.
14. M. Jakobsson and S. Wetzel. Secure Server-Aided Signature Generation. In Proc. of the International Workshop on Practice and Theory in Public Key Cryptography (PKC 2001), LNCS No. 1992, Springer, 2001.
15. N. Asokan, G. Tsudik and M. Waidners. Server-supported signatures. Journal of Computer Security, November 1997.
16. X. Ding, D. Mazzocchi and G. Tsudik. Experimenting with Server-Aided Signatures. 2002 Network and Distributed Systems Security Symposium (NDSS'02), February 2002.
17. K. Bicakci and N. Baykal. SAOTS: A New Efficient Server Assisted Signature Scheme for Pervasive Computing. In Proc. of 1st International Conference on Security in Pervasive Computing, SPC 2003, LNCS No. 2802, March 2003, Germany.
18. K. Bicakci and N. Baykal. Design and Performance Evaluation of a Flexible and Efficient Server Assisted Signature Protocol. In Proc. of IEEE 8th Symposium on Computers and Communications, ISCC 2003, Antalya, Turkey.
19. R. Gennaro and P. Rohatgi. How to Sign Digital Streams. CRYPTO 1997, LNCS No. 1294, Springer-Verlag, 1997.
20. MIRACL Multiprecision C/C++ Library. http://indigo.ie/~mscott/
21. M. Myers, R. Ankney, A. Malpani, S. Galperin, and C. Adams. Internet public key infrastructure online certificate status protocol. RFC 2560, June 1999.

Appendix A: Revocation in SAOTS Protocol

As far as digital signatures are of concern, revocation means that if a user does something that warrants revocation of his security privileges i.e. he might be fired or may suspect that his private key has been compromised, he should not generate valid digital signatures on any further messages. However, signatures generated prior to revocation may need to remain valid.

In Online Certificate Status Protocol (OCSP) [21], today's state-of-the-art approach to solve the revocation problem, to provide timely revocation information, upon verifier's query a validation server sends back a signed response showing the sender's certificate's current status. The drawback here is that it is impossible to ask a validation server whether a certificate was valid at the time of signing. Immediate revocation (the user cannot sign immediately after the revocation takes place) is possible if an online VS is employed. In order to revoke a user's public key, it is sufficient to notify the server. The server maintains a list of revoked users and it rejects signing on behalf of the user if his public key is in the list.

We now want to show a deficiency in the revocation capability of SAS protocol [16] that works in three rounds. Think of a situation where the user gets the public key signature from the VS in round 2 and postpones the execution of round 3. He then notifies the server to revoke his public key (e.g. by claiming that his private key has been stolen). Afterwards, he can cheat by executing round 3 and generating a valid signature although his public key has already been revoked. In SAOTS, this deficiency is eliminated since the protocol works in two steps in opposed to three.

Appendix B: SAOTS with Hash Chains

In SAOTS protocol, the message user sends to the server is composed of the message to be signed, its one-time signature and the public key for the next signature. We now show a variant of SAOTS to reduce the size of this message [17]. But as we will see, this reduction is possible only with a cost of off-line verification of server's signature. Now in the initialization phase of SAOTS, each user gets a certificate from a CA for the hash array of length n:

$$K_0^N, K_1^N, K_2^N, \ldots K_{n-1}^N \tag{2}$$

Where n is chosen to be large enough to encode the hashed message ($n=165$ for SHS). Each element of the array is the last element of a hash chain of length N. Then SAOTS with hash chains works in two rounds again but now the originator should receive, verify and store the signature coming from the server:

1. The originator (O) sends m and S^i to VS where
   - m is the message
   - $S^i = (K_{a_1}^i, K_{a_2}^i, K_{a_3}^i, \ldots, K_{a_p}^i)$ denotes the subset of the array that encodes the message to an OTS (composed of ith elements of hash chains). The counter i is initially set to $N - 1$ and decremented after each run.
2. Having received O's request, VS performs the followings:
   - Whether O's certificate is revoked or not.
   - Computes the encoding of $h(m)$ or in other words finds out which subset S^i would correspond to the OTS of the message.
   - Checks whether for ($q = 1$ to p) $h^{N-i}(K_{a_q}^i) = K_{a_q}^N$ or in a more efficient way ($q = 1$ to p) $h(K_{a_q}^i) = K_{a_q}^{i+1}$ if $K_{a_q}^{i+1}$ has already been received.

If these are all OK, VS signs the message and sends it back to both R and O. For secure operation, O should sign the next message only after the (off-line) verification of VS's signature otherwise the protocol can be broken. We recommend interested readers to [17] to see how the attacks work and how they can be avoided by the verification of server's signature.

In summary, with a cost of heavier computation requirement and an extra storage requirement, SAOTS with hash chains provides a total length save of $(n + 1 - p)$ hashes.

Cryptanalysis of a Zero-Knowledge Identification Protocol of Eurocrypt '95

Jean-Sébastien Coron and David Naccache

Gemplus Card International
34 rue Guynemer, 92447 Issy-les-Moulineaux, France
{jean-sebastien.coron, david.naccache}@gemplus.com

Abstract. We present a cryptanalysis of a zero-knowledge identification protocol introduced by Naccache *et al.* at Eurocrypt '95.
Our cryptanalysis enables a polynomial-time attacker to pass the identification protocol with probability one, without knowing the private key.

Keywords: Zero-knowledge, Fiat-Shamir Identification Protocol.

1 Introduction

An identification protocol enables a verifier to check that a prover knows the
private key corresponding to a public key associated to its identity. A protocol
is zero-knowledge when the only additional information obtained by the verifier is that the prover knows the corresponding private key [2]. A famous zero-
knowledge identification protocol is Fiat-Shamir's protocol [1], which is provably
secure assuming that factoring is hard. The protocol requires performing multiplications modulo an RSA modulus.

A space-efficient variant of the Fiat-Shamir identification protocol was introduced by Naccache [3] and by Shamir [5] at Eurocrypt' 94. This variant requires
only a few bytes of RAM, even for an RSA modulus of several thousands bits,
and is provably as secure as the original Fiat-Shamir protocol. This variant is
particularly interesting when the prover is implemented in a smart-card, in which
the amount of RAM is very limited.

However, the time complexity of the previous variant is still quadratic in the
modulus size, and its implementation on a low-cost smart-card is likely to be
inefficient. At Eurocrypt '95, Naccache *et al.* introduced another Fiat-Shamir
variant [4]. It uses the same idea for reducing the space-complexity, but the
prover's time complexity is now quasi-linear in the modulus size (instead of being
quadratic). As shown in [4], the new identification protocol can be executed on
a low-cost smart-card in less than a second.

In this paper, we describe a cryptanalysis of one of [4]'s time-efficient variants.
Our cryptanalysis enables a polynomial-time attacker to pass the identification
protocol with probability one, without knowing the private key. We would like to
stress that the basic quasi-linear time protocol introduced by [4] remains secure,
since it is in fact equivalent to standard Fiat-Shamir and hence to factoring.

T. Okamoto (Ed.): CT-RSA 2004, LNCS 2964, pp. 157–162, 2004.

2   The Fiat-Shamir Protocol

We briefly recall Fiat-Shamir's identification protocol [1]. The objective of the prover is to identify itself to any verifier, by proving knowledge of a secret s corresponding to a public value v, which is associated to its identity. The protocol is zero-knowledge in that it does not reveal any additional information about s to the verifier. The security relies on the hardness of factoring an RSA modulus.

Key Generation: The authority generates a k-bit RSA modulus $n = p \cdot q$, and an integer v which is a function of the identity of the prover. Using the factorization of n, it computes a square root s of v modulo n, *i.e.* $v = s^2 \mod n$. The authority publishes (n, v) and sends s to the prover.

Identification Protocol:

1. The prover generates a random $x \leftarrow Z_n$, and sends $z = x^2 \mod n$ to the verifier.
2. The verifier sends a random bit b to the prover.
3. If $b = 0$, the prover sends $y = x$ to the verifier, otherwise it sends $y = x \cdot s$ mod n.
4. The verifier checks that $y^2 = z \cdot v^b \mod n$.
5. Steps 1-4 are repeated several time to reduce the cheating probability.

3   The Space-Efficient Variant of Fiat-Shamir's Protocol

Fiat-Shamir's protocol requires to perform multiplications modulo an RSA modulus n. It has a quadratic time and linear space complexity. Therefore, the original protocol could not be implemented on low-cost smart-cards, which in 1994 contained about 40 bytes of random access memory (RAM). Naccache [3] and Shamir [5] introduced a space-efficient variant which requires only a few bytes of RAM, even for an RSA modulus of several thousands bits, and which is provably as secure as the original Fiat-Shamir protocol.

The idea is the following: assume that the prover is required to compute $z = x \cdot y \mod n$, where x and y are two large numbers which are already stored in the smart-card (*e.g.*, in its EEPROM[1]), or whose bytes can be generated on the fly. Then instead of computing $z = x \cdot y \mod n$, the prover computes

$$z' = x \cdot y + r \cdot n$$

for a random r uniformly distributed in $[0, B]$, for a fixed bound B. The verifier can recover $x \cdot y \mod n$ by reducing z' modulo n. Moreover, when computing z', the prover does not need to store the intermediate result in RAM. Instead, the

---

[1] The smart-card EEPROM is a re-writable memory, but the operation of writing is about one thousand time slower than writing into RAM, and can not be used to store fast-changing intermediate data during the execution of an algorithm.

successive bytes of z' can be sent out of the card as soon as they are generated. Therefore, a smart-card implementation of the prover needs only a few bytes of RAM (see [5] or [3] for more details).

As shown in [5], if B is sufficiently large, there is no loss of security in sending z' instead of z. Namely, from z one can generate $z'' = z + u \cdot n$ where u is a random integer in $[0, B]$. Letting $z = x \cdot y - \omega \cdot n$, we have:

$$z'' = x \cdot y + (u - \omega) \cdot n$$

Then, the statistical distance between the distributions induced by z' and z'' is equal to the statistical distance between the uniform distribution in $[0, B]$ and the uniform distribution in $[-\omega, B - \omega]$, which is equal to ω/B. Then, assuming that x and y are both in $[0, n]$, this gives $\omega \in [0, n]$, and the previous statistical distance is lesser than n/B. Therefore, by taking a B much larger than n (for example, $B = 2^{k+80}$, where k is the bit-size of n), the two distributions are statistically indistinguishable, and any attack against the protocol using z' would be as successful against the protocol using z.

The identification protocol is then modified as follows:

Space-Efficient Fiat-Shamir Identification Protocol:

1. The prover generates a random $x \leftarrow Z_n$ and a random $r \in [0, B]$, and sends $z = x^2 + r \cdot n$ to the verifier.
2. The verifier sends a random bit b to the prover.
3. If $b = 0$, the prover sends $y = x$ to the verifier, otherwise it sends $y = x \cdot s + t \cdot n$ for a random $t \in [0, B]$.
4. The verifier checks that $y^2 = z \cdot v^b \mod n$.
5. Steps 1-4 are repeated several time to reduce the cheating probability.

4   The Time-Efficient Variant of Fiat-Shamir's Protocol

The time complexity of the previous variant is still quadratic in the modulus size, and its implementation on a low-cost smart-card is likely to be inefficient. At Eurocrypt '95, Naccache *et al.* introduced yet another Fiat-Shamir variant [4]. It uses the same idea as Shamir's variant for reducing the space-complexity, but the prover's time complexity is now quasi-linear in the modulus size (instead of being quadratic). As shown in [4], the identification protocol can then be executed on a low-cost smart-card in less than a second.

The technique consists in representing the integers modulo a set of ℓ small primes p_i (usually, one takes the first ℓ primes). This is called the Residue Number System (RNS) representation. Letting $\Pi = \prod_{i=1}^{\ell} p_i$, by virtue of the Chinese Remainder Theorem, any integer $0 \le x < \Pi$ is uniquely represented by the vector:

$$(x \mod p_1, \dots, x \mod p_\ell)$$

The advantage of this representation is that multiplication is of quasi-linear complexity (instead of quadratic complexity): if x and y are represented by the vectors $(x_1, \ldots, x_\ell)$ and $(y_1, \ldots, y_\ell)$, then the product $z = x \cdot y$ is represented by:

$$(x_1 \cdot y_1 \mod p_1, \ldots, x_\ell \cdot y_\ell \mod p_\ell)$$

The size ℓ of the RNS representation is determined so that all integers used in the protocol are strictly smaller than Π; the bijection between an integer and its modular representation is then guaranteed by the Chinese Remainder Theorem. The time-efficient variant of the Fiat-Shamir protocol is the following:

Time-Efficient Variant of the Fiat-Shamir Protocol:

1. The prover generates a random $x \in [0, n]$ and a random $r \in [0, B]$, and sends $z = x^2 + r \cdot n$ to the verifier. The integers x, r and z are represented in RNS.
2. The verifier sends a random bit b to the prover.
3. If $b = 0$, the prover sends $y = x$ to the verifier, otherwise it sends $y = x \cdot s + t \cdot n$ for a random $t \in [0, B]$. The integers x, s and t are represented in RNS.
4. The verifier checks that $y^2 = z \cdot v^b \mod n$.
5. Steps 1-4 are repeated several time to reduce the cheating probability.

The only difference between this time-efficient variant and Shamir's space-efficient variant is that integers are represented in RNS. Therefore, from a security standpoint, those variants are strictly equivalent.

However, another time-efficient variant is introduced in [4], whose goal is to increase the efficiency of the verifier. The goal of this second variant is to enable the verifier to check the prover's answer in linear time when $b = 0$. In this variant, when $b = 0$, the prover also reveals r, which enables the verifier to check that $z = x^2 + r \cdot n$ by performing the computation in the RNS representation (the equality $z = x^2 + r \cdot n$ is checked modulo each of the primes p_i), which takes quasi-linear time instead of quadratic time. More precisely, this variant is the following:

Second Time-Efficient Variant of the Fiat-Shamir Protocol:

1. The prover generates a random $x \in [0, n]$ and a random $r \in [0, B]$, and sends $z = x^2 + r \cdot n$ to the verifier. The integers x, r and z are represented in RNS.

2. The verifier sends a random bit b to the prover.
3. If $b = 0$, the prover sends x and r to the verifier, in RNS representation. If $b = 1$, the prover sends $y = x \cdot s + t \cdot n$ for a random $t \in [0, B]$, where y is represented in RNS.
4. If $b = 0$, the verifier checks that $z = x^2 + r \cdot n$. The test is performed in the RNS representation. If $b = 1$, the verifier checks that $y^2 = z \cdot v \mod n$.
5. Steps 1-4 are repeated several time to reduce the cheating probability.

This second time-efficient variant is more efficient for the verifier, because when $b = 0$, the check at step 3 is performed in RNS representation, which is of quasi-linear complexity instead of quadratic complexity. Therefore, the time-complexity of this second time-efficient variant is expected to be divided by a factor of approximately two.

5   Cryptanalysis of the Second Time-Efficient Variant of Eurocrypt '95

We show that the second time-efficient variant is insecure. We describe an attacker $\mathcal{A}$ that passes the identification protocol with probability one, without knowing the private key s.

The key observation is the following: since for $b = 0$, the verifier checks that $z = x^2 + r \cdot n$ in the RNS representation, the equality checked by the verifier is actually:

$$z = x^2 + r \cdot n \quad \mod \Pi \tag{1}$$

Since the attacker can choose $x, r \in [0, \Pi]$ instead of $x \in [0, n]$ and $r \in [0, B]$, we may have $x^2 + r \cdot n > \Pi$, and therefore equation (1) does not necessarily imply that $z = x^2 + r \cdot n$ holds over the integers (or equivalently, that x is a square root of z modulo n). Therefore the zero-knowledge security proof does not apply anymore, which leads to the following attack:

Since Π is the product of small primes, it is easy to compute square roots modulo Π, as opposed to computing square roots modulo n. Therefore, the attacker can generate an integer z at step 1 so that he is guaranteed to succeed if $b = 1$. Then if $b = 0$, the attacker will also succeed by computing a square root modulo Π, which is easy.

More precisely, at step 1, the attacker generates a random $u \in \mathbb{Z}_n$ and a random $r' \in [0, B]$, and sends $z = (u^2/v \mod n) + r' \cdot n$ to the verifier. Then at step 3, if $b = 0$, the attacker generates a random $r \in [0, \Pi]$, and solves:

$$x^2 = z - r \cdot n \quad \mod \Pi$$

Since Π is the product of small primes, it suffices to take a square root of $z - r \cdot n$ modulo each of the small primes p_i. If $z - r \cdot n$ is not a square modulo a given prime p_j, it suffices to modify the value of $r \mod p_j$ without changing $r \mod p_i$ for $i \neq j$. This is possible since from the protocol, r is not required to belong to $[0, B]$. Eventually the attacker sends x and r to the verifier in RNS representation, and the attacker is successful with probability one.

Otherwise, if $b = 1$, then the attacker sends $y = u + t \cdot n$ for a random $t \in [0, B]$, and the verifier can check that $y^2 = z \cdot v \mod n$ since $u^2 = z \cdot v \mod n$.

Therefore, in both cases, the attacker passes the identification protocol with probability one, without knowing the private key..

6   Conclusion

We have shown that one of the time-efficient Fiat-Shamir variants introduced at Eurocrypt' 95 by Naccache et al. is insecure. Namely, a polynomial-time attacker can pass the identification protocol with probability one, without knowing the private key. Consequently, for practical implementations, we recommend to use [4]'s first time-efficient variant rather than [4]'s second time-efficient variant, which should be avoided. We believe that our attack illustrates the importance of careful security analysis of even apparently harmless variations of known secure protocols

References

1. A. Fiat and A. Shamir, *How to prove yourself: Practical solutions to identification and signature problems*, Proceedings of Crypto' 86, LNCS vol. 263, 1986.
2. S. Goldwasser, S. Micali and C. Rackoff, *The knowledge complexity of interactive proof-systems*, Proceedings of the 17th Annual ACM Symposium on Theory of Computing, 291–304, 1985.
3. D. Naccache, *Method, sender apparatus and receiver apparatus for modulo operation*, European patent application no. 91402958.2, November 5, 1991.
4. D. Naccache, D. MRaihi, W. Wolfowicz and A. di Porto, *Are Crypto-Accelrators really inevitable ? 20 bit zero-knowledge in less than a second on simple 8-bit microcontrollers*, Proceedings of Eurocrypt '95, Lecture Notes in Computer Science, Springer-Verlag.
5. A. Shamir, *Memory efficient variants of public-key schems for smart-card applications*, Proceedings of Eurocrypt '94, Lecture Notes in Computer Science, Springer-Verlag.

Universal Re-encryption for Mixnets

Philippe Golle[1], Markus Jakobsson[2], Ari Juels[2], and Paul Syverson[3]

[1] Stanford University
pgolle@cs.stanford.edu
[2] RSA Laboratories, Bedford, MA 01730, USA
{mjakobsson,ajuels}@rsasecurity.com
[3] Naval Research Laboratory
syverson@itd.nrl.navy.mil

Abstract. We introduce a new cryptographic technique that we call *universal re-encryption*. A conventional cryptosystem that permits re-encryption, such as ElGamal, does so only for a player with knowledge of the public key corresponding to a given ciphertext. In contrast, universal re-encryption can be done without knowledge of public keys. We propose an asymmetric cryptosystem with universal re-encryption that is half as efficient as standard ElGamal in terms of computation and storage.

While technically and conceptually simple, universal re-encryption leads to new types of functionality in mixnet architectures. Conventional mixnets are often called upon to enable players to communicate with one another through channels that are *externally anonymous*, i.e., that hide information permitting traffic-analysis. Universal re-encryption lets us construct a mixnet of this kind in which servers hold *no public or private keying material*, and may therefore dispense with the cumbersome requirements of key generation, key distribution, and private-key management. We describe two practical mixnet constructions, one involving asymmetric input ciphertexts, and another with hybrid-ciphertext inputs.

Keywords: anonymity, mix networks, private channels, universal re-encryption

1 Introduction

A *mix network* or *mixnet* is a cryptographic construction that invokes a set of servers to establish private communication channels [3]. One type of mix network accepts as input a collection of ciphertexts, and outputs the corresponding plaintexts in a randomly permuted order. The main privacy property desired of such a mixnet is that the permutation matching inputs to outputs should be known only to the mixnet, and no one else. In particular, an adversary should be unable to guess which input ciphertext corresponds to an output plaintext any more effectively than by guessing at random.

One common variety of mixnet known as a *re-encryption mixnet* relies on a public-key encryption scheme, such as ElGamal [7], that allows for re-encryption

T. Okamoto (Ed.): CT-RSA 2004, LNCS 2964, pp. 163–178, 2004.
© Springer-Verlag Berlin Heidelberg 2004

of ciphertexts. For a given public key, a ciphertext C' is said to represent a re-encryption of C if both ciphertexts decrypt to the same plaintext. In a re-encryption mixnet, the inputs are submitted encrypted under the public-key of the mixnet. (The corresponding private key is held in distributed form among the servers.) The batch of input ciphertexts is processed sequentially by each mix server. The first server takes the set of input ciphertexts, re-encrypts them, and outputs the re-encrypted ciphertexts in a random order. Each server in turn takes the set of ciphertexts output by the previous server, and re-encrypts and mixes them. The set of ciphertexts produced by the last server may be decrypted by a quorum of mix servers to yield plaintext outputs. Privacy in this mixnet construction derives from the fact that the ciphertext pair (C, C') is indistinguishable from a pair (C, R) for a random ciphertext R to any adversary without knowledge of the private key.

In this paper, we propose a new type of public-key cryptosystem that permits *universal re-encryption* of ciphertexts. We introduce the term universal encryption to mean re-encryption without knowledge of the public key under which a ciphertext was computed. Like standard re-encryption, universal re-encryption transforms a ciphertext C into a new ciphertext C' with same corresponding plaintext. The novelty in our proposal is that re-encryption neither *requires* nor *yields* knowledge of the public key under which a ciphertext was computed. (George Danezis independently discovered the same essential concept.)

When applied to mix networks, our universal re-encryption technique offers new and interesting functionality. Most importantly, mix networks based on universal re-encryption dispense with the cumbersome protocols that traditional mixnets require in order to establish and maintain a shared private key. We discuss more benefits and applications of universal mixnets in the next section. We construct a *universal mixnet* based on universal re-encryption roughly as follows. Every input to the mixnet is encrypted under the public key of the recipient for whom it is intended. Thus, unlike standard re-encryption mixnets, universal mixnets accept ciphertexts encrypted under the individual public keys of receivers, rather than encrypted under the unique public key of the mix network. These ciphertexts are universally re-encrypted and mixed by each server. The output of a universal mixnet is a set of ciphertexts. Recipients can retrieve from the set of output ciphertexts those addressed to them, and decrypt them.

Organization. The rest of the paper is organized as follows. In the next section, we give an overview of the main properties that distinguish universal mixnets from standard mixnets, and give one example of a new application made possible by universal mixnets. This is followed in section 3 by a formal definition of semantic security for universal re-encryption, as well as a proposal for creating a public-key cryptosystem with universal re-encryption based on ElGamal. In section 4, we describe our construction for an asymmetric universal mixnet. We define and prove the security properties of our system in section 5. In section 6, we propose a hybrid variant of our universal mixnet construction that combines public-key and symmetric encryption to handle long messages efficiently. We conclude in section 7.

2   Universal Mixnets: Properties and Applications

To motivate the constructions of this paper, we list here some of the main properties that set apart universal mixnets from traditional re-encryption mixnets. We also give one example of a new application made possible by universal mixnets: Anonymization of RFID tags.

Universal mixnets hold no keying material. A universal mixnet operates without a monolithic public key and thus dispenses at the server level with the complexities of key generation, key distribution, and key maintenance. This allows a universal mixnet to be set up more efficiently and with greater flexibility than a traditional re-encryption mixnet. A universal mixnet can be rapidly re-configured: Servers can enter and leave arbitrarily, even in the middle of a round of processing, without going through any setup. A mix server that crashes or otherwise disappears in the midst of the mixing process can thus be easily replaced by another server.

Universal mixnets guarantee forward anonymity. The absence of shared keys means that universal mixnets offer perfect forward-anonymity. Even if all mix servers become corrupted, the anonymity of previously mixed batches is preserved (provided that servers do not store the permutations or re-encryption factors they used to process their inputs). In contrast, if the keying material of a standard mix is revealed, an adversary with transcripts from previous mix sessions can compromise the privacy of users.

Universal mixnets do not support escrow capability. The flip-side of perfect forward-anonymity is that is that it is not possible to escrow the privacy offered by a universal mixnet in a straightforward fashion. Escrow is only achievable in a universal mix as long as every server involved in the mixing remembers how it permuted its inputs and is willing to reveal that permutation. This may be a drawback from the perspective of law enforcement. In comparison, escrow is possible in a traditional mix, provided that the shared key can be reconstructed. This requires the participation of only a quorum of servers, not all of them.

Efficiency. We present in this paper a public-key cryptosystem with universal re-encryption that is half as efficient as standard ElGamal: It requires exactly twice as much storage, and also twice as much computation for encryption, re-encryption, and decryption. In this regard, the universal mixnet constructions we propose in this paper are practical. The drawback of a universal mixnet, as we discuss in detail below, is that receivers must attempt to decrypt all output items in order to identify the messages intended for them.

2.1   Anonymizing RFID Tags

An interesting new application made possible by universal mixnets is the anonymization of radio-frequency identification (RFID) tags. An RFID tag is a small device that is used to locate and identify physical objects. RFID tags have very limited processing ability (insufficient to perform any re-encryption of data), but they allow devices to read and write to their memory [15,16]. Communication with RFID tags is performed by means of radio, and the tags themselves often obtain power by induction. Examples of uses of RFID tags include the theft-detection tags attached to consumer items in stores and the plaques mounted on car windshields for automated toll payment. Due to the projected decrease in the cost of RFID tags, their use is likely to extend in the near future to a wide range of general consumer items, including possibly even banknotes [20,12].

This raises concerns of an emerging privacy threat. Most RFID tags emit static identifiers. Thus, an adversary with control of a large base of readers for RFID tags may be able to track the movement of any object in which an RFID tag is embedded, and hence learn the whereabouts of the owner of that object. In order to prevent tracking of RFID tags, one could let some set of (honest-but-curious) servers perform re-encryption of the information that is publicly readable from RFID tags. The resulting system is similar to a mixnet, in which the permutation of ciphertexts is replaced by the movement of the RFID tags.

A traditional mix network, however, only partially solves the problem of tracking. The difficulty is that the data contained in different RFID tags may be encrypted under different public keys, depending on who possesses the authority to access that data. For example, while the data contained in tags used for automated toll payment may be encrypted under the public key of the transit agency, the data contained in tags attached to merchandise in a department store may be encrypted under the public key of that department store. To re-encrypt RFID tag data, a traditional mix network would need knowledge of the key under which that data was encrypted. The public key associated with an RFID tag could be made readable, but then the public key itself becomes an identifier permitting a certain degree of tracking. This is particularly the case if a user carries a *collection* of tags, and may therefore be identified by means of a constellation of public keys.

Universal mixnets offer a means of addressing the problem of RFID-tag privacy. If the data contained in RFID tags is encrypted with a cryptosystem that permits universal re-encryption, then this data can be re-encrypted without knowledge of the public-key. Thus universal re-encryption may offer heightened privacy in this setting by permitting agents to perform re-encryption without knowledge of public keys. While there have been previous designs using mixes for the purposes of privacy protection for low-power devices (e.g., [14]), universal re-encryption permits significant protocol and management simplification.

3 Universal Re-encryption

A conventional randomized public-key cryptosystem is a triple of algorithms, $CS = (\mathsf{KG}, \mathsf{E}, \mathsf{D})$, for key generation, encryption, and decryption. We assume, as is often the case for discrete-log-based cryptosystems, that system parameters and underlying algebraic structures for CS are published in advance by a trusted party. These are generated according to a common security parameter k. System parameters include or imply specifications of $\mathbf{M}$, $\mathbf{C}$, and $\mathbf{R}$, a message space, ciphertext space, and set of encryption factors respectively. In more detail:

- The key-generation algorithm $(PK, SK) \leftarrow \mathsf{KG}$ outputs a random key pair.
- The encryption algorithm $C \leftarrow \mathsf{E}(m, r, PK)$ is a deterministic algorithm that takes as input a message $m \in \mathbf{M}$, an encryption factor $r \in \mathbf{R}$ and a public key PK, and outputs a ciphertext $C \in \mathbf{C}$.
- The decryption algorithm $m \leftarrow \mathsf{D}(SK, C)$ takes as input a private key SK and ciphertext $C \in \mathbf{C}$ and outputs the corresponding plaintext.

A critical security property for providing privacy in a mix network is that of *semantic security*. Loosely speaking, this property stipulates the infeasibility of learning any information about a plaintext from a corresponding ciphertext [8]. For a more formal definition, we consider an adversary that is given a public key PK, where $(PK, SK) \leftarrow \mathsf{KG}$. This adversary chooses a pair (m_0, m_1) of plaintexts. Corresponding ciphertexts $(C_0, C_1) = (\mathsf{E}(m_0, r_0, PK), \mathsf{E}(m_1, r_1, PK))$ for $r_0, r_1 \in_U \mathbf{R}$ are computed, where $\in_U$ denotes uniform, random selection. For a random bit b, the adversary is given the pair (C_b, C_{1-b}), and tries to guess b. The cryptosystem CS is said to be semantically secure if the adversary can guess b with *advantage* at most negligible in k, i.e. with probability at most negligibly larger than $1/2$.

For a re-encryption mix network, an additional component known as a *re-encryption* algorithm, denoted by Re, is required in CS. This algorithm re-randomizes the encryption factor in a ciphertext. In a standard cryptosystem, this means that $C' \leftarrow \mathsf{Re}(C, r, PK)$ for $C, C' \in \mathbf{C}, r \in \mathbf{R}$, and a public key PK. Observe that re-encryption, in contrast to encryption, may be executed without knowledge of a plaintext. The notion of semantic security may be naturally extended to apply to the re-encryption operation by considering an adversary that chooses ciphertexts (C_0, C_1) under PK. The property of *semantic security under re-encryption*, then, means the following: Given respective re-encryptions (C'_b, C'_{1-b}) in a random order, the adversary cannot guess b with non-negligible advantage in k. Provided that Re yields the same distribution of ciphertexts as E (given $r \in_U \mathbf{R}$) or that the two distributions are indistinguishable, it may be seen that basic semantic security implies semantic security under re-encryption.

Bellare *et al.* [1] define another useful property possessed by the ElGamal cryptosystem. Known as "key-privacy," this property may be loosely stated as follows. Given a ciphertext encrypted under a public key randomly selected from a published pair (PK_0, PK_1), an adversary cannot determine which key corresponds to the ciphertext with non-negligible advantage. Key-privacy is one feature of the security property we develop in this paper for universal re-encryption.

As already explained, a universal cryptosystem permits re-encryption without knowledge of the public key corresponding to a given ciphertext. Let us denote such a cryptosystem by $UCS = (\mathsf{UKG}, \mathsf{UE}, \mathsf{URe}, \mathsf{UD})$, where $\mathsf{UKG}, \mathsf{UE}$, and UD are key generation, encryption, and decryption algorithms. These are defined as in a standard cryptosystem. The difference between a universal cryptosystem UCS and a standard cryptosystem resides in the re-encryption algorithm URe. The algorithm URe takes as input a ciphertext C and re-encryption factor r, but no public key PK. Thus, we have $C' \leftarrow \mathsf{URe}(C, r)$ for $C, C' \in \mathbf{C}$, $r \in \mathbf{R}$.

To define *universal semantic security under re-encryption*, i.e., with respect to URe, it is necessary to consider an adversarial experiment that is a variant on the standard one for semantic security. We define an experiment *uss* as follows for a (stateful) adversarial algorithm $\mathcal{A}$. This experiment terminates on issuing an output bit. As above, we assume an appropriate implicit parametrization of UCS under security parameter k. The idea behind the experiment is as follows. The adversary is permitted to construct universal ciphertexts under two randomly generated keys, PK_0 and PK_1. These ciphertexts are then re-encrypted. The aim of the adversary is to distinguish between the two re-encryptions. The adversary should be unable to do so with non-negligible advantage.

Experiment $\mathbf{Exp}_{\mathcal{A}}^{uss}(UCS, k)$

    $PK_0 \leftarrow \mathsf{UKG}; PK_1 \leftarrow \mathsf{UKG};$
    $(m_0, m_1, r_0, r_1) \leftarrow \mathcal{A}(PK_0, PK_1, \text{“specify ciphertexts”});$
    if $m_0, m_1 \notin \mathbf{M}$ or $r_0, r_1 \notin \mathbf{R}$ then
        output '0';
    $C_0 \leftarrow \mathsf{UE}(m_0, r_0, PK_0); C_1 \leftarrow \mathsf{UE}(m_1, r_1, PK_1);$
    $r_0', r_1' \in_U \mathbf{R};$
    $C_0' \leftarrow \mathsf{URe}(C_0, r_0'); C_1' \leftarrow \mathsf{URe}(C_1, r_1');$
    $b \in_U \{0, 1\};$
    $b' \leftarrow \mathcal{A}(C_b', C_{1-b}', \text{“guess”});$
    if $b = b'$ then output '1' else output '0'

We say that UCS is semantically secure under re-encryption if for any adversary $\mathcal{A}$ with resources polynomial in k, the probability $\mathsf{pr}[\mathbf{Exp}_{\mathcal{A}}^{uss}(UCS, k) = \text{'1'}] - 1/2$ is negligible in k.

The experiment *uss* captures the idea that the keys associated with ciphertexts are concealed by the re-encryption process in UCS. Thus, even an adversary who can compose the ciphertexts undergoing re-encryption cannot make use of differences in public keys to defeat the semantic security of the cryptosystem.

3.1   Universal Re-encryption Based on ElGamal

We present a public-key cryptosystem with universal re-encryption that may be based on the ElGamal cryptosystem implemented over any suitable algebraic group. The basic idea is simple: We append to a standard ElGamal ciphertext a second ciphertext on the identity element. By exploiting the algebraic homomorphism of ElGamal, we can use the second ciphertext to alter the encryption factor in the first ciphertext. As a result, we can dispense with knowledge of the public key in the re-encryption operation.

Let $E[m]$ loosely denote the ElGamal encryption of a plaintext m (under some key). In a universal cryptosystem, a ciphertexts on message m consists of a pair $[E[m]; E[1]]$. ElGamal possesses a homomorphic property, namely that $E[a] \times E[b] = E[ab]$ for group operator $\times$. Thanks to this property, the second component can be used to re-encrypt the first without knowledge of the associated public key. To provide more detail, let $\mathcal{G}$ denote the underlying group for the ElGamal cryptosystem; let q denote the order of $\mathcal{G}$. (Here the security parameter k is implicit in the choice of $\mathcal{G}$.) Let g be a published generator for $\mathcal{G}$. The universal cryptosystem is as follows. Note that we assume random selection of encryption and re-encryption factors in this description.

- **Key generation** (UKG): Output $(PK, SK) = (y = g^x, x)$ for $x \in_U \mathbb{Z}_q$.
- **Encryption** (UE): Input comprises a message m, a public key y, and a random encryption factor $r = (k_0, k_1) \in \mathbb{Z}_q^2$. The output is a ciphertext $C = [(\alpha_0, \beta_0); (\alpha_1, \beta_1)] = [(my^{k_0}, g^{k_0}); (y^{k_1}, g^{k_1})]$. We write $C = \mathsf{UE}_{PK}(m, r)$ or $C = \mathsf{UE}_{PK}(m)$ for brevity.
- **Decryption** (UD): Input is a ciphertext $C = [(\alpha_0, \beta_0); (\alpha_1, \beta_1)]$ under public key y. Verify $\alpha_0, \beta_0, \alpha_1, \beta_1 \in \mathcal{G}$; if not, the decryption fails, and a special symbol $\perp$ is output. Compute $m_0 = \alpha_0/\beta_0^x$ and $m_1 = \alpha_1/\beta_1^x$. If $m_1 = 1$, then the output is $m = m_0$. Otherwise, the decryption fails, and a special symbol $\perp$ is output. Note that this ensures a binding between ciphertexts and keys: a given ciphertext can be decrypted only under one given key.
- **Re-encryption** (URe): Input is a ciphertext $C = [(\alpha_0, \beta_0); (\alpha_1, \beta_1)]$ with a random re-encryption factor $r' = (k_0', k_1') \in \mathbb{Z}_q^2$. Output is a ciphertext $C' = [(\alpha_0', \beta_0'); (\alpha_1', \beta_1')] = [(\alpha_0\alpha_1^{k_0'}, \beta_0\beta_1^{k_0'}); (\alpha_1^{k_1'}, \beta_1^{k_1'})]$, where $k_0', k_1' \in_U \mathbb{Z}_q$.

Observe that the ciphertext size and the computational costs for all algorithms are exactly twice those of the basic ElGamal cryptosystem. The properties of standard semantic security and also universal semantic security under re-encryption (as characterized by experiment uss) may be shown straightforwardly to be reducible to the Decision Diffie-Hellman (DDH) assumption [2] over the group $\mathcal{G}$, in much the same way as the semantic security of ElGamal [19]. Thus, one possible choice of $\mathcal{G}$ is the subgroup of order q of $\mathbb{Z}_p^*$, where p and q are primes such that $q \mid p - 1$. Throughout the remainder of the paper, we work with the ElGamal implementation of universal re-encryption, and let g denote a published generator for the choice of underlying group $\mathcal{G}$.

4   Universal Mix Network Construction

We use the following scenario to introduce our universal mixnet construction. We consider a number of senders who wish to send messages to recipients in such a way that the communication is concealed from everyone but the sender and recipient themselves. In other words, we wish to establish channels between senders and receivers that are *externally anonymous*. We assume that every recipient has an ElGamal private/public key pair $(x, y = g^x)$ in some published group $\mathcal{G}$. We also assume that every sender knows the public key of all the receivers with whom she intends to communicate. (Alternatively, the sender may have a "blank" ciphertext for this party. By this we mean an encryption using UE of the identity element in $\mathcal{G}$ under the public key of the recipient. A "blank" may be filled in without knowledge of the corresponding public key through exploitation of the underlying algebraic homomorphism in ElGamal.) The communication protocol proceeds as follows:

1. **Submission of inputs.** Senders post to a bulletin board messages that are universally encrypted under the public key of the recipient for whom they are intended. Every entry on the bulletin board thus consists of a pair of ElGamal ciphertexts $(E[m]; E[1])$ under the public key of the recipient. Recall that the semantic security of ElGamal ensures the concealment of plaintexts. In other words, for plaintexts m and m', a universal ciphertext $(E[m]; E[1])$ is indistinguishable from another $(E[m']; E[1])$ to any entity without knowledge of the corresponding private key.

2. **Universal mixing.** Any server can be called upon to mix the contents of the bulletin board. This involves two operations: (1) The server re-encrypts all the universal ciphertexts on the bulletin board using URe, and (2) The server writes the resulting new ciphertexts back to the bulletin board in random order, overwriting the old ones. It is also desirable that a mix server be able to prove that it operated correctly. This can be done with a number of mixing schemes [6,9,11,13], and will be discussed in more detail below.

3. **Retrieval of the outputs.** Potential recipients must try to decrypt every encrypted message output by the universal mixnet. Successful decryptions correspond to messages that were intended for that recipient. The others (corresponding to decryption output '⊥') are discarded by the party attempting to perform the decryption. Recall that our construction of universal encryption based on ElGamal ensures a binding between ciphertexts and keys, so that a given ciphertext can be decrypted only under one given key.

Properties of the Basic Protocol:

1. The universal mixnet holds no keying information. Public and private keys are managed exclusively by the players providing input ciphertexts and receiving outputs from the mix.

2. The universal mixnet guarantees only external anonymity. It does not provide anonymity for senders with respect to receivers. Indeed a receiver can trace a message intended for her throughout the mixing process, since that message is encrypted under her public key. If ciphertexts are not posted anonymously, this means that the receiver can identify the players who have posted messages for her. This restriction to external anonymity is of little consequence for the applications we focus on, namely protection against traffic analysis, but should be borne in mind for other applications.

3. The chief drawback of universal mixnets is the overhead that they impose on receivers. Since the public keys corresponding to individual output ciphertexts are unknown, a receiver must attempt to decrypt each output ciphertext in order to find those encrypted under her private key. Thus the overhead for receivers is linear in the size of the input batch. (We discuss ways below and in section 6 to reduce this overhead somewhat.)

Low-volume anonymous messaging: anonymizing bulletin boards. For simplicity, we have described above the operation of a universal mixnet in which inputs are submitted, mixed and finally retrieved. This sequence of events is characteristic of all mixes. Unlike regular mixes however, universal mixes allow for repeated interleaving of the submission, mixing and retrieval steps. What makes this possible is that the decryption is performed by the recipients of the message rather than by the mixnet, so that existing messages posted to the bulletin board are at all times indistinguishable from new messages. New inputs may be constantly added to the existing content of the bulletin board, and outputs retrieved, provided there is at least one round of mixing between every submission and retrieval to ensure privacy.

This suggests a generalization of the private communication protocol described above, in which the bulletin board maintains at all times a pool of unclaimed messages. In other words, universal mixing lends itself naturally to the construction of an *anonymizing bulletin board*. Senders may add messages and receivers retrieve them at any time, provided there is always at least one round of mixing between each posting and retrieval. This protocol appears well suited to guarantee anonymity from external observers in a system in which few messages are exchanged. The privacy of the protocol relies on the existence of a steady pool of undelivered messages rather than on a constant flow of new messages. The former condition appears much easier to satisfy than the latter in cases when the total number of exchanged messages is small. This pooling of messages affords good anonymity protection, without the usual lack of verifiability of correct performance that vexes such schemes [4].

A potential drawback of a bulletin board based on universal mixing is that one must download the full contents in order to be assured of obtaining all of the messages addressed to oneself. This becomes problematic if the number of messages on the bulletin board is permitted to grow indefinitely. To mitigate this problem, it is possible to have recipients remove the messages they have received.[1] An anonymizing bulletin board based on universal mixing has the important privacy-protecting feature that removal of a particular message does not reveal which entity posted that message. Another important observation, as described in the next section, is that only a portion of each message on a

---

[1] To ensure that messages are only removed by the intended recipient, a proof of knowledge of the corresponding decryption key is required. Note that such a proof can be performed without disclosing the public key associated with the required decryption key. For ciphertext $C = [(\alpha_0, \beta_0); (\alpha_1, \beta_1)]$, this may take the form of a non-interactive zero-knowledge proof of knowledge of an exponent x such that $\alpha_1 = \beta_1^x$ – essentially a Schnorr signature [17].

bulletin board need be downloaded to allow a recipient to determine which messages are intended for her. This further restricts the work required by a receiver.

RFID-tag privacy. Universal re-encryption may be used to enhance the privacy of RFID tags. The idea is to permit powerful computing agents external to RFID tags to universally re-encrypt the tag data (recall that the tags lack the computing power necessary to do the re-encryption themselves). Thus, for example, a consumer walking home with a bag of groceries containing RFID tags might have the ciphertexts on these tags re-encrypted by computing agents provided as a public service by shops and banks along the way. In this case, the tags in the bag of groceries will periodically change appearance, helping to defeat any tracking attempt.

Application of universal mixnets to RFID-tag privacy is different in some important respects from realization of an anonymous bulletin board. As re-encryption naturally occurs for RFID tags on an individual basis, re-encryption in this setting may be regarded as realizing an *asynchronous* mixnet. There is also a special security consideration in this setting. Suppose that the ciphertext on an RFID tag is of the form $(\alpha, \beta); (1, 1)$ (where '1' represents the identity element for $\mathcal{G}$). Then the ciphertext on the tag will not change upon re-encryption. Thus, it is important to prevent an active adversary from inserting such a ciphertext onto an RFID tag so as to be able to trace it and undermine the privacy of the possessor. In particular, on processing ciphertexts, re-encryption agents should check that they do not possess this degenerate form. Of course, an adversary in this environment can always corrupt ciphertexts. Note, however, that even a corrupted ciphertext $(\alpha', \beta'); (\gamma, \delta)$ will be rendered unrecognizable to an adversary provided that $\gamma, \delta \neq 1$.

5   Security

In this section, we define two security properties of universal mixnets: *correctness* and *communication privacy*. The mixnet is correct if the set of outputs it produces is a permutation of the set of inputs. The mixnet guarantees communication privacy if, when Alice sends a message to Bob and Cathy sends a message to Dario, an observer can not tell whether Alice (resp. Cathy) sent a message to Bob or Dario.

Correctness. Correctness for universal mixnets follows directly from the definition of correctness for standard mixnets. Like standard mix servers, universal servers must prove that they have performed the mixing operation correctly. For this, we can draw on essentially any of the proof techniques presented in the literature on mixnets, as nearly all apply to ElGamal ciphertexts. For example, to achieve universal verifiability, we can use the proof techniques in [6,13,11]. A small technical consideration, which may be dealt with straightforwardly, is the form of input ciphertexts. Input ciphertexts in most mix network constructions consist of a single ElGamal ciphertext, while in our construction, an input consists of a universal ciphertext, and thus two related ElGamal ciphertexts.

Communication privacy. We define next the property of communication privacy. In order to state this definition formally, we abstract away some of the operations of the mixnet by defining them in terms of oracle operations. We do this so as to focus our exposition on our universal construction, rather than underlying primitives, particularly as our construction can make use of a broad range of choices of such primitives. We define three oracles:

• **An oracle MIX** which universally re-encrypts all ciphertexts on the bulletin board BB and outputs to BB the new set of ciphertexts in a randomly permuted order. In practice, we can substitute any mixnet with public verifiability for MIX.

• **An oracle POST** that permits message posting. POST requires a poster to submit a message, encryption factors and ciphertext. It verifies that the message, encryption factors and ciphertext are elements of the appropriate groups and permits posting if the ciphertext is a valid encryption of the message with the given encryption factors. Note that the oracle POST may be regarded as simulating a proof of knowledge of the plaintext and the encryption factor and a verification thereof. In practice, it could be instantiated with standard discrete-log-based proofs of knowledge, e.g., [5], in either their interactive or non-interactive forms.

• **An oracle RETRIEVE** that permits message retrieval. The oracle takes a private key and ciphertext from a user. The oracle verifies that the private key and ciphertext are elements of the appropriate groups. The user is allowed to remove the ciphertext if it is encrypted under the private key. Recall that our construction of universal encryption based on ElGamal ensures a binding between ciphertexts and keys, so that a given ciphertext can be decrypted only under one given key. The oracle RETRIEVE, like POST, abstracts away a proof of knowledge of the plaintext.

We define communication privacy in terms of an experiment $\mathbf{Exp}^{comm-priv}$ defined as follows. The adversary may make an arbitrary number of calls to any of the oracles RETRIEVE, MIX, or POST and may order these calls as desired. We enumerate the first several steps here for reference in our proof.

Experiment $\mathbf{Exp}_{\mathcal{A}}^{comm-priv}(UCS, k)$

    1. $PK_0 \leftarrow \mathsf{UKG}; PK_1 \leftarrow \mathsf{UKG};$
    2. $(m_0, m_1) \leftarrow \mathcal{A}(PK_0, PK_1, \text{"specify plaintexts"});$
    3. $b \in_U \{0, 1\};$
    4. $C_0' = \mathsf{UE}_{PK_b}(m_b)$ and $C_1' = \mathsf{UE}_{PK_{1-b}}(m_{1-b})$ appended to $BB;$
    5. MIX invoked;
    6. $\mathcal{A}(BB);$
    7. $L \leftarrow \{C \in BB \text{ s.t. } C \text{ is a valid ciphertext under } PK_0\};$
    8. $b' \leftarrow \mathcal{A}(L, \text{"guess } b\text{"});$
    if $b = b'$ then output '1' else output '0'

An intuitive description of this experiment is as follows. Alice and Bob wish each to transmit a single message to one of Cathy and Dario, who possess public keys PK_0 and PK_1 respectively. Our aim is to ensure that the adversary cannot tell whether Alice is sending a message to Cathy or Dario – and likewise to whom Bob is transmitting. The adversary is given the special (strong) power of determining which plaintexts, m_0 and m_1, are to be received by Cathy and Dario. The adversary observes Alice posting ciphertext C_0' and Bob posting ciphertext C_1', but does not know which ciphertext is for Cathy and which is for Dario. The bulletin board is then subjected to a mixing operation so as to conceal the communication pattern. The adversary may subsequently control when and how the mix network is invoked, and may place its own ciphertexts on the bulletin board. Finally, at the end of the experiment, the adversary is given a list L of all ciphertexts encrypted under PK_0, i.e., all the messages that Cathy retrieves. This list L will include the one such message posted by Alice or Bob in addition to all messages encrypted under PK_0 and posted by the adversary. The task of the adversary is to guess whether it was Alice who sent a message to Cathy (case $b = 0$) or Bob (case $b = 1$).

Definition 1. (Communication privacy) *We say that a universal mixnet for UCS possesses* communication privacy *if for any adversary $\mathcal{A}$ that is polynomial time in k, we have $\mathrm{pr}[\mathbf{Exp}_{\mathcal{A}}^{comm-priv}(UCS, k) = 1] - 1/2$ is negligible in k.*

Theorem 1. *Our universal mixnet possesses communication privacy provided that UCS has universal semantic security under re-encryption. For our construction involving ElGamal, privacy may consequently be reduced to the DDH assumption over $\mathcal{G}$.*

Proof. Assume we have an adversary $\mathcal{A}$ for which $\mathrm{pr}[\mathbf{Exp}_{\mathcal{A}}^{comm-priv}(UCS, k) = 1] - 1/2$ is non-negligible in k. We build a new adversary $\mathcal{A}'$ which uses $\mathcal{A}$ as a subroutine and for which $\mathrm{pr}[\mathbf{Exp}_{\mathcal{A}'}^{uss}(UCS, k) = \text{`1'}] - 1/2$ is non-negligible in k (i.e. $\mathcal{A}'$ breaks the universal semantic security of the underlying encryption scheme). $\mathcal{A}'$ operates as follows:

- At the beginning of $\mathbf{Exp}^{uss}$, $\mathcal{A}'$ is given two public keys PK_0 and PK_1. $\mathcal{A}'$ gives these two keys to $\mathcal{A}$. This simulates step 1 of $\mathbf{Exp}^{comm-priv}$.
- When $\mathcal{A}$ calls one of the oracles POST, MIX or RETRIEVE, $\mathcal{A}'$ can trivially simulate the oracle for the requested operation for $\mathcal{A}$.
- In step 2 of experiment $\mathbf{Exp}^{comm-priv}$, $\mathcal{A}$ specifies plaintexts m_0 and m_1. $\mathcal{A}'$ selects random encryption factors r_0 and r_1 and computes $C_0 = \mathsf{UE}_{PK_0}(m_0, r_0)$ and $C_1 = \mathsf{UE}_{PK_1}(m_1, r_1)$. $\mathcal{A}'$ submits these in the second step of $\mathbf{Exp}^{uss}$. $\mathcal{A}'$ then receives as input from $\mathbf{Exp}^{uss}$ two new ciphertexts C_0' and C_1'.
- In step 4 of $\mathbf{Exp}^{comm-priv}$, $\mathcal{A}'$ posts C_0' and C_1' to the bulletin board.
- In step 7 of $\mathbf{Exp}^{comm-priv}$, $\mathcal{A}'$ must identify the set of outputs encrypted under PK_0. Note that $\mathcal{A}'$ can easily identify the outputs that correspond to inputs originally submitted by $\mathcal{A}$ encrypted under PK_0, since it controls the

oracle POST and MIX. The difficulty is for $\mathcal{A}'$ to decide which of C_0' and C_1' is encrypted under PK_0 and which under PK_1. Since $\mathcal{A}'$ doesn't know that, it arbitrarily assigns C_0' to the list L of ciphertexts encrypted under PK_0.

In the last step of the simulation, $\mathcal{A}'$ assigns C_0' arbitrarily to L. We claim that if $\mathcal{A}$ can distinguish between the case where this assignment to L is correct and the case where it is incorrect, then $\mathcal{A}$ can be used to break universal semantic security in $\mathbf{Exp}^{uss}$. This may be achieved with a small modification of our simulation as follows: (1) $\mathcal{A}'$ lets $C_0' = C_0$ and $C_1' = C_1$, but invokes $\mathbf{Exp}^{uss}$ on the pair (C_0', C_1') during the mixing operation in step 5 and (2) $\mathcal{A}'$ submits to Exp^{uss} the bit b' yielded by $\mathcal{A}$ at the end of the experiment. Let us assume, therefore, that the assignment to L is correct. Given this, when $\mathcal{A}$ outputs its guess b', $\mathcal{A}'$ then outputs the same bit b' as its guess for the experiment Exp^{uss}. It is clear now that when $\mathcal{A}$ guesses correctly, so does $\mathcal{A}'$.   □

Security of UCS and chosen-ciphertext attacks. The cryptosystem UCS inherits the semantic security property of the underlying ElGamal cipher under the DDH assumption. This property is critical to our definition of communications privacy. Our model for communication privacy makes one simplifying assumption though: We assume that the adversary does not learn any information about plaintexts. For this reason, we do not require adaptive-chosen ciphertext (CCA) security of our cryptosystem. In fact, our system cannot achieve strict CCA security: In order to permit re-encryption, ciphertexts must be malleable. Note, however, that an adversary cannot repost a message or post a new message with a related plaintext since POST requires a proof of knowledge of the plaintext and encryption factors.

On the other hand, there may be circumstances in which an adversary learns information about plaintexts in our system. To show this formally, it would be necessary to modify our universal cryptosystem so as to achieve CCA security with *benign malleability*, as defined by Shoup [18]. In Shoup's terminology, we would need to require an induced *compatible relation* of plaintext equivalence by formatting plaintexts with appropriate padding. We omit detailed discussion of this topic. An adversary that can gain significant information about received messages can, after all, break the basic privacy guarantees of the system.

6   Hybrid Universal Mixing

We describe next a variant mixnet called a *hybrid universal mixnet*. This type of mixnet combines symmetric and public-key encryption to accommodate potentially very long messages (all of the same size) in an efficient manner. We refer the interested reader to [10] for definitions and examples of hybrid mixnets.

Our definition of a universal hybrid mix considers a weaker threat model than above with respect to correctness. Since hybrid mixes use symmetric encryption, we cannot verify that they execute the protocol correctly. Thus, we restrict our security model to mix servers subject only to passive adversarial corruption. Such servers are also known as *honest-but-curious*. They follow the protocol correctly but try to learn as much information as possible from its execution.

For efficiency, inputs m are submitted to a hybrid mix encrypted under an initial symmetric (rather than public) key. We denote by $\epsilon_k[m]$ the symmetric-key encryption of m under key k. Each mix server S_i re-encrypts the output of the previous mix under a new random symmetric key k_i. With k servers, the final output of the mix is $\epsilon_{k_n}[\epsilon_{k_{n-1}}[\ldots \epsilon_{k_1}[\epsilon_k[m]]\ldots]]$. The symmetric keys $k, k_1, \ldots, k_n$ must be conveyed alongside the encrypted message to enable decryption by the final recipient. These keys are themselves universally encrypted under the public key of the recipient. Universal encryption provides an efficient way of transmitting the symmetric keys without compromising privacy.

We define next our hybrid universal mixnet. Our construction imposes an upper bound n on the maximum number of times that the mixing operation is performed on any given ciphertext. The protocol consists of the following steps:

1. **Submission of inputs.** An input ciphertext takes the form

$$\epsilon_{k_0}[m], E[1], (E[k_0], E[1] \ldots E[1])$$

   where $\epsilon_{k_0}[m]$ denotes symmetric encryption of m under key k_0. This is followed by an encryption of 1, and by a vector of ciphertexts on keys, where only the first element is filled in (with k_0), leaving the remaining $n - 1$ elements as encryptions of 1.

2. **Universal mixing.** The ith server to perform the mixing operation does the following for each of the ciphertexts on the bulletin board:
   - Generates a random symmetric key k_i;
   - Adds a new layer of symmetric encryption to m under key k_i;
   - Uses the second element, $E[1]$, to compute an encryption $E[k_i]$ of k_i;
   - Rotates the elements of the vector one step leftwards, then substituting the first element with $E[k_i]$; and
   - Re-encrypts the second element and each element of the vector.

   When it has thus processed all its inputs in this manner, the server outputs them back to the bulletin board in a random order.

3. **Retrieval of the outputs.** At the end of $d \le n$ mixing operations, the final output of the mixnet assumes the form:

$$\epsilon_{k_d}[\epsilon_{k_{d-1}}[\ldots \epsilon_{k_0}[m]]\ldots], E[1], (\{E[1]\}^{n-d}, E[k_0] \ldots E[k_d]),$$

   where $\{E[1]\}^{n-d}$ denotes $n - d$ ElGamal ciphertexts on the identity element. As before, recipients try to decrypt every output of the mixnet and discard outputs for which the decryption fails. Note that a party need only decrypt the second element, $E[1]$, to determine whether a ciphertext is for her.

Remark: In principle, it is possible to use the "blank" ciphertext $E[1]$ to append ciphertexts on as many symmetric keys as desired, and thus re-encrypt indefinitely. The reason for restricting the number of "blank" ciphertexts to exactly n is to preserve a uniform length, without which an adversary can distinguish among ciphertexts that have undergone differing numbers of re-encryptions. A drawback of this approach is that a ciphertext re-encrypted more than n times will become undecipherable by the receiver.

7   Conclusion

Universal re-encryption represents a simple modification to the basic ElGamal cryptosystem that permits re-randomization of ciphertexts without knowledge of the corresponding private key. This provides a valuable tool for the construction of privacy-preserving architectures that dispense with the complications and risks of distributed key setup and management. The costs for the basic universal cryptosystem are only twice those of ordinary ElGamal. On the other hand, the problem of receiver costs in a universal mixnet presents a compelling line of further research. In our construction, a receiver must perform a linear number of decryptions to identify messages intended for her. A method for reducing this cost would be appealing from both a technical and practical standpoint.

References

1. M. Bellare, A. Boldreva, A. Desai and D. Pointcheval. Key-privacy in public-key encryption. In *Proc. of ASIACRYPT '01*, pp. 566–582. LNCS 2248.
2. D. Boneh. The Decision Diffie Hellman problem. In *Proc. of ANTS '98*, pp. 48–63. LNCS 1423.
3. D. Chaum. Untraceable electronic mail, return addresses, and digital pseudonyms. In *Communications of the ACM*, 24(2):84–88, 1981.
4. L. Cottrell. Mixmaster & remailer attacks, 1995. Available on the web at http://www.obscura.com/~loki/remailer/remailer-essay.html
5. A. de Santis, G. di Crescenzo, G. Persiano, and M. Yung. On monotone formula closure of SZK. In *Proc. of FOCS '94*, pp. 454–465. IEEE Press, 1994.
6. J. Furukawa and K. Sako. An efficient scheme for proving a shuffle. In *Proc. of CRYPTO '01*, pp. 368–387. LNCS 2139.
7. T. ElGamal. A public key cryptosystem and a signature scheme based on discrete logarithms. In *IEEE Transactions on Information Theory*, 31:469–472, 1985.
8. S. Goldwasser and S. Micali. Probabilistic encryption. In *J. Comp. Sys. Sci*, 28(1):270–299, 1984.
9. M. Jakobsson and A. Juels. Millimix: Mixing in small batches, June 1999. DIMACS Technical Report 99-33.
10. M. Jakobsson and A. Juels. An optimally robust hybrid mix network. In *Proc. of PODC '01*, pp. 284–292. ACM Press, 2001.
11. M. Jakobsson, A. Juels, and R. Rivest. Making mix nets robust for electronic voting by randomized partial checking. In *Proc. of USENIX Security'02*, pp. 339–353.
12. A. Juels and R. Pappu. Squealing euros: Privacy protection in RFID-enabled banknotes. In *Proc. of Financial Cryptography 2003*, pp. 103–121. LNCS 2742.
13. A. Neff. A verifiable secret shuffle and its application to e-voting. In *Proc. of ACM CCS'01*, pp. 116–125. ACM Press, 2001.
14. M. Reed, P. Syverson, and D. Goldschlag. Protocols using anonymous connections: mobile applications. In *Security Protocols '97*, pp. 13–23. LNCS 1361.
15. S. Sarma. Towards the five-cent tag. Technical Report MIT-AUTOID-WH-006, MIT Auto ID Center, 2001. Available from http://www.autoidcenter.org/
16. S. Sarma, S Weis, and D Engels. RFID Systems and Security and Privacy Implications. In *Proc. of CHES '02*, pp. 454–469.
17. C.-P. Schnorr. Efficient signature generation by smart cards. In *Journal of Cryptology*, 4(3):161–174, 1991.

18. V. Shoup. A proposal for an iso standard for public key encryption (version 2.1), 20 December 2001. Manuscript.
19. Y. Tsiounis and M. Yung. On the security of ElGamal-based encryption. In *Proc. of PKC '98*, pp. 117–134. LNCS 1431.
20. J. Yoshida. Euro bank notes to embed RFID chips by 2005. *EE Times*, 19 December 2001. Available at `http://www.eetimes.com/story/OEG20011219S0016`

Bit String Commitment Reductions with a Non-zero Rate

Anderson C.A. Nascimento[1], Joern Mueller-Quade[2], and Hideki Imai[1]

[1] Institute of Industrial Science, The University of Tokyo
4-6-1, Komaba, Meguro-ku, Tokyo, 153-8505 Japan
{anderson,imai}@imailab.iis.u-tokyo.ac.jp
[2] Universitaet Karlsruhe,
Institut fuer Algorithmen und Kognitive Systeme
Am Fasanengarten 5, 76128 Karlsruhe, Germany
muellerq@ira.uka.de

Abstract. We analyze a new class of primitives called weak commitments. We completely characterize when bit commitments can be reduced to these primitives. Also, we employ a new concept in cryptographic reductions, the rate of a reduction. We propose protocols achieving a nontrivial rate. We provide examples of how to implement these primitives based on oblivious transfer and on quantum mechanics. Using the theory here developed, some open problems on computationally secure quantum bit commitments are solved. Our reductions are information theoretically secure.

1 Introduction

Whenever a cryptographic primitive is implemented by a physical process there is a certain probability of failure or even the possibility that a cheater has (limited) control over system parameters of the physical process. In this paper we will introduce a new primitive, a *weak bit commitment,* where the sender (Alice) and the receiver (Bob) can cheat with a certain probability thereby reflecting the above problem. This paper will give tight bounds on the cheating probabilities for which bit commitment can be reduced to weak bit commitments in an information theoretically secure way.

But we will not only look at the possibility of reducing bit commitments to weak bit commitments, we will furthermore look at the efficiency of reductions. To do this we consider bit *string* commitments and use as a measure of efficiency the new concept of the *rate of an information theoretically secure reduction,* i. e., the length of the bit string divided by the number of weak bit commitments employed where the length of the string is going to infinity [16]. This will allow us to import methods from Shannon theory to achieve our reductions. Our bit string commitment reductions will exhibit nonzero rates.

Definition 1. *(Informal) A bit (string) commitment is a protocol consisting of two phases a commit phase and an unveil phase (which need not be entered).*

T. Okamoto (Ed.): CT-RSA 2004, LNCS 2964, pp. 179–193, 2004.

During the commit phase Alice is supposed to fix a bit (string) b which is not to be changed and will be opened to Bob in the unveil phase. We say that a protocol securely realizes bit (string) commitment if the protocol is concealing, binding, *and* sound. **Concealing:** *A bit (string) commitment scheme is* concealing *if Bob's information about the committed bit (string) before the unveil phase is negligible in the security parameter*[1]. **Binding:** *A bit (string) commitment is* binding *if the probability that Alice is able to after the commit phase unveil more than one bit (string) is negligible in the security parameter.* **Sound:** *A bit commitment (string) is sound if its probability of failure for honest Alice and honest Bob is negligible in the security parameter.*

It is important to remark that the above definition is valid only for classical bit (string) commitments. Secure quantum bit commitments are defined in a different way (see Section 6).

In a weak bit commitment Alice will be able to violate the binding property with a probability bounded by α and Bob will learn the committed bit already in the commit phase with a probability bounded by β. Two types of weak commitments will be distinguished depending on whether Alice knows in advance if her attempt to cheat will remain undetected. In a weak bit commitment of type I (see Definition 2) Alice knows if she can cheat without being detected cheating and this happens with a probability no greater than α. Bob will learn the committed bit during the commit phase with a probability no greater than β. The notion of a weak bit commitment of type II is cryptographically stronger (see Definition 3). Alice will be able to cheat without being detected with a probability no greater than α but she will not know if this is the case and she will have a risk of being detected cheating if she tries. Bob again learns the committed bit beforehand with a probability no greater than β.

Note that the actual probabilities of cheating are not specified, but bounded by α and β respectively. This is very important as it reflects an unfair primitive which might work a lot better if no-one tries to tweak system parameters. Imagine a weak bit commitment where the probability that Bob learns the committed bit beforehand is guaranteed to be exactly β, then one could perform n commitments to random bits and ask Bob to announce approximately βn bits he knows. These bits could be checked by Alice. Thereby she will be convinced that Bob knows a lot less about the remaining commitments which could then be used to implement a stronger commitment. The protocol sketched above would not work in our case as Bob can always claim to not have learned a single bit.

Later in the paper we give two examples for reductions to weak bit commitments. First, we will show how to obtain bit string commitment with a nonzero rate from *Rabin oblivious transfer* (in [11], Kilian proved that OT implies BC, but his reduction had a rate asymptotically equal to zero) and from *1 out of 2 Oblivious Transfer*. We show that 1 out of 2- OT is strictly stronger (in the sense

---

[1] As we will be interested in the asymptotic rate of reductions we will look at the limit of the rate for the length of the string going to infinity. In this situation we can identify the security parameter and the length of the string to be committed.

that better rates can be achieved) than Rabin OT when it comes to implementing string commitments. Second, we will improve a quantum bit commitment scheme based on computational assumptions by making it robust against multiple photons which could be emitted in one pulse. This was stated as an open problem in [10].

An interesting innovation in this work is the use of well known tools from Shannon theory like typical sequences, entropic inequalities and random coding arguments. Actually, we prove that when appropriately written, some reductions among cryptographic primitives are equivalent to classical problems in Shannon theory. This is a new research direction which opens interesting questions both in Shannon theory and in cryptography.

1.1   Related Work

Several researchers have been trying to base security of cryptographic protocols in somehow "weaker" primitives. However, previous works concentrated mostly on proving the existence of certain reductions, thus they, usually, did not pay attention to the question of rate. Crepeau and Kilian [3], proved that a noise channel can be used to achieve oblivious transfer and bit commitment. Crepeau [7] improved the results of [3] by using privacy amplification techniques. However, no impossibility results were proven in these papers and also there was no concern about achieving rates different than zero.

Cachin reduced OT to a weaker primitive called universal oblivious transfer in [2], but here again, the aspect of rate was not considered.

Our weak commitment is a particular case of a weak generic transfer (WGT), a primitive introduced by Damgard, Salvail and Kilian in [8]. As another particular case of a WGT they defined a weak oblivious transfer. A weak oblivious transfer is an oblivious transfer protocol where Alice and Bob can successfully cheat with non-negligible probability. They characterized when oblivious transfer can be reduced to its weaker version. However, in their reductions the achieved rates were asymptotically equal to zero. Also, their approach was different from the approach taken here which formulates reductions as Shannon theoretical problems.

In [9] and [15] string commitments achieving non trivial rates were introduced. However, these reductions were not information theoretically secure for both the sender and the receiver of a commitment.

The work which most overlaps with ours is [21], where the commitment capacity of a discrete memoryless channel (and of a class of quantum channels) was determined. Although the techniques we use are similar to the ones used in [21], there the direct part was proved by using a random code (which cannot be even efficiently stored), but here we proved that our rates are achievable by a *random linear code* (which can be efficiently stored) thus, as there is no need of efficient decodable codes in our algorithms, our solutions are practical. Also, we provide a general theory of weak commitments, whereas in [21] only the particular case of noisy channels was analyzed.

2   Preliminaries and Statement of Results

A weak bit commitment is a protocol where Alice and Bob can successfully cheat with non-negligible probability. Here, we split weak commitments in two categories according to the way they are implemented. Weak commitments of type I are the ones where the sender, with probability at most α, **knows** that she can cheat without being detected and a dishonest receiver is able to break the concealing condition of the commitment with probability of at most β.

In weak commitments of type II, the sender does not know *a priori* if she can successfully cheat or not, but if she tries to cheat, she succeeds with probability at most α. As in weak commitments of type I, the receiver can break the secrecy of the commitment before the unveiling phase with probability at most β.

We show in the following sections that the fact that the sender knows in advance if she can cheat without being detected or not makes difference in the security analysis and in the results that can be achieved.

To construct an example of a weak commitment of type I, we use a protocol proposed by Rivest in [17]. Rivest proved that if a trusted authority makes some pre-distributed data available to Alice and Bob (he called it the trusted initializer model) unconditionally secure bit commitment can be implemented. In our version, the TI, after distributing secret data to Alice and Bob, with probability α, tells Alice the secret data which was given to Bob and ensures her that she will be able to successfully cheat. With probability β, TI tells Bob the secret data which was given to Alice. In this scenario, if Alice receives Bob's secret data from the TI she knows for sure that she is able to cheat without being detected. As examples of a weak commitments of type II we cite the commitment schemes based on noisy channels such as the ones proposed in [7] and in [8]. More examples are given in the Sections 5 and 6. We formalize our definitions below.

Definition 2. *An $(\alpha, \beta) - bc(b)$ of type I, $(\alpha, \beta) - bc_I(b)$ for short, is a bit commitment protocol where additional information is provided to the participants. It consists of two phases: $\beta - Commit_I(b)$ and $\alpha - Open_I(b)$. In an $\alpha - Open_I(b)$ algorithm the committer receives extra information (which will make possible for the sender to cheat without being detected) from an oracle with probability α. In a $\beta - Commit_I(b)$ algorithm Bob, the receiver, learns the value of a committed bit with probability at most β.*

Definition 3. *An $(\alpha, \beta) - bc(b)$ of type II, $(\alpha, \beta) - bc_{II}(b)$ for short, consists of two phases: $\beta - Commit_{II}(b)$ and $\alpha - Open_{II}(b)$. In an $\alpha - Open_{II}(b)$ algorithm the committer is able to commit to something and later on unveil $b = 0$ and $b = 1$ with success probability at most α. In a $\beta - Commit_{II}(b)$ algorithm, Bob, the receiver, learns the value of a committed bit with probability at most β.*

It is clear that a bit commitment can be reduced to a wide class of weak commitments. In the following we denote sequential commitments to $b_i, 1 \leq i \leq n$. using an $(\alpha, \beta) - bc(b)$ protocol by the following notation $(\alpha, \beta) - bc(b_1 b_2 ... b_n)$.

When the same proof applies to weak commitments of type I and II, in the following we denote a generic weak commitment by $(\alpha, \beta) - bc(b)$.

Proposition 1. *A bit commitment protocol $bc(b)$ can be reduced to $(\alpha, 0) - bc_I(b)$ and $(\alpha, 0) - bc_{II}(b)$ protocols where $\alpha < 1$.*

Proof. As the same proof applies to weak commitments of type I and II, in the following we denote a generic weak commitment by $(\alpha, \beta) - bc(b)$. When the bit to be committed is b, the committer repeatedly commits to the same value b, n times, using a $(\alpha, 0) - bc(b)$ protocol, the receiver only accepts a commitment if all the committed bit are the same. Even if Alice does not stick to the protocol and commits to different bits she has in at least $n/2$ instances of the weak bit commitment protocol committed to a bit b. To unveil $\bar{b}$ she has to change no less than $n/2$ commitments. Alice's probability of successfully cheating (P_e) is no greater than $\alpha^{n/2}$ and $\lim_{n\to\infty} P_e = \lim_{n\to\infty} \alpha^{n/2} = 0$.

Proposition 2. *A bit commitment protocol $bc(b)$ can be reduced to a $(0, \beta) - bc_I(b)$ and $(0, \beta) - bc_{II}(b)$ protocols, $\beta < 1$*

Proof. Again, we denote a generic weak commitment by $(\alpha, \beta) - bc(b)$. When the committed bit is b, Alice chooses a random $n-$bit word $b_1 b_2 ... b_n$ such that $b = b_1 \oplus b_2 \oplus ... \oplus b_n$ where $\oplus$ stands for the XOR operation. Alice sequentially commits to $b_i, 1 \le i \le n$. using an $(0, \beta) - bc(b)$ protocol. We represent it by the following notation $(0, \beta) - bc(b_1 b_2 ... b_n)$ where the committed bit is equal to the XOR of $b_1 b_2 ... b_n$. If S represents the random variable associated with $b = b_1 \oplus b_2 \oplus ... \oplus b_n$ and Z is the random variable associated with the commitments observed (broken) by Bob we have that: $\lim_{n\to\infty} H(S|Z) = 1$, i.e., the scheme is concealing.

It is natural to ask when a bit commitment cannot be reduced to its weaker version. The next proposition shows us a class of weak commitments that cannot be used to achieve a bit commitment.

Proposition 3. *It is impossible to reduce a bit commitment protocol to an $(\alpha, \beta) - bc_I(b)$ when $\alpha + \beta \ge 1$.*

Proof. If $\alpha + \beta = 1$ and we assume that the events *Alice can cheat* and *Bob can cheat* are not independent, we can in the following assume a weak bit commitment where either Alice can cheat or Bob can cheat. Bit commitment cannot be implemented with such a weak primitive: All bits learned by Bob during the execution of a bit commitment based on such a weak primitive should not give away the bit Alice commits to. Hence the bit Alice is committed to is not fixed by the information Bob received, but all remaining bits can be changed by Alice. Thus no bit commitment built with a weak bit commitment with $\alpha + \beta \ge 1$ can be binding and concealing.

Propositions 1 and 2 tell us that, in principle, a "strong" bit commitment protocol can be reduced to its "weaker" versions. However, as the cheating probability goes to zero, the rate also goes to zero, since n weak commitments are used to achieve one single bit commitment. The main contribution of this paper is to show some reductions where the cheating probability goes to zero, but the rate converges to a number bounded away from zero. The key point is to look at string commitment protocols instead of single bit commitments. We state the main theorem of this paper which shows that the bound proved in the last proposition is tight. We present the proof of this theorem in the following sections. For a formal definition of rate we point at the next section.

Theorem 1. *A bit commitment protocol can be reduced to any* $(\alpha, \beta) - bc_I(b)$ *with* $\alpha + \beta < 1$ *and any* $(\alpha, \beta) - bc_{II}(b)$ *with* $\alpha < 1, \beta < 1$. *Moreover, a rate* $R = 1 - \alpha - \beta$ *is achievable in the case of commitments of type I and* $R = 1 - \beta$ *is achievable in the case of commitments of type II.*

3   String Commitment Reductions Achieving Non-trivial Rates

In this section we describe the new concept of rate in bit string commitment reductions [16]. Informally, by rate we mean the ratio between the dimension of the string which is being committed to divided by the number of weak commitments used.

Similar to error correcting codes in communication systems, we ask Alice to encode the string which she commits to into another, larger, string, before she uses the protocol $\beta - Commit$. However, differently than in a communication system, our codes not only add redundancy but also randomness to avoid Bob learning the value of the commitment before the unveiling phase. Here, we concentrate in non-interactive reductions. Thus, Alice encodes the string she wants to commit to, here called s, into another longer string x, and them commits to each bit of x by using $\beta - Commit$ protocols. Later on, to open the commitments, Alice runs the algorithm $\alpha - Open$ and opens the commitment to each bit of x and sends s to Bob. The information Bob receives during the executions of the algorithm $\alpha - Open$ is denoted by y (the domain of this information is not important in our reductions). Bob performs a test based on the information he received during the commit and opening phases. Based on the result of this test, Bob accepts or rejects Alice's commitment to s.

In the following, the *view* of a player at a certain stage of a protocol is the set of all the messages received by this player plus the random bits which were used by him during the protocol. We denote by S the random variable associated with s. X denotes the random variable associated with the encoding of s (X is the input to the algorithm $\beta - Commit$). Let Z denote the random variable associated with Bob's view of the protocol after a commitment to s is performed, that is after the $\beta - Commit$ algorithm is performed, and let Y represent Bob's view after the weak commitments are open, that is, after the protocols $\alpha - Open$ are performed.

Definition 4. *A   non-interactive   reduction   frombc($b_1 b_2 ... b_k$) to(α, β)   —   bc($x_1 x_2 ... x_n$)   consists   of   a   pair   of mappingsE : $\{0, 1\}^k \rightarrow \{0, 1\}^n$; D : $\{0, 1\}^n \rightarrow \{0, 1\}^k$ and a test $\beta(s, Y, Z) \in \{ACC, REJ\}$. The rate of this reduction is definedasR $= \frac{k}{n}$. A rate R is said to be achievable if there exists a pair (E, D) such that: $\lim_{n \to \infty} \{P[\beta(s, Y, Z) = ACC \wedge \beta(s', Y', Z) = ACC] = P_e = 0$ for any s and s' $\in \{0, 1\}^k$ such that $s \neq s'$ andlim$_{n \to \infty} I(S : Z)/k = 0$, whereI$(\cdot)$ is the mutual information function, Z is the random variable associated with Bob's view after a $\beta - Commit(E(b_1 b_2 ... b_k))$ algorithm is performed and Y represents his view after the algorithm $\alpha - Open (E(b_1 b_2 ... b_k))$ is executed (Y' denotes a possible cheating strategy for Alice)*

A reduction that achieves the supremum of all the achievable rates is called *optimal*. It is important to discuss for a while our notion of a concealing protocol $\lim_{n \to \infty} I(S : Z)/k = 0$. A stronger notion of secrecy could be proposed as in [14]: $\lim_{n \to \infty} I(S : Z) = 0$. We prove that our reduction also achieves this stronger notion of secrecy.

3.1   A Non-trivial Reduction to an $(\alpha, 0) - bc_{II}$ Protocol

Here, we compute a non-trivial achievable rate for a reduction to an $(\alpha, 0) - bc$ protocol. The mappings E and D in this reduction are given by a randomly generated linear code. Namely, a binary generating matrix G (of dimensions $k \times n$, $k = Rn$) is created at random. We know that there exists a constant ρ such that a binary matrix of size $Rn \times n$ defines a binary linear code with minimum distance at least δn except with probability not greater than $\rho^{(R-1+h(\delta))n}$, where $h(\delta) = -\delta \log \delta - (1 - \delta) \log(1 - \delta)$[7]. The mappings are made available to both Alice and Bob. They are also supposed to know the cheating probability α. To commit to a string $s = b_1 ... b_k$, uniformly chosen, Alice chooses the bit string $x = x_1 ... x_n$ associated with $b_1 ... b_k$ and then runs $Commit(x_1 ... x_n)$. To open the commitment, Alice runs $\alpha - Open(x_1 ... x_n)$. The receiver accepts the commitment to the string $b_1 ... b_k = D(x_1 ... x_n)$ iff $x_1 ... x_n$ is a valid codeword.

Proposition 4. *All   rates   of   reductions   from   bc($x_1 x_2 ... x_k$)   to $(\alpha, 0) -$ bc($y_1 y_2 ... y_n$) which are inferior to 1 are achievable.*

Proof. By definition, an $(\alpha, 0) - bc(b)$ protocol is concealing, thus a dishonest Bob can never cheat. The most general dishonest strategy for a dishonest Alice is to commit to a word **w** (which may not be a valid codeword) and later on unveil a valid codeword. The protocol is insecure if the probability that Alice can unveil two valid codewords is non-negligible. For any $R < 1$ and large n, there is a $\delta > 0$ so that the random linear code used by Alice and Bob has minimum distance equal to δn, with high probability. Suppose that Alice wants to be able to unveil two different codewords c_1 and c_2 and that she has committed to a general (not necessarily a codeword) **w**. As the minimum distance of the code is δn, the hamming distance of one of these two codewords $\{c_1, c_2\}$, w.l.o.g say

c_1, and $\mathbf{w}$ is at least $\delta n/2$. Thus, the probability that she can successfully open c_1 is upper bounded by $\alpha^{-\delta n/2}$ which goes to zero exponentially with n and our result follows

Note that, although we use random linear codes (which are not known to be efficiently decodable), are algorithms are efficient, since Bob need not decode anything in the protocol, he just checks if the word opened by Alice is a codeword or not.

3.2 A Reduction to an $(\alpha, \beta) - bc_{II}$ Protocol with $\alpha, \beta < 1$

We now show a reduction of bit commitment to $(\alpha, \beta) - bc_{II}(b)$ with $\beta \neq 0$ and $\alpha, \beta < 1$. The encoding used is the same that was used by Wyner in his seminal "wiretap channels" paper[20]. In order to commit to a string $b_1 b_2 ... b_k$, represented by a random variable S, Alice randomly selects r bits $j_1 j_2 ... j_r$, here represented by the random variable J, and concatenates these two strings forming a $q = k + r$ bits word: $b_1 b_2 ... b_k j_1 j_2 ... j_r$. Alice then proceeds with the encoding scheme described in the reduction to an $(\alpha, 0) - bc$ protocol. A binary generating matrix G of dimensions $Tn \times n$, $T = \frac{k+r}{n} = \frac{q}{n}$, where $r = \beta n$, is generated. The overall rate of the reduction is $R = \frac{k}{n}$. As in the last section, the random variable associated with the codewords is represented by X. The next theorem follows.

Proposition 5. *The above described reduction achieves unconditionally bindingness against the sender, Alice, if $T < 1 \Leftrightarrow r + k < n$.*

Proof. As we have to prove just the binding property, this proof follows directly from Proposition 4. To see this, we observe that the above reduction is equivalent (from the sender's security point of view) to a reduction to a $(\alpha, 0) - bc_{II}$. Therefore, the reduction presented in the last section works and gives us a rate $R < 1 - \beta$.

Now me must show that this reduction provides unconditionally concealment against Bob. As already defined, Z represents the random variable associated with the bits observed by Bob when a $\beta - Commit(x_1 x_2 ... x_n)$ algorithm is performed. From successive applications of the chain rule for entropy [6] we have: $H(S|Z) = H(S, Z) - H(Z) = H(S, X, Z) - H(X|S, Z) - H(Z) = H(S|X, Z) + H(X, Z) - H(X|S, Z) - H(Z) = H(S|X, Z) + H(X|Z) - H(X|S, Z)$. Here we make use of Fano's inequality. For convenience of the reader we state this theorem bellow:

Theorem 2. *Fano Inequality: Let X and Y be random variables with alphabets $\mathcal{H}$, defined a decoding probability of error P_e we have that: $h(P_e) + P_e \log(|\mathcal{H}| - 1) \geq H(X|Y)$, where $h(\cdot)$ is the entropy function.*

Define λ as the probability of wrongly guessing the random variable X when knowing Z and S. From Fano's inequality we have: $H(X|Z, S) \leq h(\lambda) + \lambda n$. Also, we know that: $H(S|Z) = H(S|X, Z) + H(X|Z) - H(X|S, Z)$, since $H(X|Z) =$

$n(1-\beta)$ (as Bob breaks $n\beta$ weak commitments in average), $H(X|S,Z) \leq h(\lambda) + \lambda n$ and $H(S|X,Z) = 0$, it follows that: $H(S|Z) \geq n(1-\beta) - h(\lambda) - \lambda n \geq n(1-\beta) - h(\lambda) - \lambda n + k - k$. As $\beta = \frac{r}{n}$, we have: $H(S|Z) \geq k + n - r - k - h(\lambda) - \lambda n \Rightarrow H(S|Z)/k \geq 1 + \frac{n-r-k}{k} - \frac{h(\lambda)}{k} - \lambda\frac{n}{k}$, but $r + k < n$, hence $H(S|Z)/k \geq 1 + \frac{h(\lambda)}{k} - \lambda/R$.

Now we must show that as k increases $\lambda \to 0$. We remember that λ is defined as the probability of wrongly guessing the random variable $X = SJ$ when knowing Z and S. We note that when the random variable S is given to Bob, only 2^r of the 2^{r+k} possible codewords are left to be guessed. Therefore, λ is the error probability of decoding a random coding scheme of rate less than $r/n = I(X:Z)/n$. From the direct part of the channel coding theorem we know that this probability goes to zero when n becomes large and hence $\lambda \to 0$. It is still necessary to prove that a stronger notion of secrecy $\lim_{n\to\infty} I(S:Z) = 0$ can also be achieved. To do so, we note that from the direct part of the noisy coding theorem, λ goes to zero exponentially with k, so if we multiply $H(S|Z)/k$ by k we are left with the following inequality: $H(S|Z) \geq k + h(\lambda) - \lambda k/R$. As λ goes to zero exponentially, we have our result: $\lim_{n\to\infty} H(S|Z) > k$. The proof of the next theorem follows from the previous arguments.

Theorem 3. *There exists a reduction of a string commitment scheme to any $(\alpha, \beta) - bc_{II}$ achieving a rate $R < 1 - \beta$ when $\alpha, \beta < 1$.*

4 A Reduction to an $(\alpha, \beta) - bc_I$ Protocol with $\alpha + \beta < 1$

Here, we prove that any $(\alpha, \beta) - bc_I$ protocol with $\alpha + \beta < 1$ can be used to implement strong string commitments achieving a rate $R = 1 - \alpha - \beta$. First we prove that string commitments can be reduced to any $(\alpha, 0) - bc_I$ protocol with $\alpha < 1$ achieving a rate $R = 1 - \alpha$. Again, the mappings E and D in this reduction are given by a randomly generated linear code. Thus, a binary generating matrix G (of dimensions $k \times n$, $k = Rn$, $R < 1 - \alpha$) is created at random. The mappings are made available to both Alice and Bob. They are also supposed to know the cheating probability α. To commit to a string $s = b_1...b_k$, uniformly chosen, Alice chooses the bit string $x = x_1...x_n$ associated with $b_1...b_k$ and then runs $Commit_I(x_1...x_n)$. To open the commitment, Alice runs $\alpha - Open_I(x_1...x_n)$. The receiver accepts the commitment to the string $b_1...b_k = D(x_1...x_n)$ iff $x_1...x_n$ is a valid codeword.

Proposition 6. *All rates of reductions from $bc(x_1x_2...x_k)$ to $(\alpha, 0) - bc_I(y_1y_2...y_n)$ which are inferior to $1 - \alpha$ are achievable.*

Proof. In average, Alice will be able to cheat in approximately αn (for large n) weak commitments. Moreover, she knows where she can cheat in advance. She will able to break the security conditions of the strong string commitment if she finds two codewords which have the same bits on exactly $(1 - \alpha)n$ positions (the ones where she will not be able to change the weak commitment). The probability that those two codewords are generated can be upper bounded by $2^{(1-\alpha)n}2^{nR}$, which can be made arbitrarily small as far as $R < 1 - \alpha$.

Observing that the arguments presented in Section 3.2 straightforwardly apply here, we are able then to prove the following theorem:

Theorem 4. *There exists a reduction of a string commitment scheme to any $(\alpha, \beta) - bc_I$ achieving a rate $R < 1 - \alpha - \beta$ when $\alpha + \beta < 1$.*

5   Weak Commitments Based on Rabin Oblivious Transfer and on 1 out of 2-Oblivious Transfer

As an example, we shall describe a weak commitment based on Rabin oblivious transfer. Our reduction, in spite of its simplicity, achieves a rate bounded away from zero thereby showing the generality of the theory developed in the previous sections. Rabin oblivious transfer is a primitive where a sender (Alice) sends a bit b to a receiver (Bob). Bob receives it with probability $1/2$ or he receives an erasure symbol $\triangle$ otherwise. The protocol is secure if Alice does not know if Bob received the bit or not. We describe a simple protocol to implement a weak bit commitment from OT. To commit to a bit b, Alice sends the bit b through the

Rabin OT. In the opening phase Alice sends the bit b to Bob in the clear. If Bob received an erasure $\triangle$ in the commit phase, Bob always accepts b. Otherwise, Bob only accepts Alice commitment if the bit she announces is in agreement with the bit he received during the commit phase.

From the point of view of this paper Rabin oblivious transfer can be seen as a $(1/2, 1/2)$-weak bit commitment of type II, since Alice does not know *a priori* if she will be able to cheat during the open phase or not and with probability $1/2$ the sender learns the commitment before the opening phase and if the bit b is erasured during the commit phase the sender can successfully open any bit to the receiver. Hence we can apply the reduction of the previous sections and any rate below $R = 1 - 1/2 = 1/2$ can be achieved by this reduction.

Now, we show another example of weak commitment based on 1 out of 2 - OT. A $(b_0, b_1) - OT(c)$ is a protocol where a sender inputs two bits b_0 and b_1 and a receiver inputs a bit c. At the end of the protocol the receiver gets b_c and the sender receives nothing. The protocol is secure if the sender learns nothing about c after an execution of it and if the receiver learns nothing about $b_{\bar{c}}$.

Our protocol is as follows: to commit to a bit b, Alice runs $(b \oplus r, r) - OT(c)$ with Bob, where r and c are chosen at random and $\oplus$ is the XOR operation. To unveil this bit, Alice sends the bits b and r to Bob in the clear. Bob accepts the commitment iff these bits are consistent with the information he has received in the commit phase. The previous protocol is obviously concealing, however Alice can cheat with probability at most $1/2$. Thus we are left with an $(1/2, 0)$-weak bit commitment of type II. Hence, by applying the reduction of last Section, any rate below $R = 1$ can be achieved by this reduction.

One can prove that a rate $R = 1/2$ is indeed optimal when reducing bit commitment to Rabin oblivious transfer[21]. Therefore, we observe that 1 out of 2-OT is *strictly more powerful* than Rabin OT when it comes to implementing string commitments.

6 Improving a Computational Secure Quantum Bit Commitment Protocol

In this section, we show how the proposed reductions can be used to turn a quantum bit commitment protocol based on any quantum one-way permutation robust against multiple-photons. This is a rather surprising application of our reductions, since they are basically classical reductions. However, in this section, we show that some of our reductions can be proven to be secure against some quantum adversaries.

Quantum cryptography came on to the scene as a possible way to achieve secure bit commitment schemes based solely on the laws of physics. These hopes were ruled out when Mayers[13] and Lo and Chau [12] proved that any bit commitment protocol can be broken if the parties involved have unlimited quantum computational power. The main idea of the proof is that the existence of a secure quantum bit commitment protocol implies the existence of a secure *purified* protocol where all actions are kept at the quantum level such that the result of the protocol is a pure state shared between Alice and Bob. If such a scheme is concealing, i. e., the state in the possession of Bob is approximately the same for a commitment to one and for a commitment to zero then Alice can change the bit she committed to on her side by applying a transformation U on her qubits. Hence no quantum bit commitment scheme can be binding and concealing. Just after the no go theorem of Mayers and Lo and Chau the research community started looking for assumptions that could be used to implement quantum bit commitment protocols. Salvail [18] proved that we can have secure bit commitment under the assumption that some kind of measurement cannot be performed by one of the participants. In the paper [10], Dumais, Salvail and Mayers generalized the idea of complexity based cryptography from the classical to the quantum world. They proposed a protocol which the security is based on any quantum one-way permutation.

We briefly introduce their notations before reviewing their protocol. In the following we use the computational $\{|0\rangle, |1\rangle\}$ and diagonal basis $\{|0\rangle_\times = \frac{1}{\sqrt{2}}(|0\rangle + |1\rangle), |1\rangle_\times = \frac{1}{\sqrt{2}}(|0\rangle - |1\rangle)\}$, here denoted by $+$ and $\times$ respectively. A bit string $y \in \{0,1\}^n$ encoded in the computational basis (diagonal basis) is represented by $|y\rangle_{\theta(0)^n}(|y\rangle_{\theta(1)^n})$. Following [10] we write the quantum state $\bigotimes_{i=1}^{n} |y_i\rangle_{\theta(w_i)}$ as $|y\rangle_{\theta(w)^n}$. The following projections are used in our discussion: $P^y_{\theta(b)^n} = |y\rangle_{\theta(b)^n \theta(b)^n} \langle y|$. $\Sigma = \{\sigma_n : \{0,1\} \to \{0,1\}^n | n > 0\}$ denotes a family of one-way permutations. Dumais et al.'s protocol is as follows: To commit to a bit w, Alice chooses $x \in \{0,1\}^n$ and computes $y = \sigma_n(x)$. Alice then sends the quantum states $|\sigma_n(x)\rangle_{\theta(w)^n}$. To unveil the bit, Alice announces w and x

to Bob. Bob measures his received state with the measurement $\{P^y_{\theta(w)^n}\}_{y \in \{0,1\}^n}$ and accepts the commitment iff the outcome of the measurement he performs is equal to $\sigma_n(x)$.

It is easy to see that the above protocol is unconditionally concealing (since the outcome of the permutation is completely random). Dumais et al. also proved that Alice can use the transformation U that changes the value of a committed bit to efficiently invert the one-way permutation σ_n, characterizing the computational bindingness of the protocol. It is important to remark that the definition of bindingness for quantum bit commitments differs from the one used for classical protocols. To require that the probability that Alice can successfully open two different commitments goes to zero is a too strong requirement for quantum protocols, since Alice can always commit to a superposition of two valid commitments. Thus, it was suggested in [10] to classify as binding, any quantum bit commitment protocol where the probability of successfully opening zero and one sum up to a value arbitrarily close to one in a security parameter. Therefore, we slightly change our definition of weak commitments of type II for quantum protocols.

Definition 5. *A quantum* $(\alpha, \beta) - qbc(b)$ *of type II,* $(\alpha, \beta) - qbc_{II}(b)$ *for short, is defined as a pair of algorithms* $\beta - qCommit_{II}(b)$ *and* $\alpha - qOpen_{II}(b)$. *In an* $\alpha - qOpen_{II}(b)$ *algorithm the committer is able to commit to something and later on unveil* $b = 0$ *and* $b = 1$ *with success probabilities which sum up to* $1 + \alpha$. *In a* $\beta - qCommit_{II}(b)$ *algorithm, Bob, the receiver, learns the value of a committed bit with probability at most* β.

Thus, if α is negligible in a security parameter, our quantum weak commitment is binding. A quantum commitment is concealing if $\lim_{m \to \infty} Tr|\rho_0 - \rho_1| = 0$, where Tr is the trace of a matrix, ρ_0 is the density matrix which represents a commitment to zero, ρ_1 is the density matrix which represents a commitment to one and m is a security parameter.

It was noted in [10] that if the photon source used in the protocol implementation is not a perfect one, i.e. if it emits multiple-photons with some probability, the protocol is not concealing anymore. Actually, it is easy to see that, given a non-perfect photon source, the probability that Bob breaks the protocol approaches one as the security factor n increases. Finding a protocol robust against multiple-photons was stated as an open problem in [10]. We use the theory of weak commitments to present a quantum bit commitment protocol based on any quantum one-way permutation robust against multiple-photons. We model a realistic source which emits multiple photons with probability p by an oracle that announces the value of the committed bit to Bob (before the unveiling phase) with probability $1 - (1 - p)^n$. Therefore, we have a $(\alpha, \beta) - qbc_{II}$ where α is negligible and $\beta = 1 - (1 - p)^n$.

Before applying our reductions, we have to prove that using repeatedly the above protocol does not compromise its security. In the following, we denote Dumais et. al. quantum bit commitment protocol performed m times with **perfect**

apparatus by $qbc(b_1...b_m)$ where $b_i, 1 \leq i \leq m$ are the bits committed to in each execution of the protocol.

Proposition 7. *Bob cannot distinguish between two commitments $qbc(b_1...b_m)$ and $qbc(w_1...w_m)$ where $b_1...b_m$ and $w_1...w_m$ are any $m-$bits long strings.*

Proof. For an honest Alice the individual bit commitments in the string commitment are performed independently where each of the states resulting from the individual commitments is on Bob's side independent of the committed bit. Then the product state which is the result of the string commitment can easily be computed to be independent of the string committed. The bit string commitment is perfectly concealing.

Proposition 8. *Let p_0 and p_1 be the probabilities that Alice can successfully open two different commitments $qbc(b_1...b_m)$ and $qbc(w_1...w_m)$ to Bob respectively. We have that $\lim_{n\to\infty} (p_0 + p_1) = 1$ where n is the dimension of the image of the one-way permutation σ_n.*

Proof. We will show that a cheating strategy for bit string commitment based on the protocol of [10] would imply a cheating strategy for one single application of the protocol which is equivalent to inverting the quantum one way function. Assume Alice is able to change at least one commitment within a string of length m with non negligible probability then Alice can instead of performing a single execution of the protocol of [10] perform a string commitment but send only one of the commitments (which she can change with a non negligible probability) to Bob. As Bob cannot detect if a bit Alice committed to is part of a string this strategy allows Alice to cheat in a single round of the protocol.

One more point is missing, could the information provided by the oracle help Bob breaking commitments where no multiple-photon is emitted? This is not the case, since the output of the one-way permutation is completely random and therefore makes each partial commitment $qbc(b_i)$ independent of the others. In other words, the density matrix of each unbroken partial commitment $qbc(b_i)$ does not change when other partial commitments are broken. Therefore, we are able to use our reduction techniques in order to cope with the multiple-photons problem in the scheme above described.

We ask Alice to use the same reduction which was used in Proposition 2: when the committed bit is b, Alice chooses a random $m-$bit word $b_1b_2...b_m$ such that $b = b_1 \oplus b_2 \oplus ... \oplus b_m$ where $\oplus$ stands for the XOR operation. Alice sequentially commits to $b_i, 1 \leq i \leq m$ using a $\beta - qCommit_{II}(b)$ protocol. By applying the reduction presented in Proposition 5 (which is basically a straightforward application of privacy amplification[1]), and Propositions 7 and 8 we see that it is possible to have a binding (α is negligible in a security parameter) and concealing ($\lim_{m\to\infty} Tr|\rho_0 - \rho_1| = 0$) quantum bit commitment protocol based on a quantum one-way permutation even when used with an imperfect source of photons. We omit the proof of the next theorem.

Theorem 5. *There exists a quantum bit commitment protocol based on any quantum one-way permutation which is binding and concealing even in the presence of multiple photons.*

Satisfactory as this result is, it has to be noted that our achieved rate goes to zero in the limit of large n. It would be desirable to have a reduction which achieves a non-trivial rate. However, the reductions achieving non-trivial rates presented in this paper do not trivially generalize to the quantum world. One of the reasons is that, it is not completely obvious how to define the binding condition for quantum string commitments, since Alice can commit to superposition of commitments to different strings. Although, recently, in a pre-print[5], Crepeau, Dumais, Mayers and Smith proposed a reasonable definition for the binding condition of quantum string commitments, the fact that Dumais et. al. protocol satisfies this definition has yet to be proven (it was conjectured to be true in [5]). If the conjectures of [5] are proven to be correct, our non-trivial reductions should apply for Dumais et. al. protocol and non-trivial rates can be achieved. However, to prove these statements is beyond the scope of this paper.

7   Conclusions

We introduced a new class of primitives called weak commitments. We completely characterized when bit commitments can be reduced to these primitives. To judge the efficiency of such a reduction we used the new concept of the rate of a reduction. The reductions presented exhibit a non-zero rate. Furthermore, we provided examples of how to implement these primitives based on oblivious transfer and on quantum mechanics. Several open questions were stated.

We thank an anonymous referee who pointed out a mistake in the proof of Theorem 3. This research was partially funded by project PROSECCO of the IST-FET programme of the EC and by a project on Research and Development on Quantum Cryptographyh of Telecommunications Advancement Organization as part of the programme Research and Development on Quantum Communication Technologyh of the Ministry of Public Management, Home Affairs, Posts and Telecommunications of Japan.

References

1. C.H. Bennett, G. Brassard, C. Crépeau, and U. Maurer, Generalized Privacy Amplification, IEEE Transaction on Information Theory , Volume 41, Number 6, pp. 1915–1923, November 1995.
2. C. Cachin, On the Foundations of Oblivious Transfer. EUROCRYPT 1998: 361–374
3. C. Crepeau and J. Kilian Achieving oblivious transfer using weakened security assumptions. In 29th Symp. on Found. of Computer Sci., pages 42–52. IEEE, 1988.
4. C. Crépeau, Efficient Cryptographic Protocols Based on Noisy Channels, Proceedings of Eurocrypt '97

5. C. Crepeau, P. Dumais, D. Mayers and L. Salvail, Apparent Collapse of Quantum State and Computational Quantum Oblivious Transfer, pre-print available at http://crypto.cs.mcgill.ca/~crepeau/PS/subqmc-3.ps
6. T. Cover and J. Thomas, Elements of Information Theory, Wiley, 1991.
7. C. Crepeau Efficient Cryptographic Protocols Based on Noisy Channels. Advances in Cryptology: Proceedings of Eurocrypt '97 , Springer-Verlag, pages 306–317, 1997.
8. I. Damgard, J. Kilian and L. Salvail, On the (Im)possibility of Basing Oblivious Transfer and Bit Commitment on Weakened Security Assumptions, proceedings of Eurocrypt'99
9. I. Damgard, T.. Pedersen, B. Pfitzmann: Statistical Secrecy and Multi-Bit Commitments; IEEE Transactions on Information Theory 44/3 (1998) 1143–1151.
10. P. Dumais, D. Mayers, and L. Salvail, Perfectly Concealing Quantum Bit Commitment from Any Quantum One-Way Permutation, Eurocypt2000, Proceedings, Lecture Notes in Computer Science, no 1807, Springer Verlag, May 2000, pp. 300–315.
11. J. Kilian Founding Cryptography on Oblivious Transfer, Proceedings of the 20th ACM Symposium on the Theory of Computing,1988
12. H.K. Lo, H. F. Chau, Is quantum Bit Commitment Really Possible?", Physical Review Letters, vol. 78, no 17, April 1997, pp. 3410
13. D. Mayers, The Trouble With Quantum Bit Commitment, http://xxx.lanl.gov/abs/quant-ph/9603015, March 1996.
14. U. Maurer and S. Wolf, Information-Theoretic Key Agreement: From Weak to Strong Secrecy for Free, Advances in Cryptology - EUROCRYPT '00, Lecture Notes in Computer Science, Springer-Verlag, vol. 1807, pp. 351–368, 2000.
15. M. Naor, Bit Commitment Using Pseudorandomness . J. of Cryptology, Volume 4, pp. 151–158
16. A. Nascimento, J. Mueller-Quade, H. Imai, Optimal Multi-Bit Commitment Reductions to Weak Commitments, ISIT 2002, pg .294
17. R.L. Rivest, Unconditionally secure commitment and oblivious transfer schemes using concealing channels and a trusted initializer, pre-print.
18. L. Salvail, , Quantum Bit Commitment From a Physical Assumption, Advances in Cryptology : CRYPTO 98, Proceedings, Lecture Notes in Computer Science, no 1462, Springer Verlag, August 1998, pp. 338–353.
19. D. Stinson ,Cryptography: Theory and Practice,CRC Press,Florida,1995
20. AD Wyner, The WireTap Channel Bell System J., 54, 1981, pp. 1355–1387.
21. A. Winter, A. Nascimento, H. Imai, Commitment Capacity of Discrete Memoryless Channels, http://xxx.lanl.gov/abs/cs.CR/0304014
22. A Yao, Security of Quantum Protocols Against Coherent Measurements, Proceedings of the 26th Annual ACM Symposium on the Theory of Computing, 1995

Improving Robustness of PGP Keyrings
by Conflict Detection*

Qinglin Jiang, Douglas S. Reeves, and Peng Ning

Cyber Defense Lab
Departments of Computer Science and Electrical and Computer Engineering
N.C. State University Raleigh, NC 27695-8207 USA
{qjiang,pning,reeves}@ncsu.edu

Abstract. Secure authentication frequently depends on the correct recognition of a user's public key. When there is no certificate authority, this key is obtained from other users using a *web of trust*. If users can be malicious, trusting the key information they provide is risky. Previous work has suggested the use of redundancy to improve the trustworthiness of user-provided key information. In this paper, we address two issues not previously considered. First, we solve the problem of users who claim multiple, false identities, or who possess multiple keys. Secondly, we show that *conflicting* certificate information can be exploited to improve trustworthiness. Our methods are demonstrated on both real and synthetic PGP keyrings, and their performance is discussed.

1   Introduction

Authentication is one of the most important objectives in information security. Public key cryptography is a common means of providing authentication. Some examples are X.509 [19] and PGP [20]. In the public key infrastructure, each user is associated with a public key, which is publicly available, and with a private key, which is kept secret by the user. A user signs something with her private key, and this signature can be authenticated using the user's public key.

The ability to exchange public keys securely is an essential requirement in this approach. Certificates are considered to be a good way to deliver public keys, and are popularly used in today's public key infrastructures. Intuitively, a certificate is an authority telling about a user's public key. In a hierarchical system, such as X.509 [19], we usually assume that the certificates contain true information, because the authority is secured and trusted. In non-hierarchical systems, such as PGP keyrings [20], each user becomes an authority. Such systems are referred to as *webs of trust*. With webs of trust, it may be risky to expect all certificates to contain true information, because not all users are fully secured and trusted. Accepting a false public key (i.e., believing it contains true

---

* This work is partially supported by the U.S. Army Research Office under grant DAAD19-02-1-0219, and by the National Science Foundation under grant CCR-0207297.

T. Okamoto (Ed.): CT-RSA 2004, LNCS 2964, pp. 194–207, 2004.

information) undermines the foundation of authentication. For this reason, a method that can be used to verify a user's public key (i.e., to detect and reject false certificates) is very much needed.

The goal of this paper is to develop a robust scheme to determine if a certificate is trustworthy. Our method uses redundant information to confirm the trustworthiness of a certificate. Previous work [16] has shown that redundancy (in the form of multiple, independent certificate chains) can be used to enhance trustworthiness. We show that in some circumstances multiple certificate chains do not provide sufficient redundancy. This is because a single malicious user may (legitimately) possess multiple public keys, or (falsely) claim multiple identities, and therefore can create multiple certificate chains which seem to be independent. Our solution to this problem is to also consider identities when determining if certificate chains are independent.

In addition, we investigate the implications of *conflicting certificates* (i.e., certificates which disagree about the public key of a user). Conflicts are simple to detect. We show that such conflicting information can be used to help identify malicious users. Based on that information, the number of certificates which can be proved to be true is increased, improving the performance of our method.

The organization of the paper is as follows. Section 2 summarizes related work. Section 3 defines our problem precisely. Section 4 presents solutions without using conflicting certificate information. Section 5 shows how conflicting certificates information may be used to improve performance. Section 6 presents our experimental results, and section 7 concludes.

2   Related Work

In addition to X.509 and PGP, discussed above, there are other public key infrastructures, such as SPKI/SDSI [6] and PolicyMaker [3]. These mainly focus on access control issues. They differ from X.509 and PGP in that they bind access control policies directly to public keys, instead of to identities.

Existing work on improving the trustworthiness of webs of trust can be classified into two categories. In the first category, *partial trust* is used to determine trustworthiness of a target public key. In [18] and [13], this trustworthiness is based on the trust value in a single certificate chain. Multiple certificate chains are used in [2] and [11] to boost confidence in the result. [9] tries to reach a consensus by combining different beliefs and uncertainties. In [15], insurance, which may be viewed as a means of reducing risk, is used to calculate the trustworthiness of a target public key.

In the second category of methods, there is no partial trust; a key is either fully trustworthy, or else it is untrustworthy. In this category, an upper bound on the number of participants that may be malicious is assumed. An example is [4], which requires a bound on the minimum network connectivity in order to reach a consensus. Another important method in this category is [16]. This method suggests using multiple public key-independent certificate chains to certify the same key. The authors showed that if at most n public keys are compromised, a public key is provably correct if there are $n + 1$ or more public key-independent certificate chains certifying it. Computation of the number of indepedent chains is accomplished by computing network flow in the certificate graph.

Methods in the first category (partial trust) are based on probabilistic models. We believe such models are more appropriate for computing the reliability of faulty systems than they are for computing trustworthiness of information provided by users. For example, such approaches require proper estimation of the trustworthiness of each user. If such estimates are incorrect, or a trusted user is compromised, then the output produced by these methods will be misleading. We believe methods in the second category are more suitable for the case of (intentionally) malicious users. That is, it should be much easier to bound the number of users who are malicious than to specify how trustworthy each user is.

A limitation of methods in the second category is that the importance of identities, as well as public keys, has not been fully considered. That is, these methods have not considered the possibility that each user may claim multiple identities, or possess multiple public keys.

All methods for distributed trust computation assume there is some initial trust between selected users. Without such initial trust, there is no basis for any users to develop trust in one another. In [5], it is shown that forging multiple identities is always possible in a decentralized system. We assume that the initial trust must be negotiated in an *out-of-band* way (such as by direct connection, or communication with a trusted third party) from the distribution of trust, and that proof of identity is available during this initial phase.

The next section presents definitions and assumptions, and a statement of the problems to be solved.

3   Problem Statement

A *user* is an entity in our system represented by an *identity* (such as the names "Bob" and "Alice"). An identity must be established when the initial trust information is negotiated between users. We assume in this work that each user legitimately has exactly one, unique *true (or valid) identity*. An identity which does not belong to a real user is a *false identity*.

We further assume each user can have, or be associated with, one or more public keys. In the case where a user has more than one public key, we assume the user further specifies each of her keys by a *key index number*. The combination of a user identity x and a key index number j is denoted x/j, and uniquely identifies a true public key. If the user with identity x only has a single public key, j will be omitted for the sake of convenience.

Our definitions of *public key certificate* and *certificate chain* follow [14]. A public key certificate is a triple $\langle x/j, k, s_{k'} \rangle$, where x is an identity, j is a key index number, k is a public key, and $s_{k'}$ is the digital signature over the combination of x/j and k. Given a certificate $C = \langle x/j, k, s_{k'} \rangle$, if (i) the identity x is a true identity, and (ii) the user with identity x says k is her public key, then C is a *true certificate* and k is a *true public key* for x. Otherwise, C is a *false certificate* and k is a *false public key* for x. If $s_{k'}$ is generated by y, we say the certificate is *issued* by y. If all certificates issued by y are true certificates, then y is a *good user*. If there exists at least one false certificate issued by y, then y is a *malicious user*.

Two certificates are said to *agree* with each other if the identities, key index numbers and public keys are the same. Two certificates are called *conflicting certificates* if the identities and key index numbers are the same, but the public keys are different. Note that the two conflicting certificates may both be true by our definition (i.e., the user with the corresponding identity says both of the two keys are her public keys, with the same index number). This may happen when a user x intentionally has two conflicting certificates issued to herself, by two separate parties, for the purpose of possessing more public keys. We expand our definition of a malicious user to also include x in such a case; the issuers of the conflicting certificates, however, are not considered to be malicious on this count, and the certificates are defined to be true.

Each user x may accumulate a set R^x of certificates about other users. Obtaining these certificates may be done in a variety of ways. We do not discuss further in this paper how certificates are distributed, which is an open problem.

In each user's set of certificates, some of them are assumed by x to be true and others are not. Denote T_0^x the set of certificates assumed by x to be true initially (i.e., they are provided by means of the initial trust distribution). Because this initial trust information is assumed to be true, the signatures on the certificates in T_0^x do not have to be further verified. A *certificate chain* is a sequence of certificates where:

1. the starting certificate, which is called the *tail* certificate, is assumed or determined to be true;
2. each certificate contains a public key that can be used to verify the digital signature associated with the next certificate in the sequence; and,
3. the ending certificate, which is called the *head* certificate, contains a desired name-to-key binding, which is called the *target*.

Each user x's set of certificates R^x may be represented by a directed *certificate graph* $G^x(V^x, E^x)$. V^x and E^x denote the set of vertexes and the set of edges in the certificate graph G^x, respectively. A vertex in V^x represents a public key and an edge in E^x represents a certificate. There is a directed edge labeled with y/j from vertex k' to vertex k in G^x if and only if there is a certificate $\langle y/j, k, s_{k'} \rangle$ in R^x. A certificate chain is represented by a directed path in the certificate graph. Two conflicting certificates are represented by two edges with the same label, but different head vertexes. In this case, the two different head vertexes are called *conflicting vertexes*.

For the sake of simplicity, we add to the certificate graph an "oracle" vertex k_0. There is a directed edge in G^x from k_0 to every key which is assumed to be true (by way of the initial trust distribution), labeled with the identity/index number bound to that key.

To depict true and false public keys in the certificate graph, we "paint" vertexes with different colors. A white vertex represents a public key that is

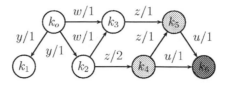

Fig. 1. User x's certificate graph

either assumed or determined to be true for the identity on the edges directed towards that vertex. A dark gray vertex represents a public key which is known to be false for the corresponding identity. If a public key is neither assumed to be true nor proved (yet) to be true for the corresponding identity, we paint it light gray.

Figure 1 is a sample certificate graph. k_0 is the oracle vertex. k_1, k_2 and k_3 are three public keys that are assumed to be true by user x. k_1 and k_2 are two conflicting vertexes because the labels on their incoming edges are the same. k_5 is z's public key, with index number 1, and k_4 is z's public key, with index number 2. k_6 is a false public key.

When there is more than one malicious user, there may be a relationship between these users. Two malicious users who cooperate with each other to falsify information are said to be *colluding*. We say that two users x and y are colluding if either: (i) there exist two false certificates, one issued by x and one by y, and they agree with each other; or, (ii) x issues a false certificate upon y's request, or vice versa.

3.1 Problem Description

We now define the problems to be solved. First we consider the case in which malicious users do not collude, followed by the colluding case. The goals are the same, that is, to maximize the number of certificates which can be proved to be true.

Problem 1 *Given a set R^x of certificates and a set $T_0^x \in R^x$ of true certificates. Assuming there are at most n malicious users, and these users do not collude, maximize the number of certificates which can be proved to be true.*

Problem 2 *Given a set R^x of certificates and a set $T_0^x \in R^x$ of true certificates. Assuming there are at most n malicious users and these users may collude, maximize the number of certificates which can be proved to be true.*

It is necessary to have a metric to evaluate the performance of proposed solutions for the above problems. Let U be the set of all users. Denote by K_a^x the set of true public keys that can be reached by at least one path from the oracle vertex k_0 in the certificate graph for user x. K_a^x is the maximum set of true public keys that any method could determine to be true by means of certificate chains in this graph. Let K_0^x be the set of public keys that are assumed to be true initially by user x, and K_M^x the set of public keys that are determined to be true by a method M. Ideally, K_a^x would be equal to $K_0^x \bigcup K_M^x$, but in practice this may be difficult to achieve. The metric definition now follows.

Definition 1 *Given a solution s to Problems 1 or 2 generated by a method M. The performance $q_x(s)$, i.e. s's performance for user x, is*

$$q_x(s) = \frac{|K_0^x| + |K_M^x|}{|K_a^x|} .$$

To capture the performance for a set of users, we propose using the weighted average of each user's performance:

$$q(s) = \frac{\sum_{x \in U} \{q_x(s) \cdot |K_a^x|\}}{\sum_{x \in U} |K_a^x|} .$$

4   Solutions to Problems 1 and 2

In this section, we present methods for solving problems 1 and 2 under two assumptions: (a) when there is only one public key per identity, and (b) when there may be multiple public keys per identity. We assume there is an upper bound on the number of users who may be malicious, and use redundancy to determine the certificates that must be true. We ignore in this section the case in which there are conflicting certificates, which is considered in section 5.

Two certificate chains are *public key-independent* if their head certificates agree, and their remaining certificates have no *public keys* in common. Two certificate chains are *identity-independent* if their head certificates agree, and their remaining certificates have no *identities* in common. We state the following theorems without proof (see [8]).

Theorem 1 *Given two identity-independent certificate chains and any number of non-colluding malicious users, the head certificates must be true.*

In the case of multiple colluding malicious users, a greater degree of redundancy is needed to verify a certificate is true:

Theorem 2 *Given $n + 1$ identity-independent certificate chains, if there are at most n colluding malicious users, then the head certificates must be true.*

Based on these results, we now present methods for maximizing the number of true certificates (when there are no conflicting certificates).

200   Q. Jiang, D.S. Reeves, and P. Ning

4.1 Maximum of One Public Key per Identity

Reiter and Stubblebine [16] considered and solved this problem. We summarize
their results for the convenience of the reader. When each identity corresponds to
only one public key, public key-independent certificate chains are also identity-
independent. In the certificate graph, public key-independent certificate chains
corresponds to paths with the same tail vertex and the same head vertex, but
which are otherwise vertex disjoint. For each vertex k_t, it is possible to use
standard algorithms for solving maximum network flow [1] in a unit capacity
network to find the maximum number of vertex-disjoint paths from k_0 to k_t, in
running time $O(|V||E|)$.

It can be shown that all certificates (edges) ending at k_t must be true if:
(i) there are any number of non-colluding malicious nodes, and the maximum
flow from k_0 to k_t is greater than or equal to 2; or, (ii) the number of (possibly
colluding) malicious users is no greater than n and the maximum flow from k_0
to k_t is greater than or equal to $n + 1$.

4.2 Multiple Public Keys Allowed per Identity

We now address the case in which an identity may be associated with multiple
public keys in x's certificate graph. For example, in figure 1 there are two vertex-
disjoint paths from k_0 to k_6. However, user z has two public keys, and the
maximum number of identity-independent paths to k_6 is only 1. Therefore, it is
not safe to conclude k_6 is u's true key if z may be malicious. This issue has not
previously been addressed.

We still wish to use the notion of redundant, identity-independent paths to
nullify the impact of malicious users. To ensure that two paths are *identity-
independent* under the new assumption, it is necessary that the two paths have
no label in common on their edges. In this case the paths are said to be *label-
disjoint*.

Suppose there exists a solution to the problem of determining the maximum
label-disjoint network flow in the graph G^x with unit capacity edges, from k_0 to
a vertex k_t. We conclude (by the reasoning previously given) that k_t is a true
public key for the identity on the edges in the maximum flow ending at k_t if:

1. the maximum flow is 2 or greater, and there is at most 1 malicious node, or
   any number of non-colluding malicious users; or,
2. the maximum flow is $n + 1$ or greater, and there are at most n colluding
   malicious users.

We now state a theorem about the complexity of finding the maximum label-
disjoint network flow in a directed graph:

Theorem 3 *Given a certificate graph G^x and a vertex k_t. If one identity may
legitimately be bound to multiple public keys, the problem of finding the maximum
number of label disjoint paths in G^x from k_0 to k_t is NP-Complete.*

The proof can be found in [8]. Therefore, to solve this problem exactly is computationally expensive in the worst case.

A heuristic for this problem follows an idea from [16]. The problem of finding the maximum number of label-disjoint paths between k_0 and k_t can be transformed to the maximum independent set (MIS)[7] problem. The MIS problem is defined as follows: Given a graph G, find the maximum subset of vertexes of G such that no two vertexes are joined by an edge. The transformation from our problem is trivial, and can be found in [8]. The size of the maximum independent set in the transformed graph is the maximum number of label-disjoint paths from k_0 to k_t in G^x. Although MIS is also a NP-complete problem, there exist several well-known approximation algorithms for it [10].

Alternatively, an exact solution may be computationally tractable if the required number of label-disjoint paths is small. Suppose we wish to solve the problem of whether there exists at least b label-disjoint paths in G^x from k_0 to k_t. The maximum number of label-disjoint paths from k_0 to k_t equals the size of the *minimum label-cut* for k_0, k_t. A label-cut is a set of labels on edges whose removal would disconnect k_t from k_0. If we enumerate all subsets of labels in G^X with b or fewer labels, and no label-cut with b or fewer labels exists, then k_t is determined to be true. The algorithm runs in $O(|E||V|^b)$ (proof omitted).

In this section, we solved problems 1 and 2, without considering the possibility of conflicting certificates. We now turn to this problem.

5   Dealing with Conflicting Certificates

We assume that conflicting certificates occur because of malicious intent, and not by accident. A malicious user may create conflicting certificates for several reasons. For example, one use is to attempt to fool user x into believing a false public key is true, by creating multiple public key-independent certificate chains to the false public key. In this case, the method of section 4.2 can first be applied to determine the set of true certificates.

However, we can exploit the existence of conflicting certificates to prove an even larger number of certificates must be true. Stubblebine and Reiter [16] pointed out that conflicting certificates represent important information, but did not suggest how they could be used. We propose below a method of doing so, based on the notion of the *suspect set*:

Definition 2 *A suspect set is a set of identities that contains at least one malicious user. A member of the suspect set is called a suspect.*

We now describe how to construct suspect sets of minimum size, and how they can be used to determine more true certificates.

5.1   Constructing Suspect Sets (Non-colluding Malicious Users)

Suppose we know or have determined a certificate is false by some means. If there is only a single malicious user, or multiple malicious users who are non-colluding, the true identity of the issuer of this false certificate must appear in every chain ending with a certificate that agrees with this false certificate. Using this insight, we propose to construct suspect sets using the following algorithm. The algorithm takes a certificate graph G^x as input.

Algorithm 1 *constructing suspect sets*

1: **for** each label y/j in the certificate graph **do**
2:    **for** each dark gray vertex k_i whose incoming edge has a label y/j **do**
3:       construct a new suspect set consisting of the set of labels, each of which is a label-cut for k_0, k_i.
4:    **end for**
5:    **for** each light gray vertex k_i with an incoming edge labeled y/j, conflicting with a white vertex with an incoming edge labeled y/j **do**
6:       construct a new suspect set consisting of {the set of labels each of which is a label-cut for k_0, k_i} $\cup y$
7:    **end for**
8:    **for** each pair k_i, k_h of light gray vertexes whose incoming edges both have a label y/j **do**
9:       construct a new suspect set consisting of {the set of labels each of which is a label-cut for either k_0, k_i or k_0, k_h} $\cup y$
10:    **end for**
11: **end for**

The intuition behind algorithm 1 is as follows. For each certificate known to be false, the malicious user's true identity must be a label-cut between k_0 and the false certificate. For each certificate conflicting with a true certificate, a malicious user has either purposely requested the conflicting certificate be issued to herself, or has issued the conflicting certificate. For each pair of conflicting but undetermined certificates (neither known to be true), any of either, both, or neither being true must be considered possible. The complexity of algorithm 1 is $O(|V|^3|E|)$ (proof omitted). We now explain how suspect set information can be useful.

5.2   Exploiting Suspect Sets (Non-colluding Malicious Users)

Consider the case where there is a single malicious user. Let L_s represent the intersection of all the suspect sets generated by algorithm 1. Clearly the single malicious user's identity is in L_s. If, on the other hand, there may be up to b non-colluding malicious users, we must determine the maximum disjoint sets (*MDS*) from all the suspect sets generated by algorithm 1. Two sets are called disjoint if they do not have any members (labels) in common. Suppose a solution to MDS consists of m suspect sets, and $m = b$. Let L_m be the union of these m sets. It is clear that all b malicious users must be in L_m.

Unfortunately, *MDS* is also NP-Complete, by transformation from the maximum independent set problem (proof omitted). As a result, the solution may only be approximated.

Given L_s or L_m, we can determine more certificates are true as follows. For each undetermined public key k_t, if there is only one malicious user, we simply test if any single label in L_s is a label-cut for k_0, k_t in the certificate graph. If not, k_t is determined to be true. If there are multiple non-colluding malicious users, we simply test if any single label in L_m is a label-cut for k_0, k_t. If not, k_t is determined to be true. For this computation, a modified breadth-first search suffices, with a complexity of $O(b \cdot |V||E|)$.

5.3   Suspect Sets (Multiple Colluding Malicious Users)

In the case of multiple colluding malicious users, we propose to use the following rules to construct suspect sets. These rules are presented in decreasing order of priority.

Suspect set rule 1 *Given a certificate chain whose head certificate is false, construct a new suspect set that contains all the identities (except the identity in the head certificate) in the certificates of the chain.*

Suspect set rule 2 *Given two certificate chains whose head certificates are conflicting with each other, if one of the head certificates is true and the other is undetermined, construct a new suspect set that contains all the identities in the certificates of the chain whose head certificate is undetermined.*

Suspect set rule 3 *Given two certificate chains whose head certificates are conflicting with each other, if both head certificates are undetermined, construct a new suspect set that contains all the identities in the certificates of the two chains.*

We do not describe an algorithm that implements these rules, due to space limitations. The algorithm is straightforward, and the rules are applied in order. To make use of the suspect sets constructed by these rules, we try to find the maximum number of disjoint sets (*MDS*), from all the suspect sets.

Suppose the number of maximum disjoint sets is found to be a. Let L_c be the union of the a sets. It is clear that at least a malicious users are included in L_c. Next, all the edges with a label in L_c are deleted from the certificate graph. By doing this, the maximum number of malicious users with certificates in the certificate graph is reduced from b to no greater than $b-a$. In this case, Theorem 2 can be applied to determine if the rest of the undetermined certificates are true, as follows. For each target public key k_t, if there exist $b-a+1$ label-disjoint paths between k_0 and k_t, the head certificates of these paths are true. The algorithm described previously for computing the minimum label-cut can be used to solve this problem.

We now present experimental evidence about the benefits of using certificate conflicts.

Table 1. Performance for PGP keyring before and after conflict detection, for a total of 10,000 public keys

# of colluding malicious users	1	2	3	4
# of provably true certificates before conflict detection	78	54	40	33
# of provably true certificates after conflict detection	9992	77	53	40

6   Experimental Results

We implemented and tested our conflict detection method to investigate its practicality and benefits. These experiments only considered the case of one legitimate public key per user. We emulated "typical" malicious user behavior, in order to contrast the performance before and after conflict detection.

For test purposes, we used actual PGP keyrings. These were downloaded from several PGP keyservers, and the keyanalyze [17] tool was used to extract strongly-connected components. In addition, we synthetically generated keyrings to have a larger number of test cases. The synthetic data was generated by the graph generator BRITE [12]. We used the default configuration file for Barabasi graphs, which we believe are similar to actual keyrings. The undirected graphs generated by BRITE were converted to directed graphs by replacing each undirected edge with two directed edges, one in each direction. The number of vertexes in each synthetic key ring was set to 100. For each data point we report, 50 problem instances were generated (using a different random number generator seed each time); the values plotted are the average of these 50 instances.

The first experiment compares our method with the method of [16] on one of the largest PGP keyrings. The graph of this keyring contains 15956 vertexes (users) and 100292 edges (certificates). For this PGP keyring we emulated the behavior of n colluding malicious user as follows. First we randomly picked a target, and then n malicious users were randomly selected. The n malicious users issued n false certificates, one per malicious user, each binding the target's identity to the same false public key.

After emulating this behavior, we applied the method of [16] to determine the maximum set of true certificates. The resulting performance is the performance *before* conflict detection. Then we applied the suspect rules of colluding malicious users (from section 5.3) to find many suspect sets, from which we constructed L_c. For each of the remaining undetermined public keys, we made use of L_c and the method of section 5.3 to determine if it was true. The resulting performance is labeled the performance *after* conflict detection. Table 1 shows the results.

For this very large PGP keyring, it is not practical to evaluate the performance for all users. Instead, we randomly picked 200 users. For each user, we randomly selected 50 public keys on which to test our method. Each user's certificate graph was the entire keyring. The figure shows how many public keys can be determined to be true when there are different numbers of malicious users. This figure shows that when there is only a single malicious user, performance is greatly improved (by two orders of magnitude) with the use of conflict detection. In these experiments, the suspect sets turned out to be quite small. The performance with conflict detection dropped dramatically, however, for the case of two or more colluding malicious users, because of lack of sufficient redundancy in the certificate graph.

For our second experiment, we synthetically generated PGP keyrings. We emulated a single malicious user's behavior, as follows. We randomly picked a target, a malicious user, and n certificate issuers. The malicious user was assumed to ask the n certificate issuers to certify n different public keys for herself. Using these n different public keys, the malicious user created n certificates, one per key, each binding the target's identity to the same false public key. After emulating this behavior, we applied the algorithm for computing the minimum label-cut for the case of $b = 2$, and determined the maximum set of true certificates. The resulting performance was the performance *before* conflict detection. Then we applied algorithm 1 to find many suspect sets, from which we constructed L_s. For each of the remaining undetermined public keys, we made use of L_s and the method of section 5.1 to try to determine if it was true. This gave the performance *after* conflict detection. Each test was run 50 times to obtain averages. The results were:

- Performance before conflict detection was 2%, regardless of the number of false certificates.
- Performance after conflict detection steadily increased from 4% (with 1 false certificate) up to 11% (with 19 false certificates).

The malicious user is faced with a dilemma (fortunately!). While increasing the number of false certificates should increase uncertainty about keys, it also becomes easier to narrow the list of "suspicious" users, thereby limiting the scope of the damage the malicious user can cause.

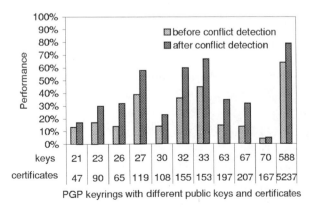

Fig. 2. Performance comparison for PGP keyrings before and after conflict detection

Our final experiment was for real PGP keyrings, and demonstrates how performance is improved by conflict detection. The results are shown in Figure 2. For all cases, performance increased when conflict detection was used.

In each PGP key ring, we emulated a single malicious user's behavior as follows. We randomly picked a target, a malicious user, and two certificate issuers. Then the malicious user asked for two different public keys, certified by the two certificate issuers. Using these two public keys, the malicious user created two public key-independent certificate chains to the target. After emulating this behavior, we again applied the method for determining if a label-cut of size b of less exists, for $b = 2$, to determine the maximum set of true certificates. The resulting performance is the performance *before* conflict detection. Then we used the method of section 5.3) to obtain the performance *after* conflict detection.

All experiments were performed on a Pentium IV, 2.0GHZ PC with 512MB memory. Running times for figure 2 ranged from 5 to 30 seconds, except for the graph with 588 vertexes, which required 10 hours of CPU time. We believe analysis of keyrings for robustness will be done infrequently, in which case these execution times should be acceptable.

7   Conclusion and Future Work

In this paper, we described the problem of proving certificates are true in webs of trust (such as PGP Keyrings). This is a difficult problem because malicious users may falsify information. Under the assumption that users may legitimately have multiple public keys, we showed that redundant *identity-independent* certificate chains are necessary. Previous methods based on *public key-independent chains* are not sufficient under this assumption.

In the case that certificate conflicts are detected, it is possible to exclude certain users from the set of possible malicious users. This allows additional certificates to be proved true. Experimental results demonstrated that (a) the web of trust is seriously degraded as the number of malicious users increases, and (b) the use of conflict detection and redundant certificates substantially improves the ability to prove certificates are true.

Our results show that current PGP keyrings are not particularly resistant to attacks by malicious users, particularly colluding users. We are currently investigating ways to increase the robustness of webs of trust, such as PGP keyrings.

References

1. R. Ahuja, T. Magnanti, and J. Orlin. *Network flows : theory, algorithms, and applications*. Prentice Hall, Englewood Cliffs, N.J., 1993.
2. Thomas Beth, Malte Borcherding, and Birgit Klein. Valuation of trust in open networks. In *Proceeding of the 3rd European Symposium on Research in Computer Security (ESORICS 94)*, pages 3–18, 1994.

3. M. Blaze and J. Feigenbaum. Decentralized trust management. In *Proceedings of the 1996 IEEE Symposium on Security and Privacy*, pages 164–173, Oakland CA USA, 6-8 May 1996.
4. M. Burmester, Y. Desmedt, and G. A. Kabatianski. Trust and security: A new look at the byzantine generals problem. In *Proceedings of the DIMACS Workshop on Network Threats*, volume 38 of *DIMACS*. American Mathematical Society Publications, December 1996.
5. John R. Douceur. The sybil attack. In *Proceedings for the 1st International Workshop on Peer-to-Peer Systems (IPTPS '02)*, MIT Faculty Club, Cambridge, MA, USA, March 2002.
6. C. Ellison, B. Frantz, B. Lampson, R. Rivest, B. Thomas, and T. Ylonen. RFC 2693: SPKI certificate theory, September 1999.
7. M.R. Garey and D.S. Johnson. *Computers and Intractability: A Guide to the Theory of NP-Completeness*. W H Freeman & Co., 1979.
8. Qinglin Jiang, Douglas S. Reeves, and Peng Ning. Improving robustness of PGP keyrings by conflict detection. Technical Report TR-2003-19, Department of Computer Science, N.C. State University, October 2003.
9. Audun Josang. The consensus operator for combining beliefs. *Artificial Intelligence*, 141(1):157–170, 2002.
10. Sanjeev Khanna, Rajeev Motwani, Madhu Sudan, and Umesh V. Vazirani. On syntactic versus computational views of approximability. In *IEEE Symposium on Foundations of Computer Science*, pages 819–830, 1994.
11. Ueli Maurer. Modelling a public-key infrastructure. In *Proceedings of the Fourth European Symposium on Research in Computer Security (ESORICS 96)*, pages 324–350, 1996.
12. Alberto Medina, Anukool Lakhina, Ibrahim Matta, and John Byers. BRITE: Universal topology generation from a user's perspective. Technical Report BU-CS-TR-2001-003, Boston University, 2001.
13. S. Mendes and C. Huitema. A new approach to the X.509 framework: Allowing a global authentication infrastructure without a global trust model. In *Proceedings of the Symposium on Network and Distributed System Security, 1995*, pages 172–189, San Diego, CA , USA, Feb 1995.
14. A. Menezes, P. Van Oorschot, and S. Vanstone. *Handbook of applied cryptography*. CRC Press, Boca Raton, Fla., 1997.
15. M. Reiter and S. Stubblebine. Toward acceptable metrics of authentication. In *IEEE Symposium on Security and Privacy*, pages 10–20, 1997.
16. M. Reiter and S. Stubblebine. Resilient authentication using path independence. *IEEE Transactions on Computers*, 47(12), December 1998.
17. M. Drew Streib. Keyanalyze - analysis of a large OpenPGP ring. http://www.dtype.org/keyanalyze/.
18. Anas Tarah and Christian Huitema. Associating metrics to certification paths. In Yves Deswarte, Gérard Eizenberg, and Jean-Jacques Quisquater, editors, *Computer Security - ESORICS 92, Second European Symposium on Research in Computer Security, Toulouse, France, November 23-25, 1992, Proceedings*, volume 648 of *Lecture Notes in Computer Science*, pages 175–189. Springer Verlag, 1992.
19. Int'l Telecommunications Union/ITU Telegraph & Tel. ITU-T recommendation X.509: The directory: Public-key and attribute certificate frameworks, Mar 2000.
20. Philip Zimmermann. *The official PGP user's guide*. MIT Press, Cambridge, Mass., 1995.

Issues of Security with the Oswald-Aigner Exponentiation Algorithm

Comodo Research Lab
10 Hey Street, Bradford, BD7 1DQ, UK
Colin.Walter@comodogroup.com
www.comodogroup.com

Abstract. In smartcard encryption and signature applications, randomized algorithms can be used to increase tamper resistance against attacks based on averaging data-dependent power or EMR variations. Oswald and Aigner describe such an algorithm for point multiplication in elliptic curve cryptography (ECC). Assuming an attacker can identify and distinguish additions and doublings during a single point multiplication, it is shown that the algorithm is insecure for repeated use of the same secret key without blinding of that key. Thus blinding should still be used or great care taken to minimise the differences between point additions and doublings.

Keywords: Addition-subtraction chains, randomized exponentiation, elliptic curve cryptography, ECC, point multiplication, power analysis, SPA, DPA, SEMA, DEMA, blinding, smartcard.

1 Introduction

Side channel attacks [6,7] on embedded cryptographic systems show that substantial data about secret keys can leak from a *single* application of a cryptographic function through data-dependent power variation and electro-magnetic radiation [12,13]. This is particularly true for crypto-systems which use the computationally expensive function of exponentiation, such as RSA, ECC and Diffie-Hellman. Early attacks required averaging over a number of exponentiations [9] to extract meaningful data, but improved techniques mean that single exponentiations using traditional algorithms may be insecure. In particular, it should be assumed that the pattern of squares and multiplies can be extracted fairly accurately from side channel leakage, perhaps by using Hamming weights to identify operand re-use. Where the standard binary "square-and-multiply" algorithm is used, this pattern reveals the secret exponent immediately.

In this context, Oswald and Aigner proposed a randomized point multiplication algorithm [10] for which there is no bijection between scalar key values and sequences of curve operations. They randomly switch to a different procedure for which multiplications appear to occur instead for zero bits but not for one bits. This alternative corresponds to a standard recoding of the input bits to

T. Okamoto (Ed.): CT-RSA 2004, LNCS 2964, pp. 208–221, 2004.
© Springer-Verlag Berlin Heidelberg 2004

remove long sequences of 1s and introduces other non-zero digits such as $\bar{1}$. On the one hand, the pattern of squares and multiplications is no longer fixed, so that averaging power traces from several exponentiations does not make sense, and, on the other hand, there is ambiguity about which digit value is associated with each multiplication.

This article analyses the set of randomized traces that would be generated by repeated re-use of the same unblinded key k. By aligning corresponding doublings in a number of traces, the possible operation sequences associated with bit pairs and bit triples of the secret key k can be extracted. With only a few traces (ten or so) this provides enough information to determine half the bits of k unequivocally, and the rest with a very high degree of certainty.

Previous work in this area includes [11] and [14]. In [11] Oswald takes a similar but *deterministic* algorithm and shows how to determine a space of possible keys from one sequence of curve operations, but not how to combine such results from different sequences. Here randomization minimises the inter-dependence between consecutive operations and so it is unclear whether or not her techniques lead to an intractable amount of computing. Okeya & Sakurai [14] treat the simple version of the randomized algorithm and succeed in combining results from different multiplications by the same key. They require the key k to be re-used $100+ \log_2 k$ times. Here we treat the more complex version of the algorithm in an extended form which might increase security. The analysis of Okeya & Sakurai is inapplicable here because it depends on a fixed finite automaton state occurring after processing a zero bit. However, using new methods we find that a) measurements from only $O(10)$ uses of the secret key reveal the key by applying theory which considers pairs of bits at a time, b) software which considers longer sequences of bits can process just two uses to obtain the key in $O(\log k)$ time, and c) for standard key lengths and perfect identification of adds and doubles, a *single* use will disclose the key in a tractable amount of time. In addition, our attack seems less susceptible to error: key bits are deduced independently so that any incorrect deductions affect at most the neighbouring one or two bits. In comparison, the attack of Okeya & Sakurai recovers bits sequentially, making recovery from errors more complex.

Although only one algorithm is studied here, a similar overall approach can be used to break most randomized recoding procedures under the same conditions. The two main properties required are: i) after a given sequence of point operations, the unprocessed part k' of the key can only have one of a small, bounded number of possible values (determined from k by the length of the operation sequence but independent of other choices); and ii) it is possible to identify an associated subset of trace suffixes for which all members correspond to the same value of k'. These also hold for the algorithm proposed by Liardet & Smart [8], which uses a sliding window of random, variable width. They seem to be the key properties required in [16] to demonstrate similar weaknesses in that algorithm.

Several counter-measures exist against this type of attack. As well as standard blinding by adding a random multiple of the group order to the exponent, different algorithms can be employed, such as [3,5]. Moreover, formulae for point additions and doublings can be made indistinguishable [1,2,4,8].

2   The Oswald-Aigner Exponentiation Algorithm

This section contains a brief outline of the Oswald-Aigner algorithm [10] in terms of the additive group of points on an elliptic curve E. Rational integers are written in lowercase while points on the curve are written in capitals and Greek characters denote probabilities. The algorithm computes the point $Q = kP$ for a given positive integer k (the secret key) and a given point P on E.

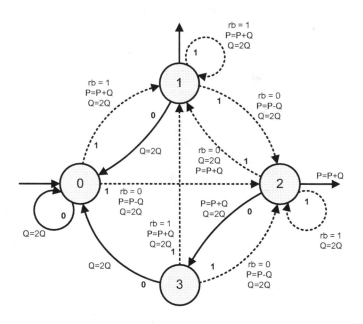

Fig. 1. Finite automaton for an extension of the algorithm. rb is a random bit.

The algorithm randomly introduces alternative re-codings to the representation of k. It can be viewed as pre-processing bits of k from right to left into a new digit set $\{-1, 0, +1, +2\}$. Then the resulting scheme for point multiplication can be performed in either direction. The conversion uses a carry bit set initially to 0. When this bit is summed with the current bit of k, the result 0, 1 or 2 can be re-coded in different ways: 0 always gives a new digit 0 with carry 0; 1 can give either new digit 1 and carry 0, or new digit $\bar{1}$ with carry 1; and 2 gives either new digit 0 and carry 1, or new digit 2 and carry 0. Fig. 1 illustrates this as a finite automaton for a slight extension of the original right-to-left algorithm. It has 4 states, numbered 0 to 3 with the carry being 1 if, and only if, the state is 2.

For the transition from state 2 to state 1, the normal order of doubling and adding is reversed. This achieves the processing for digit value 2. The extension here allows a new transition from state 0 to state 2; the original algorithm is the special case in which the random bit $rb = 1$ always for state 0. The extension also allows the random bits to be biased for each state. However, if the same distribution of random bits is used for each of the states 0, 1 and 3, the automaton simplifies to just two states, obtained by merging states 0, 1 and 3.

Figure 2 provides equivalent code for the associated right-to-left point multiplication. A left-to-right version is also possible, and can be attacked in the same way.

```
Q ← O ;                    /* O is the zero of the elliptic curve */
State ← 0 ;
While k > 0 do
{
    If (k mod 2) = 0 then
    case State of
    {
      0,1,3 : Q ← 2Q ; State ← 0 ;
      2     : P ← P+Q ; Q ← 2Q ; State ← 3 ;
    }
    else
    case State of
    {
      0,1,3 : If rb = 0 then        /* rb is a Random Bit */
                 { P ← P-Q ; Q ← 2Q ; State ← 2 }
              else
                 { P ← P+Q ; Q ← 2Q ; State ← 1 } ;
      2     : If rb = 0 then        /* rb is a Random Bit */
                 { Q ← 2Q ; P ← P+Q ; State ← 1 }
              else
                 { Q ← 2Q } ;
    } ;
    k ← k div 2 ;
} ;
If State = 2 then P ← P+Q ;
```

Fig. 2. Oswald & Aigner's randomized signed binary exponentiation (extended).

3   Efficiency Considerations

Definition 1. *Let α, β, γ and δ be the probabilities that the random bit rb is chosen to be 1 when the current state is 0, 1, 2 or 3 respectively.*

These probabilities can be chosen to improve efficiency or, as we shall see, security. For a key k whose bits are selected independently and at random from a uniform distribution, the matrix of transition probabilities between states of the automaton is then

$$
\begin{bmatrix}
\frac{1}{2} & \frac{1}{2} & 0 & \frac{1}{2} \\
\frac{\alpha}{2} & \frac{\beta}{2} & \frac{1-\gamma}{2} & \frac{\delta}{2} \\
\frac{1-\alpha}{2} & \frac{1-\beta}{2} & \frac{\gamma}{2} & \frac{1-\delta}{2} \\
0 & 0 & \frac{1}{2} & 0
\end{bmatrix}
$$

Lemma 1. *The transition matrix has an eigen-vector $(\frac{1}{2}-\mu, \frac{1}{2}-2\mu, 2\mu, \mu)$ where $\mu = \frac{2-\alpha-\beta}{12-2\alpha-4\beta-4\gamma+2\delta}$. Its elements are the probabilities associated with each state. Moreover, $0 \le \mu \le \frac{1}{4}$.*

This is an easy exercise for the reader. Taking the dot product of this eigen-vector with the vector $(\frac{1}{2}, \frac{1}{2}, 1-\frac{1}{2}\gamma, \frac{1}{2})$ of average additions associated with each state provides the expected number of additions per bit: $\frac{1}{2}+(1-\gamma)\mu$. The number of doublings is constant at one per bit. So, to minimise the total time we require $(1-\gamma)\mu = 0$, i.e. $(1-\gamma)(2-\alpha-\beta) = 0$, i.e. disallow either the transition from state 2 back to state 1, or both transitions to state 2 from states 0 and 1. Avoiding these extremes provides greater randomness. In particular, α and/or β should be kept away from 1 so that states 2 and 3 are reachable. In the limit as $\alpha\beta\gamma\delta\rightarrow1$ (which optimises efficiency), on average there is half an addition per bit of k. Thus, a typical addition chain has a little over $\frac{1}{2}\log_2 k$ additions (or subtractions). Even a modest bias towards efficiency, such as taking $\alpha = \beta = \gamma = \delta \ge \frac{3}{4}$, changes this by just 2% or less.

4   The Attack

4.1   Initial Hypotheses, Notation, and Overview of the Attack

The attack here assumes sufficiently good monitoring equipment and a sufficiently weak implementation. Specifically it is assumed that:

- Adds and doublings can always be identified and distinguished correctly in side channel leakage from a single point multiplication;   and
- Side channel traces are available for a number of different uses of the same, unblinded key value.

For ease in calculating probabilities, we assume adds and doublings can *always* be distinguished. Similar results hold if this is only usually the case.

By the first hypothesis,

- every side-channel trace *tr* can be viewed as a word over the alphabet $\{A, D\}$

where A denotes the occurrence of an addition and D that of a doubling. Here, as expected, the trace is written with time increasing from left to right. However, this is the opposite of the binary representation of the secret key k which is processed *from right to left*, so that the re-coding can be done on the fly (Fig. 2). For example, if the machine were to cycle round only states 0 and 1 giving the sequence of operations for square-and-multiply exponentiation, then the trace would be essentially the same as the binary, but reversed: every occurrence of 0 would appear as D, and every occurrence of 1 would appear as AD. So the binary representation 11001 would generate the trace $ADDDADAD$. There is one D for every bit, and we index them to correspond:

Definition 2. *The* position *of an instance of D in a trace is the number of occurrences of D to its left.*

Thus, the leftmost D of $ADDDADAD$ has position 0 and arises from processing the rightmost bit of 11001, which has index 0.

The attack consists of a systematic treatment of observations like the following. The only transition which places D before rather than after an associated occurrence of A is the transition (21). Hence, every occurrence of the substring $DAAD$ in a trace tr corresponds to traversing transition (21) then (12) or (11) in the finite automaton. This must correspond to processing a bit 1 to reach state 2, and then two further 1 bits. The trace can be split between the two adjacent As into a prefix and a suffix. There is a corresponding split in the binary representation of the secret key k such that the suffix of k has a number of bits equal to the number of Ds in the prefix of tr. This enables the position of the substring 111 to be determined in k. Moreover, by the next lemma, most occurrences of 111 can be located in this way if enough traces are available: $DAAD$ appears exactly when the middle 1 is represented by the transition (21).

Lemma 2. *If 11 occurs in the binary representation of k then the probability of the left-hand 1 being represented by transition (21) in a trace for k is $\pi = 4\mu(1-\gamma)$.*

Proof. 4μ is the probability of being in state 2 as a result of the right-hand 1 and $1-\gamma$ is the (independent) probability of selecting transition (21) next.   □

4.2   Properties of the Traces

Figure 3 lists the transitions and operation sequences which can occur for each bit pair, including the probability of each. It assumes that initial states have the probabilities determined by Lemma 1, and that neighbouring bits are unknown. The figure enables one to see which bit pairs can arise from given patterns in a trace, and to calculate their probabilities:

Lemma 3. *Let k_i denote the bit of k with index i, and μ be as in Lemma 1. Then,*
i) For a given trace, if the Ds in positions i and $i+1$ are not separated by any As, then the bit pair $k_{i+1}k_i$ is 00 with probability $(2-2\mu(1-\gamma))^{-1}$, which is at least $\frac{1}{2}$. If the Ds are separated by one or more As in any trace, then the bit pair is certainly not 00.

ii) *For a given trace, if the Ds in positions i and $i+1$ are separated by one A, then the bit pair $k_{i+1}k_i$ is 10 with probability $\frac{1}{2}$. If the Ds are separated by no As or two As in any trace, then the bit pair is certainly not 10.*

iii) *For a given trace, if the Ds in positions i and $i+1$ are separated by two As, then the bit pair $k_{i+1}k_i$ is certainly 11. The probability of two As when the bit pair is 11 is $2\mu(1-\gamma)$, assuming bit k_{i-1} is unknown.*

iv) *For a set of n traces, suppose the Ds in positions i and $i+1$ are separated by no As in some cases, by one A in some cases, and by two As in no cases. Then the bit pair $k_{i+1}k_i$ is 01 with probability $(1+(1-2\mu(1-\gamma))^n)^{-1}$.*

Bit Pair	Operation Patterns	State Sequences	Probabilities, given the bit pair
00	$D.D$	$000, 100, 300$	$1-2\mu$
	$AD.D$	230	2μ
10	$D.AD$	$001, 002$	$\frac{1}{2}-\mu$
	$D.AD$	$101, 102$	$\frac{1}{2}-2\mu$
	$AD.AD$	$231, 232$	2μ
	$D.AD$	$301, 302$	μ
01	$AD.D,\ AD.AD$	$010, 023$	$(\frac{1}{2}-\mu)\alpha,\ (\frac{1}{2}-\mu)(1-\alpha)$
	$AD.D,\ AD.AD$	$110, 123$	$(\frac{1}{2}-2\mu)\beta,\ (\frac{1}{2}-2\mu)(1-\beta)$
	$DA.D,\ \ D.AD$	$210, 223$	$2\mu(1-\gamma),\ 2\mu\gamma$
	$AD.D,\ AD.AD$	$310, 323$	$\mu\delta,\ \mu(1-\delta)$
11	$AD.AD,\ AD.AD$	$011, 012$	$(\frac{1}{2}-\mu)\alpha\beta,\ (\frac{1}{2}-\mu)\alpha(1-\beta)$
	$AD.DA,\ AD.D$	$021, 022$	$(\frac{1}{2}-\mu)(1-\alpha)(1-\gamma),\ (\frac{1}{2}-\mu)(1-\alpha)\gamma$
	$AD.AD,\ AD.AD$	$111, 112$	$(\frac{1}{2}-2\mu)\beta^2,\ (\frac{1}{2}-2\mu)\beta(1-\beta)$
	$AD.DA,\ AD.D$	$121, 122$	$(\frac{1}{2}-2\mu)(1-\beta)(1-\gamma),\ (\frac{1}{2}-2\mu)(1-\beta)\gamma$
	$DA.AD,\ DA.AD$	$211, 212$	$2\mu(1-\gamma)\beta,\ 2\mu(1-\gamma)(1-\beta)$
	$D.DA,\ \ D.D$	$221, 222$	$2\mu\gamma(1-\gamma),\ 2\mu\gamma^2$
	$AD.AD,\ AD.AD$	$311, 312$	$\mu\delta\beta,\ \mu\delta(1-\beta)$
	$AD.DA,\ AD.D$	$321, 322$	$\mu(1-\delta)(1-\gamma),\ \mu(1-\delta)\gamma$

Fig. 3. All possible operation sequences for all bit pairs, and their probabilities given the bit pair occurs. (*Bit pairs are processed right to left and operations left to right.*)

Proof. i) First, by inspection of the finite automaton, the only possible operation sequences for 00 are ADD and DD. So the Ds are always adjacent. The intervention of an A will prove that the bit pair is not 00.

Suppose there is no intervening A between the two specified Ds. Using Figure 3, if the bit pair is 00 then the probability of this is $\pi_{00} = 1$; if the bit pair is 10 then the probability is $\pi_{10} = 0$; if the bit pair is 01 then the probability is $\pi_{01} = (\frac{1}{2}-\mu)\alpha + (\frac{1}{2}-2\mu)\beta + \mu\delta$; and if the bit pair is 11 then the probability is $\pi_{11} = (\frac{1}{2}-\mu)(1-\alpha) + (\frac{1}{2}-2\mu)(1-\beta) + 2\mu\gamma + \mu(1-\delta)$. Thus, the correct deduction

of 00 is made with probability
$$\pi_{00}/(\pi_{00}+\pi_{10}+\pi_{01}+\pi_{11}) = 1/(2-2\mu(1-\gamma)) \geq \tfrac{1}{2}.$$
ii) Similarly, from Figure 3 the bit pair 10 must always include the operation A once between the two occurrences of D, but this is not the case for any other bit pair. Thus the absence of an A, or the presence of two As, guarantees the bit pair is not 10. However, suppose there is exactly one A between the specified Ds. By Figure 3, if the bit pair is 00 then the probability of this is $\pi'_{00} = 1-\pi_{00} = 0$; if the bit pair is 10 then the probability is $\pi'_{10} = 1-\pi_{10} = 1$; if the bit pair is 01 then the probability is $\pi'_{01} = 1-\pi_{01}$; and if the bit pair is 11 then the probability is $\pi'_{11} = 1-\pi_{11}-2\mu(1-\gamma)$. Thus, the correct deduction of 10 is made with probability
$$\pi'_{10}/(\pi'_{00}+\pi'_{10}+\pi'_{01}+\pi'_{11}) = \tfrac{1}{2}.$$

iii) This part is immediate from Figure 3.

iv) Finally, by parts (i) and (ii), a bit pair which includes both the possibilities of no As and of one A between the specified Ds cannot be 00 or 10; it must be 01 or 11. The probability of not having two As in any trace when the digit pair is 01 is 1, of course. By Fig. 3 the probability of not having two As in any of the n traces when the digit pair is 11 is $\pi_n = (1-2\mu(1-\gamma))^n$. Hence the probability of the pair being 01 rather than 11 is $1/(1+\pi_n)$.   □

We must be a little careful in the application of this lemma. Firstly, each part assumes no knowledge of bit k_{i-1}. Knowing it changes the probabilities. In most cases, the differences are small enough to be considered negligible; for accurate figures the table can be used to select just the cases starting in states 0 or 3 when the preceding processed bit is 0, and the cases starting in states 1 or 2 when that bit is 1. The only case where a qualitative difference occurs is for 11 when AA only occurs if $k_{i-1} = 1$. In the case of $k_{i-1} = 0$ this means we cannot distinguish 01 from 11 so easily. This is a typical problem to solve when reconstructing the whole key.

Secondly, deductions from different traces are not independent. For example, suppose all of n traces have one A between the Ds in positions i and $i+1$. From (ii) of the lemma it is tempting to deduce that the bit pair is 10 with probability $1-(\tfrac{1}{2})^n$. However, the probability of this may still only be $\tfrac{1}{2}$. In particular, this happens when the parameters $\alpha = \beta = \delta = 0$ are selected. Then the bit pairs 10 and 01 would always have exactly one A between the Ds, and bit pairs 00 and 11 would never have any As. So 01 and 10 would be equally likely with probability $\tfrac{1}{2}$ if exactly one A always occurred. The independent decisions which *can* be combined are those based on the independent choices of random bits, as in (iv).

4.3   Reconstructing the Key

For this section we assume the default values which give the original algorithm, namely $\alpha = 1$ and $\beta = \gamma = \delta = \tfrac{1}{2}$. This means $\mu = \tfrac{1}{14}$. Later we consider alternatives which might improve security. Then Figure 3 immediately yields:

Lemma 4. *For the above default values of the parameters,*
i) the bit pair 01 *has no intervening A between the associated Ds of a trace with probability* $\frac{9}{14}$ *and one intervening A with probability* $\frac{5}{14}$;
ii) the bit pair 11 *has no intervening A between the associated Ds with probability* $\frac{2}{7}$, *one intervening A with probability* $\frac{9}{14}$, *and two As with probability* $\frac{1}{14}$.

The choices which lead to the probabilities in the previous lemma are made independently for each trace. Hence, for n traces and a pair 01, there are no As in every trace with probability $(\frac{9}{14})^n$ and one A in every trace with probability $(\frac{5}{14})^n$. A similar result holds for the pair 11. By averaging:

Lemma 5. *For the default values of the parameters and n traces, in every trace a bit pair of the form* ∗1 *has:*
i) no As between the associated Ds with probability $\{(\frac{9}{14})^n + (\frac{2}{7})^n\}/2$; *and*
ii) one A with probability $\{(\frac{5}{14})^n + (\frac{9}{14})^n\}/2$.

To reconstruct the key k, first classify every bit pair as 00 if there are no intervening As in any trace, 10 if there is always one intervening A, 11 if there is an intervening AA, and, otherwise, ∗1 if there is a variable number of intervening As. This correctly classifies all pairs 00 and 10, and pairs classed as 11 or ∗1 are certainly all 11 or of the form ∗1 respectively. For $n = 10$ both probabilities in the lemma are bounded above by $\frac{1}{2}(\frac{9}{14})^{10} \approx 1/166$. Thus about 1 in 83 bits pairs 01 and 11 will be incorrectly classified as 00 or 10. Also, by the next lemma, $1-(\frac{6}{7})^{10} > \frac{3}{4}$ of pairs 11 will be located correctly by occurrences of AA when they are the left pair in triplets 111. The proof of it goes back to Lemma 2.

Lemma 6. *For the default values of the parameters and n traces, the bit pair* 11 *has at least one trace exhibiting AA with probability* $1-(\frac{6}{7})^n$ *if it has a 1 to the right and with probability* 0 *if it has a 0 to the right.*

This is now enough information to deduce almost all the bits of a standard length ECC key. Every bit which is deduced as the right member of a pair ∗1 is correctly classified as 1 since the mixture of patterns used in the classification is not possible for pairs of the form ∗0. However, about 1 in 83+1 of the bits which are deduced to be right members of a pair ∗0 is incorrectly classified as 0 because not all the possible patterns for the bit pair have occurred. In an ECC key of, say, 192 bits, about two bits will then be incorrect.

Each bit b belongs to two pairs: ∗b and b∗, say. Traces for the pair ∗b have been used to classify b. In half of all cases, there is a 0 bit to the right and the characteristic patterns of traces for the pair b0 can be used to cross-check the classification. In the other half of cases the patterns for b1 also indicate the correct value for b as a result of the ratios between the numbers of occurrences of each pattern. However, the patterns observed for overlapping bit pairs are not independent. Although unlikely, one set of patterns may reinforce rather than contradict a wrong deduction from the other set. There is no space for further detail, but the following is now clear:

Theorem 1. *Suppose elliptic curve adds and doubles can be distinguished accurately on a side channel. If the original Oswald-Aigner exponentiation algorithm is used with the same unblinded 192-bit ECC key k for 10 point multiplications then approximately half the bits can be deduced unambiguously to be 1, and the remaining bits deduced to be 0 with an average of at most about two errors.*

This theorem says that a typical ECC secret key can usually be recovered on a first attempt using a dozen traces with very little computational effort beyond extracting the add and double patterns from each trace. By checking consistency between deductions of overlapping bit pairs, most errors should be eliminated. However, it is computationally feasible to test all variants of the deduced key for up to two or three errors. The correct one from this set can surely be established by successfully decrypting some ciphertext.

4.4   Secure Parameter Choices?

From the last section, it is clear that greater security could only arise from making it less easy to distinguish between pairs of the form *0 and those of the form *1. This requires choosing parameters for which 01 and 11 are less likely to exhibit both no As and one A between the relevant Ds. From Fig. 3, the probability of no As for 01 and the probability of one A for 11 are the same, *viz.*

$$\pi = (\tfrac{1}{2}-\mu)\alpha + (\tfrac{1}{2}-2\mu)\beta + \mu\delta.$$

So this must be made close to 0 or close to 1.

For example, choosing $\alpha = \beta = 1$ makes $\mu = 0$ and so $\pi = 1$, whereas choosing $\alpha = \beta = \delta = 0$ makes $\pi = 0$. Thus both limits are possible. In general, for $\pi = 1$ (the first case) the traces match the pattern of operations for normal square-and-multiply, so we expect each A to correspond to the multiply of a 1 bit.

Although 00 and 01 are indistinguishable from the patterns, and 10 and 11 are indistinguishable (unless perhaps AA could occur), the attacker now recognises that patterns for the pairs 0* have no intervening A and patterns for the pairs 1* have one intervening A. This gives him each bit unequivocally. At the opposite extreme, if $\pi = 0$ (the second case) then 10 and 01 become indistinguishable from the patterns as do 00 and 11 (again, unless perhaps AA could occur). Now the attacker recognises pairs with equal bits from pairs with different bits. Knowing the first bit is 1, he can deduce all the bits one by one from left to right, and hence the key k.

In general the attacker can exploit the complementary frequencies of one A for the pairs 01 and 11. Either they are close enough to ensure n traces usually display both patterns (as in the previous section) or they are distinct enough for the patterns to be strongly biased in opposite directions in the trace set (as in the previous paragraph). He can then recognise either the equality of the second bits or the difference in the first bit respectively, and use the fact that each bit belongs to two pairs to cross-check the deduction of many bits. Consequently, there are no secure choices of the parameters under repeated use of the unblinded key k.

Identical working to the previous section shows that similar computations can be performed for keys of any length. With the choice of parameters there, the number of traces needed to achieve a specified degree of confidence in the determined bits is $n = O(\log \log k)$ because we want at most one error in $(\frac{14}{9})^n = O(\log k)$ bits. The same calculations apply for any π which is not 0 or 1, giving the same size order for n. For the working above in this section, mistakes are only made when too many traces record the opposite pattern to that expected from the value of π. Then, for π close enough to 0 or 1, the same bound on the size of n can be obtained for limiting the errors.

Theorem 2. *No choice of algorithm parameters is secure for a reasonable key length under the above attack if $O((\log k)^2)$ decipherings are computationally feasible and $O(\log \log k)$ traces are available from point multiplications using the same unblinded key.*

When adds and doubles are not distinguished with 100% certainty, the proportions of numbers of As can be used to assign a likelihood to the correctness of the selected bit pair. Those which are most likely to be wrong can be modified first, thereby decreasing the search time to determine the correct key.

4.5   Counter-Measures

In the absence of a secure set of parameter choices, further counter-measures are required. The most obvious counter-measure is to restore key blinding. A small number of blinding bits might still result in the attacker's desired 10 or so traces for the same key eventually becoming available. These might be identified easily within a much larger set of traces by the large number of character subsequences shared between their traces. So the size of the random number used in blinding cannot reasonably be less than the maximum lifespan of the key in terms of the number of point multiplications for which it is used. Thus 16 or more bits are needed, adding around 10% to the cost of point multiplication.

Identical formulae for additions and doublings are increasingly efficient and applicable to wider classes of elliptic curves, those of Brier and Joye [1] in particular. These should make it more difficult to distinguish adds from doubles. Another favoured counter-measure is the add-and-always-double approach. Then the pattern of adds and doubles is not key dependent. Each occurrence of DD has an add inserted to yield the pattern DAD, but the add output is discarded without having been used. This can also be done for the Oswald-Aigner algorithm provided, in addition, an extra double is performed to convert each $DAAD$ into $DADAD$. The output of this double is likewise ignored.

Alternatives algorithms exist. That described by Joye and Yen [5] is another add-and-always-double algorithm. There are also several randomized methods [3, 15] which seem to be more robust because they do not satisfy the two properties identified in the introduction as those to which the above attack can be applied.

5   One Trace

It is interesting to speculate on how much data leaks from a single point multiplication since the above counter-measures should prevent re-use of identical values for the same key. Oswald [10] noted that for some *deterministic* re-coding algorithms in which several non-zero digits generate indistinguishable As, the operation patterns resulting from numbers of up to 12 bits could only represent at most 3 keys. By breaking a standard ECC key into 12-bit sections, this means very few keys actually generate an observed patterns of operations. Moreover, these can be ordered according to their likelihood of occurrence, and this considerably reduces the average search time for the correct key. Hence the key can be recovered quite easily.

Is the same possible here? In [10] she also writes that the same attack is possible on randomized algorithms with weaker results, but provides no detail. Randomized algorithms have much weaker inter-dependencies between adjacent operation patterns. This should substantially increase the number of keys which match a specific pattern of point operations. The key Lemma 3 above does not provide certainty for many bits unless a number of traces are available; only the infrequent instances of AA seem to allow definite determination of any bits from one trace. Of course, an analysis of sub-sequences of more than two bits is possible, as in [14], but, besides better probabilities, this gives no further insight into whether it is computationally feasible to recover the key from a single trace.

Instead, software was written to enumerate all the keys which could represent a given string. On average, for the extended version of the algorithm, the trend up to 16-bit keys indicates clearly that a little over $O(\sqrt[4]{k})$ keys will match a given pattern – under 20 match a given 16-bit pattern. This would appear to ensure the strength of the algorithm when a key is used just once but only if the key has at least 2^8 bits or there is considerable ambiguity in the side channel about whether the operations are adds or doubles. The original algorithm has fewer random choices, and so has even fewer keys matching a given pattern. Thus, a standard ECC key could be recovered from a single trace in feasible time if adds and doubles are clearly distinguishable.

6   Conclusion

One of several, similar, randomized exponentiation algorithms has been investigated to assess its strength against a side channel attack which can differentiate between elliptic curve point additions and point doublings. Straightforward theory shows that at most $O(10)$ uses of the same unblinded key will enable a secret key of standard length to be recovered easily in a computationally feasible time. No choice of parameters improves security enough to alter this conclusion. Using longer bit sequences than the theory, it is also clear that software can search successfully for keys when just one side channel trace is available. However, this number may need increasing if adds and doubles might be confused or standards for key lengths are increased.

The main property which is common to algorithms which can be attacked in this way seems to be that the next subsequence of operations at a given point in the processing of the key must be chosen from a small, bounded set of possibilities which is derived from the key and the position, but is independent of previous choices. Hence, our overall conclusion is that such algorithms should be avoided for repeated use of the same unblinded key if adds and doubles can be differentiated with any degree of certainty. Furthermore, for typical ECC key lengths, a single use may be sufficient to disclose the key when adds and doubles are accurately distinguishable.

References

1. E. Brier & M. Joye, *Weierstraß Elliptic Curves and Side-Channel Attacks*, Public Key Cryptography, P. Paillier & D. Naccache (eds), LNCS **2274**, Springer-Verlag, 2002, pp. 335–345.
2. C. Gebotys & R. Gebotys, *Secure Elliptic Curve Implementations: An Analysis of Resistance to Power-Attacks in a DSP Processor*, CHES 2002, B. Kaliski, Ç. Koç & C. Paar (eds), LNCS **2523**, Springer-Verlag, 2003, pp. 114–128.
3. K. Itoh, J. Yajima, M. Takenaka & N. Torii, *DPA Countermeasures by improving the Window Method*, CHES 2002, B. Kaliski, Ç. Koç & C. Paar (eds), LNCS **2523**, Springer-Verlag, 2003, pp. 303–317.
4. M. Joye & J.-J. Quisquater, *Hessian Elliptic Curves and Side Channel Attacks*, CHES 2001, Ç. Koç, D. Naccache & C. Paar (eds), LNCS **2162**, Springer-Verlag, 2001, pp. 402–410.
5. M. Joye & S.-M. Yen, *The Montgomery Powering Ladder*, CHES 2002, B. Kaliski, Ç. Koç & C. Paar (eds), LNCS **2523**, Springer-Verlag, 2003, pp. 291–302.
6. P. Kocher, *Timing Attack on Implementations of Diffie-Hellman, RSA, DSS, and other Systems*, Advances in Cryptology – CRYPTO '96, N. Koblitz (ed), LNCS **1109**, Springer-Verlag, 1996, pp. 104–113.
7. P. Kocher, J. Jaffe & B. Jun, *Differential Power Analysis*, Advances in Cryptology – CRYPTO '99, M. Wiener (ed), LNCS **1666**, Springer-Verlag, 1999, pp. 388–397.
8. P.-Y. Liardet & N. P. Smart, *Preventing SPA/DPA in ECC Systems using the Jacobi Form*, CHES 2001, Ç. Koç, D. Naccache & C. Paar (eds), LNCS **2162**, Springer-Verlag, 2001, pp. 391–401.
9. T. S. Messerges, E. A. Dabbish & R. H. Sloan, *Power Analysis Attacks of Modular Exponentiation in Smartcards*, Proc. CHES 99, C. Paar & Ç. Koç (eds), LNCS **1717**, Springer-Verlag, 1999, pp. 144–157.
10. E. Oswald & M. Aigner, *Randomized Addition-Subtraction Chains as a Countermeasure against Power Attacks*, CHES 2001, Ç. Koç, D. Naccache & C. Paar (eds), LNCS **2162**, Springer-Verlag, 2001, pp. 39–50.
11. E. Oswald, *Enhancing Simple Power-Analysis Attacks on Elliptic Curve Cryptosystems*, CHES 2002, B. Kaliski, Ç. Koç & C. Paar (eds), LNCS **2523**, Springer-Verlag, 2003, pp. 82–97.
12. J.-J. Quisquater & D. Samyde, *ElectroMagnetic Analysis (EMA): Measures and Counter-Measures for Smart Cards*, Smart Card Programming and Security (e-Smart 2001), I. Attali & T. Jensen (eds), LNCS **2140**, Springer-Verlag, 2001, pp. 200–210.

13. J.-J. Quisquater & D. Samyde, *Eddy current for Magnetic Analysis with Active Sensor*, Proc. Smart Card Programming and Security (e-Smart 2002), Nice, September 2002, pp. 183–194.

14. K. Okeya & K. Sakurai, *On Insecurity of the Side Channel Attack Countermeasure using Addition-Subtraction Chains under Distinguishability between Addition and Doubling*, Information Security and Privacy (ACISP 2002), L. Batten & J. Seberry (eds), LNCS **2384**, Springer-Verlag, 2002, pp. 420–435.

15. C. D. Walter, *MIST: An Efficient, Randomized Exponentiation Algorithm for Resisting Power Analysis*, Proc. CT-RSA 2002, B. Preneel (ed), LNCS **2271**, Springer-Verlag, 2002, pp. 53–66.

16. C. D. Walter, *Breaking the Liardet-Smart Randomized Exponentiation Algorithm*, Proc. Cardis '02, San José, November 2002, USENIX Association, Berkeley, 2002, pp. 59–68.

Hardware Countermeasures against DPA – A Statistical Analysis of Their Effectiveness

Stefan Mangard[*]

Institute for Applied Information Processing and Communications
Graz University of Technology
Inffeldgasse 16a, A-8010 Graz, Austria
Stefan.Mangard@iaik.at
http://www.iaik.at/research/sca-lab

Abstract. Many hardware countermeasures against differential power analysis (DPA) attacks have been developed during the last years. Designers of cryptographic devices using such countermeasures to protect their devices have the challenging task to select and implement a suitable combination of countermeasures. Every device has different requirements, and so there is no universal solution to protect devices against DPA attacks.

In this article, a statistical approach is pursued to determine the effect of hardware countermeasures on the number of samples needed in DPA attacks. This approach results in a calculation method that enables designers to assess the resistance of their devices against DPA attacks throughout the design process. This way, different combinations of countermeasures can be easily compared and costly design iterations can be avoided.

Keywords: Smart cards, Side-Channel Attacks, Differential Power analysis (DPA), Hardware countermeasures

1 Introduction

During the last years, a lot of effort has been dedicated towards the research of side-channel attacks [1,9,10] and the development of corresponding countermeasures. In particular, there have been many endeavors to develop effective countermeasures against differential power analysis (DPA) [10,15] attacks.

DPA attacks are based on the fact that the power consumption of a cryptographic device depends on the internally used secret key. Since this property can be exploited with relatively cheap equipment, DPA attacks pose a serious practical threat to cryptographic devices, like smart cards.

The countermeasures that have been developed up till now against these attacks can be categorized into two groups. The first group are the so-called

---

[*] This work has been supported by the Austrian Science Fund (FWF Project No. P16110-N04).

T. Okamoto (Ed.): CT-RSA 2004, LNCS 2964, pp. 222–235, 2004.
© Springer-Verlag Berlin Heidelberg 2004

algorithmic countermeasures [4,5,7,14,22]. The basic idea of these countermeasures is to randomize the intermediate results that are processed during the execution of a cryptographic algorithm. Classical first-order DPA attacks are rendered practically impossible, if this randomization is implemented correctly. However, there are two significant drawbacks of this approach.

The first one is that the randomization is quite expensive to implement for non-linear operations as they are used in symmetric ciphers (see for example [5], [6] and [7]). The second one is that many algorithmic countermeasures do not provide sufficient protection against higher-order DPA attacks [13] or sophisticated SPA attacks [11,18]. The consequence of these facts is that algorithmic countermeasures are typically combined with hardware countermeasures [2,8,12, 16,19,20,21].

The hardware approach to counteract DPA attacks differs significantly from the algorithmic one. The intermediate results that occur during the execution of a cryptographic algorithm are not affected by this type of countermeasure. Instead, the goal of this approach is to bury the attackable part of the power consumption in different kinds of noise.

The more noise there is in the power traces recorded by the attacker, the more measurements are needed for a successful DPA attack. Although the basic idea is relatively simple, hardware countermeasures have proven to be quite effective in practice. This is why cryptographic devices are typically either protected by a combination of hardware and algorithmic countermeasures or solely by hardware countermeasures.

The decision which combination of countermeasures is implemented in a device, is made by the designers. It is their task to choose a combination of countermeasures that provides the resistance against DPA attacks that is necessary for the planned application of the device. The resistance against DPA attacks is typically specified by a number of samples: If DPA attacks with this number of samples fail, the device is resistant enough. Otherwise the requirements are not fulfilled.

Choosing a suitable combination of countermeasures is a very challenging task in practice. This is due to the fact that this decision needs to be made at a very early stage of the design process. Design iterations are costly and so the fabrication of a physical prototype to test whether a combination of countermeasures is sufficient or not, should be avoided.

In order to minimize the number of design iterations, methods are necessary to assess the effect of countermeasures on the number of samples. However, particularly for hardware countermeasures there are no publications that discuss such methods. Publications of hardware countermeasures usually just contain case studies showing that the proposed countermeasure really increases the number of samples. Yet, such case studies are only of limited use for a designer of a device who uses a different technology, a different architecture, and potentially even uses multiple countermeasures simultaneously.

In this article, a statistical approach is pursued to determine the effect of hardware countermeasures on the number of samples. This approach leads to a calculation method that allows the determination of lower bounds for the number of samples needed in DPA attacks. The presented calculation method is based on only very few parameters that can be assessed already at an early stage of the design process.

It is therefore ideally suited to help designers to choose the right combination of countermeasures already at the beginning of the design process. Of course, the presented calculation method can also be used at any time during the design process to determine whether a design fulfills certain resistance requirements or not. The more precisely the parameters of the calculation can be determined, the more precise becomes the statement on the number of samples.

This article is organized as follows: Section 2 provides a short summary of the fundamentals of DPA attacks and defines some of the notation that is used in this article. Section 3 analyzes the principles that are used by hardware countermeasures to increase the resistance against DPA attacks. The calculation of lower bounds for the number of samples is presented in section 4. In section 5, the corresponding formulas are empirically verified. Conclusions can be found in section 6.

2   Differential Power Analysis

The power consumption of a digital circuit depends on the data that the circuit processes. Thus, the power trace of a device executing a cryptographic algorithm, depends on intermediate results of this algorithm.

DPA attacks exploit the fact that in all cryptographic algorithms there occur intermediate results which are a function of the ciphertext and only few key bits. We call these key bits a subkey. In a DPA attack, one subkey after the other is attacked until the entire secret key is known or the missing rest of the key can be efficiently determined by a brute-force search. An attacker knowing the cryptographic algorithm that is executed in a device, can reveal a subkey as follows:

First, the power consumption of the device is recorded, while it encrypts S different plaintexts using the same key. In this article, we use the common assumption that these plaintexts are uniformly distributed. We refer to the power traces that are recorded during the encryptions as $P_{1...S,1...T}$, where T is the number of points that are recorded per encryption.

In the next step, the attacker chooses an intermediate result of the executed algorithm that is a function of the ciphertext and a short subkey. Based on the ciphertext and all possible values for the subkey, hypothetical values for the intermediate result are calculated. This leads to a matrix $I_{1...K,1...S}$ of hypothetical intermediate results, where K is the number of possible values for the subkey.

The subkey k_c that is actually used in the attacked device is one of the K possible values for the subkey. Hence, the values $I_{k_c,1...S}$ have actually been processed by the attacked device while it has been doing the S recorded encryptions. Consequently, the values $P_{1...S,t_c}$ depend on $I_{k_c,1...S}$, where t_c is the moment of time at which the attacked intermediate results have been processed.

The attacker determines a hypothetical power consumption value $H_{k,s}$ for every $I_{k,s}$. The absolute values of $H_{1...K,1...S}$ are of no importance for the attack—only the relative distance between the values is relevant.

Nevertheless, the calculation of $H_{1...K,1...S}$ requires some basic knowledge about how the processing of different data affects the power consumption of a device. Many devices use pre-charged buses. Such buses cause a power consumption that is proportional to the Hamming weight of the data block that is being transferred over the bus.

After having determined $H_{1...K,1...S}$, the attacker reveals the correct subkey k_c by correlating the hypothetical power consumptions with the one of the device. In this article, the Pearson correlation coefficient is used to measure this correlation.

In [10], Kocher et. al. measure this correlation by calculating the distance between means. In the context of DPA attacks, there is no significant difference between the two measures for the correlation. However, we favor the Pearson correlation coefficient because there exists a well-established theory on measuring correlations this way—the Pearson correlation coefficient is the common measure to determine the linear relationship between two variables. Equation 1 shows a definition of the correlation ρ between two variables X and Y, where $E(X)$, $E(Y)$ and $E(XY)$ are expected values, $Cov(X,Y)$ is the covariance and $Var(X)$ as well as $Var(Y)$ are the variances of the variables.

$$\rho(X,Y) = \frac{E(XY) - E(X)E(Y)}{\sqrt{Var(X)Var(Y)}} \quad \frac{Cov(X,Y)}{\sqrt{Var(X)Var(Y)}} \tag{1}$$

The definition of the Pearson correlation coefficient r is shown in equation 2. r estimates the correlation ρ between two variables based on S samples. $\bar{x}$ and $\bar{y}$ in equation 2 denote the means of the variables based on S samples.

$$r(<x_1,\ldots,x_S>,<y_1,\ldots,y_S>) = \frac{\sum_{s=1}^{S}(x_s - \bar{x})(y_s - \bar{y})}{\sqrt{\sum_{s=1}^{S}(x_s - \bar{x})^2}\sqrt{\sum_{s=1}^{S}(y_s - \bar{y})^2}} \tag{2}$$

In a DPA attack, the Pearson correlation coefficient between the values $H_{k=fixed,1...S}$ and $P_{1...S,t=fixed}$ is calculated for every fixed k and t. This leads to the matrix $\mathcal{R} = r_{1...K,1...T}$ of correlation coefficients. Since the values $P_{1...S,\forall t \neq t_c}$ and $H_{\forall k \neq k_c,1...S}$ are largely uncorrelated, the correlations $\rho_{\forall k \neq k_c,\forall t \neq t_c}$ are significantly lower than ρ_{k_c,t_c}.

If S is sufficiently large in an attack, this difference between the correlations can be detected in the matrix $\mathcal{R}$ of Pearson correlation cocfficients. In this case, one correlation coefficient of $\mathcal{R}$ is significantly larger than all other ones. The position of this peak in $\mathcal{R}$ reveals the correct subkey k_c.

The number of samples that is needed in a DPA attack to reveal k_c is mainly determined by the value ρ_{k_c,t_c}. This observation has already been made previously by Messerges et. al. in [15].

Since ρ_{k_c,t_c} is the maximum value of $\rho_{1...K,1...T}$, we refer to this correlation as ρ_{max} throughout the remainder of this article. The higher ρ_{max} is, the less samples are needed to see a significant peak at the position (k_c, t_c) of $\mathcal{R}$.

This is why it is the goal of hardware countermeasures to reduce ρ_{max} to a value that is as close to zero as possible.

3   Hardware Countermeasures

In order to increase the number of samples needed in DPA attacks, hardware countermeasures decrease the correlation between the hypothetical power consumptions and the power consumption of the device.

The hypothetical power consumptions are determined by the attacker, and therefore they cannot be controlled by the designers of a device. Yet, designers can alter the power consumption of their devices in such a way that ρ_{max} is reduced. There exist two possibilities to lower this correlation. All hardware countermeasures that have been proposed so far, rely on these two possibilities.

3.1   Reduction of the SNR

The first possibility to reduce the correlation ρ_{max} is to bury the part of the power consumption that is caused by the processing of the attacked intermediate result in a lot of noise.

The burying of this signal in noise is best measured by a signal-to-noise ratio (SNR). For the definition of this SNR, we define Q to be the power consumption caused by the attacked intermediate result and N to be additive noise. Consequently, the power consumption of a device at the time t_c can be written as $P_{s,t_c} = Q_s + N_s$.

Equation 3 shows the definition of the SNR for the signal Q. Since the DC components of N and Q are not relevant for the calculation of the correlation, only the AC components (i.e. the variances) of the signals are considered in this equation.

$$SNR = \frac{Var(Q)}{Var(N)} \tag{3}$$

The lower the SNR is, the lower is also the correlation between the correct hypothetical power consumption and the power consumption of the device.

There are several hardware countermeasures that reduce the SNR. The most prominent examples are special logic styles that minimize the data dependency of the power consumption. Such logic styles are presented by Moore et. al. in [16, 17], by Tiri et. al. in [20,21] and by Saputra et. al. in [19].

However, there are many more ways to reduce the SNR. For example, also flattening the power consumption or random charging of on-chip capacitances reduce the SNR. In fact, any processing that occurs in parallel to the execution of the cryptographic algorithm, leads to this result.

In general, the effect of a hardware countermeasure on the SNR can be determined already at an early stage of the design process. Even before the implementation of a device has started, it is possible to assess the SNR. During the implementation phase, the SNR can be determined by using tools that assess the power consumption of a device. Due to the fact that the overall power consumption of integrated circuits has become increasingly important during the last few years, several tools of this kind are available.

3.2   Random Disarrangement of t_c

The second possibility to reduce the correlation ρ_{max} is to randomly disarrange the moment of time at which the attacked intermediate result is processed. If the time t_c is different in every power trace, the correlation between the correct hypothetical power consumption and the one of the device is significantly reduced.

Random disarrangement techniques lead to the fact that there is a certain probability distribution for t_c. Clearly, the highest correlation in DPA attacks occurs at the moment of time of the power traces, where the maximum of this probability distribution is located. We refer to this moment of time as $\hat{t}_c$. The maximum probability $\hat{p}$ that is located at $\hat{t}_c$ is the decisive value determining how much the correlation is reduced in DPA attacks. The lower $\hat{p}$ is, the more samples are required in DPA attacks.

There exist many proposals for hardware countermeasures that are based on a random disarrangement of t_c. The classic countermeasure that is based on this principle is the insertion of random delays [3], which can even be implemented in software. Another approach that is also based randomizing t_c is pursued by Irwin et. al. in [8] and by May et. al. in [12]. They propose to use a non-deterministic processor to foil DPA attacks.

The countermeasure proposed by Benini et. al. in [2], also gains most of its strength by randomizing t_c. Of course, also asynchronous logic styles [16,17] are very well suited for the insertion of non-deterministic delays.

In order to determine the effect of a random disarrangement of t_c on DPA attacks early in the design process, it is necessary that $\hat{p}$ can be determined very early.

In case of the insertion of random delays, t_c is binomially distributed and so $\hat{p}$ can be calculated in a straightforward manner. In case of the other countermeasures, the distributions of t_c may be more complex. Yet, even if a direct calculation of $\hat{p}$ is not practical, it is always possible to approximate it empirically based on a software model of the countermeasure.

In this section, we have introduced the possibilities that can be used to lower the correlation ρ_{max} in DPA attacks. There are two properties of (combinations of) hardware countermeasures that largely determine the effect of the countermeasures on the number of samples: the SNR defined in equation 3 and $\hat{p}$.

Both properties can be assessed already at an early stage of the design process. The following section introduces the calculation of lower bounds for the number of samples based on these two parameters.

4   Calculation of Lower Bounds for the Number of Samples

The effect of hardware countermeasures on the number of samples is largely determined by the two parameters discussed in section 3. However, there are also some other parameters that have a certain influence of the number of samples. Throughout this article, these parameters are set to worst-case values from a designer's point of view (i.e. all unknown parameters are set in favor of a potential attacker). Hence, the calculation method introduced in this section, leads to lower bounds for the number of samples. This conservative measure is exactly what designers should use to determine the effectiveness of the hardware countermeasures in their design.

In the following subsection, first formulas are derived to calculate ρ_{max} in the presence of hardware countermeasures. Subsection 4.2 then introduces the calculation of lower bounds for the number of samples based on ρ_{max}.

4.1 ρ_{max} in the Presence of Hardware Countermeasures

The Effect of SNR on ρ_{max}: In a DPA attack on a device without random disarrangement of t_c, ρ_{max} is the correlation between the hypothetical power consumption for the correct subkey and the one of the device at the time t_c.

Equation 4 shows the calculation of ρ_{max} based on SNR. In this equation, the variable H refers to the hypothetical power consumption for the correct subkey. Q and N are used as defined in section 3: Q denotes the power consumption of the device caused by the attacked intermediate result and N denotes uncorrelated additive noise.

$$\rho(H, Q+N) = \frac{E(H(Q+N)) - E(H)E(Q+N)}{\sqrt{Var(H)(Var(Q)+Var(N))}}$$
$$= \frac{E(HQ+HN) - E(H)(E(Q)+E(N))}{\sqrt{Var(H)Var(Q)}\sqrt{1+\frac{Var(N)}{Var(Q)}}} = \frac{\rho(H,Q)}{\sqrt{1+\frac{1}{SNR}}} \quad (4)$$

The Effect of $\hat{p}$ on ρ_{max}: If t_c is randomly disarranged, the correlation ρ_{max} occurs between the correct hypothetical power consumption and the one of the device at the time $\hat{t}_c$.

In equation 5, the variable $\hat{P}$ refers to the power consumption of the device at this time $\hat{t}_c$. The probability that a power consumption at this time is caused by the processing of an attacked intermediate result is $\hat{p}$. With a probability of $(1-\hat{p})$, the power consumption at the time t_c is caused by the processing of some other data.

In equation 5, we refer to the power consumption caused by an attacked intermediate result as P. With O we refer to the one caused by the processing of other data. In practice, O is largely independent from the correct hypothetical power consumption H. This is why we set $Cov(H,O)$ to zero in equation 5.

$$\rho(H, \hat{P}) = \frac{\hat{p} * Cov(H, P) + (1 - \hat{p}) \, Cov(H, O)}{\sqrt{Var(H)Var(\hat{P})}}$$

$$= \frac{\hat{p} * Cov(H, P)}{\sqrt{Var(H)Var(\hat{P})}} = \rho(H, P) * \hat{p} * \sqrt{\frac{Var(P)}{Var(\hat{P})}} \qquad (5)$$

Calculation of ρ_{max}: The equations 4 and 5 can be combined into one formula (see equation 6) that allows to determine the effect of a given combination of hardware countermeasures on ρ_{max}.

$$\rho_{max} = \frac{\rho(H, Q)}{\sqrt{1 + \frac{1}{SNR}}} * \hat{p} * \sqrt{\frac{Var(P)}{Var(\hat{P})}} \qquad (6)$$

Besides the parameters SNR and $\hat{p}$, also the correlation $\rho(H, Q)$ and the term $F = \sqrt{\frac{Var(P)}{Var(\hat{P})}}$ influence ρ_{max}. While the correlation $\rho(H, Q)$ solely depends on how well the attacker knows the power consumption characteristics of a device, the factor F is a device specific property.

However, unlike SNR and $\hat{p}$, F is rather difficult to assess at very early stages of the design process. In order to reasonably assess F, designers need some knowledge about how the power consumption of the device looks like before and after the attacked intermediate result is processed. The range that needs to be known is the bigger, the wider the probability distribution of t_c is.

In practice, F should be set to the worst-case value 1 at the very early stages of the design process. As soon as first assessments on the power consumption of the device are available, F can be updated accordingly in the calculation of the number of samples.

Since designers should always determine the number of samples in a conservative manner, $\rho(H, Q)$ should be set to 1 throughout the design process.

Based on equation 6, ρ_{max} can be determined at any point of the design process. The better the parameters of this equation can be assessed, the better becomes the statement on the number of needed samples.

4.2   Mapping ρ_{max} to a Number of Samples

The number of samples needed in a DPA attack is the commonly used measure for the resistance of a device against these attacks. In order to reveal the correct subkey k_c, the number of samples needs to be increased in an attack until a significant peak is visible in the matrix $\mathcal{R}$.

The Pearson correlation coefficients in this matrix $\mathcal{R}$ estimate the correlations $\rho_{1...K,1...T}$ based on S samples. The sampling distribution of a Pearson correlation coefficient r is best described by transforming r to a variable z that is normally distributed. This transformation (known as Fisher's Z-Transformation) is shown in the equations 7 to 9.

$$z = \frac{1}{2} \ln \frac{1+r}{1-r} \tag{7}$$

$$\mu = \frac{1}{2} \ln \frac{1+\rho}{1-\rho} \tag{8}$$

$$\sigma^2 = \frac{1}{S-3} \tag{9}$$

Based on these formulas, the sampling distribution of each correlation coefficient r of the matrix $\mathcal{R}$ can be determined easily based on $\rho_{1...K,1...T}$ and S. The equations 7 to 9 are an approximation for $S > 30$. Yet, since the number of samples is typically much higher, this approximation is sufficient.

Calculating the exact number of samples that are needed for a DPA attack is quite difficult in practice. This has several reasons. First of all, the designers of a cryptographic device don't know to which sampling rate an attacker will set the oscilloscope in an attack, and the designers also don't know how long the recorded power traces will be. Clearly, these parameters strongly influence the size and the values of the matrix $\rho_{1...K,1...T}$.

Even if we would assume the designers knew $\rho_{1...K,1...T}$, the designers would still not know the correlation between the values of the matrix $\rho_{1...K,1...T}$. Yet, these values are correlated significantly in practice.

In order to calculate a lower bound for the number of samples only based on ρ_{max}, we use the following observation: The number of samples that is needed to see a peak in practice, is mainly determined by the distance between the sampling distributions with $\rho = 0$ and $\rho = \rho_{max}$. All values of $\mathcal{R}$ are drawn from one of these two sampling distributions. Clearly, the more overlap there is between these distributions, the less likely it is to see a significant peak in $\mathcal{R}$. An attacker can decrease this overlap by increasing the number of samples (see equation 9).

In order to measure the distance between the distributions, we calculate the probability that a value drawn from the distribution with $\rho = \rho_{max}$ is bigger than one that is drawn from the distribution with $\rho = 0$. This probability α can be calculated as shown in equation 10. This equation can be transformed to equation 11, which allows a direct calculation of the number of samples based on ρ_{max}.

$$\alpha = \Phi \left(\frac{\frac{1}{2} \ln \frac{1+\rho_{max}}{1-\rho_{max}} - \frac{1}{2} \ln \frac{1+0}{1-0}}{\sqrt{\frac{2}{S-3}}} \right) \tag{10}$$

$$S = 3 + 8 \left(\frac{Z_\alpha}{\ln \left(\frac{1+\rho_{max}}{1-\rho_{max}} \right)} \right)^2 \tag{11}$$

The quantile Z_α determines the distance between the distributions with $\rho = 0$ and $\rho = \rho_{max}$. The higher the probability α is, the bigger is the distance between the distributions and, consequently, the more likely it is to see a peak.

In practice, several values are drawn from each of these distributions. Yet, these values are not drawn independently. Therefore, it is hard to calculate the exact probability for a peak, and we have to rely on empirical results to approximate a lower bound for the number of needed samples.

Based on several practical attacks and simulations, we have determined that $\alpha = 0.9$ is a reasonable value to calculate a lower bound for the number of samples needed in a DPA attack. Setting $\alpha = 0.9999$ in equation 11 on the other hand, leads to a number of samples that reveals the attacked subkey with very high probability. Between $\alpha = 0.9$ and $\alpha = 0.9999$ there is a "gray area". The lower the value of α is, the lower is the probability of observing a significant peak in the correlation trace $r_{k_c,1...T}$. The levels $\alpha = 0.9$ and $\alpha = 0.9999$ have been chosen in a very conservative way.

In order to get more exact bounds for a particular device, the levels α may be refined as soon as simulated or measured power traces of the device are available.

Based on the formulas we have provided in this subsection, designers of cryptographic devices can determine the effect of hardware countermeasures on the number of samples as follows:

First, ρ_{max} is calculated according to equation 6. The parameters needed for this calculation, are conservatively assessed by the designers as good as it is possible at the respective stage of the design process. Based on ρ_{max}, a lower bound for the number of needed samples can then be calculated according to equation 11.

5   Empirical Verification

In order to empirically verify the formulas derived in the last section, we implemented AES-128 on an 8-bit micro controller. The micro controller was clocked with 11MHz and its power consumption was sampled with 250 MS/s during 4000 AES-128 encryptions.

We attacked an 8-bit intermediate result of AES-128 at the time it was transferred over the pre-charged bus of the micro controller. In order to verify equation 6, a different number of bits of this intermediate result were attacked. From an attacker's point of view the bits that are transferred over the bus, but are not part of the attacked intermediate result, are noise. Of course, there is also other noise in the measurement, besides the power consumption of these bits.

However, since we are not familiar with the details of the design of the micro controller, we had to assume that this noise is zero for our first calculation of ρ_{max} based on equation 12. In this equation, b is the number of bits that are attacked on the bus, and n is the variable representing the additional noise in the power traces.

Table 1. Comparison of calculated correlations with the empirically determined correlation coefficients for 4000 samples

Number of Attacked Bits	1	2	3	4	5	6	7	8
Calculated ρ_{max} $(n = 0)$	0.35	0.50	0.61	0.71	0.79	0.87	0.94	1.00
Calculated ρ_{max} $(n = 2)$	0.32	0.45	0.55	0.63	0.71	0.77	0.84	0.89
Measured r_{max} (4000 samples)	0.31	0.44	0.53	0.63	0.70	0.76	0.82	0.90

$$\rho_{max} = \frac{1}{\sqrt{1 + \frac{1}{SNR}}} = \frac{1}{\sqrt{1 + \frac{n+(8-b)}{b}}} \tag{12}$$

The first line of table 1 shows ρ_{max} for $n = 0$ and $b = 1 \ldots 8$. In the second line, the corresponding values are shown for $n = 2$. The correlation coefficients we determined empirically by performing a DPA attack with 4000 samples, can be found in line three.

The values ρ_{max} calculated based on $n = 0$, are higher than the ones determined empirically. This is a logical consequence of the fact that no noise was assumed for the calculation. However, the values in the second and third line match almost exactly—obviously setting $n = 2$ models the noise of the micro controller very well. The slight deviations between the lines two and three are a consequence of the fact that not all wires of the bus of the micro controller have the same power consumption characteristics.

Based on the micro controller, we also verified the effect of random delays on ρ_{max}. For this purpose, we disarranged the 4000 power traces using a binomial distribution with $p = \frac{1}{2}$ and $n = 50$ clock cycles for the delay—the maximum probability of this distribution is $p = \frac{1}{2^{50}}\binom{50}{25}$. However, when calculating $\hat{p}$, the fact that the micro controller processes the attacked intermediate result twice needed to be considered. The micro controller we used, processed the attacked intermediate result in two subsequent clock cycles. Consequently, $\hat{p}$ was approximated by $2 * \frac{1}{2^{50}}\binom{50}{25}$. We attacked a 4-bit intermediate result using the disarranged traces. Consequently, ρ_{max} could be determined as shown in equation 13.

$$\rho_{max} \approx \frac{1}{\sqrt{1 + \frac{6}{4}}} * 2 * \frac{1}{2^{50}}\binom{50}{25} * \sqrt{\frac{47}{260}} = 0.06 \tag{13}$$

The quotient $\frac{47}{260}$ was determined empirically. Performing 1000 attacks with different random delays based on 4000 power traces of the micro controller, lead to mean of $r_{max} = 0.063$.

In the next step, we verified the calculation of the number of samples. We calculated S based on equation 11 for the attacks without random delays which we described before. The calculated number of samples for $\alpha = 0.9$ and $\alpha = 0.9999$ are shown in table 2 (ρ_{max} was calculated using equation 12 with $n = 2$). The big difference between the number of samples for the same attack with

Table 2. The number of samples needed in a DPA attack on the micro controller for $\alpha = 0.9$ and $\alpha = 0.9999$

Number of Attacked Bits	1	2	3	4	5	6	7	8
S calculated with $\alpha = 0.9$	34	17	12	9	7	6	5	5
S calculated with $\alpha = 0.9999$	261	122	76	53	39	29	22	16

different values α, again shows that the levels $\alpha = 0.9$ and $\alpha = 0.9999$ have been chosen very conservatively.

We have performed DPA attacks with the calculated number of samples. Clearly visible peaks occurred in the attacks conducted with the numbers of samples calculated based on $\alpha = 0.9999$. In attacks with the numbers of samples shown in the first line of table 2, only some sporadic peaks occurred in hundreds of attacks.

Hence, we have been able to verify empirically all formulas presented in this article, based on attacks on an 8-bit micro controller.

6   Conclusions

Designers of cryptographic devices require methods to assess the effect of hardware countermeasures on the number of samples needed in DPA attacks. Such methods are necessary in order to avoid costly design iterations.

In this article, we have identified those properties of hardware countermeasures that affect the number of samples needed in DPA attacks. Based on these properties, we have derived formulas that allow the calculation of lower bounds for the number of samples needed in DPA attacks.

The presented formulas enable designers to assess the resistance of their devices against DPA attacks from the earliest stages of the design process onwards until the fabrication. This way designers can verify that the combination of countermeasures they have chosen to implement in their devices, indeed provides the required protection against DPA attacks.

References

1. D. Agrawal, B. Archambeault, J.R. Rao, and P. Rohatgi. The EM Side-channel(s). In *Cryptographic Hardware and Embedded Systems – CHES 2002*, Lecture Notes in Computer Science (LNCS). Springer-Verlag, 2002.
2. L. Benini, A. Macii, E. Macii, E. Omerbegovic, M. Poncino, and F. Pro. Energy-Aware Design Techniques for Differential Power Analysis Protection. In *40th Design Automation Conference – DAC 2003*. ACM, 2003.
3. C. Clavier, J.-S. Coron, and N. Dabbous. Differential Power Analysis in the presence of Hardware Countermeasures. In *Workshop on Cryptographic Hardware and Embedded Systems – CHES 2000*, volume 1965 of *Lecture Notes in Computer Science (LNCS)*, pages 252–263. Springer-Verlag, 2000.

4. J.-S. Coron. Resistance against Differential Power Analysis for Elliptic Curve Cryptosystems. In *Workshop on Cryptographic Hardware and Embedded Systems – CHES 1999*, volume 1717 of *Lecture Notes in Computer Science (LNCS)*, pages 292–302. Springer-Verlag, 1999.

5. J. Dj. Golic and C. Tymen. Multiplicative Masking and Power Analysis of AES. In *Cryptographic Hardware and Embedded Systems – CHES 2002*, Lecture Notes in Computer Science (LNCS). Springer-Verlag, 2002.

6. L. Goubin. A Sound Method for Switching between Boolean and Arithmetic Masking. In *Workshop on Cryptographic Hardware and Embedded Systems – CHES 2001*, volume 2162 of *Lecture Notes in Computer Science (LNCS)*, pages 3–15. Springer-Verlag, 2001.

7. L. Goubin and J. Patarin. DES and Differential Power Analysis – The Duplication Method. In *Cryptographic Hardware and Embedded Systems - CHES 1999*, volume 1717 of *Lecture Notes in Computer Science*, pages 158–172. Springer-Verlag, 1999.

8. J. Irwin, D. Page, and N.P. Smart. Instruction Stream Mutation for Non-Deterministic Processors. In *IEEE International Conference on Application-Specific Systems, Architectures, and Processors – ASAP 2002*, pages 286–295. IEEE, 2002.

9. P.C. Kocher. Timing Attacks on Implementations of Diffie-Hellman, RSA, DSS and Related Attacks. In *Advances in Cryptology – CRYPTO 1996*, volume 1109 of *Lecture Notes in Computer Science*, pages 104–113. Springer-Verlag, 1996.

10. P.C. Kocher, J. Jaffe, and B. Jun. Differential Power Analysis. In *Advances in Cryptology – CRYPTO 1999*, volume 1666 of *Lecture Notes in Computer Science (LNCS)*, pages 388–397. Springer-Verlag, 1999.

11. S. Mangard. A Simple Power-Analysis (SPA) Attack on Implementations of the AES Key Expansion. In *Information Security and Cryptology – ICISC 2002*, volume 2587 of *Lecture Notes in Computer Science (LNCS)*, pages 343–358. Springer-Verlag, 2002.

12. D. May, H.L. Muller, and N.P. Smart. Non-deterministic Processors. In *Information Security and Privacy – ACISP 2001*, volume 2119 of *Lecture Notes in Computer Science (LNCS)*, pages 115–129. Springer-Verlag, 2001.

13. T.S. Messerges. Using Second-Order Power Analysis to Attack DPA Resistant Software. In *Cryptographic Hardware and Embedded Systems – CHES 2000*, volume 1965 of *Lecture Notes in Computer Science (LNCS)*, pages 238–251. Springer-Verlag, 2000.

14. T.S. Messerges, E. A. Dabbish, and R. H. Sloan. Power Analysis Attacks of Modular Exponentiation in Smartcards. In *Cryptographic Hardware and Embedded Systems – CHES 1999*, volume 1717 of *Lecture Notes in Computer Science*, pages 144–157. Springer-Verlag, 2000.

15. T.S. Messerges, E.A. Dabbish, and R.H. Sloan. Examining Smart-Card Security under the Threat of Power Analysis Attacks. *IEEE Transactions on Computers*, 51(5), 2002.

16. S. Moore, R. Anderson, P. Cunningham, R. Mullins, and G. Taylor. Improving Smart Card Security using Self-timed Circuits. In *Eighth IEEE International Symposium on Asynchronous Circuits and Systems – Async 2002*. IEEE Computer Society Press, 2002.

17. S. Moore, Ross Anderson, Robert Mullins, and George Taylor. Balanced Self-Checking Asynchronous Logic for Smart Card Applications. In *Microprocessors and Microsystems Journal*, to appear.

18. E. Oswald. Enhancing Simple Power-Analysis Attacks on Elliptic Curve Cryptosystems. In *Cryptographic Hardware and Embedded Systems – CHES 2002*, volume 2523 of *Lecture Notes in Computer Science (LNCS)*, pages 82–97. Springer-Verlag, 2002.

19. H. Saputra, N. Vijaykrishnan, M. Kandemir, M.J. Irwin, R. Brooks, S. Kim, and W. Zhang. Masking the Energy Behavior of DES Encryption. In *Design, Automation and Test in Europe Conference and Exhibition – DATE 2003*, pages 84–89. IEEE, 2003.

20. K. Tiri, M. Akmal, and I. Verbauwhede. A Dynamic and Differential CMOS Logic with Signal Independent Power Consumption to Withstand Differential Power Analysis on Smart Cards. In *29th European Solid-State Circuits Conference – ESSCIRC 2002*, 2002.

21. K. Tiri and I. Verbauwhede. Securing Encryption Algorithms against DPA at the Logic Level: Next Generation Smart Card Technology. In *Cryptographic Hardware and Embedded Systems – CHES 2003*, volume 2779 of *Lecture Notes in Computer Science (LNCS)*, pages 125–136. Springer-Verlag, 2003.

22. E. Trichina, D. De Seta, and L. Germani. Simplified Adaptive Multiplicative Masking for AES and its Secure Implementation. In *Cryptographic Hardware and Embedded Systems – CHES 2002*, Lecture Notes in Computer Science (LNCS). Springer-Verlag, 2002.

Self-Randomized Exponentiation Algorithms

Benoît Chevallier-Mames

Gemplus, Card Security Group
La Vigie, Avenue du Jujubier, ZI Athélia IV,
13705 La Ciotat Cedex, France
benoit.chevallier-mames@gemplus.com
http://www.gemplus.com/smart/

Abstract. Exponentiation is a central process in many public-key cryptosystems such as RSA and DH. This paper introduces the concept of self-randomized exponentiation as an efficient means for preventing DPA-type attacks. Self-randomized exponentiation features several interesting properties:
- it is fully generic in the sense that it is not restricted to a particular exponentiation algorithm;
- it is parameterizable: a parameter allows to choose the best trade-off between security and performance;
- it can be combined with most other counter-measures;
- it is space-efficient as only an additional long-integer register is required;
- it is flexible in the sense that it does not rely on certain group properties;
- it does not require the prior knowledge of the order of the group in which the exponentiation is performed.

All these advantages make our method particularly well suited to secure implementations of the RSA cryptosystem in standard mode, on constrained devices like smart cards.

Keywords: Exponentiation, implementation attacks, fault attacks, side-channel attacks (DPA, SPA), randomization, exponent masking, blinding, RSA, standard mode, smart cards.

1   Introduction

Since the invention of the public key cryptography by Diffie and Hellman [DH76], numerous public-key cryptosystems were proposed. Amongst those that resisted cryptanalysis, the RSA cryptosystem [RSA78] is undoubtedly the most widely used. Its intrinsic security relies on the difficulty of factoring large integers. In spite of decades of intensive research, the factoring problem is still considered as a very hard problem, making the RSA cryptosystem secure for sensitive applications such as data encryption or digital signatures [PKC02].

Instead of trying to break the RSA at a mathematical level, cryptographers then turned their attention to *concrete* implementations of RSA cryptosystems.

T. Okamoto (Ed.): CT-RSA 2004, LNCS 2964, pp. 236–249, 2004.

This gave rise to fault attacks [BDL01] and side-channel attacks [Koc96,KJJ99]. Implementation attacks profoundly modified the way algorithms should be implemented.

As a general rule of thumb for preventing implementation attacks, algorithms should be randomized. In the case of the RSA cryptosystem, there are basically two approaches for randomizing the computation of $y = x^d$ (mod N). This can be achieved by:

1. randomizing the input data prior to executing the exponentiation algorithm [Koc96]; e.g., as
   a) $\hat{x} \leftarrow x + r_1 N$ for a k-bit random r_1
   b) $\hat{d} \leftarrow d + r_2 \phi(N)$ for a k-bit random r_2

   and then y is evaluated as $y = \hat{y}$ (mod N) with $\hat{y} = \hat{x}^{\hat{d}}$ (mod $2^k N$);
2. randomizing the exponentiation algorithm itself (e.g., [Wal02], [MDS99]).

The first approach, initiated by Kocher (see [Koc96, Section 10]), presents the advantage of being independent of the exponentiation algorithm. It also is worth noting that when x is the result of a probabilistic padding (e.g., OAEP [BR95] or PSS [BR96]), there is no need to further randomize x and so the exponentiation can, for example, be carried out as $y = x^{\hat{d}}$ (mod N) with $\hat{d} = d + r_2 \phi(N)$ for a random r_2. Unfortunately, such a randomization of d is restricted to CRT implementations of RSA [QC82] as the value of Euler totient function $\phi(N)$ is usually unknown to the private exponentiation algorithm in standard (i.e., non-CRT) mode.[1]

The best representative of the second approach is the MIST algorithm by Walter [Wal02]. MIST randomly generates a fresh addition chain for exponent d for performing x^d (mod N). To minimize the number of registers, the addition chain is computed on-the-fly via an adaptation of an exponentiation algorithm based on "division chains" [Wal98]. Another example is an improved version of the sliding window method proposed in [IYTT02] . Compared to the first approach, it allows to randomize the exponentiation without the knowledge of $\phi(N)$ but requires a secure division algorithm for computing the division chains or quite complicated management.

This paper presents a novel method to randomize the execution of the exponentiation, in order to prevent *Differential Power Analysis* (DPA) [KJJ99], combining the advantages of the two approaches: As in the first approach, it does not impose a particular exponentiation algorithm; and as in the second approach, it is a randomized algorithm (in particular, it does not require the knowledge of $\phi(N)$ nor of e in a private RSA exponentiation). Our method introduces the concept of *self-randomized exponentiation*, meaning that exponent d is used itself as an additional source of randomness in the exponentiation process. Self-randomized exponentiation only assumes that exponent bits are

---

[1] When the public exponent e is known and not too large, one can randomize the private exponent as $\hat{d} \leftarrow d + r(ed - 1)$. Unfortunately, in most cases, e is unknown (i.e., not available to the private exponentiation algorithm).

scanned from the most significant position and so applies to most exponentiation algorithms [MvV97, Chapter 14]. It can also be combined with most other
counter-measures such as randomizing the exponent prior to the exponentiation. Finally, our method is not restricted to exponentiation in RSA groups and
equally applies to other groups such as the group of points of an elliptic curve
over a finite field [Kob87,Mil86].

The rest of this paper is organized as follows. The next section briefly reviews exponentiation algorithms and presents the general principle behind self-
randomized exponentiation. In Section 3, two different, self-randomized exponentiation algorithms (and variants thereof) are detailed. Section 4 presents
equivalent versions but without branching instructions, so that *Simple Power
Analysis* (SPA) [KJJ99] is also prevented. It also presents a version resisting
against a powerful attacker able to "reverse" the exponentiation algorithm along
with other further optimizations. Finally, Section 5 concludes the paper.

2   Self-Randomized Exponentiation

2.1   Classical Exponentiation Algorithms

There exist two main families of exponentiation algorithms for evaluating the
value of $y = x^d \pmod{N}$, according to the direction the bits of exponent d
are scanned. This paper is only concerned with left-to-right algorithms (i.e.,
scanning d from the most significant position to the least significant position),
including the square-and-multiply algorithm and its k-ary variants, the sliding-
window algorithms, ...   (see [MvV97, Chapter 14]). Left-to-right algorithms
require fewer memory and allow the use of precomputed powers, $x^i \pmod{N}$,
for speeding up the computation of y.

2.2   General Principle

Let $d = (d_l, \ldots, d_0)_2 = \sum_{i=0}^{l} d_i 2^i$ (with $d_i \in \{0, 1\}$) denote the binary representation of exponent d. Defining

$$d_{k \to j} := (d_k, \ldots, d_j)_2 = \sum_{k \geq i \geq j} d_i 2^{i-j} \,,$$

left-to-right exponentiation algorithms share the common feature that an accumulator is used throughout the computation for storing the value of $x^{d_{l \to i}}$
$\pmod{N}$ for decreasing i's until the accumulator contains the value of $y = x^{d_{l \to 0}} = x^d \pmod{N}$.

For example, the square-and-multiply algorithm exploits the recurrence relation

$$x^{d_{l \to i}} = (x^{d_{l \to i+1}})^2 \cdot x^{d_i}$$

with $x^{d_{l \to l}} = x^{d_l}$. Therefore, writing at iteration i the value of $x^{d_{l \to i}}$ in accumulator R_0, we obtain the algorithm of Fig. 1.

Input: $x, d = (d_l, \ldots, d_0)_2$
Output: $y = x^d \pmod N$

$R_0 \leftarrow 1;\ R_1 \leftarrow x;\ i \leftarrow l$
while $(i \geq 0)$ **do**
$\quad R_0 \leftarrow R_0 \cdot R_0 \pmod N$
$\quad$ **if** $(d_i = 1)$ **then** $R_0 \leftarrow R_0 \cdot R_1 \pmod N$
$\quad i \leftarrow i - 1$
endwhile
return R_0

Fig. 1. Square-and-multiply algorithm

Building on the earlier works of [CJRR99,CJ01], we use an additive splitting of the form

$$x^d = x^{d-a} \cdot x^a$$

for a random a, as a means to mask exponent d. A straightforward application of this splitting is inefficient as it roughly doubles the running time: both x^{d-a} and x^a need to be computed.

The main idea behind self-randomized exponentiation consists in taking (part of) d as a source of randomness. So, random a in the above splitting is chosen equal to $d_{l \to i}$, for a random i, since the value of $x^{d \to i}$ is available in the accumulator and needs not to be computed. There are various ways to apply this idea. The next sections present several realizations.

3 Basic Algorithms

3.1 First Algorithm

Our first algorithm relies on the simple observation that, for any $l \geq i_j \geq 0$, we have

$$
\begin{aligned}
x^d &= x^{d_l \to 0} \\
&= x^{d_l \to 0 - d_l \to i_1} \cdot x^{d_l \to i_1} \\
&= x^{(d_l \to 0 - d_l \to i_1) - d_l \to i_2} \cdot x^{d_l \to i_1} \cdot x^{d_l \to i_2} \\
&= \ldots \\
&= x^{(((d_l \to 0 - d_l \to i_1) - d_l \to i_2) - d_l \to i_3) \cdots - d_l \to i_f} \cdot x^{d_l \to i_1} \cdot x^{d_l \to i_2} \cdot x^{d_l \to i_2} \cdots x^{d_l \to i_f} \ .
\end{aligned}
$$

If the i_j's are randomly chosen, the exponentiation process becomes probabilistic. A Boolean random variable ρ is used to determine whether or not the current loop index i belongs to the set $\{i_1, \ldots, i_f\}$. If so, exponent d is replaced with $d - d_{l \to i_j}$. This is illustrated in the next figure.

As in the classical left-to-right exponentiation algorithms, a first accumulator, R_0, is used to keep the value of $x^{d_l \to i}$. We also use a second accumulator,

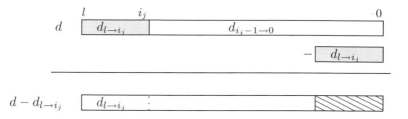

Fig. 2. Masking of exponent d (I)

R_1, to keep the value of $\prod_{i_j \geq i} x^{d_{l \to i_j}}$. To ensure the correctness of the process, the randomization step $d \leftarrow d - d_{l \to i_j}$ cannot modify the $(l - i_j + 1)$ most significant bits of d (i.e., $d_{l \to i_j}$). This latter condition is guaranteed by checking that $d_{i_j - 1 \to 0} \geq d_{l \to i_j}$ (see Fig. 2).

Applied to the classical square-and-multiply algorithm, we get the following algorithm.

Input: $x, d = (d_l, \ldots, d_0)_2$
Output: $y = x^d \pmod{N}$

$\quad R_0 \leftarrow 1;\ R_1 \leftarrow 1;\ R_2 \leftarrow x;\ i \leftarrow l$
$\quad$**while** $(i \geq 0)$ **do**
$\quad\quad R_0 \leftarrow R_0 \cdot R_0 \pmod{N}$
$\quad\quad$**if** $(d_i = 1)$ **then** $R_0 \leftarrow R_0 \cdot R_2 \pmod{N}$
$\quad\quad \rho \leftarrow_R \{0, 1\}$
$\quad\quad$**if** $((\rho = 1) \wedge (d_{i-1 \to 0} \geq d_{l \to i}))$ **then**
$\quad\quad\quad d \leftarrow d - d_{l \to i}$
$\quad\quad\quad R_1 \leftarrow R_1 \cdot R_0 \pmod{N}$
$\quad\quad$**endif**
$\quad\quad i \leftarrow i - 1$
$\quad$**endwhile**
$\quad R_0 \leftarrow R_0 \cdot R_1 \pmod{N}$
$\quad$**return** R_0

Fig. 3. Self-randomized square-and-multiply algorithm (I)

Remark 1. In Fig. 3, as at iteration $i = i_j$, the updating step, $d \leftarrow d - d_{l \to i}$, does not modify the $(l - i + 1)$ most significant bits of d, it can be equivalently replaced with $d_{i-1 \to 0} \leftarrow d_{i-1 \to 0} - d_{l \to i}$.

Analysis. We remark that the randomization step (i.e., $d \leftarrow d - d_{l \to i_j}$) modifies the $(l - i_j + 1)$ *least significant* bits of d. Furthermore, the "consistency" condition (i.e., $d_{i_j-1\to 0} \geq d_{l\to i_j}$) implies that only about the lower half of exponent d is randomized. For the RSA cryptosystem with small public exponent, this is not an issue since such a system leaks half the most significant bits of the corresponding private exponent d [Bon99, Section 4.5].

A simple variant. The previous methodology applies when the randomization step is generalized to:

$$d \leftarrow d - g \cdot d_{l\to i_j}$$

for some random g such that $d_{i_j-1\to 0} \geq g \cdot d_{l\to i_j}$. The second accumulator (say R_1, cf. Fig. 3) should then be updated accordingly as $R_1 \leftarrow R_1 \cdot R_0{}^g \pmod{N}$. Of particular interest is the value $g = 2^\tau$ as the operation $g \cdot d_{l\to i_j}$ amounts to a shifting and the evaluation of $R_0{}^g \pmod{N}$ amounts to τ squarings. Again with the example of the square-and-multiply algorithm, we have:

Input: $x, d = (d_l, \ldots, d_0)_2$
Output: $y = x^d \pmod{N}$

   $R_0 \leftarrow 1;\ R_1 \leftarrow 1;\ R_2 \leftarrow x;\ i \leftarrow l$
   while $(i \geq 0)$ **do**
      $R_0 \leftarrow R_0 \cdot R_0 \pmod{N}$
      if $(d_i = 1)$ **then** $R_0 \leftarrow R_0 \cdot R_2 \pmod{N}$
      $\rho \leftarrow_R \{0,1\};\ \tau \leftarrow_R \{0, \ldots, T\}$
      if $((\rho = 1) \wedge (d_{i-1\to\tau} \geq d_{l\to i}))$ **then**
         $d_{i-1\to\tau} \leftarrow d_{i-1\to\tau} - d_{l\to i}$
         $R_3 \leftarrow R_0$
         while $(\tau > 0)$ **do**
            $R_3 \leftarrow R_3{}^2 \pmod{N};\ \tau \leftarrow \tau - 1$
         endwhile
         $R_1 \leftarrow R_1 \cdot R_3 \pmod{N}$
      endif
      $i \leftarrow i - 1$
   endwhile
   $R_0 \leftarrow R_0 \cdot R_1 \pmod{N}$

---

   return R_0

Fig. 4. Self-randomized square-and-multiply algorithm (I')

Note that, at iteration $i = i_j$, the "consistency" condition $d_{i-1\to 0} \geq 2^\tau d_{l\to i}$ is replaced with the more efficient test $d_{i-1\to\tau} \geq d_{l\to i}$ and the updating step $d \leftarrow d - 2^\tau d_{l\to i}$ is replaced with $d_{l\to\tau} \leftarrow d_{l\to\tau} - d_{l\to i} \Leftrightarrow d_{i-1\to\tau} \leftarrow d_{i-1\to\tau} - d_{l\to i}$, as mentioned in Remark 1.

Bound T should be chosen as the most appropriate trade-off between the randomization of the most significant bits of d and the efficiency in the evaluation of τ squarings, for a τ randomly drawn in $\{0, \ldots, T\}$.

While it also randomizes the upper half of exponent d, the algorithm of Fig. 4 requires an additional register for computing $R_0^{2^\tau}$. The next section shows how to remove this drawback.

3.2   Second Algorithm

Our first algorithm (Fig. 3) only randomizes the lower half of exponent d as $d \leftarrow d - d_{l \rightarrow i_j}$; the restriction coming from the "consistency" condition imposing a half-sized masking. In order to mask the whole value of d, we use the additional trick that

$$d_{l \rightarrow i_j - c_j} = (d_{l \rightarrow i_j - c_j} - d_{l \rightarrow i_j}) + d_{l \rightarrow i_j}$$

for any $i_j \geq c_j \geq 0$. Actually, we successively apply the methodology of our first algorithm to sub-exponent $d_{l \rightarrow i_j - c_j}$.[2] Moreover, to avoid the use of additional registers, we only perform one randomization at a time. In other words, if we update exponent d as depicted in the next figure

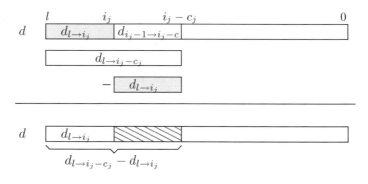

Fig. 5. Masking of exponent d (II)

a new updating step of exponent d will only be permitted after the complete evaluation of $x^{d_{l \rightarrow i_j - c_j}} \pmod{N}$. A Boolean "semaphore", σ, keeps track whether updating is permitted or not.

From Fig. 5, we observe that the $(l - i_j + 1)$ most significant bits of d (i.e., $d_{l \rightarrow i_j}$) remain unchanged by the randomization step if

$$\begin{cases} d_{i_j - 1 \rightarrow i_j - c_j} \geq d_{l \rightarrow i_j}, \\ (i_j - 1) - (i_j - c_j) \geq l - i_j \iff c_j \geq l - i_j + 1 . \end{cases}$$

We set $c_j = l - i_j + 1 + \nu_j$ for some nonnegative integer ν_j. Together with condition $i_j \geq c_j \geq 0$, this implies $2i_j \geq l + 1 + \nu_j$.

---

[2] Our first algorithm corresponds to the case $c_j = i_j, \forall j$.

Remark 2. If ν_j is equal to 0, the "consistency" condition (i.e., $d_{i_j-1\to i_j-c_j} \geq d_{l\to i_j}$) is satisfied half of time, approximating $d_{i_j-1\to i_j-c}$ and $d_{l\to i_j}$ as $(l-i_j+1)$-bit randoms. In other words, if $\nu_j = 0$, half of time randomization is possible. A larger value for ν_j increases the success probability of the consistency condition (and thus of the randomization). On the other hand, it also reduces the possible counter indexes i satisfying the condition $2i_j \geq l+1+\nu_j$.

Figure 6 presents the resulting algorithm corresponding to the square-and-multiply algorithm. For all j, the value of ν_j is taken equal to 0 (and thus $c_j = l - i_j + 1$).

Input: $x, d = (d_l, \ldots, d_0)_2$
Output: $y = x^d \pmod{N}$

$R_0 \leftarrow 1; \; R_1 \leftarrow 1; \; R_2 \leftarrow x; \; i \leftarrow l; \; c \leftarrow -1; \; \sigma \leftarrow 1$
while $(i \geq 0)$ **do**
$\quad R_0 \leftarrow R_0 \cdot R_0 \pmod{N}$
$\quad$ **if** $(d_i = 1)$ **then** $R_0 \leftarrow R_0 \cdot R_2 \pmod{N}$
$\quad$ **if** $((2i \geq l+1) \wedge (\sigma = 1))$ **then** $c \leftarrow l - i + 1$ [‡]
$\quad$ **else** $\sigma \leftarrow 0$
$\quad \rho \leftarrow_R \{0,1\}$
$\quad \epsilon \leftarrow \rho \wedge (d_{i-1 \to i-c} \geq d_{l\to i}) \wedge \sigma$
$\quad$ **if** $(\epsilon = 1)$ **then**
$\quad\quad R_1 \leftarrow R_0; \; \sigma \leftarrow 0$
$\quad\quad d_{i-1\to i-c} \leftarrow d_{i-1\to i-c} - d_{l\to i}$
$\quad$ **endif**
$\quad$ **if** $(c = 0)$ **then**
$\quad\quad R_0 \leftarrow R_0 \cdot R_1 \pmod{N}; \; \sigma \leftarrow 1$
$\quad$ **endif**
$\quad c \leftarrow c - 1; \; i \leftarrow i - 1$
endwhile
return R_0

Fig. 6. Self-randomized square-and-multiply algorithm (II)

4   Enhanced Algorithms

4.1   Side-Channel Atomicity

As presented in the previous section, our algorithms involve numerous branchings and so, although randomized, might be vulnerable to SPA-type attacks [KJJ99].

A generic yet efficient technique, called *"side-channel atomicity"* [CCJ], allows to remove branching conditions at negligible cost. As this is not the main subject of this paper and due to lack of space, we present hereafter, without any further explanation, an atomic version of our first algorithm (Fig. 3). An atomic version of our second algorithm (Fig. 6) can be found in Appendix A.

Input: $x, d = (d_l, \ldots, d_0)_2$
Output: $y = x^d \pmod{N}$
$R_0 \leftarrow 1; \ R_1 \leftarrow 1; \ R_2 \leftarrow x; \ i \leftarrow l; \ k \leftarrow 0; \ \epsilon = 0$
while $(i \geq 0)$ **do**
$\quad R_\epsilon \leftarrow R_0 \cdot R_{\epsilon + 2k} \pmod{N}$
$\quad k \leftarrow k \oplus (d_i \wedge \neg \epsilon)$
$\quad d \leftarrow d + d_{l \to i} - d_{l \to i} \times (1 + \epsilon)$
$\quad i \leftarrow i - (\neg k \wedge \neg \epsilon)$
$\quad \rho \leftarrow_R \{0, 1\}$
$\quad \epsilon \leftarrow \rho \wedge \neg k \wedge \neg \epsilon \wedge (d_{i-1 \to 0} \geq d_{l \to i})$
endwhile
$R_0 \leftarrow R_0 \cdot R_1 \pmod{N}$
return R_0

Fig. 7. Atomic self-randomized square-and-multiply algorithm (I)

4.2   Reversibility

Throughout this section, we assume that our algorithms are given in a form free of conditional branchings (e.g., by using side-channel atomicity). We will now study their respective strengths against a very powerful imaginary adversary able to distinguish the performed (modular) multiplications. Algorithms I and II involve four types of multiplication:

$$
\begin{aligned}
\mathcal{S} &: \quad R_0 \leftarrow R_0 \cdot R_0 \quad (\bmod\ N) \\
\mathcal{M} &: \quad R_0 \leftarrow R_0 \cdot R_2 \quad (\bmod\ N) \\
\mathcal{C}_1 &: \quad R_1 \leftarrow R_0 \cdot R_1 \quad (\bmod\ N) \\
\mathcal{C}_2 &: \quad R_0 \leftarrow R_0 \cdot R_1 \quad (\bmod\ N)
\end{aligned}
$$

according to the registers used for the multiplication. Provided that such an attacker makes no errors, Algorithms I and II can be reversed and the value of exponent d recovered. The reversing algorithms are presented in Fig. 8.

We insist that the assumption of recovering the exact sequence of multiplications is unrealistic for present-day cryptographic devices as they include various countermeasures to purposely prevent the distinction between $\mathcal{S}$, $\mathcal{M}$ and $\mathcal{C}_i$. Even under such a strong attack scenario, Algorithm II can be slightly modified in order to make the attack impractical.

Algorithm II (Fig. 6) is constructed by choosing parameter $\nu_j = 0$ for all j. In fact, parameter ν_j can be any nonnegative integer such that $2i_j \geq l + 1 + \nu_j$ (cf. Remark 2). Hence, the largest possible value for ν_j is $2i_j - l - 1$ and thus, since $\nu_j \geq 0$, parameter $c_j = l - i_j + 1 + \nu_j$ can take any value in the set $\{l - i_j + 1, \ldots, i_j\}$. We generalize our second algorithm by randomly picking c_j in the set $\{l - i_j + 1, \ldots, i_j\}$; i.e., by replacing Line ‡ in Fig. 6 by

$$\textbf{if } ((2i \geq l + 1) \wedge (\sigma = 1)) \textbf{ then } c \leftarrow_R \{l - i + 1, i\}$$

Doing so we obtain a third algorithm (Algorithm III). Its side-channel atomic version is fully given in Appendix A.

Input: $L = (L_{l'}, \ldots, L_0)$	**Input:** $L = (L_{l'}, \ldots, L_0)$
Output: d	**Output:** d

$u \leftarrow 0; d' \leftarrow 0; i \leftarrow l'$
while $(i \geq 0)$ **do**
   case
      $(L_i = \mathcal{S}): \quad d' \leftarrow 2d'$
      $(L_i = \mathcal{M}): d' \leftarrow d' + 1$
      $(L_i = \mathcal{C}_1): \quad u \leftarrow u + d'$
   endcase
   $i \leftarrow i - 1$
endwhile
return $(u + d')$

$d' \leftarrow 0; j \leftarrow l; i \leftarrow l'$
while $(i \geq 0)$ **do**
   case
      $(L_i = \mathcal{S}): \quad d' \leftarrow 2d'$
      $(L_i = \mathcal{M}): d' \leftarrow d' + 1$
      $(L_i = \mathcal{C}_2): \quad d' \leftarrow d' + D[\frac{l+j+1}{2}]$
   endcase
   if $(L_{i-1} = \mathcal{S})$ **then** $D[j] \leftarrow d'; j \leftarrow j - 1$
   $i \leftarrow i - 1$
endwhile
return d'

| (a) Algorithm I | (b) Algorithm II |

Fig. 8. Recovering exponent d in self-randomized exponentiation algorithms by distinguishing all the involved multiplications

Provided that multiplications can be distinguished, reversing Algorithm III translates into the successful execution of the following algorithm:

Input: $L = (L_{l'}, \ldots, L_0)$
Output: d

$d' \leftarrow 0; j \leftarrow l; j_{old} \leftarrow l; i \leftarrow l'$
while $(i \geq 0)$ **do**
   case
      $(L_i = \mathcal{S}): \quad d' \leftarrow 2d'$
      $(L_i = \mathcal{M}): d' \leftarrow d' + 1$
      $(L_i = \mathcal{C}_2): \quad$ **for** $\frac{l+j+1}{2} \leq j_{try} \leq j_{old}$, **try** $d' \leftarrow d' + D[j_{try}]; j_{old} \leftarrow j$
   endcase
   if $(L_{i-1} = \mathcal{S})$ **then** $D[j] \leftarrow d'; j \leftarrow j - 1$
   $i \leftarrow i - 1$
endwhile
return d'

Fig. 9. Exhaustive search on Algorithm III

Since the attacker does not know the random c_j chosen in the set $\{l - i_j + 1, i_j\}$, she has to to try all possible values. Such a exhaustive rapidly becomes impractical, rendering our third algorithm even secure against very powerful adversaries.

4.3   Further Optimizations

The frequency of appearance that Boolean variable $\rho = 1$ can be seen as a *tuning parameter* for choosing the best trade-off between performance and security: more randomization penalize the running time and fewer randomization eases the exhaustive search.

A good way to lower the cost of additional operations consists in slightly modifying the random generator outputting ρ so that when Hamming weight of $d - a$ (a may have several definitions according to Algorithm I, II, or III) is weaker than Hamming weight of d , ρ has a higher probability of being a 1 and conversely. By this trick, the self-randomized algorithm will tend to select the case which has the weakest Hamming weight, that is, the fastest branch. We note however that the algorithm cannot always select the fastest branch as otherwise it becomes deterministic and so is more easily reversible.

4.4   Average Timing

In the following, we give a table with complexity of different algorithms, in term of multiplications.

Table 1. Average number of modular multiplications to perform an exponentiation of length 1024

$\mathcal{S}, \mathcal{M}, \mathcal{C}_1, \mathcal{C}_2$	Square and Multiply		Our algorithms		
	naive	random exp.[3]	(I)	(II)	(III)
Multiplications	1536	$1536 + 96$	$1536 + 512 \times \bar{\rho}$	$1536 + 10$	$1536 + 10$

The overhead factor of Algorithms II and III (10) corresponds in fact to an upper bound of $\log_2 d$. This is a very small quantity but it provides an interesting entropy: the number of possible randomization for a given exponent is superior to $\binom{10}{512} > 2^{64}$.

5   Conclusion

This paper introduced the concept of self-randomized exponentiation as an efficient means for preventing DPA-type attacks. Three different such algorithms (and some SPA-protected variants thereof) were described.

Self-randomized exponentiation presents the following interesting properties:

- it is fully generic in the sense that it is not restricted to a particular exponentiation algorithm;

---

[3] By random exponent d, we mean the use of $\hat{d}$ as explained in the introduction, with a random r_2 of size 64 bits.

- it is parameterizable: a parameter allows to choose the best trade-off between security and performance;
- it can be combined with most other counter-measures;
- it is space-efficient as only an additional long-integer register is required;
- it is flexible in the sense that it does not rely on certain group properties;
- it does not require the prior knowledge of the order of the group in which the exponentiation is performed.

Of independent interest, the notion of reversibility in self-randomized exponentiation algorithms was defined and a concrete construction was given.

Acknowledgements. The author would like to thank the anonymous referees for their helpful comments that allow us to improve the readability of this paper. Thanks also go to Marc Joye for his careful attention and continuous support in this research.

References

[BDL01]   Dan Boneh, Richard A. DeMillo, and Richard J. Lipton. On the importance of eliminating errors in cryptographic computations. *Journal of Cryptology*, 14(2):101–119, 2001.

[Bon99]   Dan Boneh. Twenty years of attacks on the RSA cryptosystem. *Notices of the AMS*, 46(2):203–213, 1999.

[BR95]   Mihir Bellare and Phillip Rogaway. Optimal asymmetric encryption. In A. De Santis, editor, *Advances in Cryptology – EUROCRYPT '94*, volume 950 of *Lecture Notes in Computer Science*, pages 92–111. Springer-Verlag, 1995.

[BR96]   Mihir Bellare and Phillip Rogaway. The exact security of digital signatures - How to sign with RSA and Rabin. In U. Maurer, editor, *Advances in Cryptology – EUROCRYPT '96*, volume 1070 of *Lecture Notes in Computer Science*, pages 399–416. Springer-Verlag, 1996.

[CJ01]   Christophe Clavier and Marc Joye. Universal exponentiation algorithm: A first step towards provable SPA-resistance. In Ç.K. Koç, D. Naccache, and C. Paar, editors, *Cryptographic Hardware and Embedded Systems – CHES 2001*, volume 2162 of *Lecture Notes in Computer Science*, pages 300–308. Springer-Verlag, 2001.

[CJRR99]   Suresh Chari, Charanjit S. Jutla, Josyula R. Rao, and Pankaj Rohatgi. Towards sound approaches to counteract power-analysis attacks. In M. Wiener, editor, *Advances in Cryptology – CRYPTO '99*, volume 1666 of *Lecture Notes in Computer Science*, pages 398–412. Springer-Verlag, 1999.

[CCJ]   Benoît Chevallier-Mames, Mathieu Ciet, and Marc Joye. Low cost solutions for preventing simple side-channel power analysis: Side-channel atomicity. To appear. Preprint available on IACR ePrint.

[DH76]   Whitfield Diffie and Martin E. Hellman. New directions in cryptography. *IEEE Transactions on Information Theory*, IT-22(6):644–654, 1976.

[IYTT02]   Kouichi Itoh, Jun Yajima, Masahiko Takenaka, and Naoya Torii. Dpa countermeasures by improving the window method. In Burton S. Kaliski Jr., Ç.K. Koç, and C. Paar, editors, *Cryptographic Hardware and Embedded Systems– CHES '02*, volume 2523 of *Lecture Notes in Computer Science*, pages 303–317. Springer-Verlag, 2002.

[KJJ99]   Paul Kocher, Joshua Jaffe, and Benjamin Jun. Differential power analysis. In M. Wiener, editor, *Advances in Cryptology – CRYPTO '99*, volume 1666 of *Lecture Notes in Computer Science*, pages 388–397. Springer-Verlag, 1999.

[Kob87]   Neal Koblitz. Elliptic curve cryptosystems. *Mathematics of Computation*, 48(177):203–209, 1987.

[Koc96]   Paul Kocher. Timing attacks on implementations of Diffie-Hellman, RSA, DSS, and other systems. In N. Koblitz, editor, *Advances in Cryptology – CRYPTO '96*, volume 1109 of *Lecture Notes in Computer Science*, pages 104–113. Springer-Verlag, 1996.

[MDS99]   Thomas S. Messerges, Ezzy A. Dabbish, and Robert H. Sloan. Power analysis attacks of modular exponentiation in smartcards. In Ç.K. Koç and C. Paar, editors, *Cryptographic Hardware and Embedded Systems– CHES '99*, volume 1717 of *Lecture Notes in Computer Science*, pages 144–157. Springer-Verlag, 1999.

[Mil86]   Victor S. Miller. Use of elliptic curves in cryptography. In H.C. Williams, editor, *Advances in Cryptology – CRYPTO '85*, volume 218 of *Lecture Notes in Computer Science*, pages 417–426. Springer-Verlag, 1986.

[MvV97]   Alfred J. Menezes, Paul C. van Oorschot, and Scott A. Vanstone. *Handbook of applied cryptography*. CRC Press, 1997.

[PKC02]   PKCS #1 v2.1: RSA cryptography standard. RSA Laboratories, June 14, 2002.

[QC82]   Jean-Jacques Quisquater and Chantal Couvreur. Fast decipherment algorithm for RSA public-key cryptosystem. *Electronics Letters*, 18:905–907, 1982.

[RSA78]   Ronald L. Rivest, Adi Shamir, and Leonard M. Adleman. A method for obtaining digital signatures and public-key cryptosystems. *Communications of the ACM*, 21(2):120–126, 1978.

[Wal98]   Colin D. Walter. Exponentiation using division chains. *IEEE Transactions on Computers*, 47(7):757–765, 1998.

[Wal02]   Colin D. Walter. Mist: An efficient, randomized exponentiation algorithm for resisting power analysis. In B. Preneel, editor, *Topics in Cryptology – CT-RSA 2002*, volume 2271 of *Lecture Notes in Computer Science*, pages 53–66. Springer-Verlag, 2002.

A Side-Channel Atomic Exponentiation Algorithms

Input: $x, d = (d_l, \ldots, d_0)_2$
Output: $y = x^d \pmod N$

$R_0 \leftarrow 1;\ R_1 \leftarrow 1;\ R_2 \leftarrow x;\ i \leftarrow l;\ k \leftarrow 0;\ \epsilon = 0$
$\sigma \leftarrow 1;\ c \leftarrow -1$
while $(i \geq 0)$ **do**
    $\theta \leftarrow (c = 0)$
    $R_0 \leftarrow R_0 \cdot R_{\theta + 2k} \pmod N$
    $k \leftarrow k \oplus (d_i \wedge \neg\theta);\ i \leftarrow i - (\neg k \wedge \neg\theta)$
    $\sigma \leftarrow (\sigma \vee \theta) \wedge (2i \geq l + 1)$
    $c \leftarrow \neg\sigma(c - \neg k) + (l - i + 1) \times \sigma$
    $\rho \leftarrow_R \{0, 1\}$
    $\epsilon \leftarrow \rho \wedge \neg k \wedge \sigma \wedge (d_{i-1 \to i-c} \geq d_{l \to i})$
    $\sigma \leftarrow \sigma \wedge \neg\epsilon$
    $d_{i-1 \to i-c} \leftarrow d_{i-1 \to i-c} + d_{l \to i} - d_{l \to i} \times (1 + \epsilon)$
    $R_1 \leftarrow R_{\neg\epsilon}$
endwhile
return R_0

Fig. 10. Atomic self-randomized square-and-multiply algorithm (II)

Input: $x, d = (d_l, \ldots, d_0)_2$
Output: $y = x^d \pmod N$

$R_0 \leftarrow 1;\ R_1 \leftarrow 1;\ R_2 \leftarrow x;\ i \leftarrow l;\ k \leftarrow 0;\ \epsilon = 0$
$\sigma \leftarrow 1;\ c \leftarrow -1$
while $(i \geq 0)$ **do**
    $\theta \leftarrow (c = 0)$
    $R_0 \leftarrow R_0 \cdot R_{\theta + 2k} \pmod N$
    $k \leftarrow k \oplus (d_i \wedge \neg\theta);\ i \leftarrow i - (\neg k \wedge \neg\theta)$
    $\sigma \leftarrow (\sigma \vee \theta) \wedge (2i \geq l + 1)$
    $\gamma \leftarrow_R \{l - i + 1, i\}$
    $c \leftarrow \neg\sigma(c - \neg k) + \gamma \times \sigma$
    $\rho \leftarrow_R \{0, 1\}$
    $\epsilon \leftarrow \rho \wedge \neg k \wedge \sigma \wedge (d_{i-1 \to i-c} \geq d_{l \to i})$
    $\sigma \leftarrow \sigma \wedge \neg\epsilon$
    $d_{i-1 \to i-c} \leftarrow d_{i-1 \to i-c} + d_{l \to i} - d_{l \to i} \times (1 + \epsilon)$
    $R_1 \leftarrow R_{\neg\epsilon}$
endwhile
return R_0

Fig. 11. Atomic self-randomized square-and-multiply algorithm (III)

Flexible Hardware Design for RSA and Elliptic Curve Cryptosystems*

Lejla Batina[1], Geeke Bruin-Muurling[2], and Sıddıka Berna Örs[1]

[1] Katholieke Universiteit Leuven, ESAT/COSIC, Kasteelpark Arenberg 10,
B-3001 Leuven-Heverlee, Belgium
[2] Rembrandtlaan 45, 5261 XG, Vught, The Netherlands
{Lejla.Batina, Siddika.BernaOrs}@esat.kuleuven.ac.be
geekemuurling@hotmail.com

Abstract. This paper presents a scalable hardware implementation of both commonly used public key cryptosystems, RSA and Elliptic Curve Cryptosystem (ECC) on the same platform. The introduced hardware accelerator features a design which can be varied from very small (less than 20 Kgates) targeting wireless applications, up to a very big design (more than 100 Kgates) used for network security. In latter option it can include a few dedicated large number arithmetic units each of which is a systolic array performing the Montgomery Modular Multiplication (MMM). The bound on the Montgomery parameter has been optimized to facilitate more secure ECC point operations. Furthermore, we present a new possibility for CRT scheme which is less vulnerable to side-channel attacks.

Keywords: FPGA design, Systolic array, Hardware implementation, RSA, ECC, Montgomery multiplication, Side-channel attacks

1 Introduction

Security of communication or in general of some digital data is founded by various cryptographic algorithms. Especially implementations of Public Key Cryptography (PKC) present a challenge in vast majority of application platforms varying from software to hardware. Software platforms are cheap and a more flexible solution but it appears that only hardware implementations provide a suitable level of security especially related to side-channel attacks. Two best known and most widely used public-key cryptosystems are RSA [26] and ECC [18], [13]. When it comes to RSA, it is believed to be on its "sunset" but still keeping up with requirements. Namely, because of various factors such as well developed speed-ups in the form of Chinese Remainder Theorem (CRT) techniques and

---

* Lejla Batina and Sıddıka Berna Örs are funded by a research grants of the Katholieke Universiteit Leuven, Belgium. This work was supported by Concerted Research Action GOA-MEFISTO-666 of the Flemish Government and by the FWO "Identification and Cryptography" project (G.0141.03).

T. Okamoto (Ed.): CT-RSA 2004, LNCS 2964, pp. 250–263, 2004.
© Springer-Verlag Berlin Heidelberg 2004

its suitability for hardware, RSA is the main technology for high-speed applications in network security, financing etc. On the other hand, ECC is expected to take the lead within wireless applications. The reason is that ECC operates with higher speed, lower power consumption and smaller certificates, which are all necessities within these areas including the smartcard industry. In short, it is mostly desired to develop an architecture which can efficiently perform both RSA and ECC, RSA for VPNs, banking etc. and ECC still mostly for wireless applications.

Our contribution deals with an FPGA implementation of RSA and ECC cryptosystems over a field of prime characteristic. The architecture for Montgomery Modular Multiplication (MMM) used in this work is efficient and secure [22]. The systolic array is used for arbitrary precision in bits, hence easily bridging the gap between the bit-lengths for ECC from 160 bits to 2048 (or higher) bit long modulus for RSA. The notion of scalability we discuss includes both, freedom in choice of operand precision as well as adaptability to any desired gate complexity. To the latter is usually referred to as "flexibility". We use modular exponentiation based on Montgomery's method without any modular reduction achieving the best possible bound [29], [3]. We are first to introduce a similar bound for ECC which allows us to perform a very secure and yet efficient point addition and doubling. We show that in the case of two or more arithmetic units a high level of parallelism can be achieved altering ECC operations between those units. The eventual parallelism between more units and also between cells of the systolic array is beneficial for side-channel resistance. Moreover, in this work we introduce a new variation of Garner's scheme for CRT decryption, which has built-in countermeasure against timing and power analysis based attacks.

Since the introduced architecture was dedicated to RSA applications, it was natural to implement elliptic curve arithmetic in $GF(p)$. In this way all required components were already available as ECC in $GF(p)$ is based on ordinary modular arithmetic. Assuming one uses projective coordinates modular multiplication remains as the most time consuming operation for ECC. Hence, efficient implementation relies on efficient modular multiplication, as is the case for RSA. Nevertheless, it is also important to focus on time-constant algorithms which are less likely to leak side-channel information. To conclude, in this work we aimed to introduce a secure combined RSA-ECC implementation which as well meets high demands in speed implied by state of art for RSA hardware implementation. See for example [8].

The remainder of this paper is organized as follows. Section 2 gives a survey of previous implementations of public-key algorithms in hardware relevant for our work. In Section 3, we outline the architecture of the targeted implementation platform. Section 4 describes new options for point operations. In Section 5, the implementation results and timings are given. Section 6 introduces a new variant of Garner's scheme for CRT which is as well efficient but more resistant to side-channel attacks. Implications of the proposed changes on security of both RSA and ECC are considered in Section 7. Sections 8 concludes the paper.

2   Related Work

This section reviews some of the most relevant previous work in hardware implementations for PKC. The vast majority of published work that is considering implementations of PKC deals with software platforms. Some of the work is done on FPGAs and only very few implementations are presenting an ASIC implementation of ECC in the field of prime characteristic. Most of the work is done in binary field and some authors have considered dual field implementations i.e. ECC in prime and binary field.

Goodman and Chandrakasan proposed a domain-specific reconfigurable cryptographic processor (DSRCP) in [8]. The DSRCP performs a variety of algorithms ranging from modular integer arithmetic to elliptic curve arithmetic. They mainly discussed the arithmetic in binary field. Most recent published work is the one of Satoh and Takano [27]. They presented the dual field multiplier with the best performance so-far in both type of fields. The throughput of EC scalar multiplication is maximized by use of Montgomery multiplier and on-the-fly redundant binary converter. The great quality of their design is in scalability in operand size and also flexibility between speed and hardware area. Another hardware solution for both types of fields was presented by Wolkerstorfer in [31]. The author introduced low power design which features short critical path to enable high clock frequencies. Most operations are executed within a single clock cycle and the redundant number representation was used. The idea of unified multiplier was first introduced by Savaş et al. in [28]. The authors have discussed a scalable and unified architecture for a Montgomery multiplication module. They deployed an array of word size processing units organized in a pipeline. The same idea is the basis of work in Grosschädl [9]. The bit-serial multiplier which is introduced is performing multiplications in both types of fields. The author also modified the classical MSB-first version for iterative modular multiplication. All concepts are introduced in detail, but the actual VLSI implementation is not given. Some hardware implementations in $GF(p)$ on FPGA are also known. The ECC-only processor over fields GF(p) was proposed by Orlando and Paar [21]. They proposed so-called Elliptic Curve Processor (ECP) which is scalable in terms of area and speed. The ECP is also best suited for projective coordinates and it is using a new type of high-radix precomputation-based Montgomery multiplier. The scalability of the multiplier to larger fields was also verified in the field whose size is 521 bits. The authors have estimated eventual timing of 3 ms for computing one point multiplication in 192-bit prime field. Örs et al discussed an ECC-processor which is optimized for MMM in [23]. They described an efficient implementation of an elliptic curve processor over $GF(p)$. The processor can be programmed to execute a modular multiplication, addition/subtraction, multiplicative inversion, EC point addition/doubling and multiplication. A detailed overview of hardware implementations for PKC is given in [4] .

Still plenty of the work in ECC over GF(p) deals with software implementations, where there exist many hardware implementations over binary field. It appears that the arithmetic in characteristic 2 is easier to implement and area and power consumption are smaller than in the case of $GF(p)$. This is believed to be true, but only for platforms where specialized arithmetic coprocessors for finite field arithmetic are not available. On the other hand, an advantage of prime field is in its suitability for both RSA and ECC with an a resourceful sharing of hardware.

3   Previous Work and Background

In this paper we discuss how an FPGA implementation of Montgomery multiplication that was originally designed for RSA can efficiently be used to perform prime field ECC operations. This design consists of a Large Modular Montgomery Multiplier (MMM), designed as a systolic array. This array is one-dimensional and consists of a fixed number of Processing Cells (PCs). The MMM performs Montgomery modular multiplication that consists of the following operation: $Mont(X, Y) = XYR^{-1} \mod N$.

In the remainder we call $AR \mod N$ the Montgomery representation of A. For modular exponentiation with the MMM all intermediate results are in this form. A number can be transformed to its Montgomery representation by performing a Montgomery multiplication of that number with $R^2 \mod N$. For the transformation from Montgomery representation to the normal form a Montgomery multiplication with 1 will suffice.

3.1   Systolic Array

Figure 1 shows a schematic of the systolic array that was implemented in the MMM. A PC contains adders and multipliers that can process α bits of X and β bits of Y in one clock cycle. Here X and Y are the multiplicand and multiplier. Each PC calculates $\frac{T_j + x_i y_j + m_i N_j}{2^\alpha}$ in each clock cycle. The detailed description is given in Section 4.4.

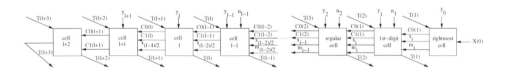

Fig. 1. Schematic of the Modular Montgomery Multiplier.

In the original notation of Montgomery after each multiplication in the exponentiation algorithm a reduction was needed [19]. The input had the restriction $X, Y < N$ and the output T was bounded by $T < 2N$. The result of this is that in the case $T > N$, N must be subtracted so that the output can be used as input of the next multiplication. To avoid this subtraction a bound for R can be calculated such that for inputs $X, Y < 2N$ also the output is bounded by $T < 2N$.

In [3] the need of avoiding this reduction after each multiplication is addressed. In practice this means that the output of the multiplication can be directly used as an input of the next Montgomery multiplication. The following theorem is proven in [3].

Theorem 1. *The result of a Montgomery multiplication $XYR^{-1} \bmod N < 2N$ when $X, Y < 2N$ and $R > 4N$.*

The final round in the modular exponentiation is the conversion to the integer domain, i.e. calculating the Montgomery multiplication of the last result and 1. The same arguments prove that this final step remains within the following bound: $Mont(T, 1) \leq N$. In practice, $A^B \bmod N = N$ will never occur since $A \neq 0$.

3.2   ECC Processor

The MMM need not only be used for fast RSA implementation but also for ECC point operations in the prime field. Due to the scalability of the design, the FPGA architecture can perform both, i.e. efficient exponentiations on large operands (for RSA) and modular multiplication on the smaller ECC operands. In Figure 2 a schematic of an FPGA implementation for ECC is given. One or two MMMs are used to perform the modular (Montgomery) multiplications. A Large Number Co-Processor (LNCP) is added to the design to perform the additions and subtractions. These units have their own RAM's and are connected with a data bus.

As already explained, the performance of an elliptic curve cryptosystem is primarily determined by the efficient realization of the arithmetic operations (addition, multiplication and inversion) in the underlying finite field. If projective coordinates are used the inversion operation becomes negligible. Therefore, coprocessors for elliptic curve cryptography are primarily designed to accelerate the field multiplication. Considering multiplication in the prime field i.e., $GF(p)$, the whole work which is done for the RSA implementation is relevant. The only difference is that shorter bit-lengths are used i.e., 160-300 bits. Scalability is again a point of concern and even more inter operability between different implementations.

4   New Implementation

In this section we present our FPGA implementation for ECC point operations for prime fields.

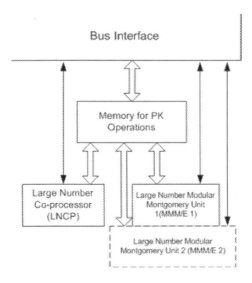

Fig. 2. Schematic of the Modular Montgomery Multiplier.

4.1 Point Addition

Point addition and doubling can be performed according to the algorithm given in [5].

Here we assume that the two points that will be added i.e., $P = (X_1, Y_1, Z_1)$ and $Q = (X_2, Y_2, Z_2)$ are already transformed to the Projective coordinates and Montgomery representation. The result point $R = P + Q = (X_3, Y_3, Z_3)$

Scheduling of point addition. Point addition can be even performed more efficient if two MMM units are used. The operations can be conveniently divided between the two units. (Modular) addition and subtraction will be done on a Large Number Co-processor. Those operations can be performed in the same time as the Montgomery multiplication. The following scheduling as shown in Table 1 can be used. Table 1 shows that the performance can almost be doubled by using two MMM units.

4.2 Point Doubling

Here we discuss a special case of point addition i.e. point doubling, where the points P and Q are respectively given as: $P = (X_1, Y_1, Z_1)$ and $R = 2P = (X_3, Y_3, Z_3)$.

Scheduling of point doubling. In Table 2 a possible schedule for point doubling over the 2 MMMs and the LNCP is given .

The difficulty in the scheduling of point doubling lies in the operations scheduled in MMM2 and the LNCP, which are all depending on the answer of the previous operation.

Table 1. Scheduling of point addition.   **Table 2.** Scheduling of point doubling.

MMM1	MMM2	LNCP
Z_2^2	Z_1^2	
$\lambda_1 = X_1 Z_2^2$	$\lambda_2 = X_2 Z_1^2$	
Z_2^3	Z_1^3	$\lambda_3 = \lambda_1 - \lambda_2$
		$\lambda_7 = \lambda_1 + \lambda_2$
$\lambda_4 = Y_1 Z_2^3$	$\lambda_5 = Y_2 Z_1^3$	
	λ_3^2	$\lambda_6 = \lambda_4 - \lambda_5$
		$\lambda_8 = \lambda_4 + \lambda_5$
λ_6^2	$\lambda_7 \lambda_3^2$	
λ_3^3	$Z_1 Z_2$	$X_3 = \lambda_6^2 - \lambda_7 \lambda_3^2$
		$\lambda_9 = \lambda_7 \lambda_3^2 - 2X_3$
$\lambda_8 \lambda_3^3$	$\lambda_9 \lambda_6$	
$Z_3 = Z_1 Z_2 \lambda_3$		$Y_3 = \dfrac{\lambda_9 \lambda_6 - \lambda_8 \lambda_3^3}{2}$

MMM1	MMM2	LNCP
$3X_1^2$	Z_1^2	
Y_1^2	Z_1^4	
$\lambda_2 = 4X_1 Y_1^2$	aZ_1^4	
$\lambda_3 = 8Y_1^4$		$\lambda_1 = 3X_1^2 + aZ_1^4$
	λ_1^2	
$Z_3 = 2Y_1 Z_1$		
		$X_3 = \lambda_1^2 - 2\lambda_2$
		$\lambda_2 - X_3$
	$\lambda_1(\lambda_2 - X_3)$	
		$Y_3 = \lambda_1(\lambda_2 - X_3) - \lambda_3$

Point multiplication can be implemented as a repeated combination of point addition and point doubling.

4.3   Modular Addition and Subtraction

Modular (i.e. Montgomery) multiplication, modular addition and modular subtraction are the basic operations for point addition. MMM is performed on our highly scalable Montgomery based multiplier. Modular addition and modular subtraction can be implemented as a repeated addition. However, the number of additions/subtractions would be data dependent. Let us take a better look at these two operations. As proven in Section 3.1, the result of an operation on our multiplier will always be smaller than twice the modulus $(2N)$. All modular additions and subtractions in the point addition scheme are with two outputs of the Montgomery multiplier.

For example:

$$\begin{aligned}
\lambda_1 &= X_1 Z_2^2 < 2p \quad \text{and} \quad \lambda_2 = X_2 Z_1^2 < 2p \\
\lambda_3 &= \lambda_1 - \lambda_2 \bmod p \\
\lambda_7 &= \lambda_1 + \lambda_2 \bmod p
\end{aligned} \tag{1}$$

The result of the modular addition and subtraction is again the input of another Montgomery multiplication and can therefore be larger than the modulus but should be positive. If it would be possible to calculate the previous calculations as "normal" i.e. non-modular addition and subtraction, this would make the operations very efficient but more importantly time constant.

Keeping in mind the "$2p$" bound for the operands as a result of the bound for the Montgomery parameter, we get:

$$\begin{aligned}
0 &< \lambda_1 + \lambda_2 &&< 4p - 1 \\
0 &< \lambda_1 + 2p - \lambda_2 &&< 4p
\end{aligned} \tag{2}$$

Our target is now to try to fix a bound for the Montgomery parameter such that we can use these non-modular addition and subtraction instead of the modular forms. To achieve this we must ensure that the inputs X and Y of the Montgomery multiplier that are smaller than $4p$ result in a Montgomery product that is smaller than $2p$.

As already mentioned, in the original implementation of our MMM the inputs of a Montgomery multiplication should be smaller than $2p$. We will use the following lemma.

Lemma 1. *If the Montgomery parameter R satisfies the following inequality $R > 16N$, then for inputs $X, Y < 4N$ the result T of the MMM will satisfy: $T < 2N$ (as required).*

Proof: The Montgomery multiplication as implemented in the MMM calculates the following:

$$T = \frac{AB + mN}{R} = \frac{AB}{R} + \frac{m}{R}N \tag{3}$$

here m is calculated modulo R. Filling in the bounds for the inputs and $R > 16N$ we get

$$T = \frac{AB}{R} + \frac{m}{R}N < \frac{4N \cdot 4N}{R} + N \leq 2N. \tag{4}$$

If n is the length of modulus N in bits then the following is valid: $N < 2^n$ and $16N < 2^{n+4}$. With $R = 2^r$, we get $r \geq n + 4$. $\square$

We have shown that for all modulus lengths, inputs smaller than $4p$ will result after a Montgomery multiplication on the MMM in a value which is smaller than $2p$. Therefore we can use the more efficient and time constant implementation of modular addition and subtraction. Furthermore, there is no any loss in efficiency caused by this enlarged bound because R is usually already bigger than this bound (especially for $\alpha, \beta > 1$.)

4.4   Montgomery Modular Multiplication

The processing cells in the systolic array shown in Figure 1 performs Equation 5. x_i and m_i have α bits, y_j and n_j have β bits. $c1_{i,j}$ and $c0_{i,j}$ denote the carry chain on the array. Because the critical path of the systolic array is the same as the critical path of one PC, the clock frequency of the Montgomery multiplier will be the same for all bit-lengths. This property gives the advantage of using the circuit for RSA and ECC.

$$2^2 \times c1_{i,j} + 2 \times c0_{i,j} + t_{i,j} = t_{i-1,j+1} + x_i \times y_j + m_i \times n_j + 2 \times c1_{i,j-1} + c0_{i,j-1} \tag{5}$$

Parameters α and β are 4 for this implementation. Table 3 shows the performance of the FPGA implementation of the Montgomery multiplier. Parameter n is the bit-length of N, l in Figure 1 is $\frac{n}{4}$.

Table 3. The performance of the FPGA implementation of the Montgomery multiplier.

Number of clock cycles	$3\frac{n}{4} + 7$
Clock period	19 ns
Clock frequency	53 MHz
Total latency	$14.25n + 133$ ns
Number of gates	4547

5   Results and Timings

In the work of Lenstra and Verheul [16], the authors made a security comparison between RSA and ECC key lengths. They introduced a table that included corresponding key bit-lengths assuring minimal security in the years to come for the two Public Key systems. In Figure 3 the performances for ECC and RSA are given according to the key sizes that were given in their paper. The figures show also that especially for the future applications the performance of ECC is more attractive than the performance of RSA.

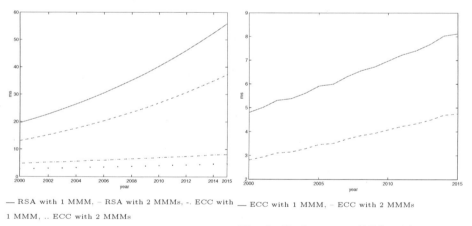

— RSA with 1 MMM, – RSA with 2 MMMs, -. ECC with 1 MMM, .. ECC with 2 MMMs

Fig. 3. Performance of RSA and ECC.

— ECC with 1 MMM, – ECC with 2 MMMs

Fig. 4. Performance ECC with 1 or 2 MMM.

Figure 4 shows the performance for an ECC implementation with one and two MMMs. The implementation with 2 MMMs is scheduled according to the schedule given in Table 1 and Table 2. Figure 4 shows a speed-up of a factor of 2 for the two MMMs variant.

For the sake of preciseness we give detailed performance results in the Table 4.

Table 4. RSA and ECC performance in ms at 53 MHz.

year [16]	RSA			ECC		
	size	1 MMM	2 MMM	size	1 MMM	2 MMM
2002	1024	22.8	15.2	139	5.3	3.1
2014	1536	52.5	35	172	8	4.7
2023	2048	90.6	60.4	197	10.5	6.1
2051	4096	350.9	234	275	11.4	6.7

6   Side-Channel Security of CRT

We will now briefly review some benefits of Montgomery's Multiplication Method, which are also evident for CRT implementations. In [11,29], $R > 4N$ is proposed which, with some savings in hardware, omits completely all reduction steps.

Especially implementations of CRT schemes are found to be very sensitive to side-channel attacks. For example, recently a new SPA-based attack was introduced by Novak [20], which is targeting the algorithm of Garner [17]. This scheme is often used in all sorts of applications, including smartcards. It is usually implemented as follows:

Algorithm 1. Garner's algorithm for CRT

INPUT: ciphertext c, $N = p \cdot q$, $(p > q)$ and precomputed values
$d_1 \equiv d \bmod (p - 1), d_2 \equiv d \bmod (q - 1)$ and $U = p^{-1}(\bmod q)$
$(C_1 \equiv C \,(\bmod p), C_2 \equiv C \,(\bmod q))$
OUTPUT: $R = M \equiv t + q \cdot (s \cdot (q^{-1} \bmod p)) \bmod p$
1. $s = M_p \equiv C_1{}^{d_1} \bmod p$,
2. $t = M_q \equiv C_2{}^{d_2} \bmod q$
3. $x = (s - t)(\bmod p)$
4. $R = t + q \cdot ((x \cdot U)(\bmod p))$

The third step is the critical one. Novak observed that if the modular subtraction is implemented in the common way it may leak information. More precisely, to perform subtraction $(\bmod p)$ one has to check the sign of $s - t$ and conditionally add p if $s - t < 0$ $(p > q$ is required). Novak managed to build a successful attack based on this observation. An implementation of the above algorithm can produce the optional pattern in a power trace as a result of the conditional addition.

We propose the following solution. Instead of the subtraction $\bmod p$, one can compute the following:

$$x = s + p - t. \qquad (6)$$

For $p > q$ the result stays within the following bounds $0 < x < 2p$ which can be handled easily if Step 4 is implemented by use of the algorithm of Montgomery. Namely, the algorithm as proposed in [11,29] for Montgomery modular multiplication takes two inputs $0 < X, Y < 2p$ and the result is also within the same interval, if the proper bound for Montgomery parameter R is chosen. This result is converted from the Montgomery domain to the usual domain by a Montgomery multiplication with 1. Changing Garner's scheme in this way the algorithm is always performing a constant execution path. We prove this in more detail.

Claim. The result of $x = s + p - t$ is always smaller than $2p$, for the parameters s, p, t defined as above, i.e. $s = M_p \equiv C_1{}^{d_1} \bmod p$, $t = M_q \equiv C_2{}^{d_2} \bmod q$ and $p > q$.

Proof: It is shown that with the use of Montgomery's algorithm and $R > 4N$, the final result of modular exponentiation is bounded by the modulus N. (See Theorem 1.)

Now, we can prove the claim. It is obvious that $0 \leq s < p$ and $0 \leq t < q$. We assume $p > q$ with which this proof does not loose its generality, the other case is almost the same. Then we get $x = s + p - t < p + p - 0 = 2p$. Hence, if the multiplication in the Algorithm 1 is implemented as the one of Montgomery, no conditional subtraction is required as in original algorithm. This concludes the proof. □

7   Security Remarks

In this section we address side-channel security i.e. resistance to timing [14], [10] and power analysis based attacks [15]. These types of attacks, together with fault-analysis based attacks [6], [12], [2] electromagnetic analysis attacks (EMA) [25], [7] and other physical attacks such as probing attacks [1] are a major concern especially for wireless applications. Mainly because of space limitation we only briefly discuss the first two, which are also believed to be the most practical .

Namely, computations performed in non-constant time i.e. computations which are time-dependent on the values of the operands, may leak secret key information. This observation is the basis for timing attacks. On the other hand, power analysis based attacks use the fact that the power consumed at any particular time during a cryptographic operation is related to the function being performed and data being processed. The attack can be usually performed easily because smartcards, for example receive the power externally and an attacker can easily get to hold on the source of this side-channel information.

In our implementation all modular reductions are excluded. The weaknesses in the conditional statements of the algorithm (used for realization of the reduction step) are time variations and therefore these should be omitted. By use of an optimal upper bound the number of iterations required in the algorithm based on Montgomery's method of multiplication can be reduced [30]. Another timing information leakage that was observed by Quisquater [10] et al. and Walter [30] was the timing difference between "square" and "multiply". This information can be used to attack RSA, even advanced exponentiation methods were used. In our architecture, this weakness is removed, because the same systolic array is performing squarings and multiplications, which are therefore indistinguishable with respect to timing.

Besides that, when considering power analysis attacks, some other precautions have also been introduced. The fact that all of the PCs operate in parallel makes these types of attacks far less likely to succeed. Both, RSA and ECC can benefit from this fact.

As already mentioned, this architecture can be an option for wireless devices, although we have chosen here to introduce a network security devoted product. Again, because of space limitation we were not able to discuss the smaller, compact implementation but that also features very secure low-power design with attractive performances.

Örs et al characterized the power consumption of a XILINX Virtex 800 FPGA in [24]. They showed that it is possible to draw conclusions about vulnerability of an ordinary ASIC in CMOS technology by performing power-analysis attacks on an FPGA-implementation. With respect to this, an FPGA design can serve as a good model for ASIC platform not just for usual hardware related properties but also for security.

8   Conclusions

We have presented the hardware implementation on systolic array architecture that is scalable in all parameters and ideally suitable for RSA and ECC algorithms.

We have also introduced a bound on Montgomery parameter R, which allows us to perform the most efficient point addition and doubling for ECC, as well as modular exponentiation. Even in the case of CRT the Montgomery's algorithm is proven to be the best option for side-channel resistance.

References

1. R. Anderson and M. Kuhn. Low cost attacks on tamper resistant devices. In B. Christianson *et al.*, editor, *Proceedings of 1997 Security Protocols Workshop*, number 1361 in Lecture Notes in Computer Science, pages 125–136. Springer-Verlag, 1997.

2. C. Aumüller, P. Bier, W. Fischer, P. Hofreiter, and J.-P. Seifert. Fault attacks on RSA with CRT: Concrete results and practical countermeasures. In B. S. Kaliski Jr., Ç. K. Koç, and C. Paar, editors, *Proceedings of Cryptographic Hardware and Embedded Systems (CHES 2002)*, Lecture Notes in Computer Science, 2002.

3. L. Batina and G. Muurling. Montgomery in practice: How to do it more efficiently in hardware. In B. Preneel, editor, *Proceedings of RSA 2002 Cryptographers' Track*, number 2271 in Lecture Notes in Computer Science, pages 40–52, San Jose, USA, February 18–22 2002. Springer-Verlag.

4. L. Batina, S. B. Örs, B. Preneel, and J. Vandewalle. Hardware architectures for public key cryptography. *Elsevier Science Integration the VLSI Journal*, 34, 2003.
5. I. Blake, G. Seroussi, and N. P. Smart. *Elliptic Curves in Cryptography*. London Mathematical Society Lecture Note Series. Cambridge University Press, 1999.
6. D. Boneh, R. A. DeMillo, and R. J. Lipton. On the importance of checking cryptographic protocols for faults (extended abstract). In W. Fumy, editor, *Advances in Cryptology: Proceedings of EUROCRYPT'97*, number 1233 in Lecture Notes in Computer Science, pages 37–51. Springer-Verlag, 1997.
7. K. Gandolfi, C. Mourtel, and F. Olivier. Electromagnetic analysis: Concrete results. In Ç. K. Koç, D. Naccache, and C. Paar, editors, *Proceedings of Cryptographic Hardware and Embedded Systems (CHES 2001)*, number 2162 in Lecture Notes in Computer Science, pages 255–265, 2001.
8. J. Goodman and A. P. Chandrakasan. An energy-efficient reconfigurable public-key cryptography processor. *IEEE Journal of Solid-State Circuits*, 36(11):1808–1820, November 2001.
9. J. Grosschädl. A bit-serial unified multiplier architecture for finite fields $GF(p)$ and $GF(2^n)$. In Ç. K. Koç, D. Naccache, and C. Paar, editors, *Proceedings of Cryptographic Hardware and Embedded Systems (CHES 2001)*, number 2162 in Lecture Notes in Computer Science, pages 206–223, 2001.
10. G. Hachez, F. Koeune, and J.-J. Quisquater. Timing attack: what can be achieved by a powerful adversary? In A. Barbé, E. C. van der Meulen, and P. Vanroose, editors, *Proceedings of the 20th symposium on Information Theory in the Benelux*, pages 63–70, May 1999.
11. G. Hachez and J.-J. Quisquater. Montgomery exponentiation with no final subtractions: Improved results. In Ç. K. Koç and C. Paar, editors, *Proceedings of Cryptographic Hardware and Embedded Systems (CHES 2000)*, number 1965 in Lecture Notes in Computer Science, pages 293–301, 2000.
12. M. Joye, A. K. Lenstra, and J.-J. Quisquater. Chinese remaindering based cryptosystem in the presence of faults. *Journal of Cryptology*, 4(12):241–245, 1999.
13. N. Koblitz. Elliptic curve cryptosystem. *Math. Comp.*, 48:203–209, 1987.
14. P. Kocher. Timing attacks on implementations of Diffie-Hellman, RSA, DSS and other systems. In N. Koblitz, editor, *Advances in Cryptology: Proceedings of CRYPTO'96*, number 1109 in Lecture Notes in Computer Science, pages 104–113. Springer-Verlag, 1996.
15. P. Kocher, J. Jaffe, and B. Jun. Differential power analysis. In M. Wiener, editor, *Advances in Cryptology: Proceedings of CRYPTO'99*, number 1666 in Lecture Notes in Computer Science, pages 388–397. Springer-Verlag, 1999.
16. A. Lenstra and E. Verheul. Selecting cryptographic key sizes. In H. Imai and Y. Zheng, editors, *Proceedings of Third International Workshop on Practice and Theory in Public Key Cryptography (PKC 2000)*, number 1751 in Lecture Notes in Computer Science, pages 446–465. Springer-Verlag, 2000.
17. A. Menezes, P. van Oorschot, and S. Vanstone. *Handbook of Applied Cryptography*. CRC Press, 1997.
18. V. Miller. Uses of elliptic curves in cryptography. In H. C. Williams, editor, *Advances in Cryptology: Proceedings of CRYPTO'85*, number 218 in Lecture Notes in Computer Science, pages 417–426. Springer-Verlag, 1985.
19. P. Montgomery. Modular multiplication without trial division. *Mathematics of Computation*, Vol. 44:519–521, 1985.

20. R. Novak. SPA-based adaptive chosen-ciphertext attack on RSA implementation. In D. Naccache and P. Pallier, editors, *Proceedings of 5th International Workshop on Practice and Theory in Public Key Cryptosystems (PKC 2002), Paris, France, February 2002*, number 2274 in Lecture Notes in Computer Science. Springer-Verlag, 2002.

21. G. Orlando and C. Paar. A scalable GF(p) elliptic curve processor architecture for programmable hardware. In Ç. K. Koç, D. Naccache, and C. Paar, editors, *Proceedings of Workshop on Cryptograpic Hardware and Embedded Systems (CHES 2001)*, number 2162 in Lecture Notes in Computer Science, pages 356–371, Paris, France, May 14-16 2001. Springer-Verlag.

22. S. B. Örs, L. Batina, B. Preneel, and J. Vandewalle. Hardware implementation of a Montgomery modular multiplier in a systolic array. In *The The 10th Reconfigurable Architectures Workshop (RAW)*, Nice, France, April 22 2003.

23. S. B. Örs, L. Batina, B. Preneel, and J. Vandewalle. Hardware implementation of an elliptic curve processor over GF(p). In *IEEE 14th International Conference on Application-specific Systems, Architectures and Processors (ASAP)*, The Hague, The Netherlands, June 24–26 2003.

24. S. B. Örs, E. Oswald, and B. Preneel. Power-analysis attacks on an FPGA – first experimental results. In C. Walter, Ç. K. Koç, and C. Paar, editors, *Proceedings of Cryptographic Hardware and Embedded Systems (CHES 2003)*, Lecture Notes in Computer Science, pages 35–50, Cologne, Germany, September 7–10 2003.

25. J. J. Quisquater and D. Samyde. Elecromagnetic analysis EMA: Measures and coutermeasures for smart cards. In *in Smart Card Programming and Security*, number 2140 in Lecture Notes in Computer Science, pages 200–210. Springer-Verlag, 2001.

26. R. L. Rivest, A. Shamir, and L. Adleman. A method for obtaining digital signatures and public-key cryptosystems. *Communications of the ACM*, 21(2):120–126, 1978.

27. A. Satoh and K. Takano. A scalable dual-field elliptic curve cryptographic processor. *IEEE Transactions on Computers, special issue on cryptographic hardware and embedded systems*, 52(4):449–460, April 2003.

28. E. Savaş, A. F. Tenca, and Ç. K. Koç. A scalable and unified multiplier architecture for finite fields GF(p) and GF(2^m). In C. Paar and Ç. K. Koç, editors, *Proceedings of Cryptographic Hardware and Embedded Systems (CHES 2000)*, number 1965 in Lecture Notes in Computer Science, pages 281–296. Springer-Verlag, 2000.

29. C. D. Walter. Precise bounds for Montgomery modular multiplication and some potentially insecure RSA moduli. In B. Preneel, editor, *Proceedings of Topics in Cryptology- CT-RSA 2002*, number 2271 in Lecture Notes in Computer Science, pages 30–39, 2002.

30. C. D. Walter and S. Thompson. Distinguishing exponent digits by observing modular subtractions. In D. Naccache, editor, *Proceedings of Topics in Cryptology - CT-RSA*, number 2020 in Lecture Notes in Computer Science, pages 192–207, San Francisco, 8-12 April 2001. Springer-Verlag.

31. J. Wolkerstorfer. Dual-field arithmetic unit for $GF(p)$ and $GF(2^m)$. In B. S. Kaliski Jr., Ç. Koç, and C. Paar, editors, *Proceedings of Cryptographic Hardware and Embedded Systems (CHES 2002)*, Lecture Notes in Computer Science, Redwood Shores, CA, USA, August 13–15 2002. Springer-Verlag.

High-Speed Modular Multiplication

Wieland Fischer and Jean-Pierre Seifert

Infineon Technologies, Secure Mobile Solutions
D-81609 Munich, GERMANY
{Wieland.Fischer,Jean-Pierre.Seifert}@infineon.com

Abstract. Sedlak's [Sed] modular multiplication algorithm is one of the first real silicon implementations to speed up the RSA signature generation [RSA] on a smartcard, cf. [DQ]. Theoretically, Sedlak's algorithm needs on average $n/3$ steps (i.e., additions/subtractions) to compute the modular product of n-bit numbers. In [FS2] we presented a theoretical algorithm how to speed up Sedlak's algorithm by an arbitrary integral factor $i \geq 2$, i.e., our new algorithm needs on average $n/(3 \cdot i)$ steps in order to compute the modular product of n-bit numbers. As an extension of [FS2] the present paper will show how this theoretical framework can be turned into a practical implementation.

Keywords: Booth recoding, Computer arithmetic, Implementation issues, Sedlak's algorithm, Modular multiplication.

1 Introduction

Without doubt it is clear that all of todays used public-key cryptography relies on modular arithmetic. Here, the most interesting operation is the modular multiplication. Thus, fast algorithms/implementations of the modular multiplication have always been in the focus of cryptographic hardware investigations. This is witnessed by a tremendous amount of literature on this constantly growing field of research, cf.[BA,Br,Gro,WQ,DJQ,Q,Mon,STK,Om,Wa].

Although Sedlak's [Sed] modular multiplication algorithm was one of the first real silicon implementations it has never received a lot of scientific attraction. While his original algorithm needs on average $n/3$ steps (i.e., additions/subtractions) to compute the modular product of n-bit numbers, it was only very recently shown by [FS2] how to speed up Sedlak's algorithm by an arbitrary integral factor $i \geq 2$. Theoretically, the new algorithm needs on average only $n/(3 \cdot i)$ steps in order to compute the modular product of n-bit numbers. As a continuation of [FS2] the present paper will show how this theoretical framework can be turned into a practical implementation. Thus, we will investigate all the subtle implementation issues to turn the former theoretical algorithm of [FS2] into a real-world algorithm. In addition to this "silicon-ready" implementation we will also present practical performance results and our silicon implementation.

The paper is organized as follows. Section 2 recapitulates the results of [FS2], i.e., we recall Booth's algorithm, explain the ZDN-reduction, then we show how

T. Okamoto (Ed.): CT-RSA 2004, LNCS 2964, pp. 264–277, 2004.

to merge both algorithms in order to get the simple as well as the double ZDN-based modular multiplication. Section 3, 4 and 5 will be concerned about the real life implementations of the two former algorithms. Section 6 will show the principle of the multiple ZDN-based modular multiplication. Finally in section 7 we give practical results for the algorithms.

2   Multiplication with Booth, the ZDN-Reduction, the Simple and Double ZDN-Based Modular Multiplication

In the following, we give a short resume of the principles of the ZDN-based modular multiplication as it was introduced in [FS2]. However, the reader is strongly encouraged to read [FS2] for more details. First, we recall the multiplication, reduction and its amalgamation to the modular multiplication. All algorithms in this chapter are purely mathematical! The variables in capital letters contain elements of $\mathbb{Q}$, or more precisely, of $\bigcup_{i \in \mathbb{N}} 2^{-i}\mathbb{Z}$. So, a division by 2 will not destroy information (which happens if one is working with physical registers of finite length).

In this paper we will be concerned with the computation of $(\alpha \cdot \beta \bmod \nu)$ where $2^{n-1} \leq \nu < 2^n$ and $0 \leq \alpha, \beta < \nu$ for some integer n of the size, e.g., 1024 or 2048. Our starting point in [FS2] was the following simple and straightforward textbook algorithm *Modular Multiplication I*, see figure, where β_i denotes i-th bit of β, i.e., $\beta = (\beta_{n-1}, \ldots, \beta_0)_2$ in binary representation. For our approach an equivalent variant *Modular Multiplication II* was used.

input: α, β, ν
 output: $\gamma := \alpha \cdot \beta \bmod \nu$
 	
 $Z := 0,\ C := \alpha,\ N := \nu$
 for $i := n - 1$ **downto** 0 **do**
 $\quad Z := Z \cdot 2$
 $\quad$ **if** $\beta_i = 1$ **then** $Z := Z + C$ **endif**
 $\quad$ /* now $Z = \alpha \cdot (\beta_{n-1}, \ldots, \beta_i)_2$ */
 endfor
 /* now $Z = \alpha \cdot \beta,\ 0 \leq Z < \nu \cdot 2^n$ */
 for $i := n - 1$ **downto** 0 **do**
 $\quad$ **if** $Z \geq N \cdot 2^i$ **then** $Z := Z - N \cdot 2^i$
 $\quad$ **endif**
 $\quad$ /* now $Z = \alpha \cdot \beta \bmod \nu \cdot 2^i$ */
 endfor
 return Z
 	
 Modular Multiplication I.

input: α, β, ν
 output: $\gamma := \alpha \cdot \beta \bmod \nu$
 	
 $Z := 0,\ C := \alpha,\ N := \nu$
 for $i := n - 1$ **downto** 0 **do**
 $\quad C := C/2$
 $\quad$ **if** $\beta_i = 1$ **then** $Z := Z + C$ **endif**
 $\quad$ /*$Z = \alpha \cdot (\beta_{n-1}, \ldots, \beta_i)_2 \cdot 2^{-n+i*}$/
 endfor
 /* $Z = \alpha \cdot \beta \cdot 2^{-n},\ 0 \leq Z < N$ */
 for $i := n - 1$ **downto** 0 **do**
 $\quad Z := Z \cdot 2$
 $\quad$ **if** $Z \geq N$ **then** $Z := Z - N$ **endif**
 $\quad$ /*now $Z = \alpha \cdot \beta \cdot 2^{-i} \bmod \nu$ */
 endfor
 return Z
 	
 Modular Multiplication II.

To enhance the average shift value 1 (over the multiplier) of the multiplication, shown in the former algorithms, the classical method of Booth [Bo] can be used. This algorithm is described in [Mac,Kor,Spa,Par] and achieves asymptotically an average shift value of 3. It requires variable shifts and the ability to subtract numbers. The method of Booth is based on representing numbers in the signed-digit notation SD2: Let β be an n-bit integer in binary notation. $\beta = (\beta_{n-1}, \dots, \beta_0)_2$, i.e., $\beta = \sum_{i=0}^{n-1} \beta_i \cdot 2^i$. Then, there exists a representation of β in the form $\beta = \sum_{i=0}^{n} \bar{\beta}_i \cdot 2^i$ (also written as $(\bar{\beta}_n, \dots, \bar{\beta}_0)_{SD2}$) where $\bar{\beta}_i \in \{-1, 0, 1\}$. Among these representations there is one with a minimal Hamming weight $H(\bar{\beta})$. For these representations one knows about the expectation value $\mathbb{E}(H(\bar{\beta})) = (n+1)/3$. With the algorithms described in[Kor,Mac,JY] such a representation can be efficiently obtained on-the fly. By virtue of this representation, one can define the two equivalent multiplication algorithms *Multiplication I* and *Multiplication II:*

<table>
<tr><td>

input: α, β
output: $\gamma := \alpha \cdot \beta$

$Z := 0$, $C := \alpha$, $m := n+1$
while $m > 0$ **do**
  LABooth$(\beta, \&m, \&s, \&v)$
  $Z := Z \cdot 2^s$
  $Z := Z + v \cdot C$

  /* now $Z = \alpha \cdot (\bar{\beta}_n, \dots, \bar{\beta}_m)_{SD2}$ */
endwhile
return Z

Multiplication I.

</td><td>

input: α, β
output: $\gamma := \alpha \cdot \beta$

$Z := 0$, $C := \alpha$, $m := n+1$, $c := 0$
while $m > 0$ **do**
  LABooth$(\beta, \&m, \&s, \&v)$
  $C := C \cdot 2^{-s}$
  $Z := Z + v \cdot C$
  $c := c + s$ /* $= n+1-m$ */
  /* $Z \cdot 2^c = \alpha \cdot (\bar{\beta}_n, \dots, \bar{\beta}_m)_{SD2}$ */
endwhile
return $Z \cdot 2^c$

Multiplication II.

</td></tr>
</table>

Here, the subroutine LABooth$(\beta, \&m, \&s, \&v)$ provides the shift value s, sign v (according to Booth) and current position m in the multiplier β for the arithmetic step. It manipulates the variables m, s and v (denoted by the &-sign).

Note that $\mathbb{E}(H(\bar{\beta})) = (n+1)/3$ implies that $\mathbb{E}(s) = 3$, which means that the above algorithm achieves asymptotically a performance factor of 3 compared to the simple multiplication. Or, one can say "one has an average shift value of 3".

Now, since the classical variable shift Booth algorithm achieves a speed-up of the factor 3 asymptotically, we have to provide a reduction algorithm of the same speed. This is accomplished by the so called ZDN algorithm ($2/3N = $ "**Z**wei **D**rittel **N**" in german) which is based on the following lemma:

Lemma 1. *Let $\nu \in \mathbb{N}$ and ζ a real number with $\zeta \in [-\nu, \nu[$. Furthermore, let $s := s_\zeta \in \mathbb{N}_0 \cup \{\infty\}$ be the unique integer such that $\zeta \cdot 2^s \in [\frac{2}{3}\nu, \frac{4}{3}\nu[$ or $\zeta \cdot 2^s \in [-\frac{4}{3}\nu, -\frac{2}{3}\nu[$. If $\zeta = 0$, we set $s_\zeta := \infty$. Then $(\zeta \cdot 2^s - \mathrm{sign}(\zeta) \cdot \nu) \in [-\frac{\nu}{3}, \frac{\nu}{3}[$.*

If $\zeta \colon \Omega \longrightarrow [-\frac{\nu}{3}, \frac{\nu}{3}[$ is a uniformly distributed random variable, then the expectation value of s_ζ is given by $\mathbb{E}(s_\zeta) = 3$.

This lemma immediately leads to the so called ZDN-reduction *Reduction II* which replaces the (somewhat) classical reduction algorithm *Reduction I* used in *Modular Multiplication II*:

input: $\zeta,\ \nu$ with $0 \le \zeta < 2^c \cdot \nu$
 output: $\gamma := \zeta \bmod \nu$
 	
 $Z := \zeta \cdot 2^{-c},\ N := \nu$
 while $c > 0$ **do**
 $\quad Z := Z \cdot 2$
 $\quad$ **if** $Z \ge N$ **then** $Z := Z - N$ **endif**
 $\quad c := c - 1$
 $\quad$ /* now $Z \cdot 2^c = (\zeta \bmod \nu \cdot 2^c)$ */
 endwhile
 	
 return Z

Reduction I.

input: $\zeta,\ \nu$ with $0 \le \zeta < 2^c \cdot \nu$
 output: $\gamma := \zeta \bmod \nu$
 	
 $Z := \zeta \cdot 2^{-c},\ N := \nu$
 while $c > 0$ **do**
 $\quad$ LARed$(Z, N, c, \&s, \&v)$
 $\quad Z := Z \cdot 2^s + v \cdot N$
 $\quad c := c - s$
 $\quad$ /* now $Z \cdot 2^c \equiv \zeta \ (\bmod\ \nu \cdot 2^c)$ */
 endwhile
 if $Z < 0$ **then** $Z := Z + N$
 return Z

Reduction II.

The subroutine LARed$(\zeta, \nu, \max, \&s, \&v)$ provides the algorithm with the shift value s and appropriate sign value $v = -\mathrm{sign}(\zeta)$ according to the lemma.

Then, the two former algorithms *Multiplication II* and *Reduction II* can be combined into one single algorithm. Starting with *Modular Multiplication III* (figure not shown here), a simple concatenation of the two algorithms, we merge the two occurring loops into one single one, resulting in *Modular Multiplication IV*. Finally, by reordering the loop we got *Modular Multiplication V* the final version of the ZDN-based modular multiplication algorithm. We also change the notation of the parameters to the obvious ones ($s_Z := s_2$, $s_\beta := s_1$).

input: α, β, ν
 output: $\gamma := \alpha \cdot \beta \bmod \nu$
 	
 $Z := 0,\ C := \alpha,\ N := \nu,$
 $m := n + 1,\ c := 0$
 while $m > 0$ **or** $c > 0$ **do**
 $\quad$ LABooth$(\beta, \&m, \&s_1, \&v_1)$
 $\quad C := C \cdot 2^{-s_1}$
 $\quad Z := Z + v_1 \cdot C$
 $\quad c := c + s_1$
 $\quad$ LARed$(Z, N, c, \&s_2, \&v_2)$
 $\quad C := C \cdot 2^{s_2}$
 $\quad Z := Z \cdot 2^{s_2} + v_2 \cdot N$
 $\quad c := c - s_2$
 endwhile
 if $Z < 0$ **then** $Z := Z + N$ **endif**
 return Z

Modular Multiplication IV.

input: α, β, ν
 output: $\gamma := \alpha \cdot \beta \bmod \nu$
 	
 $Z := 0,\ C := \alpha,\ N := \nu,$
 $m := n + 1,\ c := 0$
 while $m > 0$ **or** $c > 0$ **do**
 $\quad$ LARed$(Z, N, c, \&s_Z, \&v_N)$
 $\quad$ LABooth$(\beta, \&m, \&s_\beta, \&v_C)$
 $\quad s_C := s_Z - s_\beta$
 $\quad C := C \cdot 2^{s_C}$
 $\quad Z := Z \cdot 2^{s_Z} + v_C \cdot C + v_N \cdot N$
 $\quad c := c - s_C$
 	
 endwhile
 if $Z < 0$ **then** $Z := Z + N$ **endif**
 return Z

Modular Multiplication V.

The big advantage of this version is given by the following fact: One can now substitute the two single additions/subtractions $Z := Z + v_C \cdot C$ and $Z := Z \cdot 2^s + v_N \cdot N$ by one single 3-operand addition $Z := Z \cdot 2^s + v_C \cdot C + v_N \cdot N$. In the case of our architecture this extra-overhead is also negligible. All in all, this results in a nearly doubled performance, since the two arithmetic operations which need at least 2 clock cycles are substituted by a one-clock-cycle operation.

Remarks on behavior of the algorithm: The algorithm has an asymptotically average shift value of 3 meaning that the number of loops is $(n + 1)/3$. This would be obvious for the single LABooth, as $\mathbb{E}(s_\beta) = 3$. However, LARed undergoes the technical constraint "max = c". This especially means that the reduction is slowed down in the beginning, i.e., one cannot any longer assume that $\mathbb{E}(s_Z) = 3$ holds in general. However, we can say $\mathbb{E}(s_Z) \rightarrow 3$ for $n \rightarrow \infty$. This can be seen as follows. If $\mathbb{E}(s_Z) = 3 - \delta$, for some $\delta > 0$, then c would be increased by δ per loop. From a c, large enough, the slow down by max = c disappears implying again the result $\mathbb{E}(s_\zeta) \rightarrow 3$. In general, a modular multiplication will show the following behaviour (for large n). In the beginning LABooth dominates, thereby increasing a little bit the value c. Then, LARed will also reach its full performance, i.e., c will go up and down pretty randomly but keeping its middle position. Since the Booth algorithm is already a bit ahead, it will reach the end of the multiplication before the reduction stops. LARed has to reduce the value c down to 0, thus finishing the algorithm.

The development of the last algorithm was driven by the fact that it is much more efficient (in terms of speed) to merge two single 2-operand additions into one single 3-operand addition. This was accomplished by a slightly more complicated loop control structure. Of course, this idea can be continued in a natural (naive) fashion. Simply, merge two successive loop-iterations of *Modular Multiplication V* into one single loop-iteration. Although this will again increase the loop control structure, we start with this naive approach. Firstly, in *Loop I* (not shown here) we simply stick together two successive loop-iterations. For the following we use upper indices which are not exponents! The parts of the first resp. second loop are denoted with the upper indices 1 resp. 2. Although the second LABooth can be directly executed after the first LABooth (remember that it is independent of anything else) the situation with the second LARed is not so easy. Indeed, the second LARed depends on the result Z which is influenced by the first LARed. So, we have to do some pre-computation! This is shown in the first parameter of the second LARed in *Loop II*.

Starting to make things easier, the second LARed will not receive $Z \cdot 2^{s_Z^1} + v_C^1 \cdot C + v_N^1 \cdot N$ as input, but simply the approximation $Z \cdot 2^{s_Z^1} + v_N^1 \cdot N$. Here we assume, that the influence of C will be irrelevant in most cases. In some rare cases (when c is very small) a "wrong" reduction value can be delivered. However, the reduction value is not really wrong but only sub-optimal in terms of reduction speed. On the other side, we save the "shifting" step of Z by s_Z^2 for the next loop. This yields *Modular Multiplication VI*.

Asymptotically this algorithm needs $(n+1)/6$ loop cycles in theory. This can be seen similarly to the last paragraph and will be explained in a full version of the paper. Clearly, the "small" side computation $Z \cdot 2^{s_Z} + v_N^1 \cdot N$ needs to be done only on a small fraction of the leading register-bits. This is due to the fact that LABooth$(Z, N, \dots)$ and LABooth$(Z \cdot 2^t, N \cdot 2^t, \dots)$ return the same values s and v for any $t \in \mathbb{N}$.

$$
\begin{array}{l}
\text{LARed}(Z, N, c, \&s_Z^1, \&v_N^1) \\
\text{LABooth}(\beta, \&m, \&s_\beta^1, \&v_C^1) \\
\text{LARed}(Z \cdot 2^{s_Z^1} + v_C^1 \cdot C + v_N^1 \cdot N, \\
\qquad N, c - s_Z^1 + s_\beta^1, \&s_Z^2, \&v_N^2) \\
\text{LABooth}(\beta, \&m, \&s_\beta^2, \&v_C^2) \\
s_C^1 := s_Z^1 - s_\beta^1 \\
C := C \cdot 2^{s_C^1} \\
Z := Z \cdot 2^{s_Z^1} + v_C^1 \cdot C + v_N^1 \cdot N \\
c := c - s_C^1 \\
s_C^2 := s_Z^2 - s_\beta^2 \\
C := C \cdot 2^{s_C^2} \\
Z := Z \cdot 2^{s_Z^2} + v_C^2 \cdot C + v_N^2 \cdot N \\
c := c - s_C^2
\end{array}
$$

Loop II.

---

input: α, β, ν
output: $\gamma := \alpha \cdot \beta \bmod \nu$

$Z := 0,\ C := \alpha,\ N := \nu,$
$m := n + 1,\ c := 0$
while $m > 0$ **or** $c > 0$ **do**
$\quad$ LARed$(Z, N, c, \&s_Z, \&v_N^1)$
$\quad$ LARed$(Z \cdot 2^{s_Z} + v_N^1 \cdot N, N,$
$\qquad\qquad c - s_Z, \&s_N, \&v_N^2)$
$\quad$ LABooth$(\beta, \&m, \&s_\beta^1, \&v_C^1)$
$\quad$ LABooth$(\beta, \&m, \&s_\beta^2, \&v_C^2)$
$\quad s_C^1 := s_Z - s_\beta^1$
$\quad s_C^2 := -s_\beta^2$
$\quad c := c - s_C^1 - s_C^2$
$\quad C := C \cdot 2^{s_C^1 + s_C^2}$
$\quad Z := Z \cdot 2^{s_Z} + v_C^1 \cdot (C \cdot 2^{-s_C^2})$
$\qquad + v_N^1 \cdot N + v_C^2 \cdot C + v_N^2 \cdot (N \cdot 2^{-s_N})$
endwhile
if $Z < 0$ **then** $Z := Z + N$ **endif**
return Z

Modular Multiplication VI.

---

3 Implementing the Simple ZDN-Based Modular Multiplication

Now we want to put the algorithm *Modular Multiplication V* into a computer architecture. Therefore, we will first summarize some of the specific properties of the algorithm which are determining the architecture:
(i) Z takes positive and negative values, i.e., $Z \in [-N, N]$ in steps of 2^{-k}— for some k. (ii) C will be divided/multiplied by powers of 2. (iii) Z will only be multiplied by powers of 2. (iv) N doesn't change at all. (v) There is a 3-operand addition. (vi) The multiplier β affects the computation only through the values s_β and v_C.
These properties are mirrored by the following rough architecture description:
There is one *calculation unit* and one *control unit*. (i) The calculation unit consists of 3 registers N, C and Z of bit length $rl := n+1+k$ for some k (say, $k = 32$). There is a *virtual comma* at the bit-position k such that the interval $[-2^n, 2^n[$ is realized in steps of 2^{-k} (in two's complement representation). (ii) The register C can be shifted by the values $s_C \in \{-ShL, \ldots, +ShL\}$. (iii) There is a shifter which latches Z into the adder by shift values $s_Z \in \{0, \ldots, ShL\}$. (iv) The register N has no additional features. (v) There is a realization of the 3-operand adder. (vi) The control unit holds information about the multiplier β and parts of Z and also delivers the control signals s_Z, s_C, v_C, v_N for the adder.

To stress again one point: The three registers of length $rl = n + 1 + k$ will be used in a way such that they can contain numbers of the interval $[-2^n, 2^n[$ in steps of 2^{-k}. In other words, if $Z = (z_{n+k}, \ldots, z_0)$, then usually the contents of Z is interpreted as the binary number $(z_{n+k}, \ldots, z_0)_2$, at least modulo 2^{n+k+1}.

input: α, β, ν
output: $\gamma := \alpha \cdot \beta \bmod \nu$

$Z := 0$, $C := \alpha$, $N := \nu$,
$m := n+1$, $c := 0$, $lsb := rl - n - 1$[5)]

while $m > 0$ or $c > 0$ do
  $max := \min\{ShL, c\}$[1)]
  LARed$(Z, N, max, \&s_Z, \&v_N)$
  if $c = s_Z$ and $m > 0$ then[2)]
    $s_Z := \max\{0, s_Z - 1\}$
    $v_N := 0$
  endif
  $max := \min\{lsb, ShL\} + s_Z$[3)]
  LABooth$(\beta, max, \&m, \&s_\beta, \&v_C)$[4)]
  $s_C := s_Z - s_\beta$
  $c := c - s_C$
  $lsb := lsb + s_C$[5)]

  $C \longleftarrow C \ll s_C$
  $Z \longleftarrow Z \ll s_Z + v_C \cdot C + v_N \cdot N$

endwhile
if $Z < 0$ then $Z \longleftarrow Z + N$ endif
return Z

Modular Multiplication VII.

---

input: α, β, ν
output: $\gamma := \alpha \cdot \beta \bmod \nu$

$Z := 0$, $C := \alpha$, $N := \nu$,
$\widetilde{Z} \longleftarrow Z$, $\widetilde{N} \longleftarrow N$
(copy only the top $ShL + 4$ Bits)
$m := n+1$, $c := 0$, $lsb := rl - n - 1$
while $m > 0$ or $c > 0$ do
  $max := \min\{ShL, c\}$
  LARed$(\widetilde{Z}, \widetilde{N}, max, \&s_Z, \&v_N)$
  if $c \le ShL + 4$ and $m > 0$ then
    $s_Z := 0$
    $v_N := 0$
  endif
  $max := \min\{lsb, ShL\} + s_Z$
  LABooth$(\beta, max, \&m, \&s_\beta, \&v_C)$
  $s_C := s_Z - s_\beta$
  $c := c - s_C$
  $lsb := lsb + s_C$
  $\widetilde{Z} \longleftarrow Z$ (copy only the top
    $2 \cdot ShL + 4$ Bits)
  $C \longleftarrow C \ll s_C$
  $Z \longleftarrow Z \ll s_Z + v_C \cdot C + v_N \cdot N$
  $\widetilde{Z} \longleftarrow \widetilde{Z} \ll s_Z + v_N \cdot \widetilde{N}$
endwhile
if $Z < 0$ then $Z \longleftarrow Z + N$ endif
return Z

Modular Multiplication VIII.

However, here it is interpreted as the rational number $(z_{n+k}, \dots, z_0)_2 \cdot 2^{-k}$, modulo 2^{n+1}.

Since the value in register C will be shifted up and down, the variable lsb will keep track of its actual position such that no bits will be lost. (Hence lsb has to lie between 0 and k.) The algorithm *Modular Multiplication VII* can be now provided. Note that we now use C-style notation for shifting, i.e., multiplication with 2. Further note that negative shift values are allowed so that $C \ll s_C = C \gg -s_C$. Of course, some practical constraints are given:

1. Finite register length.
2. Finite shifter length.
3. The 2-complement representation of numbers in Z, i.e., a number that is too big could be interpreted as a negative number.

This leads to the algorithm *Modular Multiplication VII*. We will briefly comment on all of the additional technical issues which are marked by footnotes:

1. The constraint in 1) is necessary due to finite shifter length ShL of Z.
2. Due to finite shifter length the following situation is possible: the values which are returned by LARed are $v_N = 0$ and s_Z so that $(Z \ll s_Z) \approx$

$2/3N$. In addition, we may have $s_\beta = 1$ (which is the minimum during the multiplication phase) and $v_C = 1$. This means for an $\alpha \approx N$ that the new value of Z is $(Z \ll s_Z) + C + 0 \cdot N \approx 2/3 \cdot N + 1/2 \cdot N > N$. All in all this could mean that Z would erroneously be negative in the 2-complement representation. Therefore, with 2) we decrement the s_Z by one, now gaining $(Z \ll s_Z) + C + 0 \cdot N \approx 1/3 \cdot N + 1/2 \cdot N < N$.

3. Because of the finite shifter length and the fact that C must not leave the register to the right, we have $- \min\{lsb, ShL\} \leq s_C (= s_Z - s_\beta)$ and therefore $s_\beta \leq \min\{lsb, ShL\} + s_Z$ which is realized by virtue of the next point 4).

4. The maximal output value s_β of LABooth will be bounded by max in the obvious way. Therefore, we used the LABooth with an additional parameter:

$$
\begin{aligned}
&\text{LABooth}(\beta, \max, \&m, \&s, \&v) \\
\stackrel{def}{=}\; &\text{LABooth}(\beta, \&m, \&s, \&v) \\
&\textbf{if } s > \max \textbf{ then} \\
&\qquad m \longleftarrow m + s - \max \\
&\qquad s \longleftarrow \max \\
&\qquad v \longleftarrow 0 \\
&\textbf{endif}
\end{aligned}
$$

5. The lsb controls the position of the α in the register C. We assume that the algorithm starts with an lsb greater than some constant threshold value k, e.g., 32 bits.

Performance Analysis: We have seen before that this algorithm needs asymptotically $(n+1)/3$ loop cycles in theory. One loop cycle contains one 3-operand addition. In a real hardware realization this can be performed on average during 1 clock cycle. However, in reality this doesn't mean that the whole modular multiplication can be performed during $(n+1)/3$ clock cycles. Both functions LABooth and LARed have to be computed and especially LARed needs the result of the previous 3-operand addition. That means that we have to spend at least one additional clock cycle for both functions LABooth and LARed, which leads to a two-cycle loop realization. All in all we would need on average about $2n/3$ cycles to perform one modular multiplication. Nevertheless, the next section shows how to perform both functions LABooth and LARed in parallel to a simultaneously running 3-operand addition. This means that LARed must use another main input for Z—as the currently performed 3-operand addition will be ready when LARed already has to deliver its values at the latest.

4   An Enhanced Implementation

It seems that the plausibility chain forces LARed to use the result of the previously computed 3-operand addition. However, this dependence can be avoided when a good approximation of Z can be computed very fast in a way, derived from the following thoughts:

1. As s_Z is anyway bounded by the maximal shift-value ShL, it suffices to know only the top (say $ShL + 4$) bits of Z in order to compute s_Z and v_N within the LARed.

2. Assuming that c has become large enough a few steps after starting the algorithm (e.g., if $c \geq 2 \cdot ShL + 4$), the top most bits of $Z := Z \ll s_Z + v_C \cdot C + v_N \cdot N$ can be approximated fairly good by $\widetilde{Z} := Z \ll s_Z + v_N \cdot N$. And again moreover, this last computation needs to be performed only on the top most bits of Z.

3. Such an addition can be computed in parallel to the "big" addition, but much faster than the latter one.

4. Now, during the remaining time, i.e., finishing the "big" addition, LABooth and LARed can be completed. This saves one whole clock cycle at every step at the expense of using hardware parallelism together with a good approximation of the currently evaluated Z.

This leads to a fairly good description of a real existing hardware implementation of the ZDN-multiplication: *Modular Multiplication VIII*.

Considering this new algorithm, we see that we had to strengthen some constraints after the LARed. The reason is that we can ignore the influence of C only if c is large enough, i.e., only in this case we will get a good approximation for Z. We simply suppress the reduction until c has become large enough. Note that the parallelism extends over two loops: While the "big" addition at the end of one loop is computed, the control functions of the *next* loop are evaluated! A further improvement deserves its own small section:

Modulus Transformation: We have seen that computing the control values from LARed is the most expensive operation of the control part. As we have seen above we don't use the full N, rather than a good approximation $\widetilde{N}$, given by some of the top most bits of N. Thus, we can simplify LARed.

Assume that the used top most bits of N, i.e., $\widetilde{N}$ will always have a particular simple and fixed form, which is *independent* of N. E.g., if $\widetilde{N} = 0110\ldots0$, then $2/3 \cdot \widetilde{N} = 010\ldots0$. However, that means that s_Z with $2/3 \cdot N \leq (Z \ll s_Z) < 4/3 \cdot N$ can be gained by simply counting the leading zeros of $\widetilde{Z}$. A circuit for that purpose is rather small and fast.

Note that computing with such modules ν doesn't mean any restriction. This is due to the fact that it is always possible and rather simple to find a very small number t, such that $t \cdot \nu$ has the desired form, cf.[DJQ,DQ,Q]. After computing the result modulo $t \cdot \nu$ (a full RSA exponentiation, cf.[RSA]) simply reduce this result modulo ν.

5   Implementing the Doubled ZDN-Based Modular Multiplication

The manner in which the algorithm *Modular Multiplication VI* can be practically realized is very similar to that of the last two sections. Therefore we will fix a similar register architecture.

input: α, β, ν
output: $\gamma := \alpha \cdot \beta \bmod \nu$
$Z := 0, \ C := \alpha, \ N := \nu,$

$m := n + 1, \ c := 0, \ lsb := rl - n - 1$
while $m > 0$ **or** $c > 0$ **do**
   $\max := \min\{c, ShL\}$[1)]
   $\mathrm{LARed}(Z, N, \max, \&s_Z, \&v_N^1)$
   if $s_Z = c$ **and** $m > 0$ **then**[2)]

      $s_Z := \max\{0, s_Z - 1\}$
      $v_N := 0$
   endif

   $\max := \min\{c - s_Z, ShL_N\}$
   $\mathrm{LARed}(Z \ll s_Z + v_N^1 \cdot N,$
           $N, \max, \&s_N, \&v_N^2)$

   $\max := s_Z + \min\{lsb, ShL\}$[3)]
   $\mathrm{LABooth}(\beta, \max, \&m, \&s_\beta, \&v_C^1)$[4)]
   $s_C^1 := s_Z - s_\beta$
   $c := c - s_C^1$
   $lsb := lsb + s_C^1$[5)]
   $\max := \min\{lsb, ShL, ShL + s_C^1\}$[6)]
   $\mathrm{LABooth}(\beta, \max, \&m, \&s_\beta, \&v_C^2)$[4)]
   $s_C^1 := -s_\beta$
   $c := c - s_C^2$
   $lsb := lsb + s_C^2$[5)]

   $C \longleftarrow C \ll s_C^1 + s_C^2$
   $Z \longleftarrow Z \ll s_Z + v_C^1 \cdot (C \gg s_C^2) +$
      $+ v_N^1 \cdot N + v_C^2 \cdot C + v_N^2 \cdot (N \gg s_N)$

endwhile
if $Z < 0$ **then** $Z \longleftarrow Z + N$ **endif**
return Z

Modular Multiplication IX.

---

input: α, β, ν
output: $\gamma := \alpha \cdot \beta \bmod \nu$
$Z := 0, \ C := \alpha, \ N := \nu,$
$\widetilde{Z} \longleftarrow Z, \ \widetilde{N} \longleftarrow N$ (copy only the
      top $ShL + ShL_N + 4$ Bits)
$m := n + 1, \ c := 0, \ lsb := rl - n - 1$
while $m > 0$ **or** $c > 0$ **do**
   $\max := \min\{c, ShL\}$
   $\mathrm{LARed}(\widetilde{Z}, \widetilde{N}, \max, \&s_Z, \&v_N^1)$
   if $c + s_C^2 \le ShL + ShL_N + 2$ **and**
      $m_{old} > 0$ **then**
      $s_Z := 0$
      $v_N^1 := 0$
   endif
   $\widetilde{Z} \longleftarrow \widetilde{Z} \ll s_Z + v_N^1 \cdot \widetilde{N}$
   $\max := \min\{c - s_Z, ShL_N\}$
   $\mathrm{LARed}(\widetilde{Z}, \widetilde{N}, \max, \&s_N, \&v_N^2)$
   if $s_N = c$ **or** $s_N = 0$ **then**
      $s_Z := 0$
      $v_N^2 := 0$
   endif
   $m_{old} := m$
   $\max := s_Z + \min\{lsb, ShL\}$
   $\mathrm{LABooth}(\beta, \max, \&m, \&s_\beta, \&v_C^1)$
   $s_C^1 := s_Z - s_\beta$
   $c := c - s_C^1$
   $lsb := lsb + s_C^1$
   $\max := \min\{lsb, ShL, ShL + s_C^1\}$
   $\mathrm{LABooth}(\beta, \max, \&m, \&s_\beta, \&v_C^2)$
   $s_C^1 := -s_\beta$
   $c := c - s_C^2$
   $lsb := lsb + s_C^2$
   $\widetilde{Z} \longleftarrow Z$ (copy only the top
      $2 \cdot ShL + ShL_N + 4$ Bits)
   $C \longleftarrow C \ll s_C^1 + s_C^2$
   $Z \longleftarrow Z \ll s_Z + v_C^1 \cdot (C \gg s_C^2) +$
      $+ v_N^1 \cdot N + v_C^2 \cdot C + v_N^2 \cdot (N \gg s_N)$
   $\widetilde{Z} \longleftarrow \widetilde{Z} \ll s_Z + v_N^1 \cdot \widetilde{N} +$
      $+ v_N^2 \cdot (N \gg s_N)$
endwhile
if $Z < 0$ **then** $Z \longleftarrow Z + N$ **endif**
return Z

Modular Multiplication X.

The calculation unit consists again of 3 registers Z, C and N of bit length $rl = n + 1 + k$. They are used for the following operations.

1. C can be shifted by the values $-ShL, \dots, +ShL$.
2. A 5-operand addition with the following properties is realized:
$Z \longleftarrow (Z \ll s_Z) + v_C^1 \cdot (C \gg s_C^2) + v_N^1 \cdot N + v_C^2 \cdot C + v_N^2 \cdot (N \gg s_N)$. Here, C and N will enter the addition twice. Each of C and N one time directly and

a second time latched through additional shifters, where N can be shifted down by $0, \ldots, ShL_N$ bits.

Plugging in this modification into the same framework as before, the following algorithm *Modular Multiplication IX* results. Again, we will now briefly comment the modifications which we have done:

- 1) – 5) have the same reason as in the above section already explained.
- 6) The shifter limitation is chosen in such a way that $s_C^1 + s_C^2 \in [-ShL, ShL]$.

Now, for *Modular Multiplication X* we want to parallelize again the "big" addition and the computation of the look ahead values (via LARed and LABooth) for the next addition. Clearly, again some modifications have to be done on the above algorithm.

6   The Multiple ZDN-Based Modular Multiplication

The scheme which has been used for merging two successive loops into one single loop can clearly be generalized to merging $i > 2$ successive loops. As everything should be now clear we will only give a schematic sketch of the multiple ZDN algorithm (Cf. *Fig. 1*). This algorithm puts i successive loop additions into one single addition, this time using a $(2 \cdot i + 1)$-operand addition. This results in an average asymptotic shift value of $(n + 1)/(3i)$ bits per cycle.

Of course, the generation of the increasing number of control signals will become more and more complicated. So this strategy only makes sense for very long registers, and it is only practicable for small i, e.g., $i \leq 3$.

7   Practical Results

Average shift values. As all theoretically derived results are only valid for $n \to \infty$ and unlimited shifters it is interesting to consider real practical results. However, our algorithms are a mixture between multiplication and reduction, where the reduction always runs a little bit behind the multiplication: One can multiply in advance, but one cannot reduce in advance! Thus, it is difficult to give *average shift values* for the algorithms. Both algorithms have their individual shift values. Although they could be estimated taking into account a given shifter length, their interplay is hard to estimate.

Therefore, we define the average shift value as the quotient of $n + 1$ and the number of loops, i.e., the number of all necessary additions. From simulations we obtained the following shift values.

These numbers (and some which are not shown here) tell us:

- The average shift value does not depend noteworthy on the size of the k buffer bits, as long as $k \geq 20$.
- The average shift values come closer to the theoretical ones the longer the registers are.
- The average shift values come closer to the theoretical ones the longer the shifters are.

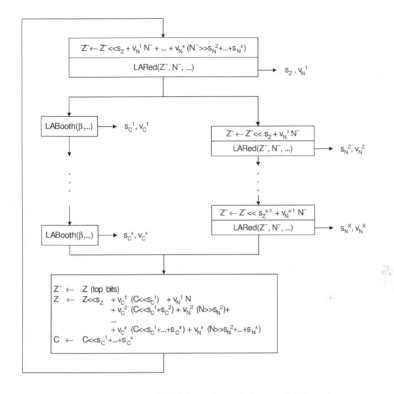

Fig. 1. The multiple ZDN-based modular multiplication.

n	ShL	MM V	MM VII	MM VIII
512	∞	2.84		
1024	∞	2.90		
2048	∞	2.92		
1024	3		2.31	2.22
	4		2.62	2.54
	5		2.77	2.71
	6		2.83	2.78
	7		2.85	2.81

Average shift values of Simple ZDN.

n	ShL	ShL_N	MM VI	MM IX	MM X
512	∞	∞	5.59		
1024	∞	∞	5.71		
2048	∞	∞	5.76		
1024	5	3		4.44	4.35
	6	4		4.98	4.90
	7	5		5.32	5.24
	8	5		5.49	5.37
	9	5		5.56	5.40

Average shift values of Double ZDN.

Silicon Realization. For the development of a new co-processor, capable to handle 2048 bit RSA, a study was done in order to compare two designs:

1. Realization of the simple ($i = 1$) ZDN-algorithm *Modular Multiplication VIII* with $n = 2048$.
2. Realization of the triple ($i = 3$) ZDN based modular multiplication, for $n = 1024$ with some additional feature described in [FS1], called *MultModDiv*.

Using the method described in [FS1] one needs 6 modular multiplications of length 1024 bit in order to emulate one modular multiplication of length 2048 bit.

Since in our architectures the complexity of a modular multiplication is linear in the bit length of the multiplier and because the algorithm in 2) is 3 times as fast as the one in 1) the performance (for RSA 2048) of these two architectures is (theoretical) comparable.

The design 2) has the advantage that here RSA for 1024 bit length is three times as fast as in design 1). On the other hand 2) is more complex, both in the implementation of the hardware as well as in the implementation of the necessary software.

Interestingly it turned out that both designs cost approximately the same silicon area: The design 2) needs a very sophisticated 7-operand-adder with more shifters than the 3-operand-adder used in 1). However in 2) only an adder of length 1024 bits has to be realized! All in all both calculation units have about the same size. In both cases, silicon area for generating the control structures are negligible compared to the adders. Finally, the design 1) was realized as a co-processor for chip card applications.

References

[Ba]   P. Barret, "Implementing the Rivest, Shamir and Adleman public-key encryption algorithm on a standard digital signal processor", *Proc. of CRYPTO '86*, Springer LNCS, vol. 263, pp. 311–323, 1987.

[Br]   E. F. Brickell, "A fast modular multiplication algorithm with application to two key cryptography', *Proc. of CRYPTO '86* Springer LNCS vol. 263, pp. 311–323, 1987.

[Bo]   A.D. Booth, "A signed binary multiplication technique", *Q. J. Mech. Appl. Math* 4(2):236-240, 1951.

[DJQ]   J.-F. Dhem, M. Joye, J.-J. Quisquater, "Normalisation in diminished-radix modulus transformation", *Electronics Letters*, vol. 33(23):1931, 1997

[DQ]   J.-F. Dhem, J.-J. Quisquater, "Recent results on modular multiplication for smart cards", *Proc. of CARDIS '98* Springer LNCS vol. 1820, pp. 336–357, 1998.

[FS1]   W. Fischer, J.-P. Seifert, "Increasing the bitlength of a crypto-coprocessor", *Proc. of CHES '02*, Springer LNCS, vol. 2523, pp. 71–81, 2002.

[FS2]   W. Fischer, J.-P. Seifert, "Unfolded modular multiplication", *Proc. of ISAAC '03*, Springer LNCS, 2003.

[Gro]   J. Großschädel, "A bit-serial unified multiplier architecture for finite fields GF(p) and GF(2^m)", *Proc. of CHES '01*, Springer LNCS, vol. 2162, pp. 206–223, 2001.

[JY]   M. Joye, S.-M. Yen, "Optimal left-to-right binary signed-digit exponent recoding", *IEEE Transactions on Computers* 49(7): 740–748, 2000.

[Kor]   I. Koren, *Computer Arithmetic Algorithms*, Brookside Court Publishers, Amherst MA, 1998.

[Mac]   O. L. MacSorley, "High-speed arithmetic in binary computers", *Proc. IRE*, 49:67–91, 1961.

[Mon]   P. L. Montgomery, "Modular multiplication without trial division", *Math. of Computation*, 44:519–521, 1985.

[NMR]   D. Naccache, D. M'Raihi, "Arithmetic co-processors for public-key cryptography: The state of the art", *IEEE Micro*, pp. 14-24, 1996.

[Om]   J. Omura, "A public key cell design for smart card chips", *Proc. of IT Workshop*, pp. 27–30, 1990.

[Par]   B. Parhami, *Computer Arithmetic*, Oxford University Press, New York, 2000.

[Spa]   O. Spaniol, *Arithmetik in Rechenanlagen*, B. G. Teubner, Stuttgart, 1976.

[Q]   J.-J. Quisquater, "Encoding system according to the so-called RSA method, by means of a microcontroller and arrangement implementing this system", U.S. Patent #5,166,979, Nov. 24, 1992.

[RSA]   R. Rivest, A. Shamir, L. Adleman, "A method for obtaining digital signatures and public-key cryptosystems", *Comm. of the ACM* **21**:120–126, 1978.

[STK]   E. Savas, A. F. Tenca, C. K. Koc, "A scalable and unified multiplier architecture for finite fields $\mathbb{F}_p$ and $\mathbb{F}_{2^k}$", *Proc. of CHES '00*, Springer LNCS, vol. 1965, pp. 277–292, 2000.

[Sed]   H. Sedlak, "The RSA cryptographic Processor: The first High Speed One-Chip Solution", *Proc. of EUROCRYPT '87*, Springer LNCS, vol. 293, pp. 95–105, 198.

[WQ]   D. de Waleffe, J.-J. Quisquater, "CORSAIR, a smart card for public-key cryptosystems", *Proc. of CRYPTO '90*, Springer LNCS, vol. 537, pp. 503–513, 1990.

[Wa]   C. Walter, "Techniques for the Hardware Implementation of Modular Multiplication", *Proc. of 2nd IMACS Internat. Conf. on Circuits, Systems and Computers*, vol. **2**, pp. 945–949, 1998.

Yet Another Sieving Device

Willi Geiselmann and Rainer Steinwandt

IAKS, Arbeitsgruppe Systemsicherheit, Prof. Dr. Th. Beth,
Fakultät für Informatik, Universität Karlsruhe,
Am Fasanengarten 5, 76 131 Karlsruhe, Germany

Abstract. A compact mesh architecture for supporting the relation collection step of the number field sieve is described. Differing from TWIRL, only isolated chips without inter-chip communication are used. According to a preliminary analysis for 768-bit numbers, with a 0.13 μm process one mesh-based device fits on a single chip of $\approx(4.9$ cm$)^2$—the largest proposed chips in the TWIRL cluster for 768-bit occupy $\approx(6.7$ cm$)^2$.
A 300 mm silicon wafer filled with the mesh-based devices is ≈ 6.3 times slower than a wafer with TWIRL clusters, but due to the moderate chip size, lack of inter-chip communication, and the comparatively regular structure, from a practical point of view the mesh-based approach might be as attractive as TWIRL.

Keywords: factorization, number field sieve, RSA

1   Introduction

Initiated by Bernstein's paper [Ber01], in the last few years several proposals for speeding-up the linear algebra step of the number field sieve (NFS) by means of specialized hardware have been put forward. While Bernstein's original proposal relied on the use of a parallel sorting algorithm, Lenstra et al. derived an improved mesh architecture that relies on a parallel routing algorithm [LSTT02]. Finally, in [GS03b] distributed variants of the proposals in [Ber01,LSTT02] are discussed where the main focus is on deriving a design that can be realized with current standard technology. In summary, building a device that performs the linear algebra step of the NFS for 768- or 1024-bit numbers within a few hours must be considered as doable with current technology.

Using the words from [LSTT02], one can "conclude that from a practical standpoint, the security of RSA relies exclusively on the hardness of the relation collection step of the number field sieve." Thus, it is no surprise that several attempts have been made to apply dedicated hardware to speed-up the relation collection step of the NFS, too. In particular, the TWINKLE device [Sha99,LS00] and the mesh-based design of [GS03a] can be seen in this context. However, none of these devices was practically capable of coping with the relation collection step of 1024-bit numbers. A significant step forward has been achieved by Shamir and Tromer [ST03] recently: the TWIRL device they describe could in principle complete the sieving part of the NFS for 1024-bit numbers in less than a year by

T. Okamoto (Ed.): CT-RSA 2004, LNCS 2964, pp. 278–291, 2004.

means of current technology. However, already for 768-bit numbers chip sizes of up to $\approx(6.7 \text{ cm})^2$ (with an irregular layout) have been proposed. Although the proposed TWIRL parameters are not optimized for chip size, actually manufacturing a TWIRL cluster seems extraordinary challenging. For 1024-bit numbers a TWIRL cluster including a wafer-sized silicon chip has been proposed; thus, from a practical point of view the question arises, whether it is possible to do with more regular and smaller chips. Also, avoiding inter-chip communication seems desirable.

In this contribution we discuss a different design which is based on a routing mesh running at 500 Mhz (instead of 1 GHz in TWIRL, where the time-critical operations are simpler). For 768-bit numbers our proposal consists of a mesh of 256×256 almost identical processing units. The layout is rather regular and the estimated silicon area for a complete mesh is about $(4.9 \text{ cm})^2$. Counting processing time per wafer, the estimated time needed for the sieving step with 768-bit numbers is about 6.3 times higher than estimated in [ST03] for TWIRL. However, due to the simpler design, manufacturing the device and applying it at least to 768-bit does not seem unrealistic. For 1024-bit numbers we cannot give a reliable answer yet, but as the design presented here allows for a rather compact storage of the factor bases—a critical point in TWIRL—exploring the 1024-bit case in more detail is certainly worthwhile.

2 Preliminaries

2.1 The Sieving Part of the NFS

A standard reference for an introduction to the number field sieve is [LHWL93]. Here we only recall those aspects of the sieving step which are relevant for describing our device.

In the first step of the NFS two univariate polynomials $f_1(x), f_2(x) \in \mathbb{Z}[x]$ are determined that share a common root m modulo n:

$$f_1(m) \equiv f_2(m) \equiv 0 \pmod{n}$$

Typically, $f_1(x)$ is of degree $d \geq 5$ and $f_2(x)$ is monic and linear (i.e., $f_2(x) = x - m$). From these two polynomials the bivariate and homogeneous polynomials $F_1(x,y), F_2(x,y) \in \mathbb{Z}[x,y]$ are derived via $F_1(x,y) := y^d \cdot f_1(x/y)$ resp. $F_2(x,y) := y \cdot f_2(x/y)$. Now everything related to $f_1(x)$ resp. $F_1(x,y)$ is said to belong to the *algebraic side*, and everything related to $f_2(x)$ resp. $F_2(x,y)$ is refered to as the *rational side*. In particular, for given smoothness bounds $B_1, B_2 \in \mathbb{N}_0$ the sets

$$P_i := \{(p,r) : f_i(r) \equiv 0 \pmod{p}, p \text{ prime}, p < B_i, 0 \leq r < p\} \subseteq \mathbb{N}^2 \quad (i = 1, 2)$$

are refered to as algebraic and rational *factor base*, respectively. Following [ST03], for the factorization of a 768-bit number, we assume $B_1 = 10^9$ and $B_2 = 10^8$.

Throughout the relation collection step, pairs of coprime integers $(a, b) \in \mathbb{Z} \times \mathbb{N}$ are to be found, such that the values $F_1(a, b)$ and $F_2(a, b)$ are *smooth*.

This means that both $F_1(a,b)$ and $F_2(a,b)$ factor over the primes $< B_1$ resp. $< B_2$, except for a small number of prime factors; the precise number of 'extra' prime factors on the rational and algebraic side is not necessarily identical. The actual computation of (a,b)-pairs where both $F_1(a,b)$ and $F_2(a,b)$ are smooth can be performed by sieving over a rectangular region $-A \leq a < A, 0 < b \leq B$ where $A, B \in \mathbb{N}$. For organizing this sieving, different techniques are available; here we focus on so-called *line sieving* which is outlined in Figure 1. At this the threshold bounds T_i correspond to the bitlength of the remaining cofactor on the algebraic resp. rational side. These bounds have to be updated several times throughout the sieving. In an actual implementation the values $\log_2(p)$ are usually replaced by an integer approximation. Also the use of base 2-logarithms is not mandatory; in analogy to [ST03], subsequently we use a 10-bit counter for summing up approximations $\lceil \log_{\sqrt{2}}(p) \rceil$. Finally, note that in the last step of the main loop in Figure 1 it is computationally too expensive to identify the factors of $F_i(a,b)$ through a simple trial-division by the primes in the respective factor base. To cope with this problem the sieving mesh described below reports prime factors of $F_i(a,b)$ that have been found.

2.2   Clockwise Transposition Routing

An important algorithmic tool used in the sieving device described below is a modification of a fast parallel routing algorithm described in [LSTT02] in the context of fast matrix-vector multiplication. We start by recalling the main ingredients of this *clockwise transposition routing*:

- the hardware platform is a mesh of (rather simple) processing units where each unit is connected to its horizontal and vertical neighbours.
- In each step of the algorithm a processing unit holds no more than *one* packet that is to be routed; only one packet can be sent and received per step.
- At the beginning of the algorithm some mesh nodes contain a data packet (the other nodes contain a *nil* value). Along with each data packet the coordinates of a processing unit in the mesh, the so-called *target node*, are stored, and the goal of the algorithm is to route all packets to the respective target.

$b \leftarrow 0$
repeat
    $b \leftarrow b + 1$
    for $i \leftarrow [1, 2]$
        $s_i(a) \leftarrow 0 \quad (\forall a : -A \leq a < A)$
        for $(p, r) \leftarrow P_i$
            $s_i(br + kp) \leftarrow s_i(br + kp) + \log_2(p) \quad (\forall k : -A \leq br + kp < A)$
    for $a \leftarrow \{-A \leq a < A : \gcd(a, b) = 1,\ s_1(a) > T_1,\ \text{and}\ s_2(a) > T_2\}$
        check if both $F_1(a,b)$ and $F_2(a,b)$ are smooth
until enough pairs (a,b) with both $F_1(a,b)$ and $F_2(a,b)$ smooth are found

---

Fig. 1. Line sieving

- The actual routing is done by repeating the following four steps until the mesh is 'empty', i.e., all packets have reached their target node (where a packet is removed from the mesh):
  1. Each node located on an *odd row* compares its packet with the node *above* it (if any). The packets are exchanged if and only if the distance-to-target of the non-nil value that vertically is farthest away from its target node is reduced in this way.
  2. Each node located on an *odd column* compares its packet with the node to its *right* (if any). The packets are exchanged if and only if the distance-to-target of the non-nil value that horizontally is farthest away from its target node is reduced in this way.
  3. Identical to the first step with 'above' replaced by 'below'.
  4. Identical to the second step with 'right' replaced by 'left'.

A theoretical analysis of this algorithm is still lacking, but experimental results demonstrate its efficiency: e.g., for the situation considered in [LSTT02], the running time of the algorithm did not exceed $2M$ steps, when dealing with an $M \times M$ mesh. In the application below, no 'packet-cancellation' (see [LSTT02]) is used and several parts of the original algorithm have been modified. Again, simulations indicate that the resulting algorithm is rather efficient. However, analogously as in [LSTT02], we cannot provide a theoretical analysis of our approach at the moment.

   Although we never encountered an infinite loop in our simulations, it can in principle even happen that our routing algorithm for certain parameter choices does not terminate. But this is of no practical concern: in the sieving procedure described later, only a certain period of time is alloted for sieving a particular range of numbers, and in the (rare) case that the routing cannot be completed within this time limit, say due to an infinite loop, only some (a, b)-candidates of that sieving interval are lost. In contrast to the linear algebra step of the NFS, where an incorrect intermediate result can have devastating consequences, the sieving process is rather robust with respect to such errors.

3   Adapting the Routing Algorithm

Subsequently, we will deal with a mesh of size $M \times M = 2^m \times 2^m$; for 768-bit numbers we can think of $m = 8$. Consequently, the largest distance a packet may have to travel during the routing is $2 \cdot (M - 1) = 2M - 2$. As a first modification, we want to 'connect' the borders of the mesh to get a torus topology and thereby reduce this maximal distance to $2 \cdot (M/2) = M$.

3.1   Using a Torus Topology

Having in mind an actual mesh of processing units, it is not desirable to install physical connections between the horizontal resp. vertical borders of the mesh, as the wires used for the 'wrap around' had to cross a length of at least $M - 2$ processing units. Instead, a standard trick can be used to derive a layout where connecting wires never cross more than one processing unit:

1. The $2^m \times 2^m$ processing units are arranged in a square, and we denote the ith column in this square by c_i ($1 \le i \le 2^m$): $c_1|c_2|c_3|c_4|\ldots|c_{2^m-2}|c_{2^m-1}|c_{2^m}$

2. Reversing the order of the columns $c_{2^{m-1}+1},\ldots,c_{2^m}$ followed by applying a perfect shuffle yields the desired column positions:

$$c_1|\mathbf{c_{2^m}}|c_2|\mathbf{c_{2^m-1}}|\ldots|\mathbf{c_{2^{m-1}+2}}|c_{2^{m-1}}|\mathbf{c_{2^{m-1}+1}}$$

3. For implementing the vertical wrap around now the same trick is applied to the rows of the resulting arrangement.

This rearrangement of the processing units is 'just' for the ease of implementation and helps to circumvent the handling of extremely unbalanced running times of signals. Thus, in the sequel when discussing algorithmic aspects and the labeling of processing units used in a computation we can ignore this implementation detail and think of an ordinary mesh architecture with wrapped-around borders.

3.2   Finding the Route to a Target Node

Each sieving line is split into subintervals of length S, and the mesh will process these subintervals one by one. For 768-bit numbers, we can think of $S = 2^{24}$, thus in a mesh of size 256×256 each processing unit will be in charge of $256 = 2^{24-16}$ consecutive sieve positions, and we focus on this parameter choice.

When preparing a packet that is to be routed in our sieving procedure, we are given a 24-bit number r that represents a non-negative integer $0 \le r < S$, and we use only this value to identify the corresponding packet's route to its target node. W. l. o g. we can choose the start s_0 of the sieving subinterval such that $256 \mid s_0$, i. e., once the packet containing r arrived at the processing unit that processes the range $\{s_0 + i \cdot 256, \ldots, s_0 + i \cdot 256 + 255\}$ (with $i \in \{0, \ldots, 2^{16} - 1\}$), the least-significant 8 bit of r determine which of the sieve positions of that processing unit has to be addressed. Thus, we want to interpret the 16-bit number i, that indicates the number of the processed subinterval, as (x, y)-coordinate ($0 \le x, y \le 2^8 - 1$) of the target node.

There are various possibilities to encode these coordinates in the remaining $24 - 8 = 16$ bit of r. For our purposes the following approach is useful: we store the x-coordinate of the target node in the odd-numbered bit positions $(23, 21, 19, 17, \ldots, 9$, where bit no. 23 is the most-significant bit) and the y-coordinate in the even bit positions $(22, 20, 18, 16, \ldots, 8)$. This interleaving of the coordinates can also be interpreted as a Kronecker/tensor product: the leading two bit of r determine in which $2^7 \times 2^7$-(sub)quadrant of the $2^8 \times 2^8$-mesh the target node is located. Similarly, the next two bit determine which (sub)quadrant (of size $2^6 \times 2^6$) inside the $2^7 \times 2^7$-quadrant has to be addressed, etc. To get a better image of the resulting pattern, Figure 2 sketches for small values of i to which processing unit the sieving range $\{s_0 + i \cdot 256, \ldots, s_0 + i \cdot 256 + 255\}$ is assigned; e. g., the processing unit at the left border of the third line in the mesh handles the 256 values $\{s_0 + 4 \cdot 256, \ldots, s_0 + 4 \cdot 256 + 255\}$.

Given an r-value we can extract the x-coordinate x_t and the y-coordinate y_t of the respective target unit by reading off the odd- resp. even-numbered bit positions. While in the original clockwise transposition routing as described

0	2	8	10	32 34	40 42
1	3	9	11	33 35	41 43
4	6	12	14	36 38	44 46
5	7	13	15	37 39	45 47
16 18	24 26	48 50	56 58
17 19	25 27	49 51	57 59
20 22	28 30	52 54	60 62
21 23	29 31	53 55	61 63

...

Fig. 2. Assigning sieving ranges to processing units

in [LSTT02] storing the target coordinates in each packet is sufficient for deciding efficiently when two packets have two be exchanged, here we also have to take care of the wrapped around borders. For this, we provide one extra bit along each axis which is set if and only if the packet wants to 'cross the border' for reaching its target. In other words, for a node with coordinates (x_0, y_0) in a $2^m \times 2^m$-mesh, the 'horizontal cross border bit' is set if and only if $|x_t - x_0| > 2^{m-1}$. Thus, with each package that is to be routed we store the two (m-bit) target coordinates (x_t, y_t) plus the two 'cross border flags'. Using an 8-bit comparer and a simple circuit that deals with the cross border flags and the most significant bit of the target coordinates, we can easily decide if two adjacent nodes have to exchange their packets. Analogously as in [LSTT02] we assume that no more than one clock cycle is to be used for such a compare/exchange operation.

3.3   Refilling the Mesh while Routing

Experimentally it turns out, that the routing algorithm performs better if the number of 'travelling' packets is not too high. In our application, a processing unit usually has several packets that have to be output on the mesh for being routed to the respective target node. In principle, a processing unit can release a new packet whenever no other packet has to be stored. However, to avoid a slow-down through congestion of the mesh, in our experiments it turned out to be more efficient to release a new packet only if in the previous two clock cycles no packet was stored in that node. In this way usually $\approx 25\%$ of the processing units are 'free', which—experimentally—allows for a quite efficient routing. Each routed packet represents a divisor of some number in the currently processed sieving range, and the next section explains this connection in more detail.

4   Organizing the Sieving

The basic organization of the sieving process is identical as in [GS03a], in particular we use line sieving, and when changing to a new line, i.e., increasing the current b-value, new data has to be loaded into the mesh. For sieving one line $-A \le a < A$ of ($\approx 3.4 \cdot 10^{13}$) a-values, only local operations inside the mesh nodes are used. As indicated above already, a sieving line is not processed 'as one piece', but divided into consecutive intervals of length $S(= 2^{24})$. Before going into the details of how these subintervals are processed, we have to say how the two factor bases are represented in the mesh—at this we want to exploit the Kronecker/tensor product arrangement explained in Section 3.2.

4.1   Storing the Factor Bases in the Mesh

Each processing unit stores[1] all elements (p, r) of the factor bases where the prime p is smaller than the size of the subinterval of the current sieving line processed by that processor—with the mentioned parameters for 768-bit numbers this translates to the condition $p \leq 2^8$. More precisely, for $p \leq 2^8$ and (p, r) being contained in a factor base, each processing stores the value $(p, (s_0 + r + 2^8 \cdot i) \bmod p)$ where s_0 is the first value in the processed sieving range of length S, and $i \in \{0, \ldots, 2^{16} - 1\}$ is the number of the subinterval of length 256 processed by that unit. The idea here is, that a processing unit will be able to test locally which 'tiny' primes divide an element in the sieving range processed in that node. Next, all pairs (p, r) with primes p that are 'up to 4 times larger'—namely with $2^8 < p \leq 2^{10}$—are stored 'once per 2×2-square' of the mesh. With the numbering from Section 3.2, this means that the prime p (along with the corresponding $(s_0 + r + 2^{10} \cdot i) \bmod p$-values) is stored in all processing units where the 'least significant tensor coordinates'—bits no. 0 and 1 in the binary representation of the number i of the subinterval of length 256—coincide. Again, the idea is to allow for a 'local' handling of prime divisors: a subquadrant of size 2×2 covers a sieving range of $2^2 \cdot 2^8 = 2^{10}$ numbers, and all prime divisors of size $\leq 2^{10}$ are available inside that square.

Next, we proceed analogously for submeshes of size 4×4 and primes $2^{10} < p \leq 2^{12}$. In other words, all processing units where the bits no. 0–3 of the binary representation of the number i of the processed subinterval coincide, store the same pairs (p, r)—where in analogy to the above r is replaced by $(s_0 + r + 2^{12} \cdot i) \bmod p$. So in each 4×4 subquadrant—which corresponds to a sieving range of length $4^2 \cdot 2^8 = 2^{12}$ all primes $\leq 2^{12}$ are 'available'. Going on in this way, in each submesh of size 8×8 we store the pairs (p, r) with $2^{12} < p \leq 2^{14}$, in each submesh of size 16×16 we take care of the primes $2^{14} < p \leq 2^{16}$, in each submesh of size 32×32 we deal with the primes $2^{16} < p \leq 2^{18}$, and in each submesh of size 64×64 we store the pairs (p, r) with $2^{18} < p \leq 2^{20}$. All pairs (p, r) with $p > 2^{20}$ are stored only once in the mesh. We do not consider subquadrants of size 128×128: due to the underlying torus topology, the horizontal or vertical distance between two nodes cannot be larger than 128 anyway.

For an actual implementation, the question of *how* to store the pairs (p, r) is crucial: with smoothness bounds $B_1 = 10^9$ and $B_2 = 10^8$, the rational factor base contains $5, 761, 456$ pairs (p, r) in total, and the algebraic factor base can be assumed to consist of $\approx 50, 850, 000$ pairs. With the prime distribution just described and leaving some leeway for multiple prime factors, we conclude that

---

[1] For the moment, we postpone a discussion of *how* to store these pairs.

each node in a 256×256 mesh has to store $\approx 1,300$ pairs (p, r). Storing them as pairs of 30-bit numbers in DRAM is extraordinary space/area-consuming, and not suitable for our purposes. Thus, before going into algorithmic details we should clarify how to store these pairs more efficiently: for storing its factor base elements, each node is equipped with three rectangular blocks of DRAM, where one DRAM block can be accessed in 'words' of 28 bit, one block can be accessed in 'words' of 31 bit, and the other block allows for access in 'words' of 7 bit. Within each block, sequential access is sufficient—random access is not required. The usage of the memory blocks depends on the size of the processed primes p: in analogy to [ST03], we call p

- **tiny**, if $2 \le p < 2^{17}$,   - **largish**, if $S < p \le B_2$.
- **smallish**, if $2^{17} \le p \le S$ - **hugish**, if $p > B_2$

For reasons of efficiency, with each tiny or smallish prime we also store the (non-negative) values $S \bmod p$ and $\lceil \log_{\sqrt{2}}(p) \rceil$. The details of the encoding used for storing the four different 'prime types' efficiently are given in [GS03c, Appendix A].

4.2   Sieving a Subinterval

Let $\tilde{a}$ be an arbitrary number from a subinterval of length $S = 2^{24}$ and $(p, r) \in P_i$ an element of a factor base. Then $\lceil \log_{\sqrt{2}}(p) \rceil$ is added to the 'length counter' $s_i(\tilde{a})$ during line sieving if and only if $\tilde{a} \equiv br \pmod{p}$ $(i = 1, 2)$. When the factor bases have been loaded into the mesh as described, the mesh is prepared to sieve the first subinterval $-A \le a < -A + S$ with $b = 1$. Each processing unit is in charge of 256 a-values (see Section 3.2), and conceptually splits into three parts:

The main part contains the DRAM with the stored factor bases along with the necessary logic to read out these elements. In particular, this logic is in charge of a flag which indicates whether currently the unique rational root (which is always stored first) or an algebraic root is processed. Also the 6-bit representation of the current $\lceil \log_{\sqrt{2}}(p) \rceil$-value is stored in this part of the processing unit.

After having retrieved an r-value from the DRAM, first we check whether it is 'relevant' for the current sieving subinterval of length S: as primes are stored repeatedly in the mesh, this 'relevance' does not depend only on r, but also on the size of p. More precisely, for $p \le 256$ all values $r < 256$ have to be considered, for $256 < p \le 1024$ all values $r < 1024$ are relevant, etc. For checking this 'relevance condition' efficiently we can make use of several OR gates that check the leading bits of r and a chain of multiplexers that is controlled by $\lceil \log_{\sqrt{2}}(p) \rceil$. If this r-value turns out to be not relevant, then we replace the old r-value in the DRAM by the new

$$r := \begin{cases} r - (S \bmod p) & , \text{ if } r - (S \bmod p) \ge 0 \\ r - (S \bmod p) + p & , \text{ otherwise} \end{cases}.$$

In this way, r is 'shifted' in the next subinterval of length S (cf. [GS03a]). Note that the value $S \bmod p$ is known already (for tiny and smallish primes it is stored in the DRAM and for largish and hugish primes we have $S = S \bmod p$), so this computation can be implemented efficiently by means of a 30-bit adder. Now the next root or prime can be processed.

On the other hand, if an r-value is identified as relevant, then the (x_t, y_t)-coordinates of the node that is in charge of this value are determined by appending the corresponding odd/even bit positions of r to the respective number of most significant bits of the node's own horizontal/vertical coordinate. In analogy to the relevance test, the precise number of bits that are to be copied from the node's own coordinates depends on p. Then the 'cross border' flags c_x, c_y are determined; for doing so, we may either use general adders with two (8-bit) inputs or an optimized component that can check whether the horizontal resp. vertical coordinate of the current node differs from x_t resp. y_t by more than 128. If the coordinates (x_t, y_t) are not identical with the node's own coordinates, then (x_t, y_t), (c_x, c_y), a 'footprint' of p, $\lceil \log_{\sqrt{2}}(p) \rceil$ (6 bit), the least significant 8 bit of r, and the flag which indicates whether the prime belongs to the algebraic or rational side are written into an output buffer which will be read by—but is not part of—the *mesh part* of the node (see below). What does 'footprint' of p mean? For promising (a, b)-pairs this footprint will be output to the processor that is in charge of the post-processing of the candidate pairs; for primes larger than some predetermined bound B_f, say $B_f = 2^{22}$, it should be possible to recover p from the footprint. In principle we could send the complete value p up to the least significant bit here, however to save some space a different footprint is preferable: we send the coordinates of the current node ($2 \cdot 8$ bit) concatenated with the bits no. 1–10 of p (i.e., the 10 least significant bits after dropping bit no. 0 which is always set). As each processing unit stores only ≈ 850 prime numbers larger than $B_f = 2^{22}$, this determines p in most cases uniquely. If a processing unit contains more than one prime with this footprint, in the postprocessing all primes with this footprint have to be taken into account. In summary, we write x_t, y_t (8 bit each), c_x, c_y (1 bit each), $\lceil \log_{\sqrt{2}}(p) \rceil$ (6 bit), the footprint (26 bit), the least significant 8 bits of r, and a one bit flag that distinguishes between the algebraic and the rational side into a 59-bit output buffer which will be read by the *mesh part* of the node (see below). According to our experiments, it is sufficient to provide space for two 59-bit entries in the output buffer; for storing the buffer entries, we can use latches which require only 4 transistors per bit.[2]

Now the currently processed prime p is added to the current r-value (with a 30-bit adder needing no more than 2 clock cycles), and we have to check as above whether the resulting new value $r := r + p$ is also 'relevant' for the processed sieving subinterval of length S. If this is not the case, then we update the old r-value in the DRAM accordingly for the next sieving subinterval, and otherwise we determine another (x_t, y_t)-pair.

Once an (x_t, y_t)-pair with the nodes own coordinates is encountered, the respective r-value has to be handled by the node itself and thus the least significant

---

[2] Actually, we could do with fewer bits: as the node's own coordinates (which are part of the 26-bit-footprint) are identical for all buffer entries, we could save some transistors by 'hardcoding' these bits.

8 bit of r, $\lceil \log_{\sqrt{2}}(p) \rceil$ (6 bit), the 26-bit footprint of p, and a 1-bit flag which indicates whether we deal with the rational or algebraic side are written in a 41-bit input buffer that is to be read by—and is part of—the *memory part* of the node (see below). Hereafter, p is added to the current r-value and checked for 'relevance' as already described. In summary, we estimate that realizing the main part requires $\approx 2{,}750$ transistors. Reading a DRAM entry takes 2 clock cycles, and the retrieved values can be processed in a pipeline structure. Thus, provided that the respective buffer is not full, basically every 4 clock cycles[3] an output is produced or the next p-value is selected. For storing the factor base elements, about $55{,}000$ bit of DRAM are needed.

The memory part of the node provides two 10-bit DRAM entries for each of the 256 a-values the node has to take care of—one 10-bit counter for the algebraic and one for the rational side. These 10-bit words are initialized with zero and used to store the sum of the $\lceil \log_{\sqrt{2}}(p) \rceil$-values that 'hit' the corresponding a-value during sieving on the algebraic resp. rational side. It is convenient to organize the DRAM for the 10-bit counters in 20-bit words, so that the algebraic and rational counter can be read simultaneously—we will exploit this when checking for simultaneous smoothness on the algebraic and the rational side.

The memory part reads from the mentioned 41-bit input buffer and uses the least significant 8 bit of r and the rational/algebraic flag to address the correct counter. To add the $\lceil \log_{\sqrt{2}}(p) \rceil$-value read from the input buffer to the respective 10-bit value in the DRAM, a 10-bit adder is used. Finally, a different part of the DRAM is needed to store footprints of prime factors larger than the already mentioned predetermined bound B_f, say $B_f = 2^{22}$. As explained above, for storing one footprint we need 26 bits. Moreover, we need 8 bits to identify the precise sieving location within the node, plus 1 bit to distinguish between the rational and the algebraic side. Thus, in total one complete entry occupies $26+8+1=35$ bit of DRAM. However, instead of equipping each individual node with DRAM for storing found prime factors, it seems more efficient to share this DRAM among two nodes that are *physical* (cf. Section 3.1) neighbours. Consequently, we add one bit to each entry to identify the processing unit. Of course, the question arises how many prime factors will be encountered per sieving subinterval. According to our experiments for $B_f = 2^{22}$ and 256 targets per node, a DRAM size of $325 \cdot 36$ bit (shared by two nodes) seems reasonable. Further on, the question of choosing the size of the input buffer arises—according to our experiments a buffer with a single 41-bit entry should already be sufficient to avoid a performance bottleneck.

Finally, we have to explain how to identify 'good' (a, b)-pairs and how to output the respective prime factors from the device. For this purpose, the (somewhat dirty) approach explained in [GS03c, Appendix B] seems reasonable. In summary, for implementing the memory part $\approx 1{,}250$ transistors should be sufficient (excluding the DRAM). For incrementing one $\lceil \log_{\sqrt{2}}(p) \rceil$-counter and checking both thresholds we allow 4 clock cycles which should provide enough

---

[3] Largish and hugish primes can usually be processed in 2 clock cycles, as most of them do not 'hit' in the current subinterval of length S.

leeway to store the (footprint of the) prime factor p into the DRAM. In addition to this, on average $\approx 11,000$ bit of DRAM (with random access) are needed for the found prime factors resp. for the $\lceil \log_{\sqrt{2}}(p) \rceil$-counters and the thresholds T_r, T_a.

In the mesh part the complete logic necessary for the clockwise transposition routing is located. In particular, this includes an 8-bit comparison unit plus some circuitry for taking care of the 'cross border flags' in each second node, which allows for an efficient (one clock cycle—cf. Section 3.2) exchange operation. The mesh part contains a register to store a complete packet as transported in the mesh. This register has the same width as the output buffer, and if a packet with (x_t, y_t) being identical to the node's own coordinates is encountered (and the input buffer of the memory part is not full), the 26-bit footprint of p, $\lceil \log_{\sqrt{2}}(p) \rceil$, the least significant 8 bits of r, and the factor base flag are copied into the input buffer of the memory part. New packets that have to be released into the mesh are read from the output buffer (which in turn is filled by the main part of the node as explained above). Implementing the mesh part should 'on average'[4] require no more than $1,100$ transistors.

4.3   Output of the Result and Moving to the Next Sieving Interval

Once a complete subinterval of size S has been sieved, we have to output the found (a, b)-pairs: for doing so, each processing unit that has set the 'done' flag during the sieving procedure, outputs the footprints of all stored prime factors along with the corresponding factor base indices (1 bit each), the coordinates ($2\times$ 8 bit) of the processing unit that found the factor, and the least significant 8 bit of the corresponding r-value. Note here, that due to the 'cleaning process' described in [GS03c, Appendix B] the end of the list of factors is marked with an 'all zero' entry. The output values are received by supporting hardware that has to perform the final smoothness testing. Using the available 59-bit bus for this purpose, reading out the results should require less than 700 clock cycles. Of course, before outputing the results, we have to be sure that the sieving of the subinterval is indeed complete. However, there is no need to use a complicated logic for this: from a simulation one can determine a reasonable upper bound for the number of clock cycles that are needed to complete the sieving of a complete subinterval—for $S = 2^{24}$ on a 256×256 mesh such a bound can be $39,500$ clock cycles (this estimation is based on simulations by means of the computer algebra system Magma [BCP97]). After that time we simply instruct each processing unit to clear its input and output buffer, to complete any missing updates of its r-values, and to output its results. In the worst case (say the routing circuit encountered an infinite loop), potentially useful (a, b)-pairs from a single subinterval of length S are lost in this way.

If the next subinterval to be sieved is in the same line, i. e., the b-value does not change, each processing unit simply has to reset its 256×2 10-bit counters

---

[4] Recall that the comparer is needed only in each second node.

to zero now, and is ready to sieve the next subinterval of length S. Analogously as in [GS03a], in this case *no new data has to be loaded into the device*, as all r-values have already been updated during the processing of the last sieving interval. Thus the change into the next sieving interval can be estimated to take no more than 500 clock cycles. Finally, passing to the next b-value requires a replacement of the stored r-values in the processing units, analogously as in [GS03a]. Using the 59-bit bus, 2000 pins per chip, and an I/O clocking rate of 133 MHz, we expect that loading the data into the DRAM takes no more than 0.02 seconds.

5   An Improvement

Before discussing the performance of the above design, one may ask about possible improvements of the discussed parameters and the design. E. g., one may think of using larger or smaller mesh sizes, of different values for S or of changing the number of primes that each processing unit deals with locally. In this paper we focus on one (which we think reasonable) parameter choice, but there is still leeway for experimenting here—lacking a theoretical analysis of the underlying routing algorithm, we cannot make reliable theoretical predictions.

In [GS03c, Appendix C] a significant improvement is described which affects the physical arrangement of the processing units as discussed in Section 3.1. The basic idea is to use four interleaving meshes instead of one. Certainly there are also other modifications of the above design, but in the estimations given in the next section we restrict to this improvement using four meshes.

6   Space Requirements and Performance of the Device

Due to the 'small' DRAM banks, we take $0.3\mu m^2$ ($0.5\mu m^2$) for a DRAM bit with sequential (random) access and $2.8\mu m^2$ per transistor into account, which is somewhat larger than the estimates in [LSTT02, Table 2] for a specialized $0.13\mu m$ DRAM process. Combined with our estimations for the sizes of the node parts, this yields an estimated total size of a 256×256-mesh of $\approx (4.9 \text{ cm})^2$. Here the use of four meshes as described in [GS03c, Appendix C] is assumed.

Having in mind the comparatively regular layout and that 90nm processes are becoming more widespread, manufacturing such chips does not seem to be unrealistic (recall that for the proposed TWIRL parameters—which are not optimized for chip size—already the algebraic sieve has an estimated size of $\approx (6.7\text{cm})^2$ [ST03, Section 4.4]). In contrast to TWIRL, in the above mesh-based design a single device handles both factor bases, and we do not require any inter-chip communication, which is in particular helpful for cooling the chips. But what about the performance of the above device which we assume to be clocked with 500 MHz (instead of 1 GHz in TWIRL, where the time-critical operations are simpler)? Assuming a sieve line width of $3.4 \cdot 10^{13}$ (see [ST03, Table 1]) and that $40,000$ clock cycles are needed per sieving subinterval of size $S = 2^{24}$, a single chip with a mesh of size 256×256 can process a sieve line in

≈ 163 seconds. This is almost a factor 20 slower than a TWIRL cluster consisting of four rational sieves (each of size $\approx(3.6\text{cm})^2$) and one algebraic sieve (of size $\approx(6.7\text{cm})^2$). However, while only six TWIRL clusters fit on a single 300 mm silicon wafer, we can fit 21 mesh-circuits of size 256×256 on such a wafer. With 21 chips we can handle a sieve line in ≈ 7.8 seconds, which is ≈ 6.3 times more than a wafer with six TWIRL clusters.

Moreover, in [ST03] the authors explain how to exclude sieve regions where a and b have a common divisor 2 or 3. The common divisor 2 can be handled easily by our device—essentially, we have to add $2p$ to r instead of p. Handling the common divisor 3 is in principle possible, but would require significant additional logic, and we do not consider such a modification here. Hence, instead of an 'essentially free' 33% time reduction in TWIRL, we assume only an 'essentially free' 25% time reduction. Thus, for completing all expected $8.9 \cdot 10^6$ sieving lines (see [ST03, Table 1]) for a 768-bit number, with one wafer we expect that ≈ 600 days are needed—roughly 6.3 times more than with TWIRL. But as smaller and regular chips are simpler to produce, and as our design does not rely on inter-chip communication, from a practical point of view the mesh-based approach might be an interesting alternative to TWIRL clusters.

7   Conclusions and Further Work

The above discussion shows that building a mesh-based sieving device for 768-bit numbers could be feasible with current technology. Depending on the number of chips one is willing to use, performing the sieving step for such numbers within a few months seems feasible. In comparison to the proposed TWIRL clusters (which are not optimized for chip size), the chips in our design are smaller, no inter-chip communication is involved, and the rather regular layout should simplify the production of a detailed hardware layout. A main drawback of the mesh-based approach is a slow-down of a factor ≈ 6.3 compared to TWIRL. However, the simpler hardware requirements might outweigh this drawback. Also, we would like to emphasize that the discussed mesh is certainly not optimal, and modifying some of the paramter choices may yield relevant speed-ups. E. g., experiments show that if one is willing to allow for larger output buffers (which of course increases the chip size), the required number of clock cycles per sieving subinterval can be reduced.

Moreover, to further reduce the chip size, one can think of using a smaller mesh with only 128×128 nodes. According to our experiments, such a design is less efficient, but of course the resulting chips are smaller and producing them can be expected to be cheaper. On the other hand, one can ask whether implementing a larger 512×512 mesh is still feasible and whether a significant speed-up is possible in this way. We have not enough experimental results to give a reliable answer here, but exploring larger meshes is certainly worthwhile, in particular in regard to 1024-bit numbers: it is a natural question to ask to what extent the above mesh-based approach can deal with 1024-bit numbers. We cannot give a satisfying answer here at the moment. However, due to the compact representation of the factor bases in our device, it is certainly worthwhile to explore the 1024-bit case in more detail: the DRAM required for storing the factor bases is one of the critical issues in TWIRL, and it seems to be a very interesting question to explore the potential of the above mesh-based approach for 1024-bit numbers.

Acknowledgements. We are indebted to Eran Tromer and an anonymous referee for several detailed and helpful comments that helped to improve the original manuscript.

References

[BCP97]   Wieb Bosma, John Cannon, and Catherine Playoust. The Magma Algebra System I: The User Language. *Journal of Symbolic Computation*, 24:235–265, 1997.

[Ber01]   Daniel J. Bernstein. Circuits for Integer Factorization: a Proposal. At the time of writing available electronically at http://cr.yp.to/papers.html #nfscircuit, 2001.

[GS03a]   Willi Geiselmann and Rainer Steinwandt. A Dedicated Sieving Hardware. In Yvo G. Desmedt, editor, *Public Key Cryptography — PKC 2003*, volume 2567 of *Lecture Notes in Computer Science*. Springer, 2003.

[GS03b]   Willi Geiselmann and Rainer Steinwandt. Hardware to Solve Sparse Systems of Linear Equations over GF(2). In Colin D. Walter, Çetin K. Koç, and Christof Paar, editors, *Cryptographic Hardware and Embedded Systems; CHES 2003 Proceedings*, volume 2779 of *Lecture Notes in Computer Science*, pages 51–61. Springer, 2003.

[GS03c]   Willi Geiselmann and Rainer Steinwandt. Yet Another Sieving Device (extended version). Cryptology ePrint Archive: Report 2003/202, 2003. At the time of writing available at http://eprint.iacr.org/2003/202/.

[LHWL93] Arjen K. Lenstra and Jr. Hendrik W. Lenstra, editors. *The development of the number field sieve*, volume 1554 of *Lecture Notes in Mathematics*. Springer, 1993.

[LS00]   Arjen K. Lenstra and Adi Shamir. Analysis and Optimization of the TWINKLE Factoring Device. In Bart Preneel, editor, *Advances in Cryptology — EUROCRYPT 2000*, volume 1807 of *Lecture Notes in Computer Science*, pages 35–52. Springer, 2000.

[LSTT02] Arjen K. Lenstra, Adi Shamir, Jim Tomlinson, and Eran Tromer. Analysis of Bernstein's Factorization Circuit. In Yuliang Zheng, editor, *Advances in Cryptology — ASIACRYPT 2002*, volume 2501 of *Lecture Notes in Computer Science*, pages 1–26. Springer, 2002.

[Sha99]   Adi Shamir. Factoring Large Numbers with the TWINKLE Device. In Çetin K. Koç and Christof Paar, editors, *Cryptographic Hardware and Embedded Systems. First International Workshop, CHES'99*, volume 1717 of *Lecture Notes in Computer Science*, pages 2–12. Springer, 1999.

[ST03]   Adi Shamir and Eran Tromer. Factoring Large Numbers with the TWIRL Device. In Dan Boneh, editor, *Advances in Cryptology — CRYPTO 2003*, volume 2729 of *Lecture Notes in Computer Science*, pages 1–26. Springer, 2003.

A Parallelizable Enciphering Mode

Shai Halevi[1] and Phillip Rogaway[2]

[1] IBM T.J. Watson Research Center, Yorktown-Heights, NY 10598, USA
shaih@watson.ibm.com,
http://www.research.ibm.com/people/s/shaih/
[2] Dept. of Computer Science,
University of California, Davis, CA 95616, USA,
Dept. of Computer Science, Fac. of Science,
Chiang Mai University, 50200, Thailand
rogaway@cs.ucdavis.edu,
http://www.cs.ucdavis.edu/~rogaway

Abstract. We describe a block-cipher mode of operation, EME, that turns an n-bit block cipher into a tweakable enciphering scheme that acts on strings of mn bits, where $m \in [1 .. n]$. The mode is *parallelizable*, but as serial-efficient as the non-parallelizable mode CMC [6]. EME can be used to solve the disk-sector encryption problem. The algorithm entails two layers of ECB encryption and a "lightweight mixing" in between. We prove EME secure, in the reduction-based sense of modern cryptography. We motivate some of the design choices in EME by showing that a few simple modifications of this mode are insecure.

1   Introduction

TWEAKABLE ENCIPHERING SCHEMES. A *tweakable enciphering scheme* is a function $\mathbf{E}$ that maps a plaintext P into a ciphertext $C = \mathbf{E}_K^T(P)$ under the control of a key K and tweak T. The ciphertext must have the same length as the plaintext and there must be an inverse $\mathbf{D}_K^T$ to $\mathbf{E}_K^T$. We are interested in schemes that are secure in the sense of a tweakable, strong pseudorandom-permutation ($\pm\widetilde{\mathrm{prp}}$): an oracle that maps (T, P) into $\mathbf{E}_K^T(P)$ and maps (T, C) into $\mathbf{D}_K^T(C)$ must be indistinguishable (when the key K is random and secret) from an oracle that realizes a T-indexed family of random permutations and their inverses. A tweakable enciphering scheme that is secure in the $\pm\widetilde{\mathrm{prp}}$-sense makes a desirable tool for solving the disk-sector encryption problem: one stores at disk-sector location T the ciphertext $C = \mathbf{E}_K^T(P)$ for plaintext P. The IEEE Security in Storage Working Group [9] plans to standardize a $\pm\widetilde{\mathrm{prp}}$-secure enciphering scheme.

OUR CONTRIBUTION. This paper specifies EME, which is a simple and parallelizable tweakable enciphering scheme. The scheme is built from a block cipher, such as AES. By making EME parallelizable we accommodate ultra-high-speed mass-storage devices to the maximal extent possible given our security goals. When based on a block cipher $E: \{0,1\}^k \times \{0,1\}^n \rightarrow \{0,1\}^n$ our mode uses a

T. Okamoto (Ed.): CT-RSA 2004, LNCS 2964, pp. 292–304, 2004.
© Springer-Verlag Berlin Heidelberg 2004

k-bit key and $2m+1$ block-cipher calls to encipher an mn-bit plaintext in a way that depends on an n-bit tweak. We require that $m \in [1 .. n]$.

The name EME is meant to suggest ECB-Mix-ECB, as enciphering under EME involves ECB-encrypting the plaintext, a lightweight mixing step, and another ECB-encryption. For a description of EME look ahead to Figs. 1 and 2.

We prove that EME is secure, assuming that the underling block cipher is secure. The proof is in the standard, provable-security tradition: an attack on EME (as a $\pm\widetilde{\mathrm{prp}}$ with domain $\mathcal{M} = \{0,1\}^n \cup \{0,1\}^{2n} \cup \cdots \cup \{0,1\}^{n^2}$) implies an attack on the underlying block cipher (as a strong PRP with domain $\{0,1\}^n$).

We go on to motivate some of the choices made in EME by showing that other choices would result in insecure schemes. Finally, we suggest an extension to EME that operates on sectors that are longer than n^2 bits.

CMC MODE. The EME algorithm is follow-on work to the CMC method of Halevi and Rogaway [6]. Both modes are tweakable enciphering schemes built from a block cipher $E\colon \{0,1\}^k \times \{0,1\}^n \to \{0,1\}^n$. But CMC is inherently sequential, as it is built around CBC, while EME overcomes this limitation, which was seen as potentially problematic for high-speed encryption devices. The change does not increase the serial complexity: both modes use about $2m$ block-cipher calls (and little additional overhead) to act on an mn-bit string.

FURTHER HISTORY. Naor and Reingold gave an elegant approach for making a strong PRP on N bits from a block cipher on $n < N$ bits [16,15]. Their approach involves a hashing step, a layer of ECB encryption (say), and another hashing step. They do not give a fully-specified mode, but they do show how to carry out the hashing step given an xor-universal hash-function that maps N bits to n bits [15]. In practice, instantiating this object is problematic: to compare well with CMC or EME one should find a hash-function construction that is computationally simpler than CBC-AES, both in hardware and software, and has a collision bound of about 2^{-128}. No such construction is known.

An early, unpublished version of the CMC paper contained buggy versions of the CMC and EME algorithms. Joux discovered the problem [10] and thereby played a key role in our arriving at a correct solution. CMC was easily fixed in response to Joux's attack, but EME did not admit a simple fix. Indeed, Section 6.1 in this paper effectively proves that no simple fix is possible for the earlier EME construction.

Efforts to construct a block cipher with a large blocksize from one with a smaller blocksize go back to Luby and Rackoff [14], whose work can be viewed as building a $2n$-bit block cipher from an n-bit one. They also put forward the notion of a PRP and a strong ("super") PRP. The first attempt to directly construct an mn-bit block cipher from an n-bit one is due to Zheng, Matsumoto, and Imai [19]. A different approach is to build a wide-blocksize block-cipher from scratch, as with BEAR, LION, and Mercy [1,4]. The definition of a tweakable block-cipher is due to Liskov, Rivest, and Wagner [13]. An earlier work by Schroeppel suggested the idea of a tweakable block-cipher, by designing a cipher that natively incorporates a tweak [18]. The concrete-security treatment of PRPs that we use begins with Bellare, Kilian, and Rogaway [2].

DISCUSSION. EME has some advantages over CMC beyond its parallelizability. First, it uses a single key for the underlying block cipher, instead of two keys. All block-cipher calls are keyed by this one key. Second, enciphering under EME uses only the forward direction of the block cipher, while deciphering now uses only the reverse direction. This is convenient when using a cipher such as AES, where the two directions are substantially different, as a piece of hardware or code might need only to encipher or only to decipher. Finally, we prove EME secure as a variable-input-length (VIL) cipher and not just as a fixed-input-length (FIL) one. This means that EME remains secure even if the adversary intermixes plaintexts and ciphertexts of various lengths during its attack.

We comment that the parallelizability goal is arguably of less utility for a $\pm\widetilde{\text{prp}}$-secure enciphering scheme than for some other cryptographic goals. This is because, parallelizable or not, a $\pm\widetilde{\text{prp}}$-secure encryption scheme cannot avoid having latency that grows with the length of the message being processed (to achieve the $\pm\widetilde{\text{prp}}$ security notion one cannot output a single bit of ciphertext until the entire plaintext has been seen). Still, parallelizability is useful even here, and the user community wants it [8]. More broadly, EME continues a tradition of trying to make modes of operation that achieve parallelizability at near-zero added computational cost compared to their intrinsically serial counterparts (examples include CTR mode, IAPM [11], and PMAC [3]).

2   Preliminaries

BASICS. We use the same notions and notations as in [6]. A *tweakable enciphering scheme* is a function $\mathbf{E}\colon \mathcal{K}\times\mathcal{T}\times\mathcal{M}\to\mathcal{M}$ where $\mathcal{M}=\bigcup_{i\in I}\{0,1\}^i$ is the *message space* (for some nonempty index set $I\subseteq\mathbb{N}$) and $\mathcal{K}\neq\emptyset$ is the *key space* and $\mathcal{T}\neq\emptyset$ is the *tweak space*. We require that for every $K\in\mathcal{K}$ and $T\in\mathcal{T}$ we have that $\mathbf{E}(K,T,\cdot)=\mathbf{E}_K^T(\cdot)$ is a length-preserving permutation on $\mathcal{M}$. The inverse of an enciphering scheme $\mathbf{E}$ is the enciphering scheme $\mathbf{D}=\mathbf{E}^{-1}$ where $X=\mathbf{D}_K^T(Y)$ if and only if $\mathbf{E}_K^T(X)=Y$. A *block cipher* is the special case of a tweakable enciphering scheme where the message space is $\mathcal{M}=\{0,1\}^n$ (for some $n\geq 1$) and the tweak space is $\mathcal{T}=\{\varepsilon\}$ (the empty string). The number n is called the *blocksize*. By $\text{Perm}(n)$ we mean the set of all permutations on $\{0,1\}^n$. By $\text{Perm}^T(\mathcal{M})$ we mean the set of all functions $\pi\colon\mathcal{T}\times\mathcal{M}\to\mathcal{M}$ where $\pi(T,\cdot)$ is a length-preserving permutation.

An *adversary* A is a (possibly probabilistic) algorithm with access to some oracles. Oracles are written as superscripts. By convention, the running time of an algorithm includes its description size. The notation $A\Rightarrow 1$ describes the event that the adversary A outputs the bit one.

SECURITY MEASURE. For a tweakable enciphering scheme $\mathbf{E}\colon\mathcal{K}\times\mathcal{T}\times\mathcal{M}\to\mathcal{M}$ we consider the advantage that the adversary A has in distinguishing $\mathbf{E}$ and its inverse from a random tweakable permutation and its inverse: $\mathbf{Adv}_{\mathbf{E}}^{\pm\widetilde{\mathrm{prp}}}(A) =$

$$\Pr\left[K \xleftarrow{\$} \mathcal{K}: \ A^{\mathbf{E}_K(\cdot,\cdot)\ \mathbf{E}_K^{-1}(\cdot,\cdot)} \Rightarrow 1\right] - \Pr\left[\pi \xleftarrow{\$} \mathrm{Perm}^{\mathcal{T}}(\mathcal{M}): \ A^{\pi(\cdot,\cdot)\ \pi^{-1}(\cdot,\cdot)} \Rightarrow 1\right]$$

The notation shows, in the brackets, an experiment to the left of the colon and an event to the right of the colon. We are looking at the probability of the indicated event after performing the specified experiment. By $X \xleftarrow{\$} \mathcal{X}$ we mean to choose X at random from the finite set $\mathcal{X}$. In writing $\pm\widetilde{\mathrm{prp}}$ the tilde serves as a reminder that the PRP is tweakable and the $\pm$ symbol is a reminder that this is the "strong" (chosen plaintext/ciphertext attack) notion of security. For a block cipher, we omit the tilde.

Without loss of generality we assume that an adversary never repeats an encipher query, never repeats a decipher query, never queries its deciphering oracle with (T, C) if it got C in response to some (T, M) encipher query, and never queries its enciphering oracle with (T, M) if it earlier got M in response to some (T, C) decipher query. We call such queries *pointless* because the adversary "knows" the answer that it should receive.

When $\mathcal{R}$ is a list of resources and $\mathbf{Adv}_{\Pi}^{\mathrm{xxx}}(A)$ has been defined, we write $\mathbf{Adv}_{\Pi}^{\mathrm{xxx}}(\mathcal{R})$ for the maximal value of $\mathbf{Adv}_{\Pi}^{\mathrm{xxx}}(A)$ over all adversaries A that use resources at most $\mathcal{R}$. Resources of interest are the running time t and the number of oracle queries q and the query complexity σ_n (where $n \geq 1$ is a number). The query complexity σ_n is measured as follows. A string X contributes $\max\{|X|/n, 1\}$ to the query complexity; a tuple of strings $(X_1, X_2, \ldots)$ contributes the sum of the contributions of each string; and the query complexity of an adversary is the sum of the contributions from all oracle queries plus the contribution from the adversary's output. So, for example, an adversary that asks oracle queries $(T_1, P_1) = (0^n, 0^{2n})$ and then $(T_2, P_2) = (0^n, \varepsilon)$ and then outputs a bit b has query complexity $3 + 2 + 1 = 6$. The name of an argument (e.g., t or σ_n) will be enough to make clear what resource it refers to.

FINITE FIELDS. We interchangeably view an n-bit string as: a string; a nonnegative integer less than 2^n (msb first); a formal polynomial over $\mathrm{GF}(2)$ (with the coefficient of x^{n-1} first and the free term last); and an abstract point in the finite field $\mathrm{GF}(2^n)$. To do addition on field points, one xors their string representations. To do multiplication on field points, one must fix a degree-n irreducible polynomial. We choose to use the lexicographically first primitive polynomial of minimum weight. For $n = 128$ this is the polynomial $\mathsf{x}^{128} + \mathsf{x}^7 + \mathsf{x}^2 + \mathsf{x} + 1$. See [5] for a list of the indicated polynomials. We note that with this choice of field-point representations, the point $\mathsf{x} = 0^{n-2}10 = 2$ will always have order $2^n - 1$ in the multiplicative group of $\mathrm{GF}(2^n)$, meaning that $2, 2^2, 2^3, \ldots, 2^{2^n-1}$ are all distinct. Finally, we note that given $L = L_{n-1} \cdots L_1 L_0 \in \{0, 1\}^n$ it is easy to compute $2L$. We illustrate the procedure for $n = 128$, in which case $2L = L{\ll}1$ if $\mathsf{firstbit}(L) = 0$, and $2L = (L{\ll}1) \oplus \mathrm{Const87}$ if $\mathsf{firstbit}(L) = 1$. Here $\mathrm{Const87} = 0^{120}10^41^3$ and $\mathsf{firstbit}(L)$ means L_{n-1} and $L{\ll}1$ means $L_{n-2}L_{n-3} \cdots L_1 L_0 0$.

Algorithm $\mathbf{E}_K^T(P_1 \cdots P_m)$	**Algorithm** $\mathbf{D}_K^T(C_1 \cdots C_m)$
100 $\quad L \leftarrow 2E_K(0^n)$	200 $\quad L \leftarrow 2E_K(0^n)$
101 $\quad$ **for** $i \in [1 \mathinner{..} m]$ **do**	201 $\quad$ **for** $i \in [1 \mathinner{..} m]$ **do**
102 $\qquad PP_i \leftarrow 2^{i-1}L \oplus P_i$	202 $\qquad CC_i \leftarrow 2^{i-1}L \oplus C_i$
103 $\qquad PPP_i \leftarrow E_K(PP_i)$	203 $\qquad CCC_i \leftarrow E_K^{-1}(CC_i)$
110 $\quad SP \leftarrow PPP_2 \oplus \cdots \oplus PPP_m$	210 $\quad SC \leftarrow CCC_2 \oplus \cdots \oplus CCC_m$
111 $\quad MP \leftarrow PPP_1 \oplus SP \oplus T$	211 $\quad MC \leftarrow CCC_1 \oplus SC \oplus T$
112 $\quad MC \leftarrow E_K(MP)$	212 $\quad MP \leftarrow E_K^{-1}(MC)$
113 $\quad M \leftarrow MP \oplus MC$	213 $\quad M \leftarrow MC \oplus MP$
114 $\quad$ **for** $i \in [2 \mathinner{..} m]$ **do**	214 $\quad$ **for** $i \in [2 \mathinner{..} m]$ **do**
$\qquad CCC_i \leftarrow PPP_i \oplus 2^{i-1}M$	$\qquad PPP_i \leftarrow CCC_i \oplus 2^{i-1}M$
115 $\quad SC \leftarrow CCC_2 \oplus \cdots \oplus CCC_m$	215 $\quad SP \leftarrow PPP_2 \oplus \cdots \oplus PPP_m$
116 $\quad CCC_1 \leftarrow MC \oplus SC \oplus T$	216 $\quad PPP_1 \leftarrow MP \oplus SP \oplus T$
120 $\quad$ **for** $i \in [1 \mathinner{..} m]$ **do**	220 $\quad$ **for** $i \in [1 \mathinner{..} m]$ **do**
121 $\qquad CC_i \leftarrow E_K(CCC_i)$	221 $\qquad PP_i \leftarrow E_K^{-1}(PPP_i)$
122 $\qquad C_i \leftarrow CC_i \oplus 2^{i-1}L$	222 $\qquad P_i \leftarrow PP_i \oplus 2^{i-1}L$
130 $\quad$ **return** $C_1 \cdots C_m$	230 $\quad$ **return** $P_1 \cdots P_m$

Fig. 1. Enciphering (left) and deciphering (right) under $\mathbf{E} = \text{EME}[E]$, where $E \colon \mathcal{K} \times \{0,1\}^n \to \{0,1\}^n$ is a block cipher. The tweak is $T \in \{0,1\}^n$ and the plaintext is $P = P_1 \cdots P_m$ and the ciphertext is $C = C_1 \cdots C_m$.

3 Specification of EME

We construct from block cipher $E \colon \mathcal{K} \times \{0,1\}^n \to \{0,1\}^n$ a tweakable enciphering scheme that we denote by $\text{EME}[E]$ or EME-E. The constructed enciphering scheme has key space $\mathcal{K}$, the key space for E, and the tweak space is $\mathcal{T} = \{0,1\}^n$. The message space $\mathcal{M} = \{0,1\}^n \cup \{0,1\}^{2n} \cup \cdots \cup \{0,1\}^{n^2}$ contains any string having any number m of n-bit blocks, where $m \in [1 \mathinner{..} n]$. The definition of EME is given in Fig. 1 and an illustration of EME is given in Fig. 2. In the figures, all capitalized variables except for K are n-bit strings (key K is an element of the key-space $\mathcal{K}$). Variable names P and C are meant to suggest *plaintext* and *ciphertext*. When we write $\mathbf{E}_K^T(P_1 \cdots P_m)$ we mean that the incoming plaintext $P = P_1 \cdots P_m$ is silently partitioned into n-bit strings $P_1, \ldots, P_m$ and when we write $\mathbf{D}_K^T(C_1 \cdots C_m)$ we mean that the incoming ciphertext $C = C_1 \cdots C_m$ is partitioned into n-bit strings $C_1, \ldots, C_m$. It is an error to provide $\mathbf{E}$ with a plaintext that is not mn bits for some $m \in [1 \mathinner{..} n]$, or to supply $\mathbf{D}$ with a ciphertext that is not mn bits for some $m \in [1 \mathinner{..} n]$.

4 Security of EME

The following theorem relates the advantage an adversary can get in attacking $\text{EME}[E]$ to the advantage an adversary can get in attacking the block cipher E.

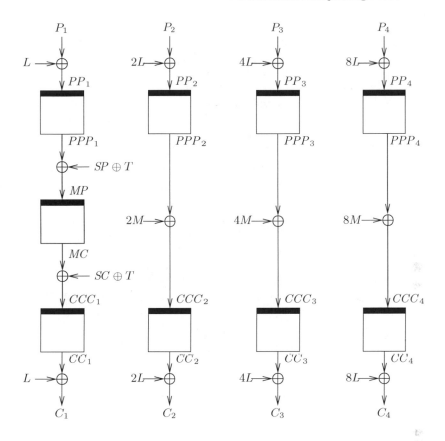

Fig. 2. Enciphering a four-block message $P_1P_2P_3P_4$ under EME. The boxes represent E_K and $L = 2E_K(0^n)$. We set $SP = PPP_2 \oplus PPP_3 \oplus PPP_4$ and $M = MP \oplus MC$ and $SC = CCC_2 \oplus CCC_3 \oplus CCC_4$.

Theorem 1. [EME security] *Fix* $n, t, \sigma_n \in \mathbb{N}$ *and a block cipher* $E \colon \mathcal{K} \times \{0,1\}^n \to \{0,1\}^n$. *Then*

$$\mathbf{Adv}^{\pm\widetilde{\mathrm{prp}}}_{\mathrm{EME}[\mathrm{Perm}(n)]}(\sigma_n) \leq \frac{7\,\sigma_n^2}{2^n} \quad and \tag{1}$$

$$\mathbf{Adv}^{\pm\widetilde{\mathrm{prp}}}_{\mathrm{EME}[E]}(t,\sigma_n) \leq \frac{7\,\sigma_n^2}{2^n} + 2\,\mathbf{Adv}^{\pm\mathrm{prp}}_{E}(t',\,\sigma_n) \tag{2}$$

where $t' = t + O(n\sigma_n)$. □

We note that the theorem does not restrict messages to one particular length: proven security is for a variable-input-length (VIL) cipher, not just fixed-input-length (FIL) one. The heart of Theorem 1 is Equation (1), whose proof is given in the full version of this paper [7]. Equation (2) embodies the standard way to pass from the information-theoretic setting to the complexity-theoretic one.

5 Proof Ideas

Since the proof of Theorem 1 is quite long, we give a brief sketch here of some of its ideas. We consider an attack against EME as a game between the attacker and the mode itself, where the cipher is replaced by a truly random permutation π and this permutation is chosen "on the fly" during this game. We give names to all of the internal blocks that occur in the game, where an internal block is any of the n-bit values PP_i, PPP_i, MP, MC, CCC_i, CC_i that arise as the game is played. For example, PPP_i^s is the PPP_i-block of the sth query of the attacker.

As usual with such modes, the core of the proof is to show that "accidental collisions" are unlikely. An accidental collision is an equality between two internal blocks that is not obviously guaranteed due to the structure of the mode. Specifically, an equality between the ith blocks in two different encipher queries $P_i^s = P_i^t$ implies that we also have the equalities $PP_i^s = PP_i^t$ and $PPP_i^s = PPP_i^t$ and so these do not count as collisions. (And likewise for decipher queries.) Most other collisions are considered accidental collisions and we show that those rarely happen.[1] Showing that accidental collisions are rare is ultimately done by case analysis (but, as usual, it takes a non-trivial argument to get there). For example, in one case we show that with high probability $PP_i^s \neq PP_j^t$; in another case we show that with high probability $PPP_i^s \neq MC^t$, etc.

The analysis of most of the cases is standard. Below we illustrate one of the more interesting cases. We show that for an encipher query P^s, the block MP^s does not collide with any of the previous MP^r blocks. This is easily seen if any of the plaintext blocks P_i^s is a "new block" (i.e., different than P_i^r for all $r < s$). But we need to show it also for the case where the plaintext P^s was obtained by "mix-and-matching" blocks from previous plaintext vectors. So let $r < s$ be the index of the last plaintext that shares some blocks with P^s, that is, $P_i^r = P_i^s$ for some index i. This means that all the blocks P_i^s appeared in queries no later than r. If queries s and r sport the same plaintext vectors, $P^r = P^s$, and differ only in the tweak values, $T^r \neq T^s$, then we clearly have $MP^r \oplus MP^s = T^r \oplus T^s \neq 0$. So assume that $P^r \neq P^s$, let I be the set of indexes where they are equal, and denote $R = [1 \mathinner{..} m^r] - I$ and $S = [1 \mathinner{..} m^s] - I$, where m^s and m^r are the lengths (in blocks) of queries r and s. That is, $P_i^r = P_i^s$ exactly for all $i \in I$, which means that all the blocks P_i^s for $i \in S$ appeared in queries before query r. This, in turn, implies that the value of PPP_i^s for any $i \in S$ depends only on things that were determined before query r.

Assume that query r was decipher (and that MC^r did not already accidentally collide with anything), so MP^r was chosen "almost at random" during the processing of query r. We show that the sum $MP^s \oplus MP^r$ can be expressed as $aMP^r + \beta$, where $a \neq 0$ is a constant and β is some expression that only depends on things that were determined before the choice of MP^r. Thus, the sum

---

[1] Actually, we only care about collisions between two values in the domain of π or between two values in its range; collisions between a domain value and a range value, such as $PP_i^s = CC_i^r$, are inconsequential and we ignore those.

$MP^s \oplus MP^r$ is rarely zero. We can write this sum as

$$MP^s \oplus MP^r = T^s \oplus \sum_{i=1}^{m^s} PPP_i^s \ \oplus T^r \oplus \sum_{i=1}^{m^r} PPP_i^r$$

$$= T^r \oplus T^s \oplus \sum_{i \in S} PPP_i^s \ \oplus \sum_{i \in R} PPP_i^r$$

$$= \text{things-that-were-determined-before-query-}r \ \oplus \sum_{i \in R} PPP_i^r$$

Assuming that R is non-empty, it is sufficient to show that we can express $\sum_{i \in R} PPP_i^r = aMP^r + \beta$ where a is non-zero and β only depends on things that were determined before the choice of MP^r. There are two cases in this proof, depending on whether $1 \in R$ or not, but they both boil down to the same point: since we use the value $2^{i-1}(MC^s \oplus MP^s)$ to mask the CCC_i block, the sum of PPP_i^r's can be written as

$$\sum_{i \in R} PPP_i^r = \text{some-expression-in-the-}CCC_i^r\text{'s-and-}MC^r \ \oplus \left(\sum_{i \in D'} 2^{i-1} \right) MP^r$$

where D' is also a non-empty set, $D' \subseteq [1 .. m^r]$, and so the coefficient of MP^r in this expression is non-zero. The case where query r is encipher is a bit longer, but it uses similar observations.

One last "trick" that is worth mentioning is the way we handle an adaptive adversary. To bound the probability of accidental collisions we analyze this probability in the presence of an augmented adversary, that can specify both the queries and their answers. That is, we let the adversary specify the entire transcript (with some minor restrictions) then choose some "permutation" π that maps the given queries to the given answers, and then consider the probability of accidental collisions. Clearly, this augmented adversary is no longer adaptive, hence the analysis becomes more tractable.

6   Some Insecure Modifications

In this section we justify two of our design choices by showing that changing them would result in insecure schemes. Specifically, we show that the block-cipher call that sits in between the two ECB layers is effectively unavoidable, and we show that that the length restriction $m < n$ also is needed.

6.1   The Middle-Layer Block-Cipher Call Is Needed

The EME construction has three block-cipher invocations in its "critical path" (that is, the construction is depth-3 in block-cipher gates). We now show that, in some sense, this is the best that you can do for a constructions of this type. Specifically, we show that for a construction of the type ECB-Mix-ECB, implementing the intermediate mixing layer by any linear transformation always results in an insecure scheme. This remains true even for an *untweakable* scheme, even when one considers only fixed-input-length inputs, even when each block-cipher call in each ECB encryption layer uses an independent key, and even if the linear transformation in the middle is key-dependent. This result implies that, as opposed to the Hash-Encrypt-Hash approach that was proven secure by Naor and Reingold [16], the "dual" approach of Encrypt-Hash-Encrypt will not be secure under typical assumptions.[2]

Formally, fix $m, n \in \mathbb{N}$ with $m \geq 2$, and let $E: \mathcal{K} \times \{0,1\}^n \to \{0,1\}^n$ be a block cipher. The scheme $\mathbf{E} = \text{BrokenEME}$ is defined on message space $\{0,1\}^{mn}$ and key space $\mathcal{K}^{2m} \times \mathcal{K}'$ where $\mathcal{K}'$ is a set of invertible linear transformations on $\{0,1\}^{mn}$. BrokenEME is keyed with $2m$ independent keys $K_1, \ldots, K_m$, $K'_1, \ldots, K'_m \in \mathcal{K}$, and with an invertible (possibly secret) linear transformation[3] $\tau: \{0,1\}^{mn} \to \{0,1\}^{mn}$. To encipher a plaintext $P = P_1 \cdots P_m \in \{0,1\}^{mn}$ we do the following:

- Set $PPP_i = E_{k_i}(P_i)$ for $i = 1 \ldots m$. Let $PPP = PPP_1 \cdots PPP_m$ be the concatenation of the PPP_i blocks ($PPP \in \{0,1\}^{mn}$).
- Apply the linear transformation τ to obtain $CCC = CCC_1 \cdots CCC_m = \tau(PPP)$.
- Set $C_i = E_{k'_i}(CCC_i)$ for $i = 1 \ldots m$. The ciphertext is the concatenation of all the C_i blocks, $C = C_1 \cdots C_m \in \{0,1\}^{mn}$.

Deciphering is done in the obvious way.

We now give an adversary A that attacks the mode, distinguishing it from a truly random permutation and its inverse using only four queries. Denote the adversary with its oracles as $A^{\mathcal{E} \mathcal{D}}$. The adversary A picks two mn-bit plaintexts that differ only in their first block, namely $P^1 = P_1 P_2 \cdots P_m$ and $P^2 = P'_1 P_2 \cdots P_m$ (with $P_1 \neq P'_1$). Then A queries its oracle as follows:

(1) Let $C^1 = C^1_1 \cdots C^1_m \leftarrow \mathcal{E}(P^1)$ and let $C^2 = C^2_1 \cdots C^2_m \leftarrow \mathcal{E}(P^2)$.
(2) Create two "complementing mixes" of the two ciphertexts, for example $C^3 = C^2_1 C^1_2 \cdots C^1_m$ and $C^4 = C^1_1 C^2_2 \cdots C^2_m$.
(3) Let $P^3 = P^3_1 \cdots P^3_m \leftarrow \mathcal{D}(C^3)$ and let $P^4 = P^4_1 \cdots P^4_m \leftarrow \mathcal{D}(C^4)$.

If the plaintext vectors P^3 and P^4 agree in all but their first block then A outputs 1 ("real") while otherwise it outputs 0 ("random"). To see that this works, we denote the intermediate variables in the four queries by PPP^i_j and CCC^i_j ($i \in [1 .. 4]$ and $j \in [1 .. m]$) and denote the "vector of differences" between PPP^1 and PPP^2 by $DP = DP_1 \cdots DP_m = PPP^1 \oplus PPP^2$. Since P^1 and P^2 agree everywhere except in their first block, it follows that also the "vector of differences" DP is zero everywhere except in the first block. Similarly, we denote the "vector of differences" between CCC^1 and CCC^2 by $DC = $

---

[2] This may seem somewhat surprising, as one may think that Encrypt-Hash-Encrypt should be at least as secure since it uses "more cryptography".

[3] In fact, it is easy to see that the attack described below works also when τ is an affine transformation.

$DC_1 \cdots DC_m \stackrel{\text{def}}{=} CCC^1 \oplus CCC^2$ and since we computed $CCC^i = \tau(PPP^i)$ and τ is a linear transformation, it follows that $DC = \tau(PPP^1) \oplus \tau(PPP^2) = \tau(PPP^1 \oplus PPP^2) = \tau(DP)$. Recall now that for any $j \in [1 .. m]$ we have either $C_j^3 = C_j^1$ and $C_j^4 = C_j^2$, or $C_j^3 = C_j^2$ and $C_j^4 = C_j^1$. It follows that for all j, $CCC_j^3 \oplus CCC_j^4 = CCC_j^1 \oplus CCC_j^2 = DC_j$, namely $CCC^3 \oplus CCC^4 = DC$. Putting this together we now compute $PPP^3 \oplus PPP^4$ as:

$$PPP^3 \oplus PPP^4 = \tau^{-1}(CCC^3) \oplus \tau^{-1}(CCC^4)$$
$$= \tau^{-1}(CCC^3 \oplus CCC^4) = \tau^{-1}(DC) = \tau^{-1}(\tau(DP)) = DP$$

This means that $PPP_j^4 = PPP_j^3$ for $j \in [2 .. m]$ and therefore also $P_j^4 = P_j^3$ for all but the first block.

6.2   The Length Restriction Is Needed

Recall that EME is defined on message space $\mathcal{M} = \bigcup_{m \in [1 .. n]} \{0,1\}^{mn}$. Here we show that the restriction $m \leq n$ is justified. In fact, we do not know whether allowing $m = n+1$ breaks the security of EME, but we can show that allowing $m = n+2$ permits easy distinguishing attacks. The details of the attack depend somewhat on the representation of the field $\text{GF}(2^n)$. Below we demonstrate it for $n = 128$, where the field $\text{GF}(2^{128})$ is represented using the polynomial $P_{128}(\mathsf{x}) = \mathsf{x}^{128} + \mathsf{x}^7 + \mathsf{x}^2 + \mathsf{x} + 1$.

Assume that $m \geq n+2$ and let J be a nonempty *proper subset* of the indexes from 2 to m, $J \subset \{2, 3, \ldots, m\}$, $J \neq \emptyset$, such that in the field $\text{GF}(2^n)$ we have $\sum_{j \in J} 2^{j-1} = 0$. For example, when $\text{GF}(2^{128})$ is represented using P_{128} we have $2^{129} + 2^8 + 2^3 + 2^2 + 2^1 = 2(2^{128} + 2^7 + 2^2 + 2^1 + 2^0) = 0$ and so we can set $J = \{130, 9, 4, 3, 2\}$. The attack proceeds as follows:

(1)   Pick an arbitrary tweak T. All the queries in the attack will use the same tweak T. (In other words, the attack works also when EME is used as an *untweakable* scheme.) Pick two plaintext vectors that differ only in their first block, $P^1 = P_1 P_2 \ldots P_m$ and $P^2 = P_1' P_2 \cdots P_m$ (with $P_1 \neq P_1'$).

(2)   Encipher both plaintext vectors to get $C^1 = \mathcal{E}(T, P^1)$ and $C^2 = \mathcal{E}(T, P^2)$.

(3)   Create a ciphertext vector C^3 such that $C_j^3 = \begin{cases} C_j^1 \text{ if } j \in J \\ C_j^2 \text{ if } j \notin J \end{cases}$.

(4)   Decipher C^3 to get $P^3 = \mathcal{D}(T, C^3)$.

Output 1 ("real") if P^3 and P^2 agree in all the blocks $j \in ([2 .. m] - J)$ and output 0 ("random") otherwise. To see that this works we denote the intermediate variables in the three queries by PPP_j^i and CCC_j^i and MP^i and MC^i and M^i (where $i \in [1 .. 3]$ and $j \in [1 .. m]$).

We note that $PPP_j^1 = PPP_j^2$ for all $j \in [2 .. m]$, and in particular for all $j \in J$. Also, from the construction of C^3 we get that $CCC_j^3 = CCC_j^1$ for $j \in J$ and $CCC_j^3 = CCC_j^2$ for $j \notin J$. Thus

$$MC^2 \oplus MC^3 = \left(T \oplus \sum_{j=1}^{m} CCC_j^2 \right) \oplus \left(T \oplus \sum_{j=1}^{m} CCC_j^3 \right)$$

$$= \sum_{j \in J} \left(CCC_j^2 \oplus CCC_j^3 \right) \;\; = \;\; \sum_{j \in J} \left(CCC_j^2 \oplus CCC_j^1 \right)$$

$$= \sum_{j \in J} \left((PPP_j^2 \oplus 2^{j-1} M^2) \oplus (PPP_j^1 \oplus 2^{j-1} M^1) \right)$$

$$= \sum_{j \in J} \left(2^{j-1} M^2 + 2^{j-1} M^1 \right) \;\; = \;\; (M^2 + M^1) \sum_{j \in J} 2^{j-1} \;\; = \;\; 0$$

So we have $MC^3 = MC^2$ and therefore also $MP^3 = MP^2$ and $M^3 = M^2$. Thus for any $j \notin J$, $j > 1$ we have $PPP_j^3 = CCC_j^3 \oplus 2^{j-1} M^3 = CCC_j^2 + 2^{j-1} M^2 = PPP_j^2$ and therefore also $P_j^3 = P_j^2$.

Acknowledgments. Phillip Rogaway was supported by NSF 0208842 and a gift from Cisco systems; thanks to the NSF and to Cisco for their kind support of my research.

References

1. R. Anderson and E. Biham. Two practical and provably secure block ciphers: BEAR and LION. In *Fast Software Encryption, Third International Workshop*, volume 1039 of *Lecture Notes in Computer Science*, pages 113–120, 1996. www.cs.technion.ac.il/~biham/.
2. M. Bellare, J. Kilian, and P. Rogaway. The security of the cipher block chaining message authentication code. *Journal of Computer and System Sciences*, 61(3):362–399, 2000. www.cs.ucdavis.edu/~rogaway.
3. J. Black and P. Rogaway. A block-cipher mode of operation for parallelizable message authentication. In L. Knudsen, editor, *Advances in Cryptology – EURO-CRYPT '02*, volume 2332 of *Lecture Notes in Computer Science*, pages 384–397. Springer-Verlag, 2002.
4. P. Crowley. Mercy: A fast large block cipher for disk sector encryption. In B. Schneier, editor, *Fast Software Encryption: 7th International Workshop*, volume 1978 of *Lecture Notes in Computer Science*, pages 49–63. Springer-Verlag, 2000. www.ciphergoth.org/crypto/mercy.
5. S. Duplichan. A primitive polynomial search program. Web document. Available at users2.ev1.net/~sduplichan/primitivepolynomials/primivitePolynomials.htm, 2003.
6. S. Halevi and P. Rogaway. A tweakable enciphering mode. In D. Boneh, editor, *Advances in Cryptology – CRYPTO '03*, volume 2729 of *Lecture Notes in Computer Science*, pages 482–499. Springer-Verlag, 2003. Full version available on the ePrint archive, http://eprint.iacr.org/2003/148/, 2003.
7. S. Halevi and P. Rogaway. A parallelizable enciphering mode. Full version available on the ePrint archive, http://eprint.iacr.org/2003/147/, 2003.
8. J. Hughes. Personal communication, 2002.
9. IEEE. Security in Storage Working Group (SISWG). See www.siswg.org and www.mail-archive.com/cryptography@wasabisystems.com/msg02102.html, May 2002.
10. A. Joux. Cryptanalysis of the EMD mode of operation. In *Advances in Cryptology – EUROCRYPT '03*, volume 2656 of *Lecture Notes in Computer Science*, pages 1–16. Springer-Verlag, 2003.

11. C. Jutla. Encryption modes with almost free message integrity. In *Advances in Cryptography – EUROCRYPT'01*, volume 2999 of *Lecture Notes in Computer Science*, pages 529–544. Springer-Verlag, 2001.
12. J. Kilian and P. Rogaway. How to protect DES against exhaustive key search. *Journal of Cryptology*, 14(1):17–35, 2001. Earlier version in CRYPTO '96. www.cs.ucdavis.edu/~rogaway.
13. M. Liskov, R. Rivest, and D. Wagner. Tweakable block ciphers. In *Advances in Cryptology – CRYPTO '02*, volume 2442 of *Lecture Notes in Computer Science*, pages 31–46. Springer-Verlag, 2002. www.cs.berkeley.edu/~daw/.
14. M. Luby and C. Rackoff. How to construct pseudorandom permutations from pseudorandom functions. *SIAM J. of Computation*, 17(2), April 1988.
15. M. Naor and O. Reingold. A pseudo-random encryption mode. Manuscript, available from www.wisdom.weizmann.ac.il/~naor/.
16. M. Naor and O. Reingold. On the construction of pseudo-random permutations: Luby-Rackoff revisited. *Journal of Cryptology*, 12(1):29–66, 1999. (Earlier version in STOC '97.) Available from www.wisdom.weizmann.ac.il/~naor/.
17. P. Rogaway, M. Bellare, J. Black, and T. Krovetz. OCB: A block-cipher mode of operation for efficient authenticated encryption. In *Eighth ACM Conference on Computer and Communications Security (CCS-8)*, pages 196–205. ACM Press, 2001.
18. R. Schroeppel. The hasty pudding cipher. AES candidate submitted to NIST. www.cs.arizona.edu/~rcs/hpc, 1999.
19. Y. Zheng, T. Matsumoto, and H. Imai. On the construction of block ciphers provably secure and not relying on any unproved hypotheses. In *Advances in Cryptology – CRYPTO '89*, volume 435 of *Lecture Notes in Computer Science*, pages 461–480. Springer-Verlag, 1989.

A   Extending EME to Longer Messages

The restriction on the message size of EME, $m \leq n$, means, for example, that when using AES as the underlying cipher one cannot encrypt messages longer than 2KB. In some applications this restriction could be problematic. We now describe EME$^+$, an extension of EME that can be used to handle message of practically any length (as long as it is an integral number of blocks).

The idea is to divide the m-block input into "chunks" of at most n blocks each and in each chunk use a construction similar to EME. Specifically, in the first chunk we use exactly the same construction as in EME while in all the other chunks we use a similar construction, where we replace the addition of $SP \oplus T$ and $SC \oplus T$ (before and after the block-cipher call in between the two ECB layers) by additions of the mask M_1 from the first chunk.

We specify in Fig. 3 the forward direction of our construction, $\mathbf{E} = \text{EME}^+[E]$. An illustration of EME$^+$ mode is given in Fig. 4. One observes that EME$^+$ is a "proper extension" of EME in that when we use it on a message of length $m \leq n$ blocks, we get back the original EME mode.

Although we have not written a proof of security for EME$^+$ we expect that such proof can be written. One would follow the arguments in the proof for the basic EME, except that one needs to analyze a few more cases in the case analysis.

Algorithm $\mathbf{E}_K^T(P_1 \cdots P_m)$

100 $L \leftarrow 2E_K(0^n)$
101 for $i \in [1 .. m]$ do
102      $PP_i \leftarrow 2^{i-1} L \oplus P_i, \quad PPP_i \leftarrow E_K(PP_i)$

110 $MP_1 \leftarrow PPP_1 \oplus PPP_2 \oplus \cdots \oplus PPP_m \oplus T$
111 $MC_1 \leftarrow E_K(MP_1), \quad M_1 \leftarrow MP_1 \oplus MC_1$
112 for $i \in [2 .. n]$ do $CCC_i \leftarrow PPP_i \oplus 2^{i-1} M_1$
113 for $j \in [2 .. \lceil m/n \rceil]$ do
114      $MP_j \leftarrow PPP_{(j-1)n+1} \oplus M_1$
115      $MC_j \leftarrow E_K(MP_j), \quad M_j \leftarrow MP_j \oplus MC_j$
116      $CCC_{(j-1)n+1} \leftarrow MC_j \oplus M_1$
117   for $i \in [(j-1)n + 2 .. jn]$ do
118          $CCC_i \leftarrow PPP_i \oplus 2^{i-1 \bmod n} M_j$
119 $CCC_1 \leftarrow MC_1 \oplus CCC_2 \oplus \cdots \oplus CCC_m \oplus T$

120 for $i \in [1 .. m]$ do
121      $CC_i \leftarrow E_K(CCC_i), \quad C_i \leftarrow CC_i \oplus 2^{i-1} L$

130 return $C_1 \cdots C_m$

Fig. 3. Enciphering under $\mathbf{E} = \text{EME}^+[E]$ with block cipher $E \colon \mathcal{K} \times \{0,1\}^n \to \{0,1\}^n$ and tweak $T \in \{0,1\}^n$ and plaintext $P_1 \cdots P_m$ and ciphertext $C_1 \cdots C_m$.

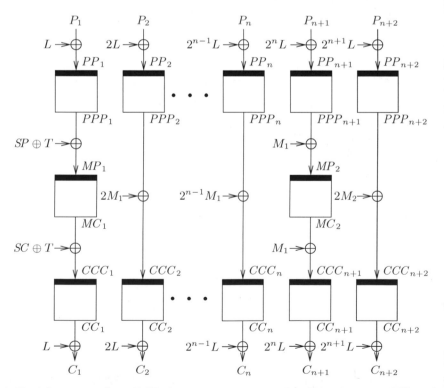

Fig. 4. Enciphering an $(n+2)$-block message under EME^+. Here $L = 2E_K(0^n)$ and $SP = PPP_2 \oplus \cdots \oplus PPP_m$ and $M_i = MP_i \oplus MC_i$ and $SC = CCC_2 \oplus \cdots \oplus CCC_m$.

Padding Oracle Attacks on the ISO CBC Mode Encryption Standard

Kenneth G. Paterson* and Arnold Yau**

Information Security Group,
Royal Holloway, University of London,
Egham, Surrey, TW20 0EX, UK
{kenny.paterson, a.yau}@rhul.ac.uk

Abstract. In [8] Vaudenay presented an attack on block cipher CBC-mode encryption when a particular padding method is used. In this paper, we employ a similar approach to analyse the padding methods of the ISO CBC-mode encryption standard. We show that, for several of the padding methods referred to by this standard, we can exploit an oracle returning padding correctness information to efficiently extract plaintext bits. In particular, for one padding scheme, we can extract all plaintext bits with a near-optimal number of oracle queries. For a second scheme, we can efficiently extract plaintext bits from the last (or last-but-one) ciphertext block, and obtain plaintext bits from other blocks faster than exhaustive search.

Keywords: padding oracle attack, CBC-mode encryption, ISO standard

1 Introduction

1.1 Background

In [8] Vaudenay presented an attack on block cipher CBC-mode encryption when a particular padding method is used. The attack requires an oracle which on receipt of a ciphertext, decrypts it and replies to the sender whether the padding is valid or not. The attack model assumes the attacker to have intercepted some such padded then CBC-mode encrypted ciphertext under some key K, and have access to the aforementioned padding validity oracle (operating using the same key K). The result is that the attacker can recover the plaintext corresponding to any block of ciphertext using an average of $128b$ oracle calls, where b is the number of bytes in a block and a byte is eight bits.

Further research has been done by Black and Urtubia [1], who generalised Vaudenay's attack to other padding schemes and modes of operations, and presented a padding method which prevents the attack. In [2], Canvel *et al* demonstrated the practicality of padding oracle attacks and showed how subtleties in

---

* This author supported by the Nuffield Foundation, NUF-NAL02

** This author supported by EPSRC and Hewlett-Packard Laboratories Bristol through CASE award 01301027. Also supported by EU Fifth Framework Project IST-2001-324467 "CASENET".

T. Okamoto (Ed.): CT-RSA 2004, LNCS 2964, pp. 305–323, 2004.

security protocol implementation can lead to flaws. First of all they realised an SSL/TLS padding oracle by exploiting timing information that is available upon submission of correctly and incorrectly padded ciphertexts. Secondly an attack against the IMAP protocol when used over SSL/TLS was implemented. In a typical setting, the attack recovers the IMAP password within one hour. Klíma and Rosa [7] applied idea of a "format correctness oracle" (of which padding is a special case) to construct a PKCS#7 validity oracle and were able to decrypt one PKCS#7 formatted ciphertext byte with on average 128 oracle calls.

1.2   ISO Standards

The current ISO standard for modes of operation of a block cipher is the second edition of ISO/IEC 10116 [4] (the third edition [5] is under development at the time of writing). It does not, however, specify any padding methods for the modes of operation (including CBC) that require one. In Section 5: Requirements it indicates that padding methods are beyond its scope and instead refers to ISO/IEC 9797-1 [3] (MACs using a block cipher) and 10118-1 [6] (general hash functions) where a few such methods are defined. Using a similar approach to [8], we have found attacks of various severity against some of those methods when used with CBC-mode encryption. Thess attacks do not, however, entail any security implications for those padding methods when they are used within their proper contexts (i.e. MACs and hash functions).

Note that in Annex B.2.3 of ISO/IEC 10116, ciphertext-stealing and another method are described for the special treatment of the last two blocks when encrypting under CBC-mode, when padding the plaintext is *not* acceptable. The standard does not prescribe that these methods be used, only that they *can* be used instead of padding. We emphasise that we are not attacking these two methods, but rather the padding methods in ISO/IEC 9797-1 and 10118-1 that are recommended for use in ISO/IEC 10116.

1.3   Our Contribution

We assume that an attacker has access to a padding oracle operating under the fixed key K and has intercepted a ciphertext encrypted in CBC-mode under that key. The attacker's aim is to recover the plaintext for that ciphertext. We further assume that the attacker is able to choose the initialisation vector when submitting ciphertexts to the oracle. This assumption prevents our attack from working when secret IVs are used; this is permitted in [4]. Some or all of these assumptions may be unwarranted when one is attacking a real system.

Under the above assumptions, our main results are as follows:

1. Attacking against padding method 3 of [3], the attacker can recover the plaintext for every ciphertext block with $n + O(\log_2 n)$ oracle calls for each block, where n is the block size.
2. There are two attacks against padding method 3 of [6], though they are to some degree interdependent. The padding method requires a parameter r to

be chosen where $1 \leq r \leq n$. In the first of our two attacks, the attacker can recover all plaintext bits to all ciphertext blocks with a complexity of $O(2^{r-1})$ oracle calls per block when $r < n$. When $r = n$ the complexity increases to $O(2^n)$. In our second attack, depending on which of two possible states the padding is in, the attacker either recovers the whole of the last plaintext block with $n + O(\log_2 n)$ oracle calls, or recovers some u bits of the last-but-one plaintext block which then speeds up the first attack by a factor of 2^{u-1} in recovering the remaining $n - u$ bits of the block.

We will first introduce some notation used throughout the paper, followed by a review of CBC-mode encryption. Then we present in turn each padding method in [3] and [6] and, if applicable, our attack against it. We conclude with a few remarks about the need for careful cryptographic design to prevent side-channel attacks.

2 Symbols, Notation, and CBC-Mode Review

2.1 Symbols and Notation

Each symbol and notation will be introduced on their first use, but we find it convenient to gather them here for reference purposes.

C : ciphertext output after CBC-mode encryption and ciphertext the attacker is trying to decrypt

C' : ciphertext to be submitted to an oracle during an attack

$d_K(Y)$: decryption of ciphertext block Y under key K

$e_K(X)$: encryption of plaintext block X under key K

D : unpadded data string to be CBC-mode encrypted

I_j : the jth intermediate block during CBC-mode encryption, i.e. $D_j \oplus C_{j-1}$, or in the case $j = 1$, it is $D_1 \oplus IV$

I'_j : the jth intermediate block during the attack, i.e. $d_k(C'_j)$

IV : the initialisation vector used in CBC-mode

L_D : the length (in bits) of the data string D

n : the block size (in bits) of the block cipher

P : the result of applying a given padding method to D

P' : data string computed by the padding oracle in the course of verifying padding

q : the number of blocks in data string P after padding

VALID and INVALID: oracle responses to, respectively, correct and incorrect padding after receipt and decryption of some ciphertext

$X||Y$: the result of concatenation of strings X and Y

$X \oplus Y$: the result of exclusive-or (XOR) of strings X and Y

X_2 : the binary representation of the value X

X_j : the jth block of the plaintext or ciphertext X

$X_{j,k}$: the kth bit of the plaintext or ciphertext block X_j

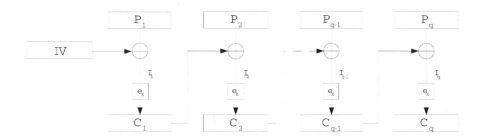

Fig. 1. CBC-mode encryption

2.2   Review of CBC-Mode Encryption

Cipher Block Chaining (CBC) is a mode of operation for an n-bit block cipher for encrypting data of arbitrary length. It has been standardised in second edition of ISO/IEC 10116 [4] and it is, quite naturally, included in the lastest draft of the third edition of that standard [5].

Let the encryption operation of the block cipher under key K be e_K, and the data we wish to encrypt be D. CBC-mode encryption (Figure 1) operates as follows:

1. A *padding method* is applied to D to make a padded message P of bitlength a multiple of n.
2. P is divided into n-bit blocks $P_1, P_2 \ldots P_q$.
3. An n bit number is chosen, at random or in a specified way, as the initialisation vector IV.
4. Compute ciphertext block $C_1 = e_K(IV \oplus P_1)$ and then
5. $C_i = e_K(C_{i-1} \oplus P_i)$, for $2 \leq i \leq q$
6. The resulting $C = IV||C_1||C_2|| \ldots ||C_q$ is the CBC-encrypted ciphertext.

We assume that IV is always prepended to the ciphertext. This allows a more concise notation for our attacks to follow and means that IV effectively plays the role of the "zeroth" ciphertext block; we write $C_0 = IV$.

Let d_K denote the inverse operation to e_K. To decrypt a block C_i (Figure 2) we simply have to compute $D_i = d_k(C_i) \oplus C_{i-1}$ for $2 \leq i \leq q$, and $D_1 = d_k(C_1) \oplus IV$.

Some security properties of CBC-mode are outlined in Section 2 of [8].

3   Attacking the Padding Methods of ISO/IEC 9797-1

3.1   The Standard

The standard [3] specifies six algorithms to compute an m-bit MAC using an n-bit block cipher with a secret key. The algorithms themselves are essentially

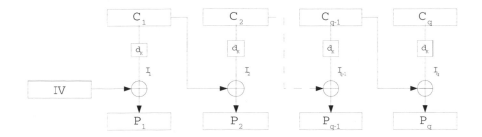

Fig. 2. CBC-mode decryption

instances of the CBC-MAC method or variants of it. Padding is applied, as with CBC-mode encryption, when the plaintext is not of length (in bits) a multiple of n, the block cipher size. For some methods it is always applied, regardless of the plaintext length.

3.2   Padding Method 1

The method is described as follows:

> "The data string D to be input to the [...] algorithm shall be right-padded with as few (possibly none) '0' bits as necessary to obtain a data string whose length (in bits) is a positive integer multiple of n."

Notice that this method is many-to-one: different data strings may be padded to yield the same result, which means that padding cannot be removed unambiguously if the length of the plaintext is not known. Consequently, given a padded data string, one cannot even tell where the data/padding boundary is, let alone check for padding validity. In fact, without data length information, every plaintext P is a validly padded version of at least one data string D. This of course limits the applicability of the padding technique to cases where the plaintext is of a fixed length, or where the proper length is somehow otherwise conveyed to the recipient.

No attack can be based on information returned from a padding oracle because any ciphetext submitted to such an oracle will decrypt to give a correctly padded plaintext.

3.3   Padding Method 2

The method:

> "The data string D to be input to the [...] algorithm shall be right-padded with a single '1' bit. The resulting string shall then be right-padded with as few (possibly none) '0' bits as necessary to obtain a data string whose length (in bits) is a positive integer multiple of n."

This method has been analysed in [1] (it is called OZ-PAD in that paper). The key result of [1] is that the method appears to resist padding oracle attacks. This is because practically all data strings are correctly padded, with the only exception being when a block contains all '0' bits. However this padding mechanism still lacks what is known as "semantic security" — an INVALID reply from the padding oracle would tell the attacker that the decrypted plaintext block is not a particular bit string. See [1] for details.

3.4   Padding Method 3

The method (Figure 3):

> "The data string D to be input to the [...] algorithm shall be right-padded with as few (possibly none) '0' bits as necessary to obtain a data string whose length (in bits) is a positive integer multiple of n. The resulting string shall then be left-padded with a block L. The block L consists of the binary representation of the length (in bits) L_D of the unpadded data string D, left-padded with as few (possibly none) '0' bits as necessary to obtain an n-bit block. The right-most bit of the block L corresponds to the least significant bit of the binary representation of L_D."

Fig. 3. ISO/IEC 9797-1 padding method 3

We have an attack against this padding scheme that decrypts, a block at a time, arbitrary ciphertexts $C_1||C_2||\ldots||C_q$. This attack takes $n + O(\log_2 n)$ oracle calls per block. There are two phases to this attack: determining L_D and the actual decryption.

Phase 1: Determining L_D. We want to find L_D, the content of the first block, which indicates the length of the unpadded data. To do that we use the padding oracle to determine the number of '0' bits that have been appended to the last block, if any. This is performed as follows (Figure 4).

Firstly notice that in CBC-mode decryption, flipping (complementing) any single bit at position i in block C_j would flip also the decrypted plaintext bit at position i in block P_{j+1} (whilst corrupting the whole of plaintext block P_i).

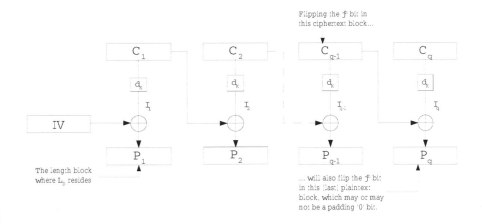

Fig. 4. Attack phase 1 — obtaining L_D

This allows us to flip arbitrary bits within a block in the decrypted plaintext by appropriately altering the ciphertext. This observation is in fact the basis of all of our attacks.

A padded data string consists of $q \geq 2$ blocks. Here we consider the case $q \geq 3$; the case $q = 2$ is handled separately below. The string is right-padded with some '0' bits and left-padded with the length block containing the binary representation of L_D. L_D is effectively a pointer to the last bit of the unpadded data, all the bits after which should be '0.' Let's now see what happens if we flip a single bit in P_q, the last plaintext block of the data string (by flipping a bit in C_{q-1}, the last-but-one ciphertext block). This change does not affect the decryption of C_1 (since $q \geq 3$) so the length block is left intact. So one of two things might happen:

1. The bit flipped is part of the original unpadded data. The padding is therefore still intact and correct and the oracle returns VALID.
2. The bit flipped is one of those '0' bits padded. The oracle therefore detects a '1' bit where it should have been '0,' and thus returns INVALID.

This means that after flipping a single bit in D_q, a VALID oracle response implies the padding boundary is to the right of the current position, and to the left otherwise. So we now can work out the exact location of the boundary by flipping the last plaintext block one bit at a time, say from right to left. The transition point of oracle response from INVALID to VALID tells us the location of the boundary we are after. This can be made more efficient by using a binary search similar to that presented in Section 3 of [1]. Once we have the boundary it is trivial to compute the value L_D from the number of blocks in the ciphertext and the position of the boundary within the last block.

This phase is presented in Algorithm 9797-1-m3-get-L_D-general below. The notation $X_{a,b}$ denotes the bit at position b of the ciphertext or plaintext block X_a. We number the positions in a block from 0 to $n-1$, going from left to right.

This method of obtaining L_D does not work when the unpadded data string consists only of a single block (this includes the case of the data being the null string). Here, the padded data string consists of two blocks $P_1||P_2$. Now flipping bits in the last (second) plaintext block would require changes in the first ciphertext block C_1, which in turn would corrupt the first plaintext block where L_D is supposed to reside.

Fortunately, there is a way to circumvent this problem at least for block sizes $n = 2^m$, $m \geq 1$, the most common situation in practice. Let $C = IV||C_1||C_2$ be the ciphertext for which we wish to determine L_D. It is not hard to see that if $IV' = IV \oplus \underbrace{0\ldots0}_{n-m-1} 1\underbrace{0\ldots0}_{m}$, then $C' = IV'||C_1||C_1||C_2$ is also a valid ciphertext unless $L_D = 0$ or $L_D = n$ (in which cases the padding oracle will return INVALID on submission of C'). In the situation where C' is valid then we can simply apply the method described above to C' to obtain $L'_D = L_D + 2^m$, and hence L_D.

We need to apply a further trick to distinguish the remaining cases, i.e. when $L_D = 0$ or $L_D = n$. Now we set $IV'' = IV \oplus \underbrace{0\ldots0}_{n-m-2} 11\underbrace{0\ldots0}_{m}$ and submit $C'' = IV''||C_1||C_1||C_2$ to the padding oracle. If $L_D = 0$, then C'' will, on decryption, contain a length field L''_D with $L''_D = 3n$. Since the unpadded data in C'' is of length at most $2n$, the padding oracle will output INVALID. On the other hand, if $L_D = n$, then C'' will yield $L''_D = 2n$ and C'' accepted as VALID. Hence one futher oracle query on a carefully chosen C'' is sufficient to decide whether $L_D = 0$ or $L_D = n$.

The special case $q = 2$ is presented in Algorithm 9797-1-m3-get-L_D-special below.

Phase 2: Decrypting. We now have L_D, the binary encoding of which is the content of the first plaintext block. We can deduce that I_1, the first intermediate block, is equal to $L_D \oplus IV$. Note that by manipulating IV, we can change the content of the first block to indicate a data length of any desired value. If L'_D is the desired value, we can take $IV' = (L'_D)_2 \oplus I_1 = (L'_D)_2 \oplus (L_D)_2 \oplus IV$.

We are now ready to decrypt an arbitrary ciphertext block C_k from the ciphertext $IV||C_1||C_2||\ldots||C_q$, where $2 \leq k \leq q$ (Figure 5). Note that there is no need to decrypt C_1 as it just encrypts the value L_D. The decryption is done in a bit-by-bit fashion, starting from the rightmost bit. So to begin with we submit to the oracle the ciphertext $C' = IV'||C_1||R||C_k$ where

$$IV' = (2n-1)_2 \oplus (L_D)_2 \oplus IV,$$

and R is a random n-bit block.

After decryption, L'_D, the length field in the resulting plaintext $P'_1||P'_2||P'_3$ points to the last-but-one bit of P'_3, the last block. Now the padding oracle outputs VALID for C' if the last bit of P'_3 is equal to '0,' and INVALID if $P'_{3,n-1}$

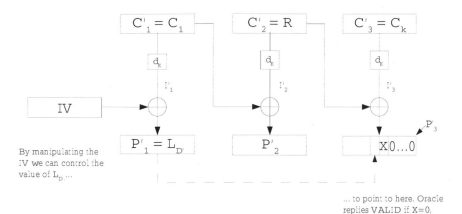

Fig. 5. Attack phase 2 — decrypting

Algorithm
9797-1-m3-get-L_D-general

Input: $IV||C_1||C_2||\ldots||C_q$
Output: L_D

Ensure: $q \geq 3$
  $C := IV||C_1||C_2||\ldots||C_q$
  $l := 0$
  $u := n - 1$
  repeat
    $h := \lceil (l + u)/2 \rceil$
    $C_{q-1,h} := C_{q-1,h} \oplus 1$
    if oracle(C) = VALID **then**
      $l := h$
    else if oracle(C) = INVALID **then**
      $u := h\text{-}1$
    end if
    $C_{q-1,h} := C_{q-1,h} \oplus 1$
  until $l = u$
  return $L_D := (q-1)n + l + 1$

Algorithm
9797-1-m3-get-L_D-special

Input: $IV||C_1||C_2$
Output: L_D

Ensure: $n = 2^m, m \geq 1, q = 2$
  $IV' := IV \oplus \underbrace{0\ldots0}_{n-m-1} 1 \underbrace{0\ldots0}_{m}$
  $C' := IV'||C_1||C_1||C_2$
  if oracle(C') = VALID **then**
    $L'_D = 9797\text{-}1\text{-}m3\text{-}get\text{-}L_D\text{-}general(C')$
    return $L_D := L'_D - 2^m$
  else
    $IV'' := IV \oplus \underbrace{0\ldots0}_{n-m-2} 11 \underbrace{0\ldots0}_{m}$
    $C'' := IV''||C_1||C_1||C_2$
    if oracle($C'' = $ VALID) **then**
      return $L_D := n$
    else
      return $L_D := 0$
    end if
  end if

is equal to '1.' We then have $I'_{3,n-1} = P'_{3,n-1} \oplus R_{n-1}$ and this block I'_3 is equal to the original intermediate block I_k. So we can obtain $P_{k,n-1} = I'_{3,n-1} \oplus C_{k-1,n-1}$.

To decrypt the next bit, we construct a new ciphertext for the oracle. We want, after decryption, the value in L'_D to decrement by one, and to ensure that $P'_{3,n-1}$ is '0'. We can achieve the former by altering IV appropriately and the latter by keeping/flipping last bit of C'_2 if the previous response was VALID/INVALID. Submitting the resulting ciphertext to the oracle, a VALID response indicates $P'_{3,n-2}$ equals '0,' and '1' otherwise. We can then compute $I'_{3,n-2} = P'_{3,n-2} \oplus R_{n-2}$. Note that the random block R at this iteration which may (or may not) have changed from the last iteration. We can now obtain $P_{k,n-2} = I'_{3,n-2} \oplus C_{k-1,n-2}$.

The process is repeated, decrementing L'_D by one per iteration while making sure the bit positions in P'_3 corresponding to those we have obtained stay at '0.' One bit of I'_3 and one bit of P_k are obtained at each iteration and we stop after $n-1$ iterations when the $n-1$ rightmost bits of those blocks are determined. We cannot get the leftmost bit of the block I'_3 (hence $P_{k,0}$) using this approach because at the next step L'_D would indicate a length $2n$, a multiple of the block size, and according to the standard, we would never append a new block in such cases.

Instead, we extract this leftmost bit $P_{k,0}$ by using a different approach. We assume that standard binary encoding is used for length information, with least significant bit in the rightmost position. (A similar attack can be mounted in the opposite situation too, but we omit the details.) Consider the ciphertext $C' = IV'||C'_1||R$ where $IV' = C_{k-1} \oplus 0P_{k,1}P_{k,2}\ldots P_{k,n} \oplus (n)_2$, $C'_1 = C_k$ and R is a random n-bit block. This ciphertext is constructed in such a way that the length field is equal to $P_{k,0}0\ldots0 \oplus (n)_2$, indicating a length of either n or $n+2^{n-1}$ depending on the value of $P_{k,0}$. So if C' is submitted to the oracle, then an output of VALID(INVALID) tells us that $P_{k,0} = 0$ ($P_{k,0} = 1$, respectively.)

We summarise the decryption phase as the pair of algorithms 9797-1-m3-decrypt and 9797-1-m3-decrypt-last-bit below. In these algorithms, Ω is the function which takes as input a ciphertext C and is defined as:

$$\Omega(C) = \begin{cases} 0 & \text{if the padding oracle returns VALID for input } C, \\ 1 & \text{if the padding oracle returns INVALID for input } C. \end{cases}$$

Complexity. Phase 1, in the general case ($q \geq 3$), should take no more than $\log_2 n$ oracle calls using binary search. To decrypt many messages encrypted under a fixed key K, this phase only needs to be performed once. Phase 2 takes one oracle call per plaintext bit, thus n calls per plaintext block. For the special case $q = 2$, one further oracle call is required in situations where $L_D = 0$ or $L_D = n$ to distinguish between them (no further oracle calls are needed otherwise).

Fewer than $\log_2(n) + 1 + (q-1)n$ oracle calls are needed to recover all the bits of plaintext from a q block ciphertext (remember that the first block contains L_D which is not part of the unpadded data string, and its value is already known after phase 1 anyway).

Optimality. The oracle returns one bit of information per use, so $(q-1)n$ is information theoretically the smallest number of oracle calls needed to recover

Algorithm
9797-1-m3-decrypt

Input: L_D, IV, C_1, C_k
Output: $P_{k,1} P_{k,2} \ldots P_{k,n-1}$,   the rightmost $n-1$ bits of P_k

$R :=$ a random n-bit block
for $j := n-1$ **to** 1 **do**
   $IV' := IV \oplus L_D \oplus (n+j)_2$
   $b := \Omega(C')$
   $C' := IV' || C_1 || R || C_k$
   $P_{k,j} := b \oplus R_j \oplus C_{k-1,j}$
   $R := R \oplus \underbrace{0 \ldots 0}_{j} b \underbrace{0 \ldots 0}_{n-j-1}$
end for
return $P_{k,1} P_{k,2} \ldots P_{k,n-1}$

Algorithm
9797-1-m3-decrypt-last-bit

Input: $C_{k-1}, C_k, P_{k,1} P_{k,2} \ldots P_{k,n}$
Output: $P_{k,0}$, the leftmost bit of P_k

$R :=$ a random n-bit block
$IV' := C_{k-1} \oplus 0 P_{k,1} P_{k,2} \ldots P_{k,n} \oplus (n)_2$
$C' := IV' || C^k || R$
$P_{k,0} := \Omega(C')$
return $P_{k,0}$

$(q-1)n$ bits of plaintext entropy. Hence our attack makes nearly optimal use of the padding oracle, especially when many ciphertexts are decrypted for the same key K.

4   Attacking the Padding Methods of ISO/IEC 10118-1

4.1   The Standard

ISO/IEC 10118 is a standard for hash functions, Part 1 [6] of which describes the general construction of a hash function.

Padding methods 1 and 2 in this standard are identical to the respective methods in ISO/IEC 9797-1 which were already discussed in the previous section. We focus instead on padding method 3 of [6].

4.2   Padding Method 3

In the standard, L_1 is used to denote the block size. It will henceforth be replaced by our usual notation n to be consistent with the rest of this paper. The method is as follows (Figure 6):

> "This padding method requires the selection of a parameter r (where $r \leq n$), e.g. $r = 64$, and a method of encoding the bit length of the data D, i.e. L_D as a bit string of length r. The choice for r will limit the length of D, in that $L_D < 2^r$.
> "The data D [...] is padded using the following procedure.
>   1. D is concatenated with a single '1' bit.

Fig. 6. ISO/IEC 10118-1 padding method 3

2. The result of the previous step is concatenated with between zero and $n - 1$ '0' bits, such that the length of the resultant string is congruent to $n - r$ modulo n. The result will be a bit string whose length will be r bits short of an integer multiple of n bits (in the case $r = n$, the result will be a bit string whose length is an exact multiple of n bits).
3. Append an r-bit encoding of L_D using the selected encoding method, yielding the padded version of D."

The above description can be summarised as "pad a '1' followed by the smallest number of '0's needed to push the r bits of L_D right to the end of a block." Using this method, the padded bits for data string D are appended in one of two ways:

Same-block $(L_D \bmod n) \leq (n - r - 1)$. The last block has enough space after the last plaintext bit to contain at least a single '1' bit and the r bits of L, the length block that holds L_D. The number of padded bits is between $r + 1$ and $n - 1$.

Cross-block $(L_D \bmod n) \geq (n - r)$. The last block does not have enough space to contain a '1' bit and the r bits of L. The number of padded bits is between n and $n + r$ and the padding extends over two blocks. Note that this will always be the case when $r = n$.

We have identified two attacks against this method, though they are to some degree dependent on each other. Note that no encoding method (for L_D) is specified in the standard. Our attacks work no matter which encoding method is used, though the attacker needs to know this method. We expect that base 2 encoding will be used in most cases and it will be used for illustrative purposes henceforth.

Attack 1: Directed IV search. This attack works against any block C_k of the ciphertext $IV||C_1||C_2||\ldots||C_q$ and recovers the corresponding plaintext block using on average $2^{r-1} + 2^{2r-n+1}$ and at most $2^r + 3 \cdot 2^{2r-n-1}$ padding oracle queries, provided $r \leq n - 1$. When $r = n$, the attack requires on average 2^n and at most 2^{n+1} oracle queries.

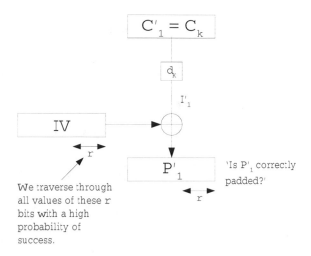

Fig. 7. Directed IV search

We first consider the case $r \leq n - 1$. We submit to the oracle strings of the form

$$IV' \| C_1'$$

where IV' is a specially selected initialisation vector and $C_1' = C_k$, hoping that the oracle returns **VALID**. This situation is depicted in (Figure 7). If it does, then the plaintext block P_k can be extracted using Attack 2 below on the ciphertext $IV' \| C_1'$. The overall complexity will be the sum of the two attacks' complexities, and will be dominated by the complexity of this first phase.

How then should IV' be selected? Notice that there is a probability of $1 - 2^{r-n}$ that there is a '1' somewhere in the leftmost $n - r$ bits of P_k. Thus, if we traverse through all 2^r possible settings of the rightmost r bits of IV', then with probability $1 - 2^{r-n}$ we will obtain (at least) one **VALID** reply from the padding oracle. The expected number of oracle queries in this situation is therefore 2^{r-1}. But with probability 2^{r-n}, all replies will be **INVALID**. Now if we flip the bit in position $n - r - 1$ of IV' and repeat the above process, it is easy to see that we are guaranteed to obtain at least one **VALID** response. A simple analysis shows that the number of oracle queries needed is equal to $2^{r-1} + 2^{2r-n+1}$ on average and is always at most $2^r + 3 \cdot 2^{2r-n-1}$.

An algorithm for Attack 1 in the case $r \leq n - 1$ is given in Algorithm 10118-1-m3-a1-general above.

Next we consider the case $r = n$. Here a valid plaintext must be at least two blocks in length and a three-block ciphertext $IV' \| R \| C_k$ is required to perform the attack. Instead of modifying only the initialisation vector as before, we now

**Algorithm
10118-1-m3-a1-general**

Input: C_k, n, r
Output: IV' s.t. $IV'||C_k$ is a valid ciphertext

Ensure: $1 \le r < n$
$\quad IV_0 :=$ a random n-bit block
$\quad IV' := \underbrace{0 \ldots 0}_{n}$

$\quad i := 0$
$\quad$**repeat**
$\qquad IV' := IV_0 \oplus \underbrace{0 \ldots 0}_{n-r-1} \underbrace{i_2}_{r+1}$
$\qquad C := IV'||C_k$
$\qquad i := i+1$
$\quad$**until** oracle(C) = VALID
$\quad$**return** IV'

**Algorithm
10118-1-m3-a1-special**

Input: C_k, n, r
Output: IV', R s.t. $IV'||R||C_k$ is a valid ciphertext

Ensure: $r = n$
$\quad IV_0 :=$ a random n-bit block
$\quad R_0 :=$ a random n-bit block
$\quad$**for** $i := 0$ to $2^n - 1$ **do**
$\qquad R := R_0 \oplus \underbrace{i_2}_{n}$
$\qquad$**for** $j := 0$ to 1 **do**
$\qquad\quad IV' := IV_0 \oplus \underbrace{0 \ldots 0 j}_{n}$
$\qquad\quad C := IV'||R||C_k$
$\qquad\quad$**if** oracle(C) = VALID **then**
$\qquad\qquad$**return** IV', R
$\qquad\quad$**end if**
$\qquad$**end for**
$\quad$**end for**

also change the random block R at each iteration. The most likely valid two-block plaintext to obtain at random is

$$\underbrace{x_0 x_1 \ldots x_{n-2} 1}_{n} \, || \, \underbrace{(L_D = n - 1)_2}_{n}$$

where each x_i can be either '0' or '1.' A valid two-block plaintext is guaranteed to occur if we traverse through all 2^{n+1} possible settings of the second plaintext block along with rightmost bit of the first plaintext block (by, respectively, changing R and the rightmost bit of IV'), so on average this strategy has a complexity of 2^n oracle calls.

This special case is illustrated in Algorithm 10118-1-m3-a1-special above.

DECRYPTING. Once we have a valid padding we can employ Attack 2 below with input a valid ciphertext of the form $IV'||C_k$ (when $r \le n-1$) or $IV'||R||C_k$ (when $r = n$).

We consider first the case $r \le n - 1$. Here the plaintext corresponding to the ciphertext submitted to Attack 2 will always be same-block padded (because it only contains one block). Then Attack 2 will efficiently recover the entire last plaintext block for this ciphertext, which we denote by P_1'. P_1' will in general consist of data bits, padding bits and length information. From P_1', it is trivial to recover P_k (the plaintext block that we are actually after). We have:

$$P_k = P_1' \oplus IV' \oplus C_{k-1}.$$

For the case $r = n$, the first phase of Attack 2 below efficiently recovers the length of the unpadded data for the valid ciphertext $IV'||R||C_k$. This information is contained in the length field which occupies all of P_2', the second plaintext block for this ciphertext. Thus after the first phase of Attack 2, P_2' is known. Now P_k can be recovered from:

$$P_k = P_2' \oplus R \oplus C_{k-1}.$$

COMPLEXITY. Obtaining a valid plaintext block takes on average $2^{r-1} + 2^{2r-n+1}$ oracle calls when $r \le n - 1$ and on average 2^n oracle calls in the case $r = n$. Our use of Attack 2 below has a complexity of $n + O(\log_2 n)$ oracle calls for all values of r (recall that for $r = n$, only the first phase of Attack 2 is needed, while for $r \le n - 1$, the plaintext is same-block padded in which case Attack 2 is efficient). Thus our use of Attack 2 does not contribute significantly to the overall complexity to decrypt a single block.

IMPACT. This attack applies to any ciphertext block and all n bits within the block are recovered. For many choices of r this attack is many orders faster than an exhaustive key search, and for a small enough r this attack will be practical whenever a padding oracle is available. When $r = n$, our attack is still better than an exhaustive key search for block ciphers whose key size is greater than the block length. It is interesting to note that the parameter r seemingly innocent of any security implications turns out not to be so at all.

Attack 2: Attacking the last block(s). This attack is conceptually similar to the one against padding method 3 of ISO/IEC 9797-1 given above: there are two phases, the first of which determines L_D and the second of which recovers any plaintext that is found in a "mixed" block – that is, a block that consists of both data and padding bits. There is obviously at most one such block in any plaintext padded using this padding method, which is either the last block or the one that immediately precedes it. If the padding ends exactly on a block boundary, then our attack does not recover any (unpadded) plaintext.

OBTAINING L_D. We want to know L_D, the data length. For ease of presentation we first examine the case $r \le n - 2$, but our algorithm to follow handles all values of r. Here, in the same-block padded case, the last plaintext block P_q corresponding to the last ciphertext block C_q in the ciphertext $IV||C_1||C_2||\dots||C_q$ has a format as follows:

$$\underbrace{[DATA]}_{t} \underbrace{10\dots0}_{p} \underbrace{(L_D)_2}_{r}$$

where $t + p + r = n$ and $p \ge 1$. In the cross-block padded case, the above format spans the last two blocks P_{q-1} and P_q and we put $t + p + r = 2n$. We note that the attacker does not at first know which of the cases he is faced with.

Given a q-block ciphertext, we want to flip the plaintext bit $P_{q,n-r-2}$, the rightmost position at which a data bit could ever reside, given q and our assumption on r. We submit to the padding oracle the ciphertext

$$IV||C_1||C_2||\ldots||C_{q-1} \oplus \underbrace{0\ldots0}_{n-r-2}10\underbrace{0\ldots0}_{r}||C_q.$$

(Recall that $C_0 = IV$, so the case where $q = 1$ is included here.)

Upon submission of the above ciphertext, the oracle will return:

- VALID meaning the padding has not been disturbed so the bit flipped is a data bit. Since this bit is at the rightmost possible data bit position, we can deduce the data length $L_D = (q-1)n + n - r - 1$. Or else,
- INVALID meaning a padding bit has been flipped so the padding is no longer valid. Therefore the padding boundary is somewhere to the left of this bit, so we continue by resetting this bit and flipping the bit immediately to the left, and test the resulting ciphertext for padding correctness. We repeat this, flipping bits further and further to the left (and into the previous block if necesssary) until the first time the oracle returns VALID. This indicates that the tested bit is the last data bit, and L_D is determined accordingly.

One might worry about instances when cross-block padding arises, where flipping bits in the last plaintext block (by flipping bits in the last-but-one ciphertext block) would turn the last-but-one plaintext block into "garbage" and along with it, potentially, any padding bits within it, so the oracle might report INVALID for the wrong bits. On closer inspection, however, this turns out not to be an issue because all we want to know is whether the padding boundary is to the left or right of the bit in question. Even if the oracle does report INVALID for the wrong bits, it does still imply the boundary is to the left, and VALID would just mean that unpadded data bits have been corrupted so the boundary is still to the right.

A binary search can also be applied here: for any single flipped bit, a VALID response means the start of the padding is to the right of this bit, whereas INVALID means it is to the left. This speed-up is made in Algorithm10118-1-m3-a2-get-L_D below.

We are now ready for the decryption stage. Same-block and cross-block padded messages are treated differently; recall that knowledge of L_D indicates which case the attacker is faced with.

DECRYPTING: SAME-BLOCK. Recall the structure of the last plaintext block: t data bits, followed by p padding bits in the form $10\ldots0$ and finally r bits of an encoding of data length L_D. We can recover the remaining t bits of the plaintext in the last block, again using a similar method to decryption phase of the attack on ISO/IEC 9797-1 method 3. We submit to the oracle $IV'||C_1'$ where $C_1' = C_q$ and

$$IV' = C_{q-1} \oplus \underbrace{0\ldots0}_{n-r}\underbrace{(L_D)_2}_{r} \oplus \underbrace{0\ldots0}_{t}\underbrace{10\ldots0}_{p}\underbrace{(t-1)_2}_{r}.$$

Algorithm 10118-1-m3-a2-get-L_D

Input: $IV||C_1||C_2||\ldots||C_q, n, r$
Output: L_D

$C := IV||C_1||C_2||\ldots||C_q$
$l := (q-2)n + n - r$
$u := (q-1)n + n - r - 1$
repeat
   $h := \lfloor (l+u)/2 \rfloor$
   $C_{\lfloor h/n \rfloor, h \bmod n} := C_{\lfloor h/n \rfloor, h \bmod n} \oplus 1$
   if oracle(C) = VALID **then**
      l := h+1
   else if oracle(C) = INVALID **then**
      u := h
   end if
   $C_{\lfloor h/n \rfloor, h \bmod n} := C_{\lfloor h/n \rfloor, h \bmod n} \oplus 1$
until $l = u$
return $L_D := l$

After decryption the length block in the plaintext block P_1' should have the value $t - 1$ which points to the last-but-one bit of the original data sub-block, with the middle padding sub-block being all '0's. A VALID response means the last (t^{th}) data bit in P_1' is a '1,' and '0' otherwise.

By decrementing the length field sub-block in P_1' one by one whilst keeping all recovered bit positions '0,' a single bit is revealed at each iteration until the whole block is recovered. We can now compute the intermediate block I_1' by XORing the final IV with D_1', and then by XORing I_1' with C_{q-1} we get the original last plaintext block.

This decryption procedure is presented in Algorithm 10118-1-m3-decrypt-same-block below.

DECRYPTING: CROSS-BLOCK. For cross-block padded plaintexts, P_q is determined completely by L_D and the padding. However, the padding extends into the penultimate plaintext block P_{q-1}. Suppose u bits of padding are present in P_{q-1}. Then we show how to decrypt C_{q-1} using Attack 1 above, but with a speed-up factor of 2^{u-1}.

Let $v = L_D \bmod n$, then the number of known plaintext bits u is equal to $n - v$ and those bits are of the form $\underbrace{10\ldots0}_{u}$. If we submit the ciphertext $IV'||C_1'$ to the oracle where

$$IV' = C_{q-2} \oplus \underbrace{0\ldots0}_{n-u}\underbrace{10\ldots0}_{u} \oplus \underbrace{0\ldots0}_{n-r}\underbrace{(n-r-1)_2}_{n-r}$$

Algorithm 10118-1-m3-decrypt-same-block

Input: $L_D, IV, C_{q-1}, C_q, r, n$
Output: $P_q := P_{q,0} P_{q,1} \dots P_{k,t-1} \underbrace{10 \dots 0}_{p} \underbrace{(L_D)_2}_{r}$

Ensure: L_D indicates that the plaintext is same-block padded
$\quad C_1' = C_q$
$\quad t := L_D \bmod n$
$\quad$ **for** $j := t - 1$ to 0 **do**
$\quad\quad IV' := C_{q-1} \oplus \underbrace{0 \dots 0}_{n-r} \underbrace{(L_D)_2}_{r} \oplus \underbrace{0 \dots 0}_{t} \underbrace{10 \dots 0}_{p} \underbrace{(j)_2}_{r}$
$\quad\quad C' := IV' \| C_1'$
$\quad\quad b := \Omega(C') \oplus 1$
$\quad\quad P_{q,j} := b \oplus IV_j' \oplus C_{q-1,j}$
$\quad\quad IV_j' := IV_j \oplus b$
$\quad$ **end for**
$\quad$ **return** $P_q := P_{q,1} P_{q,2} \dots P_{k,t} \underbrace{10 \dots 0}_{p} \underbrace{(L_D)_2}_{r}$

and $C_1' = C_{q-1}$, then we only need to go through all 2^{r-u+1} settings of the $r-u+1$ bits to the left of the u known bits (by changing IV') to guarantee a valid plaintext. This strategy takes on average 2^{r-u} oracle calls which is a fraction $2^{-(u-1)}$ of the original 2^{r-1} oracle calls for Attack 1 without the knowledge of the u padding bits.

COMPLEXITY. It takes $\log_2 n$ oracle calls to find L_D. For same-block padded plaintexts, it takes one call per bit for decrypting. So to recover the t data bits of the last block, $t + \log_2 n$ oracle calls are required.

For cross-block padded plaintexts, on average 2^{r-u} oracle calls are needed to recover the whole of the penultimate plaintext block P_{q-1}, where u is the number of known bits from finding L_D.

IMPACT. The attack is highly efficient in terms of oracle queries at extracting plaintext bits from the last plaintext block P_q. A maximum of $n - r - 1$ bits of data can be recovered in this way and the attack is therefore significant for short messages, especially in combination with a small r. One might argue that $r = n$ is a natural choice for the implementor. In this case, the padding is always cross-block and the attacker must resort to the speeded-up version of Attack 1.

5   Conclusions

We argue that, at least for the CBC-mode of operation for a block cipher standard, it is not good enough just to standardise the mode; an entire specification handling bit-level computations is needed, which necessarily includes padding issues. Padding methods devised for hashing or MACs, as we have shown, may not be suited to encryption operations where a different adversarial model may be applicable.

We also make the point that there is a need for careful consideration of the potential for side-channel cryptanalysis for cryptographic primitives and security protocols in their design phase. Designs should be fully specified so as to allow as little room as possible for the implementor to take potentially weak approaches during implementation.

We agree with the argument in Section 7 of [1] for the practice of the encryption being accompanied by strong integrity checks when possible and appropriate. Such "authenticated encryption" would, within the context of this paper, prevent any practical attempts at constructing a valid ciphertext which in turn precludes the existence of a padding oracle, and hence all the associated attacks that we have discovered.

Acknowledgement. We thank Alain Hiltgen for useful comments on the paper and for showing us how to extract the leftmost bits of plaintext in Section 3.4.

References

1. J. Black and H. Urtubia. Side-Channel Attacks on Symmetric Encryption Schemes: The Case for Authenticated Encryption. *Proceedings of the 11th USENIX Security Symposium, San Francisco, CA, USA, August 5-9, 2002,* pp. 327–338, 2002.
2. B. Canvel, A. Hiltgen, S. Vaudenay, and M. Vuagnoux. Password Interception in a SSL/TLS Channel. In *Proc. CRYPTO 2003,* D. Boneh (ed.), LNCS Vol. 2729, pp. 583–599, 2003.
3. ISO/IEC 9797-1: Information technology — Security tehniques — Message Auhentication Codes (MACs) — Part 1: Mechanisms using a block cipher. 1999.
4. ISO/IEC 10116 (2nd edition): Information technology — Security techniques — Modes of operation for an n-bit block cipher. 1997.
5. ISO/IEC 3rd CD 10116 (3rd edition): Information technology — Security techniques — Modes of operation for an n-bit block cipher (Commitee Draft). 2002.
6. ISO/IEC FDIS 10118-1: Information technology — Security techniques — Hashfunctions — Part 1: General (Final Draft). 2000
7. V. Klima and T. Rosa. Side Channel Attacks on CBC Encrypted Messages in the PKCS#7 Format. Cryptology ePrint Archive, Report 2003/098, 2003.
8. S. Vaudenay. Security Flaws Induced by CBC Padding — Applications to SSL, IPSEC, WTLS ... . In *Proc. EUROCRYPT'02,* LNCS Vol. 2332, pp. 534–545, 2002.

A 1 Gbit/s Partially Unrolled Architecture of Hash Functions SHA-1 and SHA-512

Roar Lien, Tim Grembowski, and Kris Gaj

ECE Department, George Mason University, 4400 University Drive, Fairfax, VA 22030
kgaj@gmu.edu

Abstract. Hash functions are among the most widespread cryptographic primitives, and are currently used in multiple cryptographic schemes and security protocols, such as IPSec and SSL. In this paper, we investigate a new hardware architecture for a family of dedicated hash functions, including American standards SHA-1 and SHA-512. Our architecture is based on unrolling several message digest steps and executing them in one clock cycle. This modification permits implementing majority of dedicated hash functions with the throughput exceeding 1 Gbit/s using medium-size Xilinx Virtex FPGAs. In particular, our new architecture has enabled us to speed up the implementation of SHA-1 compared to the basic iterative architecture from 544 Mbit/s to 1 Gbit/s using Xilinx XCV1000. The implementation of SHA-512 has been sped up from 717 to 929 Mbit/s for Virtex FPGAs, and exceeded 1 Gbit/s for Virtex-E Xilinx FPGAs.

1 Introduction

Hash functions are very common and important cryptographic primitives. Their primary application is their use for message authentication, integrity, and non-repudiation as a part of the Message Authentication Codes (MACs) and digital signatures [1].

The current American federal standard, FIPS 180-2, recommends the use of one of the four hash functions developed by National Security Agency (NSA) and approved by NIST. By far the most widely used of these four functions is SHA-1 (Secure Hash Algorithm-1), a revised version of the standard algorithm introduced in 1993. The best attack against this algorithm is in the range of 2^{80} operations, which makes its security equivalent to the security of Skipjack and the Digital Signature Standard (DSS). After introducing a new secret-key encryption standard, AES (Advanced Encryption Standard), with three key sizes, 128, 192, and 256 bits, the security of SHA-1 did not any longer match the security guaranteed by the encryption standard. Therefore, an effort was initiated by NSA to develop three new hash functions, with the security equivalent to the security of AES with 128, 192, and 256 bit key respectively. This effort resulted in the development and standardization of three new hash functions referred to as SHA-256, SHA-384, and SHA-512 [1].

T. Okamoto (Ed.): CT-RSA 2004, LNCS 2964, pp. 324–338, 2004.

All four standardized algorithms have a similar internal structure and operation. All of them are based on sequential processing of consecutive blocks of data, and therefore cannot be easily sped up by using pipelining or parallel processing (at least when only one stream of data is being processed).

The majority of reported implementations of SHA-1 based on the current generation of FPGA devices, such as Virtex [2], can only reach the throughputs up to 500 Mbit/s [3-9]. The higher speeds can only be accomplished by using more expensive FPGA devices, such as Virtex-E or Virtex II (see Table 1). Similarly, the FPGA implementations of SHA-512 based on the medium cost Virtex devices reach the speeds in the range of 700 Mbit/s [3, 4].

Significantly higher speeds might be required for applications such as High Definition Television (HDTV), videoconferencing, Virtual Private Networks, etc. [10]. Our goal was to propose, implement, and verify a new architecture of standard hash functions that would allow them to be executed with the throughputs in the range of 1 Gbit/s using medium cost FPGA devices, such as Xilinx Virtex 1000.

2 Hardware Architectures of Hash Functions

A general block diagram common for all four SHA standards and many other dedicated hash functions is shown in Fig. 1. An input message passes first through the preprocessing unit which performs padding and forms message blocks of the fixed length, 512 or 1024 bits, depending on the hash function. The preprocessing unit passes message blocks to the message scheduler unit. Message scheduler unit generates message dependent words, W_t, for each step of the message digest. The message digest unit performs actual hashing. In each step, it processes a new word generated by the message scheduler unit. The message digest is the most critical part of the implementation, as it determines both the speed and area of the circuit.

The most straightforward implementation of the message digest, most often used in practice is shown in Fig. 2a. It is called the basic iterative architecture (or just basic architecture). In this architecture, registers R and H are first both initialized with a value of the constant initialization vector, IV. Subsequently, the architecture executes one step of the message digest per one clock period. In each step t, the message digest accepts a different message dependent word, W_t, and a different step dependent constant, K_t. After executing all steps, the result of the last step, stored in the register R, is added to the previous value of the register H. Then, the processing of the message di-

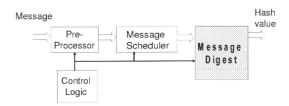

Fig. 1. General block diagram of the hardware implementation of a dedicated hash function, such as SHA-1 and SHA-512

a)
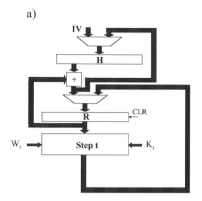

b)
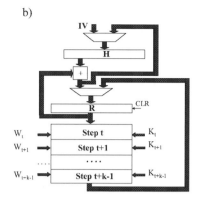

Fig. 2. General diagrams of the message digest units for a) basic architecture, b) partially unrolled architecture with k steps unrolled

gest resumes for a new set of the message dependent words, W_t, corresponding to the new block of the message.

Two straightforward ways of speeding up hardware implementations of hash functions (and any other logic functions) are parallel processing using multiple instantiations of the basic architecture, and pipelining. Out of these two methods, pipelining is more attractive because of the smaller area penalty. Nevertheless, both of these architectures are able to improve an average circuit throughput only under the assumption that multiple independent streams of data are processed simultaneously. If a single long message needs to be hashed, none of these architectures offers *any* improvement in terms of the execution time.

A new architecture of the dedicated hash functions investigated in this paper is shown in Fig. 2b. It is called *partially unrolled architecture*. In this architecture, k steps have been "unrolled" and are executed in the same clock cycle. As a result, the total number of clock cycles necessary to compute one iteration of the message digest has been reduced by a factor of k. At the same time, the critical path through k steps is likely to be significantly shorter than k times the path through a single step. This is because in hash functions, the critical path through a step of the message digest is different for each word of the step input (see Fig. 3).

3   Previous Work

Fully and partially unrolled architectures of dedicated hash functions have been investigated by several authors in the past, but no definite conclusions have been made. In [11] a fully unrolled architecture of MD5 has been compared with a basic iterative architecture. Unrolling of all 64 rounds resulted in a throughput increase by a factor of 2.1, while at the same time the circuit area increased by a factor of 5.4. In [12] a partially unrolled architecture of SHA-1, with the number of rounds unrolled k=5, has been investigated. A high level architecture presented in this paper was very similar to the one proposed in this paper. Nevertheless, the reported results were rather discouraging, with only 11% gain in the circuit throughput and a 43% penalty in the aging, with only 11% gain in the circuit throughput and a 43% penalty in the circuit area for the partially unrolled architecture over the basic iterative architecture.

All other hardware implementations of dedicated hash functions reported in the literature [9, 13, 14] or available as commercial IP cores [3-8] have followed the basic iterative architecture with only one step of hash function executed in each clock cycle.

4 Details of the Hardware Architectures

4.1 Internal Structure of the Message Digests of SHA-1 and SHA-512

Internal structures of the message digests for SHA-1 and SHA-512 are shown in Fig. 3. In both functions, input registers are initialized with the constant initialization vector, and are updated with the new value in each round. In SHA-1, four out of five words (A, B, C, and D) remain almost unchanged by a single round. These words are only shifted by one position down. The last word, E, undergoes a complicated transformation equivalent to multioperand addition modulo 2^{32}, with five 32-bit operands dependent on all input words, the round-dependent constant K_t, and the message dependent word W_t. The internal structure of the message digest of SHA-512 is similar. The primary differences are as follows: The number of words processed by each round is 8, each word is 64 bits long, and the longest path is equivalent to addition of seven 64-bit operands modulo 2^{64}. These operands depend on seven out of eight input words (all except D), the round-dependent constant K_t, and a message dependent word W_t. Six out of eight input words remain unchanged by a single round.

4.2 Basic Architecture of SHA-1

From Fig. 3a, the critical path of a single SHA-1 round involves the calculation of the chaining variable A at the moment t+1, given by the following formula:

$$A_{t+1} = A_t <<<5 + f_t(B_t, C_t, D_t) + E_t + K_t + W_t + HA'_t$$

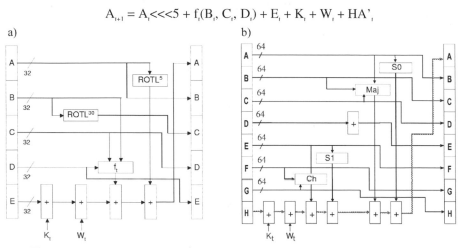

Fig. 3. Internal structure of a single message digest round of a) SHA-1, b) SHA-512

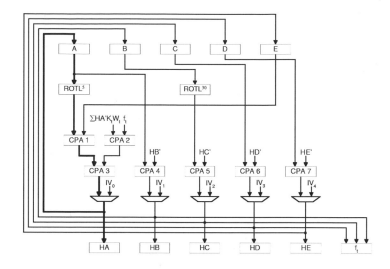

Fig. 4. Our implementation of the message digest unit of SHA-1 in the basic iterative architecture

where X_t is a value of the variable X in the step t, and $HA'_t = HA$ when t=79, otherwise 0. HA is a word A of the register H in Fig. 2a.

Additionally, we know that

$$B_t = A_{t-1}, \quad C_t = B_{t-1} <<< 30, \quad D_t = C_{t-1}.$$

None of these operations involve any logic, consequently, the expression

$$f_t(B_t, C_t, D_t) = f_t(A_{t-1}, B_{t-1} <<< 30, C_{t-1})$$

can be precomputed in the previous clock cycle, t-1, and will not contribute to the critical path. Similarly, the sum

$$\Sigma_{HA'\,Kt\,Wt} = K_t + W_t + HA'_t$$

can be precomputed by the message scheduler unit, because all values are known already in the previous clock cycle.

As a result, the critical path reduces to the addition of four operands

$$A_{t+1} = A_t <<< 5 + E_t + \Sigma_{HA'\,Kt\,Wt} + f_t(A_{t-1}, B_{t-1} <<< 30, C_{t-1}).$$

All aforementioned optimizations lead to the schematic of the basic architecture of SHA-1 shown in Fig. 4. The lowest level multiplexers choose initialization vectors IV_0 to IV_4 only in the first clock cycle of computations for any new message. The variables HB'.. HE' are equal to HB..HE only in the last step of the message digest computations for a given message block, i.e., only when t=79; otherwise, they are equal to zero.

4.3 Partially Unrolled Architecture of SHA-1

The optimization of the unrolled message digest is relatively straightforward. The general technique employed is to precalculate sums at the earliest possible stage using either regular carry propagate adders (CPAs) or carry save adders (CSAs) (see Fig. 5). The calculations in the critical path follow a sequence of computations described by the equations below:

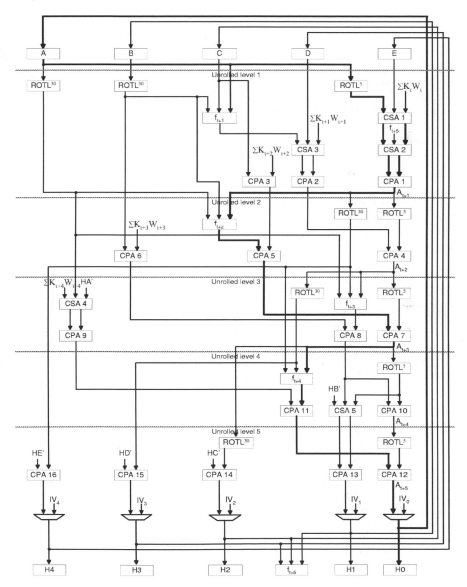

Fig. 5. Our implementation of the message digest unit of SHA-1 in the partially unrolled architecture with 5 steps unrolled

$A_{t+1} = A_t{<<<}5 + f_t(B_t, C_t, D_t) + E_t + K_t + W_t = A_t{<<<}5 + f_t(B_t, C_t, D_t) + E_t + \sum K_t W_t$

$A_{t+2} = A_{t+1}{<<<}5 + f_{t+1}(B_{t+1}, C_{t+1}, D_{t+1}) + E_{t+1} + K_{t+1} + W_{t+1} =$

$\qquad = A_{t+1}{<<<}5 + [f_{t+1}(A_t, B_t{<<<}30, C_t) + D_t + \sum K_{t+1} W_{t+1}]$

$A_{t+3} = A_{t+2}{<<<}5 + [f_{t+2}(A_{t+1}, A_t{<<<}30, B_t{<<<}30) + [C_t + \sum K_{t+2} W_{t+2}]]$

$A_{t+4} = A_{t+3}{<<<}5 + [f_{t+3}(A_{t+2}, A_{t+1}{<<<}30, A_t{<<<}30) + [B_t{<<<}30 + \sum K_{t+3} W_{t+3}]]$

$A_{t+5} = A_{t+4}{<<<}5 + [f_{t+4}(A_{t+3}, A_{t+2}{<<<}30, A_{t+1}{<<<}30) + [A_t{<<<}30 + \sum K_{t+4} W_{t+4} + HA'_{t+4}]]$.

At each stage two paths are critical. One is a calculation of the new value of A_{t+i} (i=1..5), which involves rotation by five positions and a single addition. The second is the precalculation of the value of $[f_{t+i} + [E_{t+i} + \sum K_{t+i} W_{t+i}]]$ to be used in the next stage. This precalculation involves the calculation of f_{t+i} and a single addition of a precalculated value $[E_{t+i} + \sum K_{t+i} W_{t+i}]$.

In the first stage of computations (computing A_{t+1}), precalculated values do not exist, so the computations must be performed from scratch. In every second stage starting from stage two, the precomputation of the sum $[f_{t+i} + [E_{t+i} + \sum K_{t+i} W_{t+i}]]$ is the most time consuming operation. Finally, in every second stage starting from stage three, the only contribution to the critical path is a single addition.

4.4 Basic Architecture of SHA-512

From Fig. 3b, the critical path of a single SHA-512 round involves the calculation of the chaining variable A at the moment t+1, given by the following formula:

$$A_{t+1} = S0(A_t) + Maj(A_t, B_t, C_t) + S1(E_t) + Ch(E_t, F_t, G_t) + K_t + W_t + H_t + HA_t'$$

where X_t is a value of the variable X in the step t; S0, Maj, S1, Ch are the logic functions defined in the SHA-512 standard, and $HA'_t = HA$ when t=79, otherwise 0.

Additionally, we know that

$$H_t = G_{t-1}.$$

The functions S0 and Maj execute in parallel in approximately the same amount of time. The same holds true for functions S1 and Ch.

The sum

$$KWHA_t = K_t + W_t + G_{t-1} + HA'_t$$

can be precomputed in the previous clock cycle, t-1.

As a result, the critical path reduces to the addition of five operands

$$A_{t+1} = S0(A_t) + Maj(A_t, B_t, C_t) + S1(E_t) + Ch(E_t, F_t, G_t) + KWHA_t.$$

All aforementioned optimizations lead to the schematic of the basic architecture of SHA-512 shown in Fig. 6. The registers HA-HH are set to the initialization vectors IV_0 to IV_7 only in the first clock cycle of computations for any new message. The multiplexers selecting between HB and '0', HC and '0', etc. choose non-zero values only in the last step of the message digest computations for a given message block, i.e., only when t=79.

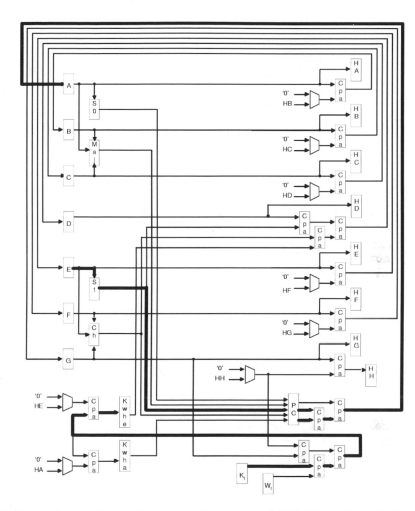

Fig. 6. Our implementation of the message digest unit of SHA-512 in the basic iterative architecture (PC – a 5-to-3 parallel counter, see [9])

4.5   Unrolled Architecture of SHA-512

The unrolled architecture of SHA-512 is shown in Fig.7. Because of the dependence of E_{t+1} on E_t, and A_{t+1} on A_t and E_t (see Fig. 3b), three major critical paths (A0 to A0, E0 to A0 and E0 to E0) exist in the circuit. These paths are marked in Fig. 7 with thicker lines. Values of variables A_{t+i}, and E_{t+i} are denoted as "Ai" and "Ei" respectively, e.g., "E2" denotes E_{t+2}. Precomputations in the previous clock cycle are used to reduce the number of operands in the first four stages of the unrolled architecture. Recall that in the basic architecture, the KWHA$_t$ sum is computed based on the equation H$_t$ = G$_{t-1}$. In the unrolled architecture with k=5, t changes by 5 every clock cycle As a result, H$_t$ = G$_{t-1}$ = F$_{t-2}$ = E$_{t-3}$ = E$_{t+2-5}$ = "E2" in the previous clock cycle.

332   R. Lien, T. Grembowski, and K. Gaj

On the far left side of Fig. 7, "E2" is used to precompute KWH0 (notation for KWHA$_{t+0}$) for the next clock cycle.

$$KWH0 = KWHA_t = K_t + W_t + H_t + HA'_t$$

This method is repeated in stages two to four in order to compute KWHA$_{t+i}$ (denoted in Fig. 7 as KWHi, i=1..3). In stage 5, H$_{t+4}$ = E$_{t+1}$ = "E1", so this value is computed in the same clock cycle, and as a result is not included in the earlier precomputed KWH4 = KWHA$_{t+4}$, which reduces to KWHA$_{t+4}$ = K$_{t+4}$ + W$_{t+4}$. Please, note that in Fig. 7, the sum K$_{t+i}$ + W$_{t+i}$ is denoted as KWi.

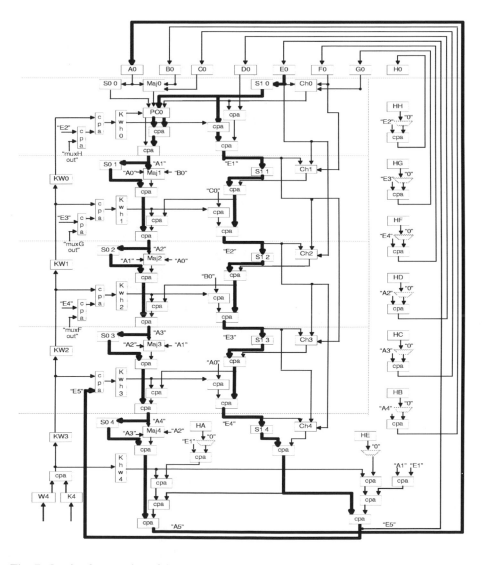

Fig. 7. Our implementation of the message digest unit of SHA-512 in the partially unrolled architecture with 5 steps unrolled

Further reductions in critical paths were accomplished in each stage by adding values of logic functions S1 and Ch as early as possible, reusing values of S1 + Ch, and by selective routing to balance the number of slices in various critical paths.

5   Design Methodology and Results

Our target FPGA device was the Xilinx Virtex XCV1000-6. This device is composed of 12,288 basic logic cells referred to as CLB (Configurable Logic Block) slices, includes 32 4-kbit blocks of synchronous dual-ported RAM, and can achieve synchronous system clock rates up to 200 MHz [2]. XCV1000 was chosen because of the availability of a general purpose PCI board, SLAAC-1V, based on three FPGA devices of this type [10]. Additionally, a new family of Virtex-E Xilinx devices was targeted as well.

All hardware architectures were first described in VHDL, and their operation verified through functional simulation using Active HDL, from Aldec, Inc. Test vectors and intermediate results from the reference software implementations based on the Crypto++ library [15] were used for debugging and verification of VHDL codes. The revised VHDL code became an input to logic synthesis performed using FPGA Compiler II from Synopsys. Tools from Xilinx ISE 4.2 were used for mapping, placing, and routing. These tools generated reports describing area and speed of implementation, a netlist used for timing simulation, and a bitstream used to configure an actual FPGA device. All designs were fully verified through behavioral, post-synthesis, and timing simulations.

The experimental testing of our cryptographic modules was performed using the SLAAC-1V hardware accelerator board, including three Virtex 1000 FPGAs as the primary processing elements. Only one of the three FPGA devices was used to implement hash core.

Test program written in C used the SLAAC-1V APIs and the SLAAC-1V driver to communicate with the board. Our testing procedure is composed of three groups of tests. The first group verifies the circuit functionality at a single clock frequency. The goal of the second group is to determine the maximum clock frequency at which the circuit operates correctly. Finally, the purpose of the third group is to determine the limit on the maximum encryption and decryption throughput, taking into account the limitations of the PCI interface.

In Fig. 8, the minimum clock periods of SHA-1 and SHA-512 obtained using static timing analysis and the experiment are given. For the unrolled architecture, the effective clock period is the minimum time necessary for the data signals to pass the critical path. Since in both our unrolled designs, the data signal is traveling through the critical path over multiple clock periods, the effective clock period is a multiple of the actual clock period. In case of the unrolled architecture for SHA-1 the multiplication factor is 2, in case of the SHA-512 architecture, the multiplication factor is 5.

Based on the knowledge of the minimum clock period, the maximum data throughput has been computed according to the equation:

Throughput=Message_block_size / (Effective_clock_period * Number_of_rounds/k)

The maximum throughput values calculated based on the minimum clock periods obtained using static timing analysis and experiment are shown in Fig. 9. In the same figure, these results are compared with the experimentally measured data throughputs that take into account the delay contributions and the bandwidth limit of the PCI interface. This comparison demonstrates that the PCI interface is capable of operating with a constant uninterrupted data flow up to about 960-990 Mbit/s, and has a negligible influence on the data throughput below this communication rate.

The number of CLB Slices used by our implementations of SHA-1 and SHA-512 are shown in Tables 1 and 2. In SHA-512, four 4 kbit block RAMs are used to store 80 64-bit constants K_t.

Out of the two analyzed hash standards, SHA-1 offers much better potential for loop unrolling. As a result of loop unrolling, the throughput of SHA-1 increased by a factor of almost two (1.9 times), while at the same time its area grew only by a factor of three. SHA-512 is much less suitable for loop unrolling, as its observed speed-up was only 30%, and the area increase 48%.

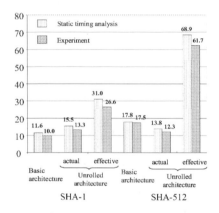

Fig. 8. Minimum clock periods of SHA-1 and SHA-512 in the basic iterative architecture and partially unrolled architecture

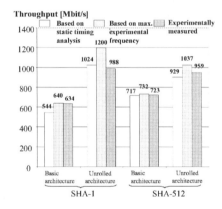

Fig. 9. Maximum throughputs of SHA-1 and SHA-512 in the basic iterative architecture and partially unrolled architecture

6   Comparison with Other Hash Cores

There exist multiple commercial IP cores implementing SHA-1 [3-8]. In Table 2, we present the comparison of our designs for SHA-1 with the most representative IP cores with equivalent functionality. For the Xilinx Virtex family of FPGA devices, our core for SHA-1 in the basic iterative architecture outperforms the second best core (from Helion Technology Ltd) by 13%, using 30% less CLB slices. Our core for the partially unrolled architecture of SHA-1 with 5 rounds unrolled, outperforms all reported Virtex cores by a factor of at least two in terms of throughput, and uses about two times more area. The similar advantages exist for the implementations using

Virtex-E devices, where our core for the unrolled architecture approaches the throughput of 1.2 Gbit/s.

Table 1. Comparison of our designs for SHA-1 with the representative commercial IP cores with equivalent functionality

Source	Clock frequency [MHz]	Throughput [Mbit/s]	Area [CLB Slices]
Xilinx Virtex			
Our, basic	**85**	**544**	**480**
Our, unrolled (k=5)	**64[1]**	**1024**	**1480**
ALMA Technologies	70	442	686
Helion Technology Ltd.	76	480	689
Ocean Logic Pty Ltd	56	352	612
Xilinx Virtex-E			
Our, basic	**103**	**659**	**484**
Our, unrolled (k=5)	**72.5**	**1160**	**1484**
ALMA Technologies	87	549	686
Bisquare Systems Private Limited	66	422	579
Helion Technology Ltd.	95	600	689
Intron, Ltd.	71	449	716
Ocean Logic Pty Ltd	71.5	452	612
Xilinx Virtex-II			
ALMA Technologies	102	644	686
Amphion Semiconductor	99	626	854
Helion Technology Ltd.	103.5	654	569
Ocean Logic Pty Ltd	79	498	612

Table 2. Comparison of our designs for SHA-512 with the representative commercial IP cores with equivalent functionality

Source	Clock frequency [MHz]	Throughput [Mbit/s]	Area [3] [CLB Slices]
Xilinx Virtex			
Our, basic	**56**	**717**	**2384 Slices**
Our, unrolled (k=5)	**67[2]**	**929**	**3521 Slices**
ALMA Technologies	56	707	2690 Slices
Xilinx Virtex-E			
Our, unrolled (k=5)	**72[2]**	**1034**	**3517 Slices**
ALMA Technologies	68	859	2690 Slices
Xilinx Virtex-II			
ALMA Technologies	72	910	2507 Slices
Amphion Semiconductor	50	626	2403 Slices

[1] multi-cycle clock used in the critical path, critical path $\leq 2\,T_{CLK} = 2/f_{CLK}$, 5 steps executed in 2 clock cycles

[2] multi-cycle clock used in the critical path, critical path $\leq 5\,T_{CLK} = 5/f_{CLK}$, 5 steps executed in 5 clock cycles;

[3] each circuit contains additionally 4 Block RAMs

At this point, there are relatively few cores available for the new standard, SHA-512 (see Table 2) [3, 4]. Our implementation of the basic iterative architecture slightly outperforms the equivalent core from ALMA Technologies in terms of throughput, using a smaller amount of FPGA resources. Our partially unrolled architecture is the fastest core for the Virtex family of FPGA devices outperforming the second best core by 30% at the cost of only 31% increase in the circuit area. For the Virtex-E family of FPGA devices our core is the only currently available SHA-512 core that exceeds the throughput of 1 Gbit/s.

7   Comparison with Software Implementations

Efficient software implementations of hash functions have been extensively studied in the literature [17-20]. In [17], basic recommendations on developing an efficient and portable implementation of SHA-1 in C have been formulated. In [18], a close to optimum implementations of dedicated hash functions using Pentium's superscalar architecture have been presented. In [19], software parallelism of all major dedicated hash functions have been studied. Finally, in [20], optimizations targeting Pentium III have been investigated. These optimizations made use of MMX registers and instructions available in Pentium III.

In this paper, we used for comparison, software implementations of SHA-1 and SHA-512, available as a part of the Crypto++ library [15]. Although Crypto++ is not the fastest of the reported software implementations, the reason for using this library was its portability, availability in public domain, and wide practical deployment.

A PC with 2.2 GHz clock, 1 GByte RAM, and cache size 512KB, running Windows XP was used in our measurements. The Crypto++ implementation of hash functions written in C++ was compiled using MS Visual Studio with Service Pack 5. The obtained throughput was 40.5 Mbit/s for SHA-1 and 30.4 Mbit/s for SHA-512. These throughputs were respectively 25 times and 31 times smaller than the throughputs of our partially unrolled hardware implementations of SHA-1 and SHA-512 for Xilinx Virtex 1000-6 FPGAs.

8   Summary

A new *partially unrolled* architecture has been proposed for a family of dedicated hash functions, including four American standard algorithms SHA-1, SHA-256, SHA-384, and SHA-512. The unrolled architecture has been designed, optimized, and experimentally verified for the most widely used hash algorithm, SHA-1, and one of the new hash standard algorithms SHA-512. For the purpose of comparison, the basic iterative architecture has been implemented for both functions as well.

The new architecture appeared to be particularly suitable for the implementation of SHA-1. For the number of rounds unrolled equal to k=5, it allowed to almost double the throughput of SHA-1 compared to the basic iterative architecture, at the cost of increasing circuit area by a factor of three. The similar design for SHA-512 appeared to have much less benefit; the increase in the circuit throughput was only 30%, and the area of the circuit increased by 48%.

This different behavior of two hash algorithms could be easily explained by analyzing the structure of both algorithms. In the unrolled architecture of SHA-1, many message digest steps could be substantially sped up by preprocessing partial results of a given step in the previous steps. The same optimization was not possible in SHA-512 due to sequential dependencies present in the algorithm.

Our partially unrolled implementation of SHA-1 reached the target throughput of 1 Gbit/s in Virtex XCV1000, and outperformed all known to the authors commercial IP cores with equivalent functionality by at least a factor of two. Our implementation of SHA-512 also compared favorably with commercial IP cores, and reached a target throughput of 1 Gbit/s using Virtex-E family of Xilinx FPGAs. To our best knowledge, our implementations of SHA-1 and SHA-512 are the only FPGA implementations of these hash functions available to date that can sustain a throughput over 1 Gbit/s for a single stream of data.

References

1. NIST Cryptographic Toolkit, available at http://csrc.nist.gov/CryptoToolkit/
2. Xilinx, Inc.: Virtex 2.5 V Field Programmable Gate Arrays, available at www.xilinx.com
3. ALMA Technologies web page, available at http://www.alma-tech.com
4. Amphion Semiconductor web page, available at http://www.amphion.com
5. Bisquare Systems Private Limited web page, available at http://www.bisquare.com
6. Helion Technology Limited web page, available at http://www.heliontech.com
7. Intron, Ltd. Web page, available at http://www.intron.lviv.ua
8. Ocean Logic Pty Ltd web page, available at http://www.ocean-logic.com
9. Grembowski T., Lien R., Gaj K., Nguyen N., Bellows P., Flidr J., Lehman T., Schott B.: Comparative Analysis of the Hardware Implementations of Hash Functions SHA-1 and SHA-512, LNCS 2433, 5th International Conference, ISC 2002, Sao Paulo, Brazil, Sep./Oct. 2002, 75–89
10. Bellows P., Flidr J., Gharai L., Perkins C., Chodowiec P., and Gaj K.: IPsec-Protected Transport of HDTV over IP, LNCS 2778, 13th International Conference on Field Programmable Logic and Applications, FPL 2003, Lisbon, Portugal, Sep. 2003, 869–879
11. Deepakumara J., Heys H.M., and Venkatesan R.: FPGA Implementation of MD5 Hash Algorithm, Proc. IEEE Canadian Conference on Electrical and Computer Engineering (CCECE 2001), Toronto, Ontario, May 2001, available at http://www.engr.mun.ca/~howard/PAPERS/ccece_2001.pdf
12. Hoare R., Menon P., and Ramos M.: 427 Mbits/sec Hardware Implementation of the SHA-1 Algorithm in an FPGA, International Association of Science and Technology for Development (IASTED) Journal 2002
13. Ting K.K., Yuen S.C.L., Lee K.H., and Leong P.H.W.: An FPGA Based SHA-256 Processor, Proc. 12th International Conference, FPL 2002, Montpellier, France September 2–4, 2002
14. Kang K.Y., Kim D.W., Kwon T.W., and Choi J.R.: Hash Function Processor Using Resource Sharing for IPSec, Proc. 2002 International Technical Conference On Circuit/Systems, Computers and Communications
15. Crypto++, free C++ class library of cryptographic schemes, available at http://www.eskimo.com/~weidai/cryptlib.html
16. Digital Signature Standard Validation System (DSSVS) User's Guide available at http://csrc.nist.gov/cryptval/shs.html
17. McCurley K.S.: A Fast Portable Implementation of the Secure Hash Algorithm, Sandia National Laboratories Technical Report SAND93–2591

18. Bosselaers A., Govaerts R. and Vandewalle J.: Fast Hashing on the Pentium, in N. Koblitz (Ed.): Advances in Cryptology–CRYPT0 '96, LNCS 1109, Springer-Verlag Berlin Heidelberg 1996, 298–312
19. Bosselaers A., Govaerts R. and Vandewalle J.: SHA: A Design for Parallel Architectures?, in W. Fumy (Ed.): Advances in Cryptology–EUROCRYPT '97, LNCS 1233, Springer-Verlag Berlin Heidelberg 1997, 348–362
20. Nakajima J. and Matsui M.: Performance Analysis and Parallel Implementation of Dedicated Hash Functions, in L.R. Knudsen (Ed.): EUROCRYPT 2002, LNCS 2332, Springer-Berlin Heidelberg 2002, 165–180

Fast Verification of Hash Chains

Marc Fischlin[*]

Department of Computer Science & Engineering,
University of California, San Diego, USA
mfischlin@,cs.ucsd.edu
http://www-cse.ucsd.edu/ mfischlin/

Abstract. A hash chain is a sequence of hash values $x_i = \mathsf{hash}(x_{i-1})$ for some initial secret value x_0. It allows to reveal the final value x_n and to gradually disclose the pre-images $x_{n-1}, x_{n-2}, \ldots$ whenever necessary. The correctness of a given value x_i can then be verified by re-computing the chain and comparing the result to x_n. Here we present a method to speed up the verification by outputting some extra information in addition to the chain's end value x_n. This information allows to relate the verifier's workload to a variably chosen security bound. That is, on input a putative chain value the verifier determines a security level (i.e., security against adversaries with at most T steps and success probability ϵ) and performs only a fraction $p = p(T, \epsilon)$ of the original work by using the additional information. We also show lower bounds for the length of this extra information.

Keywords. Certificate, hash chain, Hash function, Hash tree.

1 Introduction

A hash chain, introduced by Lamport [12], is a sequence of hash values $x_i = \mathsf{hash}(x_{i-1})$ for a seed x_0 where hash is some collision-intractable hash function (or some other publicly computable one-way function). Such a chain allows the owner of the seed to publish the chain's end value x_n and to stepwise release the pre-images $x_{n-1}, x_{n-2}, \ldots$ such that revealing x_{n-i} at step i does not help to find some of the values $x_{n-i-1}, \ldots, x_0$. The receiver can check the validity of some received x_{n-i} by re-calculating the chain starting with x_{n-i} up to x_n.

Hash chains have numerous applications. One of the best known is Micali's suggestion to use them as certificate chains [14]. Roughly, for the user's public key pk of a signature or encryption scheme the certification authority (CA) publishes a certificate of x_n and pk (and possibly further information). For the i-th time period of some pre-determined length the CA hands the pre-image x_{n-i} to the user who can then provide this value as a certificate for his public key during this time period. To revoke the certificate the CA stops delivering

[*] This work was supported by the Emmy Noether Programme Fi 940/1-1 of the German Research Foundation (DFG).

T. Okamoto (Ed.): CT-RSA 2004, LNCS 2964, pp. 339–352, 2004.

the pre-images. Since it is infeasible to find the pre-image of the previously given value, forgery of certificates for future time periods is unlikely.

Other applications areas of hash chains include the design of micropayments schemes [10,17], the S/KEY one-time authentication (RFC 1760) [5,6], securing routing information (e.g., [9,7,8]) and spam-fighting protocols [4]. Similarly, one-way chains —where the hash function is replaced by a one-way function— have been deployed for the BiBa signature scheme [16] and for multicast authentication [15].

In some of the aforementioned areas the verification procedure can be shortened significantly. Namely, if the verifier stores a previously verified chain value x_m for $m < n$, then the next time a value is presented, the verifier merely has to re-calculate the chain up to the stored value x_m. However, considering certificates for example, the owner of the seed may visit some sites only sporadically. Similarly, for routing protocols the information may be passed unfrequently. Or, the verifier may not be able to store previous chain values due to memory limitations or other restrictions. Finally, in some solutions, like the anti-spam solution of Dwork et al. [4], the values are not released gradually but rather require the verifier to re-compute a full hash chain. Hence there are cases where large parts of the chain may still have to be verified.

Related Results. The need for faster verification of hash chains has immediately lead to so-called hash trees [13]. Such constructs condense the long chains to tree-like structures such that the path from any value to the published root shrinks to logarithmic length. Unfortunately, in order to verify a given value the user has to supply logarithmically inner nodes of the tree as a proof of correctness. Hence, two of the advantages of hash chains, low communication complexity and structural simplicity, vanish and are traded for faster verification.

Interestingly, quite a few efforts have dealt with the problem of fast *computation* of intermediate values x_i, for both chains and trees [3,11,18]. That is, if the user only stores the seed x_0 then, in order to release x_i of a hash chain, he needs to re-calculate the chain starting with x_0 up to x_i. It is preferrable for the seed owner, of course, to keep some intermediate values $x_{i_1}, \ldots, x_{i_k}$ confidentially, and to recover any x_i from these values much faster.

The results in [3,11,18] give constructions for storing and recovering intermediate values of chains and trees. They also give lower bounds showing that the constructions are optimal with respect to time/storage trade-offs. However, none of these solutions improves the *verification* time. This is especially true for the hash chains, and which would lessen the disadvantage of chains versus trees.

Our Results. Our solution is to let the owner of the seed generate some supplementary information which is published together with the chain's end value x_n. This extra information then allows to improve the verification time when the verifier is presented an allegedly correct chain value. Specifically, for security bound T and ϵ on the adversarial running time and success probability, respectively, our construction allows to decide correctness after roughly a fraction $p = (\log T + \log \frac{1}{\epsilon})/100$ of the original workload. Here, the workload is the number of hash function evaluations (i.e., it is equal to i if x_{n-i} is given).

The interesting property of our construction is that the two security parameters T and ϵ can be chosen individually by any verifier, even differently for each verification run. Once the verifier has selected "his" security level this determines the fraction $p = p(T, \epsilon)$ of hash chain computations. In other words, the more liberal the verifier chooses the security level the less work he has to carry out.

We emphasize that the security level (T, ϵ) for the fast verification should not be confused with security of the hash functions against collision-finders. As for collisions we know that, by the birthday paradox, collisions for hash functions with n-bits output can be generated with probability more than $1/2$ within $2^{n/2}$ steps. Once such a collision is found the complete hash chain becomes disaffected. In our model, we simply assume that finding such collision is beyond feasible attacks. Security here refers to attacks in which the verifier should be forced to perform more than a fraction $p(T, \epsilon)$ of the work; even if the adversary overcomes this security bound the verifier can may still raise the level for the next verification.

In our solution the extra information the seed owner attaches to x_n is called a check-bit vector. As explained, this check-bit vector is a universal parameter enabling different security/workload levels for the verifiers. Another interesting characteristic, in addition to the time improvement, is the length of such check-bit vectors: very long vectors may outweigh the gain in verification time. We therefore investigate lower bounds for this length.

The bounds on the length of check-bit vectors vary with the way the vectors are created. In the most simple case the extra information is chosen according to the time period i and merely consist of some fixed number of the bits of the intermediate value x_{n-i}. For this type of schemes, under which our construction falls, we show that approximately $(\log T - \log n + \log \frac{1}{\epsilon}) \log n$ bits are required. In comparison, our solution produces check bit vectors of about $100 \log_2 n$ bits, which for $n = 1,024$, $T = 2^{40}$ and $\epsilon = 2^{-20}$ and $p = 60\%$ for example, yields respectable $1,000$ bits. Still, this is slightly better than the usual $160 \log_2 n = 1,600$ bits to communicate the inner nodes of a tree, and the communication of the public check-bit vector amortizes over the time periods. Yet, hash trees are usually much faster verifiable, in particular with respect to "standard" security levels.

Organization. In Section 2 we define check-bit schemes and their security formally. We present our lower bounds in Section 3 and our construction appears in Section 4. We conclude with a brief discussion in Section 5.

2   Definition

In the most simple form, a hash chain (for a given length parameter n) can be described by two algorithms $\mathcal{G}$ and $\mathcal{V}$, the generator and verification algorithm. The former algorithm simply chooses a random x_0 and computes the chain up to x_n, and the verifier on input x_n and some putative chain value x for time period i merely checks that $\mathsf{hash}^i(x) = x_n$.

Check-Bit Schemes. Here we augment the basic hash chain generation and verification. Algorithm $\mathcal{G}$, when generating the chain for seed x_0, repeatedly runs a deterministic selection algorithm $\mathcal{S}$ as subroutine for each hash function iteration. For each such execution, for $i = n-1$ down to 0, algorithm $\mathcal{S}$ produces a string cb_i (possibly the empty string λ), which is determined by the time period number i, the intermediate value $x_{n-i} = \mathsf{hash}^{n-i}(x_0)$ and the preceding strings $\mathsf{cb}_{i+1}, \ldots, \mathsf{cb}_{n-1}$.

The so-called check-bit vector cb is the concatenation of all strings $\mathsf{cb}_0, \mathsf{cb}_1$, $\ldots$, cb_{n-1}, ordered according to the release time. We assume that the position of cb_i within cb and its length are recoverable from cb; this clearly inhibits lossy encodings and we thus call the constructions allowing to recover cb_i *schemes with lossless encoding*. Nonetheless, since we only deal with such schemes throughout the paper we often drop this appendix. Let $\mathsf{cb}_{\geq i}$ be the string $\mathsf{cb}_i || \ldots || \mathsf{cb}_{n-1}$ and set $\mathsf{cb}_{>i} = \mathsf{cb}_{\geq i+1}$ for $i \leq n-1$ (where $\mathsf{cb}_{>n-1} = \lambda$).

As before, the verifier $\mathcal{V}$ takes a value x and integer i together with the chain's end value x_n as input, and verifies that x is the correct pre-image for time period i. This time, however, the verifier also gets the check bit vector cb as extra input and uses this value to shorten the verification: For each hash function iteration in time period j the verifier now also calls $\mathcal{S}(j, \mathsf{hash}^{j-i}(x), \mathsf{cb}_{>j})$ and compares the result to the given cb_j. If a mismatch occurs then reject, else continue (possibly up to the chain's end).

Moreover, the verifier gets two parameters T and ϵ representing the bounds on the adversarial running time and success probability (both characteristics are specified below). Instructively, one may think of these two variable security parameters as determined by $\mathcal{V}$ before starting the verification, although for ease of notation we sometimes set these parameters instead and then provide these fixed values to the verifier.

Definition 1. *A check-bit scheme with lossless encoding and for parameter n is a triple $(\mathcal{G}, \mathcal{V}, \mathcal{S})$ of algorithms (of which $\mathcal{G}$ is probabilistic) such that*

Algorithm $\mathcal{G}$:

     – *picks a seed x_0 according to some efficiently samplable distribution,*
     – *computes $x_i = \mathsf{hash}(x_{i-1})$ for $i = 1, 2, \ldots, n$,*
     – *computes $\mathsf{cb}_i = \mathcal{S}(i, x_{n-i}, \mathsf{cb}_{>i})$ for $i = n-1, \ldots, 0$,*
     – *outputs (x_0, x_n, cb).*

Algorithm $\mathcal{V}$:

     – *gets inputs x_n, cb and x, an integer i as well as T and ϵ,*
     – *repeats the following until $i = 0$ or halt:*
         • *if $\mathsf{cb}_i \neq \mathcal{S}(i, x, \mathsf{cb}_{>i})$ then reject and stop*[1]

---

[1] Here $\mathcal{V}$ recovers $\mathsf{cb}_i, \mathsf{cb}_{>i}$ from cb.

   • *else set $i \leftarrow i - 1$ and $x \leftarrow \mathsf{hash}(x)$*
– *if $x = x_n$ then accept, else reject.*

Algorithm $\mathcal{S}$:

   – *takes an integer i, a value x and a string $\mathsf{cb}_{>i}$ as input,*
   – *computes and returns $\mathsf{cb}_i = \mathcal{S}(i, x, \mathsf{cb}_{>i})$.*

In addition, the scheme is complete, i.e., the verifier never rejects a valid input x_n, cb, i and $x = x_{n-i}$ produced by $\mathcal{G}$, independently of T and ϵ.

Note that the selection algorithm $\mathcal{S}$ is defined to be deterministic. On one hand, this simplifies the definition and analysis significantly. On the other hand, it does not weaken the model too much. Namely, for a given hash function hash define $\mathsf{hash}'(x_i \| r) = \mathsf{hash}(x_i) \| r$ such that r remains unchanged during the iterations. If $x_0 \| r$ is chosen at random by $\mathcal{G}$ then $\mathcal{S}$ can use r as externally provided random coins. This corresponds, of course, to public coins, as the right part of the chain's end value $x_n \| r$ is output, too. However, public randomness ensures that *any* verifier can re-calculate the selection algorithm's output and compare it to the given check-bit vector.

Attacks. In order to define security we have to specify the attack mode first. We measure the running time T of the adversary by counting the hash function evaluations only. Formally, we therefore provide the attacker with an oracle $\mathsf{hash}(\cdot)$ which she can access, but for which "guessing" images, i.e., generating images without querying the oracle, is infeasible. The next parameter $\epsilon \in [0, 1)$ basically represents a bound on the adversary's success probability. We also introduce a parameter $p \in [0, 1)$ which bounds the fraction of the original work the verifier performs. We define the following experiment for a check-bit scheme $(\mathcal{G}, \mathcal{V}, \mathcal{S})$ with parameter n:

Experiment $\mathrm{Exp}_{\mathcal{A}}(T, \epsilon, p)$:

   – Algorithm $\mathcal{G}$ generates (x_0, x_n, cb)
   – The adversary $\mathcal{A}$ gets as input (x_n, cb). The adversary also gets access to an oracle $\mathsf{Release}(\cdot)$ which takes integers j as input and returns x_{n-j}. Let r denote the minimum over all queries to $\mathsf{Release}$ (where $r = n$ if $\mathcal{A}$ has never queried the oracle).
   – In addition to oracle queries the adversary performs internal computations and finally outputs (x, k).
   – The verifier $\mathcal{V}$ is invoked on $(x_n, \mathsf{cb}, x, k, T, \epsilon)$ and returns the decision after V hash function evaluations.

We say that adversary $\mathcal{A}$ wins experiment $\mathrm{Exp}_{\mathcal{A}}(T, \epsilon, p)$,

   – if the adversary makes at most T hash function evaluations, and
   – if the verifier makes $V \geq \lceil pk \rceil$ hash function evaluations, and
   – if the adversary has queried the oracle $\mathsf{Release}$ only about values larger than k, i.e., if $k < r$.

Security. Informally, a check-bit scheme is (T, ϵ, p)-verifiable if no adversary running in time T can cause the verifier to perform a fraction p or more of the work with probability more than ϵ. Here, the work refers to the number k of hash function evaluations required to verify the correct value x_{n-k} at time period k. As explained above, we usually envision the security bound as chosen by the verifier, and that this bound then determines the required fraction of the work. In this sense, $p = p(T, \epsilon)$ is a function of the security level, and we call a check-bit scheme p-verifiable if for any (T, ϵ) it is $(T, \epsilon, p(T, \epsilon))$-verifiable. More formally,

Definition 2. *A check-bit scheme $(\mathcal{G}, \mathcal{V}, \mathcal{S})$ with parameter n is called (T, ϵ, p)-verifiable if, for any adversary $\mathcal{A}$ running in time at most T, the probability of $\mathcal{A}$ winning experiment $\mathrm{Exp}_{\mathcal{A}}(T, \epsilon, p)$ is at most ϵ. The scheme is p-verifiable if, for any adversary $\mathcal{A}$ and any T, ϵ, the probability of $\mathcal{A}$ winning experiment $\mathrm{Exp}_{\mathcal{A}}(T, \epsilon, p(T, \epsilon))$ is at most ϵ.*

We have chosen a relative bound to measure the work to be performed, i.e., if $p = 1/2$ then at time period $3n/4$ the verifier needs $3n/8$ hash evaluations, at time period n the verifier has to compute $n/2$ hash values etc. Alternatively, one may define an absolute bound saying that the verifier has to do $w = w(T, \epsilon)$ (or less) hash function evaluations, independently of the time period. But first note that such an absolute bound easily follows if we set $w = pn$. Second, some applications may bear in mind that verification is faster for the first time periods. In this case, it is preferrable to have a relative work reduction saying that you save up to 50%, for instance, at any time period.

3   Lower Bounds

We first show a lower bound for special check-bit procedures in Section 3.1. This bound holds for arbitrary security parameters T, ϵ and thus even yields a bound for the more liberal case of (T, ϵ, p)-verifiable schemes. The bound says that the selection algorithm $\mathcal{S}$ must essentially generate check-bit vectors of $(\log T - \log hn + \log \frac{1}{\epsilon}) \log n$ bits, where h is the maximum number of hash function evaluations for each of the n iterations (including the ones for the computation of $\mathcal{S}$).

The bound above holds for selection algorithms where the length of the output cb_i may depend on the position i but not the intermediate value. We call such schemes *position-driven* selection algorithms:

Definition 3. *Let $(\mathcal{G}, \mathcal{V}, \mathcal{S})$ be a check-bit scheme (for parameter n). Algorithm $\mathcal{S}$ is position-driven if for any two seeds x_0, y_0 we have $|\mathsf{cb}_i(x_0)| = |\mathsf{cb}_i(y_0)|$.*

In general, the length of cb_i may depend on the preceding values or check bits as well, and thus $|\mathsf{cb}_i(x_0)|$ can be different from $|\mathsf{cb}_i(y_0)|$. In this case, the generator $\mathcal{G}$ possibly outputs some seeds x_0 with very short check-bit vectors. For such schemes we yet show in Section 3.2 that check-bit vectors with only a slightly smaller length than above must still be produced with high probability.

For both bounds, i.e., even if $|\mathsf{cb}_i(x_0)| \neq |\mathsf{cb}_i(y_0)|$, we make the following assumption which basically says that the adversary will find matching check bits for random samples with at least the guessing probability.

Assumption 1. *Let $(\mathcal{G}, \mathcal{V}, \mathcal{S})$ be a check-bit scheme with lossless encoding and parameter n. Then, for any i, we assume that for random x_0, y_0 the probability that $\mathsf{cb}_i(x_0) = \mathsf{cb}_i(y_0)$ is at least $2^{-\min\{|\mathsf{cb}_i(x_0)|, |\mathsf{cb}_i(y_0)|\}}$. The probability is over the choice of x_0 and y_0.*

3.1   Position-Driven Selection Algorithms

Throughout this section we use the following notation (visualized in Figure 1): We let $[1, n]$ be the set of integers between 1 and n. Each integer represents the number of hash function evaluations that are required to verify a given value x at time period i.

We divide $[1, n]$ into disjoint intervals. For this, let $\alpha_0, \ldots, \alpha_I$ be a sequence of increasing values with $\alpha_0 = 0$ and $\alpha_I = 1$ for an appropriate integer I (which we will specify later). For $\ell = 1, \ldots, I$ define the ℓ-th interval $\mathcal{I}_\ell$ to be $[\alpha_{\ell-1}n + 1, \alpha_\ell n]$, where we assume for simplicity that all $\alpha_\ell n$'s are integers.

Let $(\mathcal{G}, \mathcal{V}, \mathcal{S})$ be a (T, ϵ, p)-verifiable check-bit scheme with a position-driven selection algorithm. For a seed x_0 chosen by $\mathcal{G}$ let c_ℓ be the number of check-bit positions in the interval $\mathcal{I}_\ell$. Note that, by assumption, c_ℓ does not depend on x_0. The sum over all c_ℓ's is therefore the total number of check bits for which we prove our lower bound.

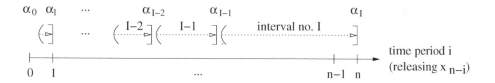

Fig. 1. Idea of Lower Bound

In the sequel we set $q = 1 - p$ for $p > 0$ of the (T, p, ϵ)-verifiable scheme and we let $\alpha_\ell = q\alpha_{\ell+1}$ for $\ell = I - 1, \ldots, 1$. Then $\alpha_\ell = q^{I-\ell}$ for $\ell \geq 1$ and each interval $\mathcal{I}_\ell$ is by a factor $1/q$ larger than the previous one. Recall that we also assume that $\alpha_\ell n$ is an integer for all ℓ, thus n must be a power of $1/q$ and we must have $I = \log_{1/q} n$ for the number I of intervals. For instance, for $p = 1/2$ we have $\log_2 n$ intervals, each one half the size of the following one.

Let h be the maximum of $\mathcal{G}$'s hash function evaluations when computing x_i and cb_i in some i-th step. Then h includes the single evaluation to derive the next chain value and at most $h - 1$ hash function computations of $\mathcal{S}$.

Lemma 1. *We have $c_\ell \geq \log_2 T - \log_2 hn - \log_2 \ln \frac{1}{1-\epsilon}$ for all $\ell = 1, \ldots, I$.*

Note that, for very small ϵ, the latter term $\log_2 \ln \frac{1}{1-\epsilon}$ becomes roughly $\log_2 \epsilon$. Hence, the smaller the error should be the more check bits are requried.

Proof. Suppose that for some interval $\mathcal{I}_\ell$ the number c_ℓ is strictly less than the given bound. We show how to construct an adversary $\mathcal{A}$ then that runs at most $T = 2^t$ steps and succeeds with probability more than ϵ in making the verifier evaluate a fraction p or more of the $k = \alpha_\ell n$ iterations for $x_{n-\alpha_\ell n}$.

Adversary $\mathcal{A}$ repeats the following at most $r = T/hn$ times. $\mathcal{A}$ selects a random seed y_0 and iterates the hash function until all check bits c_ℓ in interval $\mathcal{I}_\ell$ have been computed. If these check bits match the original ones, then output $x = \mathsf{hash}^{n-\alpha_\ell n}(y_0)$ and stop, else repeat.

Note that the adversary's running time is certainly bounded above by T. This holds since the computation of the c_ℓ check bits via the position-driven selection algorithm in each round requires at most hn hash function iterations, and since the number of repetitions is at most r.

It remains to calculate the success probability. In each loop the probability of $\mathcal{A}$ finding a value x for which the check bits match is, by Assumption 1, at least 2^{-c_ℓ}. Hence, the probability that $\mathcal{A}$ does not find a suitable x during all r rounds is at most:

$$
\begin{aligned}
\left(1 - 2^{-c_\ell}\right)^r &\leq \exp\left(-r 2^{-c_\ell}\right) = \exp\left(-2^{t - \log_2 hn - c_\ell}\right) \\
&< \exp\left(-2^{t - \log_2 hn - \left(t - \log_2 hn - \log_2 \ln \frac{1}{1-\epsilon}\right)}\right) \\
&= \exp\left(-2^{\log_2 \ln \frac{1}{1-\epsilon}}\right) = \exp\left(-\ln \frac{1}{1-\epsilon}\right) \\
&= 1 - \epsilon
\end{aligned}
$$

The probability of $\mathcal{A}$ finding such an x is therefore strictly more than ϵ, contradicting the security of the scheme. Therefore, the assumption about c_ℓ falling below the bound must be false. $\qquad\qquad\qquad\qquad\qquad\qquad\qquad\qquad\qquad\square$

We immediately get from the previous lemma:

Theorem 2. *Let $(\mathcal{G}, \mathcal{V}, \mathcal{S})$ be a (T, ϵ, p)-verifiable check-bit scheme with lossless encoding and parameter n. Let $\mathcal{S}$ be a position-driven selection algorithm and assume that Assumption 1 holds. Presume further that the computation of a chain of length n requires at most hn hash function evaluations. Then the length of the check-bit vector is at least*

$$
\left(\log_2 T - \log_2 hn - \log_2 \ln \tfrac{1}{1-\epsilon}\right) \log_{1/(1-p)} n
$$

Proof. According to the lemma, for each interval $\mathcal{I}_\ell$ we have for the number of check bits:

$$
c_\ell \geq t - \log_2 hn - \log_2 \ln \tfrac{1}{1-\epsilon}
$$

It follows for the overall number of check bits:

$$
\sum_{\ell=1}^{I} c_\ell \geq \left(t - \log_2 hn - \log_2 \ln \tfrac{1}{1-\epsilon}\right) \log_{1/q} n
$$

This proves the lower bound.   □

For example, if $n = 1,024$, $h = 1$ and the verifier chooses a security level of $T = 2^{40}$ and $\epsilon = 2^{-20}$, then for $p = 1/2 = q$ we need approximately $(40 - 10 + 20) \cdot 10 = 500$ bits.

3.2 General Check-Bit Schemes

For non-position-driven selection algorithms the size of the output cb_i may vary with the intermediate values. Luckily, we can modify the proof above to obtain a slightly relaxed bound.

Take all the values α_ℓ, $\mathcal{I}_\ell$ etc. as in the previous section and let $(\mathcal{G}, \mathcal{V}, \mathcal{S})$ be a check-bit scheme, not necessarily with a position-driven selection algorithm. Let c_ℓ denote again the number of check bits in interval $\mathcal{I}_\ell$ —which now is a random variable over $\mathcal{G}$'s choice. In addition, fix some constant $a \in (0, 1)$.

Lemma 2. *The probability that $\mathcal{G}$ picks a seed x_0 such that*

$$c_\ell \geq \log_2 T - \log_2 hn - \log_2 \ln \tfrac{1}{1 - \epsilon^{1-a}} \quad \text{for all } \ell = 1, \ldots, I$$

is at least $1 - I\epsilon^a$.

Substituting $\log_2 \ln \frac{1}{1 - \epsilon^{1-a}}$ by the approximation $\log_2 \epsilon^{1-a}$ again, the success probability now enters as $(1 - a) \log_2 \epsilon$. Hence, the smaller a the larger the vector length —but the smaller the probability of outputting such a long vector as well.

Proof. Suppose for sake of contradiction that this probability is less than $1 - I\epsilon^a$. Then there exists a fixed ℓ_0 such that the probability of $\mathcal{G}$ picking a seed x_0 such that

$$c_{\ell_0} < \text{bound}_{\ell_0} := \log_2 T - \log_2 hn - \log_2 \ln \tfrac{1}{1 - \epsilon^{1-a}}$$

is at least ϵ^a.

Next, as in the previous case, we construct an adversary $\mathcal{A}$ trying to cause more than a fraction p of the work for interval $\mathcal{I}_{\ell_0}$ with probability more than ϵ. $\mathcal{A}$ repeats the following $r = T/hn$ times. $\mathcal{A}$ selects a random seed y_0 and computes the chain for this seed up to time period $\alpha_{\ell_0} n$. Let x be $\text{hash}^{n - \alpha_{\ell_0} n}(y_0)$. The adversary continues to iterate the hash function $(\alpha_{\ell_0} - \alpha_{\ell_0 - 1})n$ times. If the check bits do not match the given ones then repeat the process. Else return x.

The running time of $\mathcal{A}$ is bounded above by T since the adversary makes at most hn hash function iterations for each of the r tries. As for the success probability, condition on the event that $\mathcal{G}$ outputs some x_0 for which $c_{\ell_0} < \text{bound}_{\ell_0}$. This happens with proability at least ϵ^a. Next note that the adversary succeeds if the at most bound_{ℓ_0} bits of the attempt match. This happens with probability at least $2^{-\text{bound}_{\ell_0}}$ according to Assumption 1.

Hence, under the condition that $\mathcal{G}$'s seed x_0 causes c_{ℓ_0} to be less than the bound, it follows as before that $\mathcal{A}$ fails with probability

$$\left(1 - 2^{-\text{bound}_{\ell_0}}\right)^r < 1 - \epsilon^{1-a}$$

The probability that $\mathcal{A}$ succeeds in the experiment is therefore more than ϵ^{1-a} times the probability that $c_{\ell_0} < \text{bound}_{\ell_0}$ for $\mathcal{G}$'s output. Multiplying these two probabilities we obtain a successful attack with probability more than ϵ. Thus the initial assumption must have been wrong.   □

Theorem 3. *Let $(\mathcal{G}, \mathcal{V}, \mathcal{S})$ be a (T, ϵ, p)-verifiable check-bit scheme with lossless encoding and parameter n. Presume that the computation of a chain of length n requires at most hn hash function evaluations and let $a \in (0, 1)$ be a constant. Then, under Assumption 1, with probability at least $1 - \epsilon^a \log_{1/(1-p)} n$ (over $\mathcal{G}$'s seed choice) the check-bit vector has at least*

$$\left(\log_2 T - \log_2 hn - \log_2 \ln \tfrac{1}{1-\epsilon^{1-a}} \right) \log_{1/(1-p)} n$$

bits.

4   Constructions of Check-Bit Schemes

In this section we present our check-bit scheme. We start with an elementary attempt which provides an *absolute* work bound of $w = \log_2 T + \log_2 \frac{1}{\epsilon}$ hash function evaluations for the desired security parameter. However, the *relative* performance (relative to the time period and the original number of hash function evaluations) is rather bad, so we elaborate on a construction with relative bound $p = (\log_2 T + \log_2 \frac{1}{\epsilon})/100$. This, unfortunately, comes with an increase in the length of check-bit vector.

4.1   Construction with Absolute Bound

In our construction with absolute bound the selection algorithm $\mathcal{S}$ simply outputs the least siginficant bit of intermediate value x_{n-i} for each percent of computation (i.e., if $i = \lfloor jn/100 \rfloor$ for some j). Here, the value 100 is chosen rather arbitrarily; any other granularity may be selected as well. The verifier, when checking some input x, i, then merely compares the least significant bits of the intermediate values when re-calculating the chain, and stops if a mismatch occurs.

Construction 4. *The check-bit scheme $(\mathcal{G}_{abs}, \mathcal{V}_{abs}, \mathcal{S}_{abs})$ with parameter $n > 100$ is described by the following selection algorithm:*

Algorithm $\mathcal{S}_{abs}(x, i)$:

>   *if $i = \lfloor jn/100 \rfloor$ for some $j \in \{1, \ldots, 100\}$*
>   *then   output $\mathsf{cb}_i = $ [least signifcant bit of x]*
>   *else   output $\mathsf{cb}_i = \lambda$*

Note that the length of the check-bit vector is constant and adds 100 bits to the public chain's end value x_n of typically 160 or 256 bits.

The idea of the scheme is as follows. Suppose that the distribution of the bits is approximately uniform, and that the adversary cannot do better than computing chains for randomly chosen seeds. Then, for such a seed the probability of hitting $w = \log_2 T + \log_2 \frac{1}{\epsilon}$ of the given check bits is at most $2^{-w} = \epsilon T^{-1}$. Hence, the overall success probability of the adversary making T or less steps is at most ϵ.

The scheme, as is, does not provide a reasonable *relative* security level, though. For instance, consider the time period k which is $\log_2 \frac{1}{\epsilon} - 1$ percent from the end value n. An adversary that outputs a random x together with k makes $\mathcal{V}$ evaluate the whole hash function till the end with probability 2ϵ (because there are at most $\log_2 \frac{1}{\epsilon} - 1$ check bits in this interval). Hence, for any given ϵ the verifier performs 100% of the original computation with probability more than ϵ for some point k. Otherwise the length of the vector could not go below our lower bound.

Because of our interest in relative bounds we omit a formal security statement and analysis of this scheme here and turn to the next construction instead.

4.2   Construction with Relative Bound

The problem with the approach in the previous subsection is that the check bits are distributed equidistantly over the chain of length n. Yet, the workload of the verifier varies with the distance to the end value and is thus relative to the position. The idea is now to increase the density of check bits towards the end of the chain such that the number of check bits compensates for the reduced work towards the first time periods.

We partition the chain of length n into $I = \log_2 n$ intervals of length $1, 2, 4, \ldots, n/4, n/2$. For ease of notation we presume that n is a power of 2. For $\ell = 1, \ldots, I$ interval $\mathcal{I}_\ell$ ranges from $2^{\ell-1} + 1$ to 2^ℓ. In interval $\mathcal{I}_I$ we let $\mathcal{S}_{rel}$ output the least significant bit of the intermediate values at positions $jn/100$. Again, any other base instead of 100 may be chosen. In interval $\mathcal{I}_{I-1}$ we double the check bits by outputting the bits of each value at position $jn/200$. In general, we output the least significant bit of value x_{n-i} for $i \in \mathcal{I}_\ell$ if $i = jn/(100 \cdot 2^{I-\ell})$.

Another refinement is to return the b least significant bits instead of a single one only. This improves the error detection probability. We thus define our check-bit scheme with respect to a parameter b which can be an arbitrary integer but which is fixed for a specific instance.

Construction 5. *The check-bit scheme* $(\mathcal{G}_{rel,b}, \mathcal{V}_{rel,b}, \mathcal{S}_{rel,b})$ *with parameter* $n > 100$ *is described by the following selection algorithm:*

Algorithm $\mathcal{S}_{rel,b}(x, i)$:

> *if* $i \in \mathcal{I}_\ell$ *and* $i = \lfloor \frac{jn}{100 \cdot 2^{I-\ell}} \rfloor$ *for some* $j \in \{\lfloor \frac{100}{2n} \rfloor + 1, \ldots, \lceil \frac{100}{n} \rceil\}$
> *then* *output* $\mathsf{cb}_i = [b$ *least signifcant bits of* $x]$
> *else* *output* $\mathsf{cb}_i = \lambda$

For each interval $\mathcal{I}_\ell$ the variable j runs through 50 values, for each such values producing b bits output. Hence, the overall length of a check-bit vector is given by:

$$50b \cdot I = 50b \cdot \log_2 n$$

If we choose $b = 2$, for instance, then we get a check bit vector of $100 \log_2 n$ bits, and for $n = 1,024$ the check-bit vector is $1,000$ bits.

In order to show security we first need to specify the assumption about the hash function, or more precisely, about the bits we output:

Assumption 6. *For any two seeds $x_0 \neq y_0$ we assume that the check-bit vectors $\mathsf{cb}(x_0)$ and $\mathsf{cb}(y_0)$ generated by $\mathcal{S}_{rel,b}$ are uniformly and independently distributed strings of the corresponding length (where the probability is over the choice of the hash function hash).*

This assumption is (almost) satisfied if hash is for example modelled as a random oracle [2]. In this case, the bits are uniformly and independently distributed as long as no intermediate collisions occur. Such collisions are, however, very unlikely and happen only with negligible probability.

Also note how this assumption captures the adaptive queries of the adversary to $\mathsf{Release}(\cdot)$ in experiment $\mathrm{Exp}_A(T, \epsilon, p)$. Specifically, the assumption quantifies over all seeds and thus, even if the adversary knows a seed x_0 generated by $\mathcal{G}_{rel,b}$, it is infeasible to find another seed complying with (parts of) $\mathsf{cb}(x_0)$ better than with trial-and-error. Also, even if given pre-images of x_n *and* the check-bit vector $\mathsf{cb}(x_0)$, it remains infeasible to find another preceding pre-image.

From a practical point of view, well-known hash functions like SHA-1 and RIPEMD-160 seem to approximate this assumption quite well. To best of our knowledge the distribution of the least significant bits is not known to be biased significantly. Similarly, providing very few bits of a pre-image is not known to substantially help inverting the hash function. See [1] for results.

Theorem 7. *Under Assumption 6 the check-bit scheme $(\mathcal{G}_{rel,b}, \mathcal{V}_{rel,b}, \mathcal{S}_{rel,b})$ in Construction 5 constitutes a p-verifiable check-bit scheme for*

$$p = \frac{\frac{2}{b}(\log_2 T + \log_2 \frac{1}{\epsilon})}{100}.$$

For chains of length n the scheme generates check-bit vectors of length $50b \cdot \log_2 n$.

Proof. Assume that the adversary's final output is a pair (x, k). Let ℓ be the interval number in which k lies, i.e.,

$$\frac{n}{2 \cdot 2^{I-\ell}} < k \leq \frac{n}{2^{I-\ell}}.$$

Then, one precent of the work to verify the pair (x, k) corresponds to at least $\frac{1}{2} \cdot \frac{1}{2^{I-\ell}}$ percent of the work to verify the whole chain. By construction, on the other hand, we have $b2^{I-i} \geq b2^{I-\ell}$ check bits in each interval $\mathcal{I}_i$ for $i \leq \ell$. Hence, if we perform $100p$ percent of the verification work for (x, k), then we consult at least $100p \cdot \frac{1}{2} \cdot \frac{1}{2^{I-\ell}} \cdot b2^{I-\ell} = 100bp/2$ check bits in total.

By the bound on the running time the adversary can probe at most T values (x, i) during the experimental phase. The probability that a specific of these values matches the first c given check bits $\mathsf{cb}_{\geq i}$ is by assumption at most 2^{-c}. Hence, the probability that any of the at most T samples matches those bits is bounded above by $T \cdot 2^{-c}$. Together with the fact above the probability of the adversary finding a value matching $100bp/2$ check bits is at most

$$T \cdot 2^{-100bp/2} = T \cdot 2^{-\log_2 T - \log_2 \frac{1}{\epsilon}} = \epsilon.$$

The length of the check-bit vector has already been discussed above. □

Returning to our example with $n = 1,024$ and $b = 2$, for $T = 2^{40}$ and $\epsilon = 2^{-20}$ the verifier requires about $p = 60\%$ of the original workload. For $b = 3$ and vectors of $1,500$ bits the work in this case even reduces to 40%.

5   Discussion

We have presented constructions to improve the verification time of hash chains. Our solutions enable the verifier to select a flexible security level and to relate the work to be done to this security level. Our constructions and lower bounds rely on so-called check-bit schemes where basically some bits of the intermediate values are output. Fortunately, such schemes are very simple and can be integrated quite easily; they preserve the simplicity of hash chains and are applicable in general. Disadvantageously, as we have shown, those schemes cannot go below certain bounds when it comes to the length of the check-bit vectors.

It remains an open problem to provide other check-bit schemes with shorter vectors, e.g., by using lossy encoding techniques. Yet, those schemes should have comparable simplicity as the basic scheme in this paper, otherwise the running time may be dominated by the additional effort, invalidating the benefits of faster verification. Similarly, it would be interesting to show lower bounds for more general check-bit schemes.

Acknowledgment. We thank the anonymous reviewers of RSA-CT 2004 for valuable comments.

References

1. M. Bellare and T. Kohno. *Hash Function Balance and its Impact on Birthday Attacks*. Number 2003/65 in Cryptology eprint archive. eprint.iacr.org, 2003.
2. M. Bellare and P. Rogaway. *Random Oracles are Practical: A Paradigm for Designing Efficient Protocols*. Proceedings of the Annual Conference on Computer and Communications Security (CCS). ACM Press, 1993.
3. D. Coppersmith and M. Jakobsson. *Almost Optimal Hash Sequence Traversal*. Financial Cryptography (FC) 2002, Volume 2357 of Lecture Notes in Computer Science. Springer-Verlag, 2002.
4. C. Dwork, A. Goldberg, and M. Naor. *On Memory-Bound Funtions for Fighting Spam*. Advances in Cryptology — Crypto 2003, Volume 2729 of Lecture Notes in Computer Science. Springer-Verlag, 2003.
5. N. Haller. *The S/KEY One-Time Password Scheme*. Symposium on Network and Distributed Systems Security, pages 151–157. Internet Society, 1994.

6. N. Haller. *The S/KEY One-Time Password Scheme*, 1995.
7. Y.-C. Hu, D. Johnson, and A. Perrig. *SEAD: Secure Efficient Distance Vector Routing in Mobile Wireless Ad Hoc Networks*. Workshop on Mobile Computing Systems and Applications (WMCSA) 2002. IEEE Computer Society Press, 2002.
8. Y.-C. Hu, A. Perrig, and D. Johnson. *Efficient Security Mechanisms for Routing Protocols*. Annual Symposium on Network and Distributed System Security (NDSS) 2003. Internet Society, 2003.
9. R. Hauser, A. Przygienda, and G. Tsudik. *Reducing the Cost of Security in Link State Routing*. Annual Symposium on Network and Distributed System Security (NDSS)'97. Internet Society, 1997.
10. R. Hauser, M. Steiner, and M. Waidner. *Micro-Payments Based on iKP*. Proceedings of SECURICOM'96, Worldwide Congress on Computer and Communications Security and Protection, pages 67–82. ???, 1996.
11. M. Jakobsson, T. Leighton, S. Micali, and M. Szydlo. *Fractal Merkle Tree Representation and Traversal*. Topics in Cryptology — Cryptographer's Track, RSA Conference (CT-RSA) 2003, Volume 2612 of Lecture Notes in Computer Science, pages 314–326. Springer-Verlag, 2003.
12. L. Lamport. *Password Authentication with Insecure Communication*. Communications of the ACM, 24(11):770–772, 1981.
13. R. Merkle. *A Digital Signature Based on a Conventional Encryption Function*. Advances in Cryptology — Crypto'87, Volume 293 of Lecture Notes in Computer Science, pages 369–378. Springer-Verlag, 1988.
14. S. Micali. *Efficient Certificate Revocation*. Technical Report MIT/LCS/TM-542b, MIT Laboratory for Computer Science, 1996.
15. A. Perrig, R. Canetti, D. Song, and D. Tygar. *The TESLA Broadcast Authentication Protocol*. CryptoBytes, Volume 5, pages 2–13. RSA Security, 2002.
16. A. Perrig. *The BiBa One-Time Signature and Broadcast Authentication Protocol*. Proceedings of the Annual Conference on Computer and Communications Security (CCS), pages 28–37. ACM Press, 2001.
17. R. Rivest and A. Shamir. *PayWord and MicroMint: Two Simple Micropayment Schemes*. Security Protocols, Volume 1189 of Lecture Notes in Computer Science, pages 69–87. Springer-Verlag, 1997.
18. Y. Sella. *On the Computation-Storage Trade-Offs of Hash Chain Traversals*. Financial Cryptography (FC) 2003, Lecture Notes in Computer Science. Springer-Verlag, 2003.

Almost Ideal Contrast Visual Cryptography with Reversing

Duong Quang Viet[1] and Kaoru Kurosawa[2]

[1]Tokyo Institute of Technology,
2-12-1 O-okayama, Meguro-ku, Tokyo 152-8552, Japan
viet@crypt.ss.titech.ac.jp
[2] Department of Computer and Information Sciences,
Ibaraki University
4–12–1 Nakanarusawa, Hitachi, Ibaraki 316-8511, Japan
kurosawa@cis.ibaraki.ac.jp

Abstract. A drawback of visual cryptography schemes (VCS) is much loss of contrast in the reconstructed image. This paper shows that no loss of contrast can be almost achieved if we are allowed to use a very simple non-cryptographic operation, reversing black and white. Many copy machines have this function these days. Therefore, our VCS is very attractive.

Keywords: Visual cryptography, ideal contrast, perfect black

1 Introduction

Visual cryptography schemes (VCS) were introduced by Naor and Shamir [9] and have been studied by many researchers [1,3,4,5,7,11]. A (k, n)-VCS is a method to encode a secret image I into n transparencies, where each participant receives one transparency. In the reconstruction phase, any k participants can recover the secret image by superimposing their transparencies. However, any $k - 1$ participants have no information on I. This can be done without any knowledge of cryptography and without performing any cryptographic operations.

A drawback of these schemes is much loss of contrast in the reconstructed image. In a $(2, 2)$-VCS [9], a black pixel is translated into a black region but a white pixel is translated into a grey region (half black and half white).

On the other hand, Blundo et al. showed how to construct a perfect black (k, n)-VCS for any $2 \leq k \leq n$, where the reconstruction of black region is perfect [6,3]. However, the reconstruction of white region is very dark.

In this paper, we show that no loss of contrast can be almost achieved if we are allowed to use a very simple non-cryptographic operation, reversing black and white. That is, all the black region is reversed to white and all the white region is reversed to black. Many copy machines have this function these days. Therefore, our VCS is very attractive. We call our construction a (k, n)-VCS with *reversing*.

T. Okamoto (Ed.): CT-RSA 2004, LNCS 2964, pp. 353–365, 2004.

We first show a perfect black (k, n)-VCS with *reversing* such that the reconstruction of white region is almost perfect. This means that the contrast is almost ideal. The cost we have to pay is the size of shares. If the size of shares is c times larger, then the grey level of white region converges to zero exponentially.

We next show how to convert a perfect *black* (k, n)-VCS (with reversing) to a perfect *white* (k, n)-VCS with reversing. Perfect *white* VCSs are much more preferable than perfect *black* VCSs because the white region is much larger than the black region in usual images. From our first result, we can obtain a perfect *white* (k, n)-VCS with reversing such that the reconstruction of black region is almost perfect.

We finally show a perfect black VCS for any monotone access structure. This means that we can obtain a VCS with *reversing* for any monotone access structure such that the contrast is almost ideal. (Perfect black VCSs have been known only for (k, n)-threshold cases so far.)

It will be a further work to find another simple non-cryptographic operation which can achieve almost ideal contrast.

Related work: Naor and Shamir showed an improved scheme in [10]. However, it works only for $(2, 2)$-VCS.

2   Preliminaries

For a random variable X, $E[X]$ denotes the expected value and $\mathrm{Var}[X]$ denotes the variance. We sometimes use $+$ to express OR.

2.1   Model

A (k, n)-visual cryptography scheme (VCS) consists of a distribution phase and a reconstruction phase. Let I be a secret image which consists of black and white pixels P.

In the distribution phase, a dealer $\mathcal{D}$ encodes each pixel P into n shares $s_1, \cdots, s_n$, one for each transparency. $\mathcal{D}$ then gives s_i to participant $\mathcal{P}_i$ for $i = 1, \cdots, n$.

In the reconstruction phase, any k participants $\mathcal{P}_{i_1}, \cdots, \mathcal{P}_{i_k}$ reconstruct I by superimposing their transparencies. That is, the reconstructed pixel is given by

$$\tilde{P} = s_{i_1} + s_{i_2} + \cdots + s_{i_k},$$

where $+$ means OR. However, any $k - 1$ participants have no information on I.

Each s_i consists of m sub-pixels, where m is called the *expansion rate*. Hence s_i is described by a Boolean vector of length m

$$v_i = (c_{i,1}, \cdots, c_{i,m}),$$

where $c_{i,j} = 1$ if the j-th sub-pixel in s_i is black. Let $C = [c_{i,j}]$ be the $n \times m$ Boolean matrix which consists of $v_1, \cdots, v_n$. We say that C is the *encoding matrix* of P.

Usually, the dealer $\mathcal{D}$ computes the encoding matrix C of a pixel P from two matrices M_0 and M_1 as follows: C is obtained by randomly permuting the columns of M_0 if P is white and by randomly permuting the columns of M_1 if P is black. M_0 and M_1 are called the *basis matrices*.

$\tilde{P}$ is interpreted as black if $w_H(\tilde{P})$ is large, and as white if $w_H(\tilde{P})$ is small, where $w_H(\tilde{P})$ denotes the Hamming weight of $\tilde{P}$. We define the grey level of a pixel P as

$$\mathsf{GREY}(P) = w_H(\tilde{P})/m,$$

where $P =$ white or black.

Therefore, $\mathsf{GREY}(white)$ should be close to zero and $\mathsf{GREY}(black)$ should be close to one. The *contrast* is ideal if

$$\mathsf{GREY}(white) = 0 \text{ and } \mathsf{GREY}(black) = 1.$$

2.2   Naor-Shamir (2, 2)-VCS

Naor and Shamir showed the first (k, n)-VCS [9]. Fig 1 illustrates their $(2, 2)$-VCS.

In the distribution phase, each pixel P is split into two sub-pixels in each of the two shares s_1 and s_2. If P is white, then the dealer $\mathcal{D}$ randomly chooses one of the first two rows of Fig 1. If P is black, then $\mathcal{D}$ randomly chooses one of the last two rows of Fig 1. $\mathcal{D}$ then gives s_1 to participant $\mathcal{P}_1$ and s_2 to participant $\mathcal{P}_2$.

In other words, the basis matrices are

$$M_0 = \begin{pmatrix} 1 & 0 \\ 1 & 0 \end{pmatrix}, \quad M_1 = \begin{pmatrix} 1 & 0 \\ 0 & 1 \end{pmatrix} \tag{1}$$

The dealer $\mathcal{D}$ computes the encoding matrix C of a pixel P by randomly permuting the columns of M_0 if P is white and by randomly permuting the columns of M_1 if P is black.

In the reconstruction phase, the two participants superimpose s_1 and s_2. If P is black, then they get two black sub-pixels; if P is white, then they get one black sub-pixel and one white sub-pixel. Therefore,

$$\mathsf{GREY}(black) = 1, \quad \mathsf{GREY}(white) = 1/2.$$

2.3   Perfect Black VCS

We say that a (k, n)-VCS is *perfect black* if

$$\mathsf{GREY}(black) = 1 \text{ and } \mathsf{GREY}(white) < 1.$$

The (n, n)-VCS shown by Naor and Shamir [9] is perfect black. The expansion rate is $m = 2^{n-1}$ and they showed that it is optimum.

For any $2 \le k \le n$, Blundo et al. showed a perfect black (k, n)-VCS such that

$$\mathsf{GREY}(white) = 1 - 1/m$$

for some expansion rate m [6].

3   Basic Construction

In this section, we show a basic construction of our schemes. We present a perfect black $(2,2)$-VCS with reversing such that $\mathsf{GREY}(white) = 1/4$. Since $\mathsf{GREY}(white) = 1/2$ in the Naor-Shamir $(2,2)$-VCS, the contrast is improved.

Definition 1. *We say that an image I is reversed if all black pixels are reversed into white and all white pixels are reversed into black.*

Let $\overline{P}$ denote the reversed pixel of P. Our scheme is illustrated in Fig 2 and Fig 3.

(Distribution phase) A dealer $\mathcal{D}$ runs the distribution phase of Naor-Shamir $(2,2)$-VCS twice independently. Let (s_1, s_2) be the shares of the first run and (s_1', s_2') be the shares of the second run. Then the share of participant $\mathcal{P}_1$ of our VCS is (s_1, s_1') and that of participant $\mathcal{P}_2$ is (s_2, s_2'). See Fig 2.

(Reconstruction phase)

Step 1. Two participants superimpose s_1, s_2 and obtain $T = s_1 + s_2$. Similarly, they superimpose s_1', s_2' and obtain $T' = s_1' + s_2'$. They are illustrated in the last columns of Fig 2(a) and Fig 2(b).

Step 2. They next reverse T, T' and obtain $\overline{T}$ and $\overline{T'}$ as shown in Fig 3.

Consider a pixel P.

- If P is black, then T and T' all black. Therefore, $\overline{T}$ and $\overline{T'}$ are all white.
- If P is white, then T and T' are grey such that a half region is black and the other half is white in each one of the four cases. Therefore, $\overline{T}$ and $\overline{T'}$ are also grey such that a half region is white and the other half is black in each one of the four cases.

Step 3. The two participants superimpose $\overline{T}$, $\overline{T'}$ and obtain $\overline{T} + \overline{T'}$.

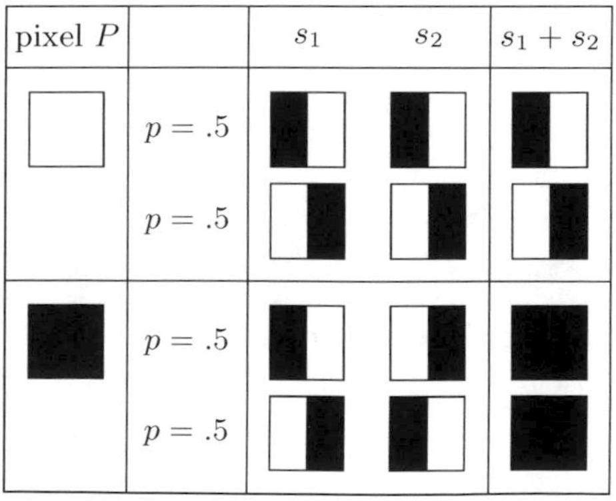

Fig. 1. Naor-Shamir 2-out-of-2 visual cryptography scheme

(a) First run

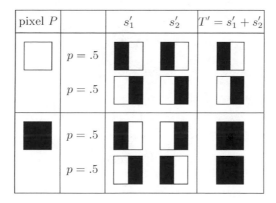

(b) Second run

Fig. 2. Proposed $(2,2)$-VCS (1)

From Fig 3, we can see that:

- If P is black, then $\overline{T} + \overline{T'}$ is always white.
- If P is white, then $\overline{T} + \overline{T'}$ is black with probability $1/2$ and grey (half black and half white) with probability $1/2$. This is because (s_1, s_2) and (s'_1, s'_2) are generated independently and randomly.

Step 4. Finally the two participants reverse $\overline{T} + \overline{T'}$ and obtain $\overline{\overline{T} + \overline{T'}}$.

It is clear that:

- If P is black, then $\overline{\overline{T} + \overline{T'}}$ is always black.
- If P is white, then $\overline{\overline{T} + \overline{T'}}$ is all white with probability $1/2$ and it is grey (half black and half white) with probability $1/2$.

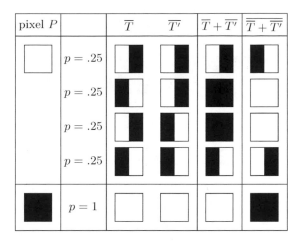

Fig. 3. Proposed $(2,2)$-VCS (2)

(Contrast): In our $(2,2)$-VCS with reversing, we obtain that $\mathsf{GREY}(black) = 1$ and
$$E[\mathsf{GREY}(white)] = (1/2) \times 0 + (1/2) \times (1/2) = 1/4.$$

4   General Construction: (k, n)-Threshold Case

In this section, we show a general construction of our (k, n)-VCS with *reversing*. The reconstruction of black region is perfect and the reconstruction of white region is almost perfect.

The cost we have to pay is the size of shares. If the size of shares is c times larger, then the grey level of white region converges to zero exponentially.

4.1 c-Run (k, n)-VCS with Reversing

Suppose that there exists a perfect black (k, n)-VCS. We then construct a "c-run (k, n)-VCS with *reversing*" as follows. (Remember that there exists a perfect black (k, n)-VCS for any $2 \le k \le n$.)

Let P a secret pixel to be distributed.

(Distribution phase) In the distribution phase, the dealer $\mathcal{D}$ runs the distribution phase of the perfect black (k, n)-VCS c times independently. Let $(s_{1,i}, \cdots, s_{n,i})$ be the set of shares in the i-th run for $i = 1, \cdots, c$. The share of participant $\mathcal{P}_j$ of our VCS is then $(s_{j,1}, \cdots, s_{j,c})$.

(Reconstruction phase) Any k participants, say $\mathcal{P}_{j_1}, \cdots, \mathcal{P}_{j_k}$, reconstruct P as follow.

1. For $i = 1, \cdots, c$, they superimpose their shares and obtain
$$T_i = s_{j_1,i} + \cdots + s_{j_k,i}$$

2. They reverse T_i and obtain $\overline{T}_i$ for $i = 1, \cdots, c$.
3. They superimpose $\overline{T}_1, \cdots, \overline{T}_c$ and obtain $U = \overline{T}_1 + \cdots + \overline{T}_c$.
4. We reverse U and obtain $\tilde{P}$, where

$$\tilde{P} = \overline{U} = \overline{\overline{T}_1 + \cdots + \overline{T}_c}.$$

On the other hand, it is clear that any $k-1$ participants have no information on P from the property of the original (k,n)-VCS.

4.2   Contrast

We show that the contrast is almost ideal in our construction. It is easy to see that $\mathsf{GREY}(black) = 1$ because the original VCS is perfect black. We now show that both $E[\mathsf{GREY}(white)]$ and $\mathrm{Var}[\mathsf{GREY}(white)]$ converge to zero.

Theorem 1. *Suppose that* $\mathsf{GREY}(white) = q < 1$ *in the original perfect black VCS. Then in our c-run VCS with reversing,*

(1) $E[\mathsf{GREY}(white)] = q^c$.
(2) $\mathrm{Var}[\mathsf{GREY}(white)] \le q^c(1 - q^c)$.

Proof. (1) Let P be a white pixel. Each T_i is described by a Boolean vector of length m

$$A_i = (a_{i,1}, \cdots, a_{i,m}),$$

where m is the expansion rate. Similarly, the reconstructed pixel $\tilde{P}$ is described by a Boolean vector

$$W = (w_1, \cdots, w_m).$$

Now since

$$w_j = \overline{\overline{a}_{1,j} + \cdots + \overline{a}_{c,j}},$$

it holds that

$$w_j = a_{1,j} \times \cdots \times a_{c,j}$$

from De Morgan's law. Therefore,

$$
\begin{aligned}
E[w_H(W)] = E[\sum_j w_j] &= \sum_j E(w_j) = \sum_j E[a_{1,j} \times \cdots \times a_{c,j}] \\
&= \sum_j \Pr(a_{1,j} = \cdots = a_{c,j} = 1) \\
&= \sum_j \Pr(a_{1,j} = 1) \times \cdots \times \Pr(a_{c,j} = 1) \\
&= \sum_j q^c \\
&= m q^c.
\end{aligned}
$$

Consequently, $E[\mathsf{GREY}(white)] = E[w_H(W)]/m = q^c$.

(2) It is easy to see that $(w_1 + \cdots + w_m) \leq m$ because $w_j = 0$ or $w_j = 1$. Therefore,

$$(w_1 + \cdots + w_m)^2 \leq m(w_1 + \cdots + w_m) = m \sum_{j=1}^{m} w_j$$

Hence

$$\text{Var}[w_H(W)] = E[w_H(W)^2] - E[w_H(W)]^2 = E[(\sum_j w_j)^2] - m^2 q^{2c}$$

$$\leq mE[\sum_j w_j] - m^2 q^{2c}$$

$$= mE[w_H(W)] - m^2 q^{2c}$$

$$= m^2 q^c(1 - q^c)$$

Consequently, $\text{Var}[\text{GREY}(white)] = \text{Var}[w_H(W)]/m^2 \leq q^c(1 - q^c)$.

$\square$

Therefore,

$$\lim_{c\to\infty} E[\text{GREY}(white)] = 0 \text{ and } \lim_{c\to\infty} \text{Var}[\text{GREY}(white)] = 0.$$

This means that we can obtain asymptotically ideal contrast by letting c large.

4.3 Complexity

The reconstruction phase of the c-run (k,n)-VCS with reversing requires $c + 1$ reversing operations and superimposing $kc - 1$ transparencies. The size of shares become c times larger than that of the original VCS.

4.4 Corollaries

From the previous result [6], we obtain the following corollary.

Corollary 1. *For any $2 \leq k \leq n$, there exists a perfect (k,n)-VCS with reversing such that*

$$E[\text{GREY}(white)] = (1 - 1/m)^c$$
$$\text{Var}[\text{GREY}(white)] \leq (1 - 1/m)^c \{1 - (1 - 1/m)^c\}$$

for some expansion rate m, where c is any positive integer.

If we use the Naor-Shamir $(2,2)$-VCS, we obtain the following corollary.

Corollary 2. *There exists a perfect black $(2,2)$-VCS with reversing such that*

$$E[\text{GREY}(white)] = (1/2)^c$$
$$\text{Var}[\text{GREY}(white)] \leq (1/2)^c \{1 - (1/2)^c\}$$

with the expansion rate $m = 2$, where c is any positive integer.

As an example, we present a 3-Run $(2, 2)$-VCS.

(Distribution phase) The dealer $\mathcal{D}$ runs the distribution phase of Naor-Shamir $(2, 2)$-VCS three times independently. Let (s_1, s_2) be the shares of the first run, (s_1', s_2') be the shares of the second run and (s_1'', s_2'') be the set of shares of the third run.

Then the share of participant $\mathcal{P}_1$ is then (s_1, s_1', s_1''). and that of participant $\mathcal{P}_2$ is (s_2, s_2', s_2'').

(Reconstruction phase)

1. We superimpose s_1 and s_2, and then obtain $T = s_1 + s_2$. Similarly, we obtain $T' = s_1' + s_2'$ and $T'' = s_1'' + s_2''$.
2. We reverse T, T' and T'', and obtain $\overline{T}, \overline{T'}$ and $\overline{T''}$.
3. We superimpose $\overline{T}, \overline{T'}, \overline{T''}$ and obtain $U = \overline{T} + \overline{T'} + \overline{T''}$.
4. We reverse U and obtain $\tilde{P}$.

(Contrast): We can then see that $\mathsf{GREY}(black) = 1$ and

$$E[\mathsf{GREY}(white)] = (1/4) \times (1/2) + (3/4) \times 0 = 1/8.$$

5   Perfect White VCS

5.1   Conversion from Perfect Black VCS

We say that a (k, n)-VCS is *perfect white* if

$$\mathsf{GREY}(white) = 0 \text{ and } \mathsf{GREY}(black) > 0.$$

In usual pictures, the white region is much larger than the black region. Therefore, perfect *white* VCSs are much preferable than perfect *black* VCSs. However, no perfect white VCS has been known.

In this section, we show that a perfect *white* (k, n)-VCS with reversing is easily obtained from a perfect *black* (k, n)-VCS (with reversing).

Theorem 2. *Suppose that there exists a perfect black (k, n)-VCS such that $E[\mathsf{GREY}(white)] = p$. Then there exists a perfect white (k, n)-VCS such that $E[\mathsf{GREY}(black)] = 1 - p$.*

Proof. We describe a perfect *white* (k, n)-VCS.

In the distribution phase,

1. the dealer $\mathcal{D}$ first reverses the original image I and obtains $\overline{I}$.
2. $\mathcal{D}$ then applies the distribution phase of the perfect *black* (k, n)-VCS to $\overline{I}$.

In the reconstruction phase,

1. a qualified subset of participants apply the reconstruction phase of the perfect *black* (k, n)-VCS and obtains a reconstructed image $\overline{I}'$.
2. They finally reverse $\overline{I}'$ and obtain $\overline{\overline{I}'}$.

Then it is easy to see that the above scheme is a perfect *white* (k, n)-VCS such that $E[\mathsf{GREY}(black)] = 1 - p$.

□

As the original perfect *black* (k, n)-VCS with reversing, we can use our construction shown in the previous section.

5.2 Example

As an example, we show how to convert the perfect black $(2, 2)$-VCS of Sec.2.2 into a perfect white $(2, 2)$-VCS with reversing. (See Fig 4.)

In the distribution phase,

1. the dealer $\mathcal{D}$ first reverses the original image I. Hence each white pixel is reversed into black and each back pixel is reversed into white.
2. $\mathcal{D}$ then applies the distribution phase of the perfect *black* $(2, 2)$-VCS. Participant $\mathcal{P}_1$ obtains a share s_1 and participant $\mathcal{P}_2$ obtains a share s_2.

In the reconstruction phase,

1. the two participants superimpose s_1 and s_2 and obtains $s_1 + s_2$.
2. They finally reverse $s_1 + s_2$ and obtain $\overline{s_1 + s_2}$.

From Fig 4, we see that a perfect white $(2, 2)$-VCS is obtained such that $\mathsf{GREY}(black) = 1/2$.

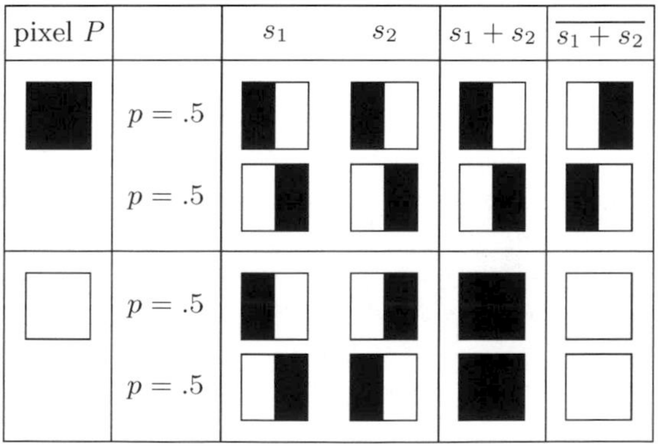

Fig. 4. Perfect white 2-out-of-2 visual cryptography scheme

6   Perfect Black VCS for General Access Structure

Perfect black VCSs have been known only for (k, n)-threshold cases so far although VCS itself can be constructed for general access structures [1], In this section, we show a perfect black VCS for any monotone access structure. This means that we can obtain a VCS with *reversing* for any monotone access structure such that the contrast is almost ideal.

Let $(M_{n,0}, M_{n,1})$ be the basis matrices of a perfect black (n, n)-VCS of Naor and Shamir [9]. Let

$$M_{n,0} = \begin{pmatrix} e_{n,1} \\ \vdots \\ e_{n,n} \end{pmatrix}, \quad M_{n,1} = \begin{pmatrix} e'_{n,1} \\ \vdots \\ e'_{n,n} \end{pmatrix}$$

where $e_{n,i}$ and $e'_{n,i}$ are binary vectors of length 2^{n-1}.

6.1   Access Structure

Let $\mathcal{P} = \{1, \ldots, n\}$ be a set of participants. (k, n)-threshold secret sharing schemes are generalized to secret sharing schemes with monotone access structures Γ [8,2], where Γ is a set of all subsets of participants which can determine the secret.

In a secret sharing scheme, on input a secret s, a dealer $\mathcal{D}$ computes $(v_1, \ldots, v_n)$ and gives v_i to participant i so that only qualified subsets of participants (access subset) can recover the secret. Let

$$\Gamma \overset{\triangle}{=} \{A \subseteq \mathcal{P} \mid A \text{ can determine } s\}.$$

Then Γ is called an access structure and $A \in \Gamma$ is called an access set. Let

$$\Gamma_0 \overset{\triangle}{=} \{A \subseteq \mathcal{P} \mid A \text{ is a minimal access set.}\}.$$

Definition 2. Γ *is said to be monotone if*

$$A \in \Gamma, A \subseteq A' \Rightarrow A' \in \Gamma.$$

We require that any $B \notin \Gamma$ has no information on s. Then it is known that a secret sharing scheme for Γ is exists if and only if Γ is monotone.

6.2   General Construction

We show a pair of basis matrices (L_0, L_1) of a perfect black VCS for any monotone access structure Γ_0. For Γ_0, there exists a usual type secret sharing scheme as follows. Suppose that $\Gamma_0 = \{A_1, \cdots, A_t\}$, where $A_j = \{j_1, \cdots, j_{|A_j|}\}$. For $1 \leq j \leq t$, the dealer $\mathcal{D}$ chooses random bits such that

$$s = b_{j,1} \oplus \cdots \oplus b_{j,|A_j|}.$$

and gives $b_{j,u}$ to participant j_u.

Now let $G = (g_{i,j})$ be an intermediate dummy $n \times t$ matrix.

(**Construction of L_0**): If $i = j_u$ for some $1 \le u \le |A_j|$, then replace $g_{i,j}$ with $e_{|A_j|,u}$. Else, replace $g_{i,j}$ with $\underbrace{(1, \cdots, 1)}_{2^{|A_j|-1}}$. Then we obtain L_0.

(**Construction of L_1**): If $i = j_u$ for some $1 \le u \le |A_j|$, then replace $g_{i,j}$ with $e'_{|A_j|,u}$. Else, replace $g_{i,j}$ with $\underbrace{(1, \cdots, 1)}_{2^{|A_j|-1}}$. Then we obtain L_1.

In this construction, the expansion rate is $m = 2^{|A_1|-1} + \cdots + 2^{|A_t|-1}$ and $\mathsf{GREY}(white) = 1/m$.

We show an example for $\Gamma_0 = \{\{1,2\}, \{2,3,4\}\}$ below.

$$L_0 = \begin{pmatrix} 101111 \\ 100011 \\ 110101 \\ 110110 \end{pmatrix}, \quad L_1 = \begin{pmatrix} 101111 \\ 010011 \\ 110101 \\ 111001 \end{pmatrix}$$

This is a perfect black VCS for Γ_0. The expansion rate is $m = 2^{2-1} + 2^{3-1} = 6$ and $\mathsf{GREY}(white) = 1/m = 1/6$.

6.3   Construction for Special Cases

In this subsection, we present a better construction for $\Gamma_0 = \{\{1,2\}, \{2,3\}\{3,4\}\}$. For Γ_0, there exists a usual type secret sharing scheme as follows. Let $\{0,1\}$ be the set of secrets. The dealer $\mathcal{D}$ chooses random bits $b_1, \cdots, b_4$ such that

$$s = b_1 \oplus b_2 = b_3 \oplus b_4$$

The set of shares are $v_1 = b_1$, $v_2 = b_2$, $v_3 = (b_1, b_3)$, $v_4 = b_4$.

Now as an intermediate dummy matrix, let

$$G = \begin{pmatrix} b_1, \; x \\ b_2, \; x \\ b_1, \; b_3 \\ x, \; b_4 \end{pmatrix}$$

(**Construction of L_0**): In G, replace b_1 and b_3 with $e_{2,1}$. Replace b_2 and b_4 with $e_{2,2}$. Replace x with $(1,1)$.

(**Construction of L_1**): In G, replace b_1 and b_3 with $e'_{2,1}$. Replace b_2 and b_4 with $e'_{2,2}$. Replace x with $(1,1)$.

Then we obtain (L_0, L_1) as follows.

$$L_0 = \begin{pmatrix} 1011 \\ 1011 \\ 1010 \\ 1110 \end{pmatrix}, \quad L_1 = \begin{pmatrix} 1011 \\ 0111 \\ 1010 \\ 1101 \end{pmatrix}$$

The expansion rate is $m = 4$ and $\mathsf{GREY}(white) = 1/4$. If we use the general construction, then the expansion rate is $m = 6$ and $\mathsf{GREY}(white) = 1/6$.

We can apply the same technique to $\Gamma_0 = \{\{1,2\}, \{1,3\}, \{2,3,4\}\}$, but not to $\Gamma_0 = \{\{1,2\}, \{2,3\}, \{3,4\}, \{2,4\}\}$ nor $\Gamma_0 = \{\{1,2\}, \{1,3\}, \{1,4\}, \{2,3,4\}\}$. The details will be given in the final paper.

References

1. G. Ateniese, C. Blundo, A. D. Santis and D. R. Stinson, "Visual cryptography for general access structures", in: Information and Computation, vol. 129, pp. 86–106 (1996).

2. J. C. Benaloh and J. Leichter, "Generalized secret sharing and monotone functions", in: Proc. of Crypto'88, Lecture Notes on Computer Science, LNCS vol. 403, pp. 27–36 (1990).

3. C. Blundo and A. De. Santis, "Visual cryptography schemes with perfect reconstruction of black pixels", in: Computer and Graphics, vol. 22, no. 4, pp. 449–455 (1998).

4. C. Blundo, A. De. Santis and D. R. Stinson, "On the contrast in visual cryptography schemes", in: Journal of Cryptology, vol. 12, no. 4, pp. 261–289 (1999).

5. C. Blundo, P. D'Arco, A. De. Santis and D. R. Stinson, "Contrast optimal threshold visual cryptography schemes", to appear in: SIAM Journal on Discrete Mathematics
(Available from http://cacr.math.uwaterloo.ca/dstinson/papers/COTVCS.ps).

6. C. Blundo, A. De. Bonis and A. De. Santis, "Improved schemes for visual cryptography", in: Designs, Codes, and Cryptography, vol. 24, pp. 255–278 (2001).

7. P. A. Eisen and D. R. Stinson, "Threshold Visual Cryptography Schemes With Specified Whiteness Levels of Reconstructed Pixels", in: Designs, Codes, and Cryptography, vol. 25, no. 1, pp. 15–61 (2002).

8. M. Itoh, A. Saito and T. Nishizeki, "Multiple assignment scheme for sharing secret", in: Journal of Cryptology, vol. 6, no. 1, pp. 15–20 (1993)

9. M. Naor and A. Shamir, "Visual cryptography", in: Proc. of Eurocrypt'94, Lecture Notes on Computer Science, LNCS vol. 950, pp. 1–12 (1995).

10. M. Naor and A. Shamir, "Visual cryptography II: Improving the Contrast Via the Cover Base", in: Proc. of Security Protocols'96, Lecture Notes on Computer Science, LNCS vol. 1189, pp. 197–202 (1997). Full version available from http://www.wisdom.weizmann.ac.il/naor/PAPERS/new_cov.ps.

11. E. R. Verheul and H. C. A. van Tilborg, "Constructions and properties of k out of n visual secret sharing schemes", in: Designs, Codes, and Cryptography, vol. 11, no. 2, pp. 179–196 (1997).

Weak Fields for ECC

Alfred Menezes[1], Edlyn Teske[1], and Annegret Weng[2]

[1] University of Waterloo, Canada
{ajmeneze,eteske}@uwaterloo.ca
[2] University of Essen, Germany
weng@exp-math.uni-essen.de

Abstract. We demonstrate that some finite fields, including $\mathbb{F}_{2^{210}}$, are weak for elliptic curve cryptography in the sense that any instance of the elliptic curve discrete logarithm problem for *any* elliptic curve over these fields can be solved in significantly less time than it takes Pollard's rho method to solve the hardest instances. We discuss the implications of our observations to elliptic curve cryptography, and list some open problems.

1 Introduction

Elliptic curve cryptography (ECC) is being standardized by accredited standards organizations and governments around the world. The security of elliptic curve systems is based on the hardness of the *elliptic curve discrete logarithm problem* (ECDLP): given an elliptic curve E defined over a finite field $\mathbb{F}_q$, a point $P \in E(\mathbb{F}_q)$ of order r, and a second point $Q \in \langle P \rangle$, determine the integer $l \in [0, r - 1]$ such that $Q = lP$. Elliptic curve systems are especially attractive because Pollard's rho method [34], the best algorithm known for the solving the general ECDLP, has a fully-exponential expected running time of $\sqrt{\pi r}/2$ point additions.

For a given underlying field $\mathbb{F}_q$, maximum resistance to Pollard's rho method can be attained by selecting an elliptic curve E for which r is prime and is as large as possible. The most favourable situation arises when $\#E(\mathbb{F}_q)$ is prime or almost prime, i.e., $\#E(\mathbb{F}_q) = dr$, where r is prime and the co-factor d is small (e.g., $d \in \{1, 2, 3, 4\}$). In this case, since $\#E(\mathbb{F}_q)$ lies in the Hasse interval $[(\sqrt{q} - 1)^2, (\sqrt{q} + 1)^2]$, we have $r \approx q$ and we say that the elliptic curve has a security level of $\frac{1}{2} \log_2 q$ bits.

Some ECC standards recommend or mandate a small selection of finite fields and elliptic curves. Among these, the most influential has been the FIPS 186-2 standard [8] for the elliptic curve digital signature algorithm (ECDSA) which recommends five prime fields $\mathbb{F}_p$ for specified primes p of bitlengths 192, 224, 256, 384, and 512, and the five characteristic two finite fields $\mathbb{F}_{2^{163}}$, $\mathbb{F}_{2^{233}}$, $\mathbb{F}_{2^{283}}$, $\mathbb{F}_{2^{409}}$, and $\mathbb{F}_{2^{571}}$. The recommended elliptic curves over these fields have security levels of approximately 80, 112, 128, 192, and 256 bits, which match the security levels of the SKIPJACK, Triple-DES, AES-Small, AES-Medium, and AES-Large symmetric-key encryption schemes. Fixing a small set of allowable fields has the advantages of facilitating interoperability, and permitting the optimization of

T. Okamoto (Ed.): CT-RSA 2004, LNCS 2964, pp. 366–386, 2004.
© Springer-Verlag Berlin Heidelberg 2004

hardware and software implementations by exploiting properties of the chosen fields.

It is therefore reasonable to expect that commercial deployments of ECC will converge upon a small selection of finite fields. This does not appear to be a serious limitation for the following reasons. First, there are an enormous number of elliptic curves to choose from; more precisely, there are roughly $2q$ isomorphism classes of elliptic curves over $\mathbb{F}_q$. Second, the orders of these curves are roughly uniformly distributed over the Hasse interval in the case of prime fields, and over the even integers in the Hasse interval in the case of characteristic two finite fields. Consequently, elliptic curves of almost prime orders are plentiful and can be easily found. Finally, there are very few elliptic curves of almost prime order over a field $\mathbb{F}_q$ for which the ECDLP can be solved in subexponential (or faster) time—those that succumb to the Weil and Tate pairing attacks [12,29], and the attack on prime-field anomalous curves [35,36,38]. It is easy to recognize these curves, and thus the aforementioned attacks can readily be circumvented.

Nonetheless, the possibility still remains that algorithms will subsequently be discovered for efficiently solving any instance of the ECDLP for *any* elliptic curve over a selected field. If ECC solutions employing that field were widely deployed (especially in hardware), then the consequences of such a discovery would be more drastic than if an attack were discovered on a special class of curves because a change in the underlying field would be required. Determining whether such finite fields exist is therefore an important problem in elliptic curve cryptography.

Definition 1. *A finite field* $\mathbb{F}_q$ *is said to be* bad *for elliptic curve cryptography if the following conditions are satisfied:*

1. *for some elliptic curves E over* $\mathbb{F}_q$, *solving the ECDLP in* $E(\mathbb{F}_q)$ *using Pollard's rho method (and its parallelized versions [31]) is intractable using existing computer technology; and*
2. *algorithms are known that can feasibly solve (using existing computer technology) any ECDLP instance for any elliptic curve over* $\mathbb{F}_q$.

No bad fields for ECC are presently known. The contribution of this paper is the observation that some finite fields are weak in the following sense.

Definition 2. *A finite field* $\mathbb{F}_q$ *is said to be* weak *for elliptic curve cryptography if the following conditions are satisfied:*

1. *for some elliptic curves E over* $\mathbb{F}_q$, *solving the ECDLP in* $E(\mathbb{F}_q)$ *using Pollard's rho method (and its parallelized versions [31]) is intractable using existing computer technology; and*
2. *algorithms are known for which any ECDLP instance for any elliptic curve over* $\mathbb{F}_q$ *can be solved in significantly less time than it takes Pollard's rho method to solve the hardest ECDLP instances over* $\mathbb{F}_q$.

While the ECDLP for elliptic curves over a weak field may in fact be intractable in general, demonstrating that a field is weak provides some evidence that the field may be bad, and therefore unsuitable for elliptic curve cryptography.

Of course our definition of a weak field is not precise since "significantly less" has not been quantified. We remark that the discovery [16,45] of a $\sqrt{N}$-speedup of Pollard's rho method for solving the ECDLP in the group of $\mathbb{F}_{2^N}$-rational points on a Koblitz curve[1] caused some to view the security of these curves with suspicion. For Koblitz curves over $\mathbb{F}_{2^{163}}$ and $\mathbb{F}_{2^{283}}$, the speedup is by a factor of only 13 and 17, respectively. In this paper, we present reasonable arguments that the finite fields $\mathbb{F}_{2^N}$, where $N \in [185, 600]$ is divisible by 5, are weak fields for ECC. In particular, we show that the ECDLP for all elliptic curves over $\mathbb{F}_{2^{210}}$ (respectively, one-quarter of all elliptic curves over $\mathbb{F}_{2^{210}}$) can be solved 2^{13} times faster (respectively, 2^{20} times faster) than it takes Pollard's rho method to solve the hardest instances. These speedups are significantly greater than the aforementioned speedups for Koblitz curves, and moreover are applicable to *all* (respectively, one-quarter of all) elliptic curves over $\mathbb{F}_{2^{210}}$. While upto now it was believed that an elliptic curve over $\mathbb{F}_{2^{210}}$ whose group order is twice a prime offers a security level of 104 bits, our results show it can have a security level of at most 91 bits, that is, the same as a curve over $\mathbb{F}_{2^{183}}$ is able to offer. The field $\mathbb{F}_{2^{210}}$ is interesting because its arithmetic can be efficiently implemented by successive extensions, e.g., $\mathbb{F}_{2^2} \subseteq \mathbb{F}_{2^6} \subseteq \mathbb{F}_{2^{30}} \subseteq \mathbb{F}_{2^{210}}$. As another example, we show that the ECDLP for all elliptic curves over $\mathbb{F}_{2^{600}}$ can be solved about 2^{69} times faster than it takes Pollard's rho method to solve the hardest instances. Hence an elliptic curve over $\mathbb{F}_{2^{600}}$ can have a security level of at most 230 bits.

Organization. The remainder of this paper is organized as follows. In Section 2, we summarize the recent work on the Weil descent attack on the ECDLP. Our detailed arguments that the fields $\mathbb{F}_{2^N}$, where $N \in [185, 600]$ is divisible by 5, are weak are presented in Section 3. In Section 4, we examine the fields $\mathbb{F}_{2^N}$, where N is divisible by 4, for weakness. In Section 5, we further explore the special case $N = 210$. We draw our conclusions in Section 6 and list some interesting open problems.

2   Weil Descent Attack on the ECDLP

Frey [11] first proposed using Weil descent as a means to reduce the ECDLP in elliptic curves over finite fields $\mathbb{F}_{q^n}$ to the discrete logarithm problem in the jacobian variety of a curve of larger genus over the proper subfield $\mathbb{F}_q$. If a subexponential-time algorithm is known for the DLP for the resulting curve, then this could lead to an algorithm that solves the original ECDLP instance faster than Pollard's rho method.

---

[1] A *Koblitz curve* is an elliptic curve defined over $\mathbb{F}_2$. There are two such curves: $y^2 + xy = x^3 + 1$ and $y^2 + xy = x^3 + x^2 + 1$. These curves admit fast point multiplication algorithms (see [40]) and are therefore favoured over other curves defined over $\mathbb{F}_{2^N}$.

Let l and n be positive integers, and let $N = ln$. Let $q = 2^l$, and let $k = \mathbb{F}_q$ and $K = \mathbb{F}_{q^n}$. Consider the (non-supersingular) elliptic curve E defined over K by the equation

$$E \ : \ y^2 + xy = x^3 + ax^2 + b, \quad a \in K, b \in K^*.$$

We assume that $\#E(K) = dr$ where d is small and r is prime, whence $r \approx q^n$. Let $b_i = \sigma^i(b)$, where $\sigma : K \to K$ is the Frobenius automorphism defined by $\alpha \mapsto \alpha^q$. The *magic number* for E relative to n is defined to be

$$m = m(b) = \dim_{\mathbb{F}_2}(\mathrm{Span}_{\mathbb{F}_2}\{(1, b_0^{1/2}), (1, b_1^{1/2}), \ldots, (1, b_{n-1}^{1/2})\}). \tag{1}$$

Assume now that either n is odd, or $m(b) = n$, or $\mathrm{Tr}_{K/\mathbb{F}_2}(a) = 0$. Gaudry, Hess and Smart [18] showed how Weil descent can be used to reduce instances of the ECDLP in the subgroup of order r of $E(K)$ to instances of the hyperelliptic curve discrete logarithm problem (HCDLP) in a subgroup of order r of the jacobian $J_C(k)$ of a hyperelliptic curve C of genus $g = 2^{m-1} - 1$ or 2^{m-1} defined over k. One first constructs the Weil restriction $W_{E/k}$ of scalars of E, which is an n-dimensional abelian variety over k. Then, $W_{E/k}$ is intersected with $n - 1$ hyperplanes to eventually obtain the hyperelliptic curve C from an irreducible reduced component in the intersection. The reduction algorithm, together with the fastest known algorithm for solving the HCDLP in $J_C(k)$, is called the *GHS attack* on the ECDLP.

Since subexponential-time algorithms are known for the HCDLP for large genus hyperelliptic curves [1], it is possible that the GHS attack can solve the original ECDLP instance faster than Pollard's rho method. In [30], it was shown that for all elliptic curves over $\mathbb{F}_{2^N}$ where $N \in [160, 600]$ is prime, the genus g of C is either too small (whereby the attack fails because $J_C(\mathbb{F}_2)$ is too small to yield any non-trivial information about the ECDLP in $E(\mathbb{F}_{2^N})$), or is too large ($g \geq 2^{16} - 1$, whereby the attack fails because the HCDLP in $J_C(\mathbb{F}_2)$ is intractable). In [23], the GHS attack was used to solve an instance of the ECDLP over $\mathbb{F}_{2^{124}}$ (which is infeasible to solve using Pollard's rho method) by reducing it to an instance of the HCDLP in a genus 31 hyperelliptic curve over $\mathbb{F}_{2^4}$ and solving the latter using the Enge-Gaudry algorithm [17,7]. A convincing argument was presented that the GHS attack could also be used to solve instances of the ECDLP for a certain class of elliptic curves over $\mathbb{F}_{2^{155}}$ by reducing them to instances of the HCDLP in genus 31 hyperelliptic curves over $\mathbb{F}_{2^5}$. The effectiveness of the GHS attack for elliptic curves over $\mathbb{F}_{2^N}$ where $N \in [100, 600]$ is composite was extensively analyzed in [28], where the elliptic curves most susceptible were identified and enumerated. In Section 3 we examine in greater detail the effectiveness of the GHS attack on the ECDLP over fields $\mathbb{F}_{2^N}$ where N is a multiple of 5. In Section 4, we study the case where N is a multiple of 4. The special case $N = 210$ is further examined in Section 5.

3   The Fields $\mathbb{F}_{2^N}$ with $N = 5l$

In this section, we argue that the fields $\mathbb{F}_{2^N}$ with $N = 5l$ are weak for ECC. We restrict our attention to $l \in [32, 120]$ (equivalently $N \in [160, 600]$), since these are the values of interest for cryptographic applications. We draw our conclusions by analyzing Pollard's rho method and the GHS attack for solving instances of the ECDLP over these fields. We emphasize that our conclusions are meaningful for practice because our analyses are exact—that is, they do not involve any asymptotics, crude approximations, or hidden constants.

3.1   Exact Analysis of Pollard's Rho Method

The instances of the ECDLP over $\mathbb{F}_{2^N}$ most resistant to Pollard's rho method (using the random walk of Teske [41]) are for elliptic curves E that have almost prime order $\#E(\mathbb{F}_{2^N}) = 2r$ for some prime r. Since $r \approx 2^{N-1}$, Pollard's rho method has an expected running time of $\sqrt{\pi 2^{N-1}}/2 \approx 2^{(N-1)/2}$ steps, where the dominant operation in each step is an addition in $E(\mathbb{F}_{2^N})$. We note that even though the expression $\sqrt{\pi r}/2$ for the running time of Pollard's rho method is an asymptotic one (as $r \to \infty$), it has been proven under reasonable assumptions [42, Corollary 5.1] that the actual running time for any fixed value of r is within a very small constant multiple of the asymptotic time. Thus

$$T_\rho = 2^{(N-1)/2} = 2^{2.5l-0.5} \tag{2}$$

is indeed a very accurate approximation for the running time of Pollard's rho method for solving the hardest instances of the ECDLP over $\mathbb{F}_{2^N}$.

3.2   Exact Analysis of the GHS Attack

Let E be an elliptic curve defined over $\mathbb{F}_{2^N}$, and let $P \in E(\mathbb{F}_{2^N})$ have prime order r. The GHS attack first uses the GHS reduction to yield an explicit group homomorphism $\Phi : \langle P \rangle \to J_C(\mathbb{F}_{2^l})$, where C is a hyperelliptic curve defined over $\mathbb{F}_{2^l}$, and then uses the Enge-Gaudry index-calculus algorithm to solve the resulting HCDLP instance. For these parameters, the GHS reduction algorithm takes less than a minute on a workstation. Thus we do not include the running time of the GHS reduction in our analysis of the GHS attack.

If the coefficients of E belong to $\mathbb{F}_{2^l}$, then $\#E(\mathbb{F}_{2^l})$ divides $\#E(\mathbb{F}_{2^N})$ and hence r has bitlength at most $N - l$. In this case, Pollard's rho algorithm can solve each ECDLP instance in at most $T'_\rho = 2^{(N-l)/2}$ steps, which is significantly less than T_ρ. For example, if $(N, l) = (160, 32)$ then $T_\rho = 2^{79.5}$ and $T'_\rho = 2^{64}$, and if $(N, l) = (600, 120)$ then $T_\rho = 2^{299.5}$ and $T'_\rho = 2^{240}$. Therefore, we will henceforth assume that E is not isomorphic to an elliptic curve defined over $\mathbb{F}_{2^l}$. In particular the magic number m (defined in (1)) is not 1, and hence it follows from [30, Corollary 9] that $m = 5$. Therefore C has genus $g = 15$ or 16. In fact, the vast majority of the $2^{N+1} - 2^{l+1}$ isomorphism classes of elliptic curves defined over $\mathbb{F}_{2^N} \setminus \mathbb{F}_{2^l}$ yield a genus 16 curve.

Theorem 3 *The GHS reduction yields a genus 15 hyperelliptic curve C defined over $\mathbb{F}_{2^l}$ for exactly $2^{4l+1} - 2$ isomorphism classes of elliptic curves defined over $\mathbb{F}_{2^N} \setminus \mathbb{F}_{2^l}$.*

Proof. Let $q = 2^l$. Let $E : y^2 + xy = x^3 + ax^2 + b$ with $b \in \mathbb{F}_{2^N} \setminus \mathbb{F}_{2^l}$, and let $\mathrm{Ord}_b(x)$ denote the unique monic polynomial $f \in \mathbb{F}_2[x]$ of least degree such that $f(\sigma)(b) = 0$. (If $f(x) = \sum_{i=0}^{r} f_i x^i$, then $f(\sigma)(b) = \sum_{i=0}^{r} f_i \sigma^i(b)$.) Let $t(x) = x^4 + x^3 + x^2 + x + 1$. Since $b \notin \mathbb{F}_{2^l}$, $\mathrm{Ord}_b(x) = x^5 - 1$ or $\mathrm{Ord}_b(x) = t(x)$. Now, by [20, Corollary 6] we have $g = 15$ if and only if $\mathrm{Tr}_{\mathbb{F}_{2^N}/\mathbb{F}_{2^l}}(b^{1/2}) = 0$, which is the case if and only if $\mathrm{Tr}_{\mathbb{F}_{2^N}/\mathbb{F}_{2^l}}(b) = 0$. On the other hand, since $\mathrm{Tr}_{\mathbb{F}_{2^N}/\mathbb{F}_{2^l}}(b) = t(\sigma)(b)$ it is easy to see that $\mathrm{Tr}_{\mathbb{F}_{2^N}/\mathbb{F}_{2^l}}(b) = 0$ if and only if $\mathrm{Ord}_b(x) \mid t(x)$, which is the case if and only if $\mathrm{Ord}_b(x) = t(x)$. By [30, Corollary 8], the latter is true for exactly $2^{4l} - 1$ elements $b \in \mathbb{F}_{2^N} \setminus \mathbb{F}_{2^l}$.   □

In our analysis, we restrict ourselves to the case $g = 16$, for which the HCDLP is harder to solve than for $g = 15$.

The Enge-Gaudry algorithm [17,7] for finding the logarithm of a divisor D_2 to the base D_1 in $J_C(\mathbb{F}_{2^l})$ has three stages. First, a smoothness bound $t \in [1, g]$ is selected and a *factor base* $\{P_1, P_2, \dots, P_w\}$ is constructed which contains exactly one of D and $-D$ for each prime divisor D of degree less than or equal to t. In the second *relation generation* stage, a random walk is performed in the set of reduced divisors equivalent to divisors of the form $\alpha D_1 + \beta D_2$. Each t-smooth divisor encountered in this walk yields a relation $\alpha_i D_1 + \beta_i D_2 \sim R_i = \sum_j e_{ij} P_j$. After slightly more than w such relations have been generated and stored, one can find by linear algebra modulo r a non-trivial linear combination $\sum_i \gamma_i (e_{i1}, e_{i2}, \dots, e_{iw}) = (0, 0, \dots, 0)$. Thus $\sum_i \gamma_i R_i = 0$, and then $\log_{D_1} D_2 = -(\sum_i \gamma_i \alpha_i)/(\sum_i \gamma_i \beta_i) \bmod r$ can be easily computed.

There is a one-to-one correspondence between points in $C(\mathbb{F}_{2^l})$ and degree one divisors in $J_C(\mathbb{F}_{2^l})$. The divisor corresponding to a point $(x, y) \in C(\mathbb{F}_{2^l})$ is ramified if and only if $h(x) = 0$ where $v^2 + h(u)v = f(u)$ is the Weierstrass equation of C; otherwise the divisor splits. According to the Hasse-Weil bound, $\#C(\mathbb{F}_{2^l}) = 2^l + 1 - \gamma$, where $|\gamma| \leq 2g\sqrt{2^l}$. Hence 2^l is a very good approximation for the number of degree one divisors in $J_C(\mathbb{F}_{2^l})$. We select the smoothness bound $t = 1$, and then the size of the factor base is $w \approx 2^{l-1}$. Creating the factor base takes negligible time compared to the relation generation and matrix stages, so we ignore that stage in our running time analysis.

The number of 1-smooth divisors in $J_C(\mathbb{F}_{2^l})$ is approximately $(2^l)^g/g!$ [17, Proposition 4]. In fact, the exact number can be efficiently computed.

Lemma 4 *Let C be a hyperelliptic curve of genus g over $\mathbb{F}_q$, and suppose that there are A_1 split degree one divisors and B_1 ramified degree one divisors in $J_C(\mathbb{F}_q)$ (so $A_1 + B_1 = \#C(\mathbb{F}_q)$, and $w = A_1/2 + B_1$). Then the number of 1-smooth divisors in $J_C(\mathbb{F}_q)$ is*

$$M(1) = \sum_{i=1}^{g} \left([x^i] \left(\frac{1+x}{1-x} \right)^{A_1/2} (1+x)^{B_1} \right),$$

where $[\,]$ denotes the coefficient operator.

Proof. Similar to the proof of Lemma 2 in [23].   □

Now, $A_1 \in [2^l + 1 - 2g\sqrt{2^l}, 2^l + 1 + 2g\sqrt{2^l}]$, and $B_1 \in [0, g]$. Using either the maximum possible values for A_1 and B_1, or the minimum values for A_1 and B_1, we verified that $M(1) \approx (2^l)^g/g!$ is indeed a very good approximation for each $l \in [32, 120]$. By Weil's theorem, the size of $J_C(\mathbb{F}_{2^l})$ satisfies $(\sqrt{2^l} - 1)^{2g} \leq \#J_C(\mathbb{F}_{2^l}) \leq (\sqrt{2^l} + 1)^{2g}$. Thus $\#J_C(\mathbb{F}_{2^l}) \approx 2^{lg}$ is a very good approximation when $l \in [32, 120]$. Hence the expected number of random walk steps before w relations are obtained is $T_1 = w \cdot \#J_C(\mathbb{F}_{2^l})/M(1) \approx 2^{l-1}g!$. For $g = 16$, we have

$$T_1 \approx 2^{l+43}.$$

The two dominant operations in a random walk step are an addition in $J_C(\mathbb{F}_{2^l})$ and a smoothness testing. A polynomial $a(u)$ can be tested for 1-smoothness by first removing repeated factors (by performing a squarefree factorization) and then checking whether the resulting polynomial divides $u^{2^l} - u$. If $a(u)$ is found to be 1-smooth, then it can be factored using the Cantor-Zassenhaus algorithm [5]. We ignore the running time of the factorization step in our estimates because 1-smooth divisors are encountered relatively infrequently—once every $g! = 16! \approx 2^{44}$ random walk steps.

The system of linear equations has dimension slightly more than w and about g non-zero coefficients per equation. It can be solved using Lanczos's algorithm [6], whose running time is closely approximated by $T_2 = gw^2$ arithmetic operations modulo r. We thus have

$$T_2 \approx 2^{2l+2}.$$

3.3   Comparisons

In order to compare the cost of Pollard's rho method for solving the ECDLP in $E(\mathbb{F}_{2^N})$ with the cost of the Enge-Gaudry algorithm for solving the HCDLP in $J_C(\mathbb{F}_{2^l})$, we need to estimate the relative cost of the basic operations in these algorithms. Let c_E denote the time to perform an elliptic curve addition in $E(\mathbb{F}_{2^N})$, c_J the time to perform an addition in $J_C(\mathbb{F}_{2^l})$ (where C has genus 16), c_S the time to test whether a monic polynomial $a \in \mathbb{F}_{2^l}[u]$ of degree 16 is 1-smooth, and c_r the time to perform a multiplication modulo r. Then the expected cost of Pollard's rho method is

$$R_\rho \approx c_E T_\rho = c_E 2^{2.5l-0.5},$$

the expected cost of the random walk stage of the Enge-Gaudry algorithm is

$$R_1 \approx (c_J + c_S)T_1 = (c_J + c_S)2^{l+43},$$

and the expected cost of the matrix stage of the Enge-Gaudry algorithm is

$$R_2 \approx c_r T_2 = c_r 2^{2l+2}. \tag{3}$$

A deficiency in the above comparison is that it only considers the total time taken, and not other scarce resources consumed such as memory, number or processors, and communications between processors. Pollard's rho method can be effectively parallelized (see [31]) so that its expected running time on a network of S processors is T_ρ/S steps. Moreover, the processors do not communicate with each other, and only occasionally transmit data to a central server. The amount of data stored at the server can be controlled without any noticeable impact on the running time (see [26]). Thus time is the only scarce resource consumed by (parallelized) Pollard's rho method.

The relation generation stage in the Enge-Gaudry algorithm can also be effectively parallelized with a speedup that is linear in the number of processors employed, and where the processors do not communicate with each other and only occasionally transmit data to a central server. However, it is not known whether the matrix stage can be parallelized in this way. Moreover, the matrix may have large storage requirements. Thus, in practice, the matrix stage may be the bottleneck in an application of the Enge-Gaudry algorithm. Note, however, that Bernstein [3] and Wiener [44] have recently shown that the full cost[2] of solving a D-dimensional system of sparse linear equations over $\mathbb{F}_2$ can be reduced from $D^{3+o(1)}$ to $D^{7/3+o(1)}$. Consequently, we are of the opinion that our comparisons of Pollard's rho method and the Enge-Gaudry algorithm that only consider running times are adequate and meaningful for determining the effectiveness of the GHS attack on elliptic curve cryptographic schemes. This reasoning is more sound when the time cost $c_r T_2$ is significantly less than $c_E T_\rho$.

To complete the comparisons, we need relative estimates for c_E, c_J, c_S and c_r. When mixed affine-projective coordinates are employed, an elliptic curve operation in $E(\mathbb{F}_{2^N})$ requires 8 multiplications in $\mathbb{F}_{2^N}$. Thus $c_E \approx 8c_N$, where c_N is the time to perform a multiplication in $\mathbb{F}_{2^N}$, and we have

$$R_\rho \approx c_N 2^{2.5(l+1)}. \tag{4}$$

The dominant computation in smoothness testing is the evaluation of $u^{2^l} \bmod a$, where a is a monic polynomial of degree (at most) 16.[3] First, one iteratively computes and stores $u^{2^i} \bmod a$ for $i \in [9, 15]$; this can be done with 224 multiplications in $\mathbb{F}_{2^l}$. Then, one can compute $u^{2^i} \bmod a$ for $5 \le i \le l$ by successive squarings; this can be done with $128(l-4)$ multiplications in $\mathbb{F}_{2^l}$. Thus $c_S = (128l - 288)c_l$, where c_l is the time to perform a multiplication in $\mathbb{F}_{2^l}$.

The fastest algorithm known for performing the jacobian arithmetic in a genus 16 hyperelliptic curve appears to be NUCOMP [24].[4] The precise operation count of NUCOMP has not been worked out. However, Jacobson [22] has reported that the cost c_J of a jacobian addition using NUCOMP is less than the

---

[2] The *full cost* of an algorithm is its running time multiplied by the number of processors employed.

[3] In practice, the squarefree factorization may not be performed. In that case, Gaudry's algorithm only considers 1-smooth divisors the points in whose supports all have coefficient 1. This does not significantly affect the expected number of random walk steps.

[4] Experiments carried out by Jacobson [22] indicate that NUCOMP is faster than Cantor's algorithm and its variants [4,33] for hyperelliptic curves of genus ≥ 7.

cost c_S of computing u^{2^l} mod a. For example, he reports that $c_S \approx 2.3c_J$ when $l = 37$. The ratio c_S/c_J grows with l because the number of $\mathbb{F}_{2^l}$-multiplications for smoothness testing increases with l, while the number of $\mathbb{F}_{2^l}$-multiplications for jacobian addition is independent of l. Thus the approximation $R_1 \approx c_S T_1$ is justified, and hence

$$R_1 \approx (128l - 288)2^{l+43}c_l. \tag{5}$$

Finally, we need to estimate the relative costs c_N, c_l and c_r of a multiplication in $\mathbb{F}_{2^N}$, $\mathbb{F}_{2^l}$, and modulo r, respectively. We use the relative timings on a Pentium II 400 MHz reported by Hankerson [19] for his optimized implementation of multiplication in $\mathbb{F}_{2^N}$ and $\mathbb{F}_{2^l}$ using the methods of [27], and for integer multiplication with Barrett reduction [2]. Table 1 shows these costs for some selected fields.

Table 1. Time estimates for Pollard's rho method for solving an ECDLP instance in $E(\mathbb{F}_{2^N})$, and for the relation generation and matrix stages of the Enge-Gaudry algorithm for solving an HCDLP instance in $J_C(\mathbb{F}_{2^l})$ where C is a genus 16 hyperelliptic curve. c_N, c_l and c_r are the relative times for a multiplication in $\mathbb{F}_{2^N}$, $\mathbb{F}_{2^l}$, and modulo an N-bit prime, respectively. The estimates for R_ρ, R_1, R_2 were derived using formulas (4), (5), (3), respectively. In columns 3, 4, 5, the time units are $\mathbb{F}_{2^N}$-multiplications, $\mathbb{F}_{2^l}$-multiplications, and modulo-r multiplications, respectively. In columns 9, 10, 11, the time unit is an $\mathbb{F}_{2^l}$-multiplication.

N	l	R_ρ/c_N	R_1/c_l	R_2/c_r	c_N	c_l	c_r	R_ρ	R_1	R_2
160	32	$2^{82.5}$	2^{87}	2^{66}	7.7	1.0	5.8	$2^{85.5}$	2^{87}	2^{69}
185	37	2^{95}	2^{92}	2^{76}	7.9	1.0	5.8	2^{98}	2^{92}	2^{79}
210	42	$2^{107.5}$	2^{97}	2^{86}	10.3	1.0	8.0	$2^{110.5}$	2^{97}	2^{89}
255	51	2^{130}	2^{107}	2^{104}	11.0	1.0	7.7	2^{133}	2^{107}	2^{107}
385	77	2^{195}	2^{133}	2^{156}	15.0	1.0	9.4	2^{199}	2^{133}	2^{160}
515	103	2^{260}	2^{160}	2^{208}	15.3	1.0	11.1	2^{264}	2^{160}	2^{212}
600	120	$2^{302.5}$	2^{177}	2^{242}	17.7	1.0	12.5	$2^{306.5}$	2^{177}	2^{246}

Remark 5 *(miscellaneous notes on Table 1)*

(i) *The relative times for c_N, c_l and c_r are, of course, dependent on the choice of algorithms, platform, and implementation. Nevertheless, we do not expect that these relative times will differ by large factors (e.g., greater than 4) from the "correct" times. For example, the ratios c_N/c_l for the seven (N,l) pairs of Table 1 that we obtained using the routines for field arithmetic in Victor Shoup's NTL package [37] are 2.3, 2.1, 3.8, 3.5, 6.0, 8.7, and 10.5.*

(ii) *We use the estimates for R_ρ, R_1 and R_2 to justify our main conclusion that the fields $\mathbb{F}_{2^N}$, where $N \in [185, 600]$ is divisible by 5, are weak for ECC. This statement becomes stronger as N increases. In particular, it is debatable whether our estimates justify the conclusion that the field $\mathbb{F}_{2^{185}}$ is weak for*

ECC. This field is of special interest because it is explicitly included in the
OAKLEY key agreement protocol that was proposed for Internet applications
[32].

(iii) *By selecting only a proportion of degree one divisors in the factor base, one*
can decrease the cost of the matrix stage at the expense of increasing the
cost of the relation generation stage. More precisely, if the factor base size
is reduced by a factor of 2^d, then R_2 decreases by a factor of 2^{2d}, while R_1
increases by a factor of $2^{d(g-1)}$. For example, if we select $d = 4$ for the case
$N = 600$, then we obtain $R_1 = 2^{237}$ and $R_2 = 2^{238}$. We can then derive our
claim made at the end of Section 1 that the ECDLP for all elliptic curves
over $\mathbb{F}_{2^{600}}$ can be solved about 2^{69} times faster than it takes Pollard's rho
method to solve the hardest instances. Similarly, for $(N, d) = (385, 1.5)$ we
have $(R_1, R_2) = (2^{155.5}, 2^{157})$, and for $(N, d) = (515, 3)$ we have $(R_1, R_2) =
$(2^{205}, 2^{206})$.

4   The Fields $\mathbb{F}_{2^N}$ with $N = 4l$

Smart [39] presented some experimental evidence that the fields $\mathbb{F}_{2^N}$, where
N is divisible by 4, are weak for ECC. In this section we repeat our analysis
from Section 3 of the GHS attack for the ECDLP over these fields, and precisely
quantify the weakness of the fields. We conclude that the fields $\mathbb{F}_{2^{4l}}$ exhibit some
signs of being weak, but are not as weak as the fields $\mathbb{F}_{2^{5l}}$.

Let $N = 4l$. If E is an elliptic curve defined over $\mathbb{F}_{2^N} \setminus \mathbb{F}_{2^{2l}}$, then it follows
from [30, Theorems 5, 6] that the magic number m (defined in (1)) is 3 or 4. By
[20, Corollary 6], the GHS reduction yields a hyperelliptic curve C of genus 4 or
8, respectively, over $\mathbb{F}_{2^l}$. The genus of C is 8 in a vast majority of the cases. Since
this yields the worse running time for the Enge-Gaudry algorithm, we focus on
this case.

Arguing as in Section 3, we have

$$R_\rho \approx c_N 2^{2l+2.5}, \qquad R_1 \approx c_l(32l - 48)2^{l+14}, \qquad R_2 \approx c_r 2^{2l+1}.$$

Hence the running time of Pollard's rho algorithm is very close to the running
time of the matrix stage. However, if the factor base size is reduced by a factor
of 2^d, then R_2 decreases by a factor of 2^{2d}, while R_1 increases by a factor of 2^{7d}
(cf. Remark 5(iii)). Table 2 list the costs R_ρ, R_1, R_2 for some selected fields and
choices of d that roughly balance R_1 and R_2.

5   The Field $\mathbb{F}_{2^{210}}$

In this section we argue that the field $\mathbb{F}_{2^{210}}$ is particularly weak for ECC. Recall
from Section 3.3 that $R_\rho \approx c_N 2^{107.5}$, $R_1 \approx c_l 2^{97}$, and $R_2 \approx c_r 2^{86}$ for the
parameters $(N, l) = (210, 42)$. We next consider the GHS attack with parameters
$(N, n, m) = (210, 6, 5)$.

Table 2. Time estimates for Pollard's rho method for solving an ECDLP instance in $E(\mathbb{F}_{2^N})$, and for the relation generation and matrix stages of the Enge-Gaudry algorithm for solving an HCDLP instance in $J_C(\mathbb{F}_{2^l})$ where C is a a genus 8 hyperelliptic curve. The factor base in the Enge-Gaudry algorithm is comprised of $1/2^d$ of all degree-one prime divisors. c_N, c_l and c_r are the times for a multiplication in $\mathbb{F}_{2^N}$, $\mathbb{F}_{2^l}$, and modulo an N-bit prime, respectively.

N	l	d	R_ρ/c_N	R_1/c_l	R_2/c_r
160	40	2	$2^{82.5}$	2^{78}	2^{77}
192	48	3	$2^{98.5}$	2^{94}	2^{91}
224	56	4	$2^{114.5}$	2^{109}	2^{104}
256	64	5	$2^{130.5}$	2^{124}	2^{119}
384	96	8	$2^{194.5}$	2^{178}	2^{177}
512	128	12	$2^{258.5}$	2^{238}	2^{233}
600	150	14	$2^{302.5}$	2^{274}	2^{273}

5.1   Exact Analysis of the GHS Attack with $(N, n, m) = (210, 6, 5)$

About 2^{175} isomorphism classes of elliptic curves over $\mathbb{F}_{2^{210}}$ have magic number $m = 5$ relative to $n = 6$, as a consequence of which there exists an even more effective ECDLP solver for at least about 25% of all elliptic curves over $\mathbb{F}_{2^{210}}$. We first analyze the GHS attack in this case, and then discuss how to extend it beyond these 2^{175} isomorphism classes.

Let $E : y^2 + xy = x^3 + ax^2 + b$ be an elliptic curve over $\mathbb{F}_{2^{210}}$ with magic number $m = 5$ relative to $n = 6$. First note that by [28, Lemma 8] and [20] we require that $\mathrm{Tr}_{\mathbb{F}_{2^N}/\mathbb{F}_2}(a) = 0$ for the GHS reduction to yield a group homomorphism $\Phi : E(\mathbb{F}_{2^{210}}) \to J_C(\mathbb{F}_{2^{35}})$ into the jacobian of a hyperelliptic curve C defined over $\mathbb{F}_{2^{35}}$. We thus restrict ourselves to curves of the form $E : y^2 + xy = x^3 + b$. Then $m = 5$ only if $b \in \mathbb{F}_{2^{210}} \setminus \mathbb{F}_{2^{35}}$, while if $b \in \mathbb{F}_{2^{35}}$ we have $m = 1$ and $\#E(\mathbb{F}_{2^{35}}) \mid \#E(\mathbb{F}_{2^{210}})$. Again, the GHS reduction takes only a few seconds. The resulting hyperelliptic curve has genus 15 or 16. Similar to Theorem 3, this time using $t(x) = x^4 + x^2 + 1$ and taking into account that $\mathrm{Tr}_{\mathbb{F}_{2^{210}}/\mathbb{F}_2}(a) = 0$ we can show that there are exactly $2^{140} - 2^{70}$ isomorphism classes of elliptic curves defined over $\mathbb{F}_{2^{210}} \setminus \mathbb{F}_{2^{35}}$ for which the GHS attack yields a hyperelliptic curve C defined over $\mathbb{F}_{2^{35}}$ of genus $g = 15$. For exactly $(2^{140} - 2^{70})(2^{35} - 1) \approx 2^{175}$ isomorphism classes, a genus 16 curve is obtained.

Using the Enge-Gaudry index-calculus algorithm with a factor base of $w \approx 2^{34}$ degree-one prime divisors, it takes an expected number of $T_1 \approx 2^{34+35g}/M(1)$ random walk steps in the jacobian to complete the relation generation stage. For the case $g = 16$, we get $R_1 \approx c_{35}2^{90}$ and $R_2 \approx c_r 2^{72}$.

5.2   Extended GHS Attack

We next discuss how to extend the GHS attack beyond the set of elliptic curves with magic number 5 relative to $n = 6$. We first classify the curves over $\mathbb{F}_{2^{210}}$ with $(n, m) = (6, 5)$.

Theorem 6 *Let $N \equiv 0 \pmod 6$, and let $E : y^2 + xy = x^3 + ax^2 + b$ be an elliptic curve over $\mathbb{F}_{2^N}$ with magic number $m = 5$ relative to $n = 6$. Then $\mathrm{Tr}_{\mathbb{F}_{2^N}/\mathbb{F}_2}(b) = 0$.*

Proof. Let $N = n \cdot l$, $q = 2^l$, and let $\sigma : \mathbb{F}_{2^N} \to \mathbb{F}_{2^N}$ denote the power-q Frobenius $\alpha \mapsto \alpha^q$. Since $x^6 - 1 = (x-1)^2(x^2 + x + 1)^2$ over $\mathbb{F}_2$, there are two possibilities to obtain magic number 5, namely $\mathrm{Ord}_b(x) = (x-1)^j(x^2 + x + 1)^2$ with $j = 0$ or $j = 1$. If $j = 0$, then $0 = \mathrm{Ord}_b(\sigma)(b) = b^{q^4} + b^{q^2} + b$. Thus, $\mathrm{Tr}_{\mathbb{F}_{2^N}/\mathbb{F}_2}(b) = \sum_{i=0}^{N-1} b^{2^i} = \sum_{i=0}^{2l-1}(b + b^{q^2} + b^{q^4})^{2^i} = 0$. If $j = 1$, then $0 = \mathrm{Ord}_b(\sigma)(b) = b^{q^5} + b^{q^4} + b^{q^3} + b^{q^2} + b^q + b = 0$. Hence $\mathrm{Tr}_{\mathbb{F}_{2^N}/\mathbb{F}_2}(b) = \sum_{i=0}^{l-1}(b + b^q + b^{q^2} + b^{q^3} + b^{q^4} + b^{q^5})^{2^i} = 0$. $\qquad\square$

The following result is probably well known. We include a proof since we could not find it elsewhere.

Lemma 7 *Let $E : y^2 + xy = x^3 + b$ be an elliptic curve over $\mathbb{F}_{2^N}$, where $N \geq 3$. Then $\mathrm{Tr}_{\mathbb{F}_{2^N}/\mathbb{F}_2}(b) = 0$ if and only if $\#E(\mathbb{F}_{2^N}) \equiv 0 \pmod 8$.*

Proof. Since the 2^s-torsion group ($s \in \mathbb{N}$) of E over the algebraic closure $\overline{\mathbb{F}}_{2^N}$ is cyclic, we equivalently show that $\mathrm{Tr}_{\mathbb{F}_{2^N}/\mathbb{F}_2}(b) = 0$ if and only if $E(\mathbb{F}_{2^N})$ has a point of order 8. The 8th division polynomial of E is given by

$$f_8(x) = x^{28} + (b^2 + b)x^{20} + (b^4 + b^3)x^{12} + b^6 x^4 = (x^6 + bx^2)^2(x^{16} + bx^8 + b^4),$$

where $x^6 + bx^2$ is the 4th division polynomial. In the very last term, we substitute x^8 by z. Then E has a point of order 8 with x-coordinate defined over $\mathbb{F}_{2^N}$ if and only if $z^2 + bz + b^4$ factors over $\mathbb{F}_{2^N}$, which is the case if and only if $\mathrm{Tr}_{\mathbb{F}_{2^N}/\mathbb{F}_2}(b^2) = \mathrm{Tr}_{\mathbb{F}_{2^N}/\mathbb{F}_2}(b) = 0$. It remains to show that if $\mathrm{Tr}_{\mathbb{F}_{2^N}/\mathbb{F}_2}(b) = 0$, then the y-coordinate is also defined over $\mathbb{F}_{2^N}$. So assume $\mathrm{Tr}_{\mathbb{F}_{2^N}/\mathbb{F}_2}(b) = 0$, let $\beta \in \mathbb{F}_{2^N}$ such that $b = \beta^2 + \beta$, and let $t \in \mathbb{F}_{2^N}$ such that $\beta = t^8$. Then

$$z^2 + bz + b^4 = (z + t^{32} + t^{24} + t^{16} + t^8)(z + t^{32} + t^{24}),$$

and the eighth root $t^4 + t^3$ of $t^{32} + t^{24}$ is the x-coordinate of the 8-division point. It is easily verified that $y = t^8 + t^5 \in \mathbb{F}_{2^N}$ is an appropriate y-coordinate. $\qquad\square$

Corollary 8 *Let $E : y^2 + xy = x^3 + ax^2 + b$ be an elliptic curve over $\mathbb{F}_{2^{210}}$ with magic number $m = 5$ relative to $n = 6$ and for which the GHS reduction yields a group homomorphism $\Phi : E(\mathbb{F}_{2^{210}}) \to J_C(\mathbb{F}_{2^{35}})$ into the jacobian of a hyperelliptic curve C defined over $\mathbb{F}_{2^{35}}$ (of genus 15 or 16). Then $\#E(\mathbb{F}_{2^{210}}) \equiv 0 \pmod 8$.*

Proof. By [28, Lemma 8] and [20], we require that $\mathrm{Tr}_{\mathbb{F}_{2^{210}}/\mathbb{F}_2}(a) = 0$ for the GHS reduction to work when $(n, m) = (6, 5)$. The statement then immediately follows from Theorem 6 and Lemma 7. $\qquad\square$

Extending the GHS attack to the entire isogeny class of a weak curve over $\mathbb{F}_{2^{210}}$. We argue that for all but a few elliptic curves E with $\#E(\mathbb{F}_{2^{210}}) \equiv 0$ (mod 8), any ECDLP instance can be solved essentially in running time $R_1 \approx c_{35}2^{90}$.

An isogeny between two elliptic curves E and E' over a field K is a non-constant morphism $\Psi : E \rightarrow E'$ such that the neutral element of E is mapped to the neutral element of E'. The curves E and E' are called isogenous over K if Ψ is defined over K; we write $E \sim E'$. If K is a finite field, then $E \sim E'$ if and only if $\#E(K) = \#E'(K)$. The equivalence classes with respect to isogeny are called isogeny classes.

Let E be a non-supersingular elliptic curve over $\mathbb{F}_{2^N}$. We call $t = 2^N + 1 - \#E(\mathbb{F}_{2^N})$ its trace and $\Delta = t^2 - 4 \cdot 2^N$ its discriminant; note that $\Delta < 0$ and $\Delta \equiv 1$ (mod 8). The endomorphism ring $\text{End}(E)$ of E is an order in the maximal order $\mathcal{O}$ of the imaginary quadratic number field $\mathbb{Q}(\sqrt{\Delta})$. More precisely, $\mathbb{Z}[\pi] \subseteq \text{End}(E) \subseteq \mathcal{O}$, where $\pi : E \rightarrow E$ is the 2^N-th power Frobenius map on E. The endomorphism class of E, denoted by $\mathcal{C}(E)$, is the set of all isogenous, non-isomorphic curves E' with $\text{End}(E) = \text{End}(E')$. There exists a one-to-one correspondence between the Picard group (denoted $\text{Cl}(\text{End}(E))$) of the order $\text{End}(E)$ and $\mathcal{C}(E)$ ([10, Th. 3.4.6]).

For any elliptic curve E over $\mathbb{F}_{2^{210}}$ we can use an algorithm of Kohel [25] to compute a chain of isogenies defined over $\mathbb{F}_{2^{210}}$ from E to an elliptic curve E' with $\text{End}(E') = \mathcal{O}$. This takes running time $O(s^3)$, where s is the largest prime dividing the conductor $c = [\mathcal{O} : \text{End}(E)]$ of $\text{End}(E)$. Note that c divides $[\mathcal{O} : \mathbb{Z}[\pi]]$. In practice, $[\mathcal{O} : \mathbb{Z}[\pi]]$ is small and smooth so that Kohel's algorithm takes negligible time compared to R_1. For the following, we therefore may assume that $\text{End}(E)$ is maximal. Then $\text{Cl}(\text{End}(E))$ is the ideal class group of the maximal order $\mathcal{O}$, which we simply denote by Cl.

Now, there exist 2^{209} isomorphism classes of elliptic curves $E_{0,b}$ over $\mathbb{F}_{2^{210}}$ with $\text{Tr}_{\mathbb{F}_{2^{210}}/\mathbb{F}_2}(b) = 0$ (i.e., with group order divisible by 8), and $2^{175} - 2^{105}$ elliptic curves $E_{0,b}$ over $\mathbb{F}_{2^{210}}$ with $(n, m) = (6, 5)$. It is therefore reasonable to expect that a randomly chosen elliptic curve over $\mathbb{F}_{2^{210}}$ with group order divisible by 8 has magic number 5 relative to $n = 6$ with probability approximately $2^{175}/2^{209} = 2^{-34}$. Moreover, we make the heuristic assumption that the same is true when E is chosen randomly from a fixed endomorphism class.

Assumption A. Let $E = E_{0,b}$ an elliptic curve over $\mathbb{F}_{2^{210}}$ with $\#E_{0,b}(\mathbb{F}_{2^{210}}) \equiv 0$ (mod 8) and such that $\mathcal{C}(E)$ contains a curve with magic number $m = 5$ relative to $n = 6$. Then any curve E' that is randomly chosen from $\mathcal{C}(E)$ (with respect to the uniform distribution) has magic number $m = 5$ relative to $n = 6$ with probability 2^{-34}.

Remark 9 *(further justification of Assumption A) For arbitrary $N \equiv 0$ (mod 6), let $N = 6l$ and $q = 2^l$. By [30, Theorem 5], there exist $q^5 - q^3 = 2^{5N/6} - 2^{N/2}$ isomorphism classes of elliptic curves $E_{0,b}$ over $\mathbb{F}_{2^N}$ with $(n, m) = (6, 5)$, while there exist 2^{N-1} isomorphism classes $E_{0,b}$ with $\text{Tr}_{\mathbb{F}_{2^N}/\mathbb{F}_2}(b) = 0$.*

Thus, the probability in the above Assumption A generalizes to $2^{-(N/6-1)}$. *For* $36 \leq N \leq 84$, *this has been confirmed in extensive experiments.*

Remark 10 (*restriction of Assumption A*) *Of course, Assumption A is not accurate if* Cl *is very small, in the order of* $210 \cdot 2^{34} \approx 2^{42}$ *and smaller. This can happen only if* $\#E(\mathbb{F}_{2^{210}})$ *lies at the extreme ends of the Hasse interval and thus* Δ *is significantly smaller than its expected value* 2^{212}, *or if* Δ *has a very large square factor. But note that* $\Delta < 2^{150}$ *if and only if* $|t| > \sqrt{2^{212} + 2^{150}}$, *which affects only a very small fraction of at most* $1/2^{63}$ *of the elliptic curves over* $\mathbb{F}_{2^{210}}$; *the proportion of elliptic curves over* $\mathbb{F}_{2^{210}}$ *that have* $\Delta < 2^{100}$ *is at most* $1/2^{113}$. *If* $\Delta > 2^{150}$ *and* $\Delta = f^2 d$ *with* $d \equiv 1 \pmod{8}$ *and squarefree, then* $\#\text{Cl} \leq 2^{42}$ *only if* f *is (roughly) at least* 2^{30}, *which is most unlikely for non-subfield curves.*

Remark 11 (*the exceptional set in Assumption A*) *Our reasoning below does not apply when* $\mathcal{C}(E)$ *does not contain a curve with* $(n, m) = (6, 5)$. *We expect this to be a very rare case, that, again, should only happen when* Cl *is very small. As with the previous remark, this affects only a tiny proportion of elliptic curves over* $\mathbb{F}_{2^{210}}$ *with group order divisible by 8.*

Given a curve E over $\mathbb{F}_{2^{210}}$ with group order divisible by 8, it is now possible to compute a curve E' over $\mathbb{F}_{2^{210}}$, isogenous to E and with $(n, m) = (6, 5)$ along with a chain of low-degree isogenies from E to E'. This is based on ideas from [15] to simulate a random walk in the endomorphism class of E, exploiting the above one-to-one correspondence between Cl and $\mathcal{C}(E)$. This works as follows: Let $E = E_{0,b}$, let $j(E) = b^{-1}$ be its j-invariant, and let p be a prime with $\left(\frac{\Delta}{p}\right) = 1$. Then p splits in $\mathcal{O}$, $(p) = \mathfrak{p}_1 \mathfrak{p}_2$, and the modular polynomial $\Phi_p(j(E), X)$ has two roots j_1 and j_2 in $\mathbb{F}_{2^{210}}$ [13]. These roots can be computed by a probabilistic algorithm using $O(210p^2)$ operations in $\mathbb{F}_{2^{210}}$. The two isogenies mapping E to $E_{0,j_1^{-1}}$ and $E_{0,j_2^{-1}}$ correspond to the multiplication of a fixed ideal, say $\mathcal{O}$, by the two prime ideals $\mathfrak{p}_1$ and $\mathfrak{p}_2$ lying over p. As explained in [15], it is easy to determine whether j_1 corresponds to $\mathfrak{p}_1$ or $\mathfrak{p}_2$. Now, let $\mathcal{P}$ be the set of the 30 smallest primes p such that $\left(\frac{\Delta}{p}\right) = 1$, and such that the pairs of ideal classes corresponding to the prime ideals lying over p are pairwise distinct in Cl. A pseudo-random walk (E_i) in $\mathcal{C}(E)$ is defined as follows: Let $E_0 = E_{0,b}$ and $b_0 = b$ and $\mathfrak{a}_0 = \mathcal{O}$. For $i = 1, 2, \ldots$, let $p \in_R \mathcal{P}$ and $j = b_{i-1}$, and compute the two roots in $\mathbb{F}_{2^{210}}$ of $\Phi_p(j, X)$; let j' be one of these roots, and let $b_i = (j')^{-1}$. Simultaneously a chain $(\mathfrak{a}_i)$ of ideals in Cl is computed such that for each index k, the ideal $\mathfrak{a}_k$ corresponds to the isogeny mapping E to E_k.

The set $\mathcal{P}$ has been chosen such that the walk (E_i) indeed simulates a random walk in the endomorphism class of E. Experimentally, we found that $\max\{p \in \mathcal{P}\} \in [190, 530]$, where we considered 5000 randomly chosen discriminants, with only 2 discriminants yielding maximum values > 500 (and we obtained $\max\{p \in \mathcal{P}\} \in [150, 380]$ if we required only $\#\mathcal{P} = 20$). Thus, each random-walk step takes up to about $210 \cdot 500^2 \approx 2^{26}$ operations in $\mathbb{F}_{2^{210}}$, given that computing the roots of the modular polynomial is by far the most time-consuming step.

Now, under Assumption A, after expected 2^{34} random-walk steps in $\mathcal{C}(E)$ an elliptic curve E_k over $\mathbb{F}_{2^{210}}$ is encountered that is isogenous to E and whose magic number relative to $n = 6$ is $m = 5$. Thus, altogether it takes something on the order of 2^{60} operations in $\mathbb{F}_{2^{210}}$ to find a curve with $(n, m) = (6, 5)$ isogenous to a given curve over $\mathbb{F}_{2^{210}}$, along with an ideal $\mathfrak{a}$ that represents the isogeny between the two curves. We note that this running time is negligible compared to R_1 and R_2. Also, this step can be efficiently parallelized.

The remaining steps to compute the explicit isogeny between E and E_k are identical with Stages 2 and 3 of [15]: index-calculus techniques are used to represent $\mathfrak{a}$ as a product of just a few ideals of small norm, and finally Vélu's formulae are applied. This can be accomplished in time $O(2^{N/4+\varepsilon}) = const \cdot 2^{53}$, which also is negligible when compared to R_1 and R_2.

5.3   Further Extension to Elliptic Curves with $\mathrm{Tr}_{\mathbb{F}_{2^{210}}/\mathbb{F}_2}(b) \neq 0$

We further extend the set of elliptic curves over $\mathbb{F}_{2^{210}}$ for which any ECDLP instance can be solved potentially faster than applying Pollard's rho method to the hardest ECDLP instances over $\mathbb{F}_{2^{210}}$. For this, we use Hess's recent generalization [20] of the GHS attack to reduce instances of the ECDLP to instances of a discrete logarithm problem in the divisor class group of a curve C over $\mathbb{F}_{2^{35}}$. Note that this curve C is in general not hyperelliptic. Nevertheless, subexponential-time methods for discrete logarithm computation are available for such curves of large genus (see [20] and the references given there). However we do not have an exact analysis of their running times.

Let $N = nl$ for some integers n and l. Consider the elliptic curve $E : y^2 + xy = x^3 + ax^2 + b$ over $\mathbb{F}_{2^N}$ with $b \neq 0$, and let $\langle P \rangle$ be a subgroup of $E(\mathbb{F}_{2^N})$ of prime order r.

Let $q = 2^l$, and for $\gamma \in \mathbb{F}_{2^N}$ let $\mathrm{Ord}_\gamma(x)$ denote the unique monic polynomial $f \in \mathbb{F}_2[x]$ of least degree such that $f(\sigma)(\gamma) = 0$ where σ is the power-q Frobenius.

Let $\gamma_1, \gamma_2 \in \mathbb{F}_{2^N}$ such that $b = (\gamma_1 \gamma_2)^2$. Let $c = 1/\gamma_1$; then $\gamma_2 = b^{1/2}c$. Let $s_i = \deg(\mathrm{Ord}_{\gamma_i})$ $(i = 1, 2)$ and $t = \deg(\mathrm{lcm}(\mathrm{Ord}_{\gamma_1}, \mathrm{Ord}_{\gamma_2}))$. Via a birational transformation the defining equation of E can be brought into the form $y^2 + y = 1/(cx) + a + b^{1/2}cx$. Then Hess's generalization [20, Theorems 4,5,7] of the GHS attack allows one to effectively reduce the ECDLP in $\langle P \rangle$ to the DLP in a subgroup of order r of the divisor class group of an explicitly computable curve C over $\mathbb{F}_{2^l}$ of genus $g = 2^t - 2^{t-s_1} - 2^{t-s_2} + 1$, provided that $\mathrm{Tr}_{\mathbb{F}_{2^N}/\mathbb{F}_2}(a) = 0$ or $\mathrm{Tr}_{\mathbb{F}_{q^n}/\mathbb{F}_q}(\gamma_1) \neq 0$ or $\mathrm{Tr}_{\mathbb{F}_{q^n}/\mathbb{F}_q}(\gamma_2) \neq 0$. The following theorem, due to Hess [21], gives an effective method to obtain curves C of relatively small genus, which applies to all but a small exceptional set of elliptic curves over $\mathbb{F}_{2^{210}}$. Here, Φ denotes the Euler phi function for polynomials, i.e., for $m(x) \in \mathbb{F}_2[x]$, $\Phi(m(x))$ is the number of elements $\gamma \in \mathbb{F}_{2^N}$ with $\mathrm{Ord}_\gamma(x) = m(x)$.

Theorem 12 *Let $n = n_1 n_2$, $K = \mathbb{F}_{q^n}$, $k = \mathbb{F}_q$ and $K_1 = \mathbb{F}_{q^{n_1}}$. Let $\beta \in K$.*

(i) There exist $\gamma_1, \gamma_2 \in K$ with $\beta = \gamma_1 \gamma_2$ and

$$\mathrm{Ord}_{\gamma_1}(x) \mid (x^{n_1} - 1) \quad \text{and} \quad \mathrm{Ord}_{\gamma_2}(x) \mid \frac{(x-1)(x^n - 1)}{x^{n_1} - 1}. \tag{6}$$

(ii) If $\mathrm{Tr}_{K/K_1}(\beta) \neq 0$ then $\gamma_1 = \mathrm{Tr}_{K/K_1}(\beta)$ and $\gamma_2 = \beta/\gamma_1$ satisfy (6).

(iii) If $\mathrm{Tr}_{K/K_1}(\beta) = 0$ then $\gamma_1 = 1$ and $\gamma_2 = \beta$ satisfy (6).

(iv) Let $m_1, m_2 \in \mathbb{F}_2[x]$ such that $m_1 \mid (x^{n_1} - 1)$ and $m_2 \mid (x-1)(x^n-1)/(x^{n_1}-1)$. Let $\mathrm{Tr}_{K/K_1}(\beta) \neq 0$. The number of $\beta = \gamma_1\gamma_2$ such that $\mathrm{Ord}_{\gamma_1}(x) = m_1(x)$ and $\mathrm{Ord}_{\gamma_2}(x) = m_2(x)$ is $\Phi(m_1(x))\Phi(m_2(x))/(q-1)$.

(v) In the case of (ii) we have:
    (a) $k(\beta) = k(\gamma_1, \gamma_2)$.
    (b) $\mathrm{Tr}_{K/k}(\gamma_2) \neq 0$ if and only if n_1 odd.
    (c) $\mathrm{Tr}_{K/k}(\gamma_1) \neq 0$ if and only if $v_{x+1}(\mathrm{Ord}_{\gamma_1}(x)) = 2^{v_2(n)}$.
       In general, $v_{x+1}(\mathrm{Ord}_{\gamma_1}(x)) \leq 2^{v_2(n)}$.

Proof. Let us first note that for $\gamma \in K$, $\mathrm{Ord}_\gamma(x)$ divides $(x^{n_1} - 1)$ if and only if $\gamma \in K_1$. Further, $\mathrm{Ord}_\gamma(x)$ divides $(x-1)(x^n - 1)/(x^{n_1} - 1)$ if and only if $\mathrm{Tr}_{K/K_1}(\gamma + \gamma^q) = 0$, and the latter implies $\mathrm{Tr}_{K/K_1}(\gamma) \in k$.

(i) Follows from (ii) and (iii).

(ii) Let $\gamma_1 = \mathrm{Tr}_{K/K_1}(\beta) \neq 0$ and $\gamma_2 = \beta/\gamma_1$. Then $\gamma_1 \in K_1$. Also, $\mathrm{Tr}_{K/K_1}(\gamma_2) = \mathrm{Tr}_{K/K_1}(\beta)/\gamma_1 = 1$, and by the additivity of the trace function $\mathrm{Tr}_{K/K_1}(\gamma_2 + \gamma_2^q) = 0$. Now, for $g(x) = (x^n - 1)/(x^{n_1} - 1) = 1 + x^{n_1} + x^{2n_1} + \cdots + x^{(n_2-1)n_1}$ we have $g(\sigma)(\alpha) = \mathrm{Tr}_{K/K_1}(\alpha)$ for all $\alpha \in K$. Thus,

$$(\sigma - 1)\left(\frac{\sigma^n - 1}{\sigma^{n_1} - 1}\right)(\gamma_2) = \left(\frac{\sigma^n - 1}{\sigma^{n_1} - 1}\right)(\gamma_2 + \gamma_2^q) = 0.$$

(iii) If $\mathrm{Tr}_{K/K_1}(\beta) = 0$ then, for any $\gamma_1 \in K_1$ and $\gamma_2 = \beta/\gamma_1$, we also have $\mathrm{Tr}_{K/K_1}(\gamma_2) = 0$.

(iv) Suppose $\mathrm{Tr}_{K/K_1}(\beta) \neq 0$. For $i = 1, 2$, there are $\Phi(m_i(x))$ elements γ_i such that $\mathrm{Ord}_{\gamma_i}(x) = m_i(x)$. Now, let $\beta = \gamma_1\gamma_2 = \gamma_1'\gamma_2'$ where $\mathrm{Ord}_{\gamma_i}(x) = \mathrm{Ord}_{\gamma_i'}(x) = m_i(x)$. Since $m_1(x)$ divides $x^{n_1} - 1$, we have $\gamma_1' = \gamma_1/\lambda$ for some $\lambda \in K_1$. Now,

$$\begin{aligned}
0 &= \mathrm{Tr}_{K/K_1}(\gamma_2' + (\gamma_2')^q) = \mathrm{Tr}_{K/K_1}(\lambda\gamma_2 + (\lambda\gamma_2)^q) \\
&= \mathrm{Tr}_{K/K_1}(\lambda\gamma_2 + (\lambda\gamma_2)^q) + 2\mathrm{Tr}_{K/K_1}(\lambda\gamma_2^q) \\
&= \lambda\mathrm{Tr}_{K/K_1}(\gamma_2 + \gamma_2^q) + (\lambda + \lambda^q)\mathrm{Tr}_{K/K_1}(\gamma_2^q) = (\lambda + \lambda^q)\mathrm{Tr}_{K/K_1}(\gamma_2^q).
\end{aligned}$$

Since $\mathrm{Tr}_{K/K_1}(\beta) \neq 0$, also $\mathrm{Tr}_{K/K_1}(\gamma_2^q) \neq 0$. Thus, $\lambda + \lambda^q = 0$, and therefore $\lambda \in \mathbb{F}_q$, $\lambda \neq 0$. On the other hand, since $f(\sigma)(\lambda\gamma) = \lambda f(\sigma)(\gamma)$ for any $\lambda \in \mathbb{F}_q$ and $f \in \mathbb{F}_2[x]$, if $\lambda \in \mathbb{F}_q$, $\lambda \neq 0$, then $\mathrm{Ord}_\gamma(x) = \mathrm{Ord}_{\lambda\gamma}(x)$.

(v) (a) We have $\beta \in k(\gamma_1, \gamma_2)$, since $\beta = \gamma_1\gamma_2$. Further, $\gamma_1, \gamma_2 \in k(\beta)$ since $\gamma_1 = \mathrm{Tr}_{K/K_1}(\beta)$ and $k(\beta)$ is Galois.
    (b) We have

$$\begin{aligned}
\mathrm{Tr}_{K/k}(\gamma_2) = \mathrm{Tr}_{K/k}(\beta/\gamma_1) &= \mathrm{Tr}_{K_1/k}\left(\mathrm{Tr}_{K/K_1}(\beta/\gamma_1)\right) \\
&= \mathrm{Tr}_{K_1/k}(1) = [K_1 : k] = n_1.
\end{aligned}$$

(c) We have $v_{x+1}(x^n - 1) = 2^{v_2(n)}$. If $v_{x+1}(\operatorname{Ord}(\gamma_1)) < 2^{v_2(n)}$ then $\operatorname{Ord}(\gamma_1) \mid \frac{x^n-1}{x-1}$, hence $\operatorname{Tr}_{K/k}(\gamma_1) = 0$. The result follows.   □

Corollary 13 *For any elliptic curve* $E : y^2 + xy = x^3 + ax^2 + b$ *over* $\mathbb{F}_{q^6}$ *with* $\operatorname{Tr}_{\mathbb{F}_{q^6}/\mathbb{F}_{q^3}}(b) \neq 0$, *Hess's generalization [20] of the GHS attack can be used to reduce the ECDLP in* $E(\mathbb{F}_{q^6})$ *to the DLP in the divisor class group of a curve over* $\mathbb{F}_q$ *of genus at most 14.*

Proof. Let $\beta = b^{1/2}$. We apply Theorem 12 with $n = 6$ and $n_1 = 3$. With $\gamma_1 = \operatorname{Tr}_{\mathbb{F}_{q^6}/\mathbb{F}_{q^3}}(\beta)$ and $\gamma_2 = \beta/\gamma_1$, Hess's generalization of the GHS attack applies and yields a curve over $\mathbb{F}_q$ of genus $g = 2^t - 2^{t-s_1} - 2^{t-s_2} + 1$, where s_1, s_2, t are as defined above. Table 3 lists all possible values for g for the various choices of $\operatorname{Ord}_{\gamma_1}$ and $\operatorname{Ord}_{\gamma_2}$.   □

Table 3. Genera in Hess's generalization of the GHS attack (see Corollary 13) for fields $\mathbb{F}_{q^6}$.

$\operatorname{Ord}_{\gamma_1}(x)$	s_1	$\operatorname{Ord}_{\gamma_2}(x)$	s_2	t	g
$x + 1$	1	$x + 1$	1	1	1
$x + 1$	1	$(x + 1)^2$	2	2	2
$x + 1$	1	$x^2 + x + 1$	2	3	3
$x + 1$	1	$(x + 1)(x^2 + x + 1)$	3	3	4
$x + 1$	1	$(x + 1)^2(x^2 + x + 1)$	4	4	8
$x^2 + x + 1$	2	$(x + 1)^2$	2	4	9
$x^2 + x + 1$	2	$x^2 + x + 1$	2	3	5
$x^2 + x + 1$	2	$(x + 1)(x^2 + x + 1)$	3	3	6
$x^2 + x + 1$	2	$(x + 1)^2(x^2 + x + 1)$	4	4	12
$x^3 - 1$	3	$(x + 1)^2$	2	4	11
$x^3 - 1$	3	$(x + 1)(x^2 + x + 1)$	3	3	7
$x^3 - 1$	3	$(x + 1)^2(x^2 + x + 1)$	4	4	14

From Theorem 12(iv) and [30, Theorem 5] we see that the vast majority of elliptic curves over $\mathbb{F}_{q^6}$ yield a genus 14 curve over $\mathbb{F}_q$.

Remark 14 *(application of Theorem 12 to fields* $\mathbb{F}_{2^{4l}}$*) For* $n = 4$ *we can use the same technique to show that Hess's generalization of the GHS attack can be used to produce curves over* $\mathbb{F}_{2^l}$ *of genus at most 6. In fact, for most such curves the genus is equal to 6 and in general the resulting curve will be non-hyperelliptic.*

5.4   Comparisons

Table 4 shows the costs R_ρ, R_1, R_2 for the attacks on the ECDLP for elliptic curves defined over $\mathbb{F}_{2^{210}}$ as discussed in this section.

Consequently, for all elliptic curves over $\mathbb{F}_{2^{210}}$, the ECDLP can be solved about 2^{13} times faster than it takes Pollard's rho method to solve the hardest

Table 4. Time estimates for Pollard's rho method for solving an ECDLP instance in $E(\mathbb{F}_{2^{210}})$, and for the relation generation and matrix stages of the Enge-Gaudry algorithm for solving an HCDLP instance in $J_C(\mathbb{F}_{2^{42}})$ and $J_C(\mathbb{F}_{2^{35}})$ where C is a genus 16 hyperelliptic curve. c_{210}, c_l and c_r are the relative times for a multiplication in $\mathbb{F}_{2^{210}}$, $\mathbb{F}_{2^l}$, and modulo an 210-bit prime, respectively.

N	n	l	R_ρ/c_{210}	R_1/c_l	R_2/c_r	c_{210}	c_l	c_r	R_ρ	R_1	R_2
210	5	42	$2^{107.5}$	2^{97}	2^{86}	10.3	1.0	8.0	$2^{110.5}$	2^{97}	2^{89}
210	6	35	$2^{107.5}$	2^{90}	2^{72}	10.3	1.0	8.0	$2^{110.5}$	2^{90}	2^{75}

instances. For about a quarter of all curves over $\mathbb{F}_{2^{210}}$ (those with $\mathrm{Tr}_{\mathbb{F}_{2^{210}}/\mathbb{F}_2}(a) = \mathrm{Tr}_{\mathbb{F}_{2^{210}}/\mathbb{F}_2}(b) = 0$) the ECDLP can be solved about 2^{20} times faster than with Pollard's rho method. As argued in Section 5.3, for essentially all elliptic curves over $\mathbb{F}_{2^{210}}$, the ECDLP presumably can be solved significantly faster than with Pollard's rho method, although an exact analysis has not been conducted.

6   Conclusions

We have argued that the fields $\mathbb{F}_{2^N}$, where $N \in [185, 600]$ is divisible by 5, are weak for ECC. The fundamental open problem is to determine whether there are any fields that are bad for ECC. We have provided some evidence that the field $\mathbb{F}_{2^{210}}$ is a prime candidate for being bad.

Another candidate for a bad field is $\mathbb{F}_{2^{161}}$. For 2^{94} of the 2^{162} isomorphism classes of elliptic curves E over $\mathbb{F}_{2^{161}}$, the GHS reduction yields a hyperelliptic curve C of genus (7 or) 8 over $\mathbb{F}_{2^{23}}$, where the HCDLP is feasible. In our notation, we have $R_\rho = c_E 2^{80}$, $R_1 = (c_J + c_S)2^{37}$, and $R_2 = c_r 2^{47}$, where c_E denotes the time to perform an elliptic curve addition in $E(\mathbb{F}_{2^{161}})$, c_J is the time to perform an addition in $J_C(\mathbb{F}_{2^{23}})$, c_S is the time to test whether a monic polynomial $a \in \mathbb{F}_{2^{23}}[u]$ of degree 8 is 1-smooth, and c_r is the time to perform a multiplication modulo a 160-bit prime. If an arbitrary ECDLP instance over $\mathbb{F}_{2^{161}}$ can be efficiently mapped to an ECDLP instance for an isogenous elliptic curve that belongs to the aforementioned class of 2^{94} curves, then one would conclude that $\mathbb{F}_{2^{161}}$ is bad for ECC (see also [15] and [28, Remark 20]). No such mapping is known so far (see also [43]).

An important open question in hyperelliptic curve cryptography is whether there are algorithms for solving the HCDLP curve that are faster than the Enge-Gaudry algorithm. Because of the relevance to solving the ECDLP, improvements by a constant factor would be of interest. For example, the possibility of using sieving (see [9]) to generate relations needs to be further explored.

Galbraith [14] has shown that Weil descent can be used to attack the HCDLP over some low genus hyperelliptic curves defined over characteristic two finite fields of composite extension degrees. An open problem is to determine whether there are any weak fields for genus two hyperelliptic curve cryptography.

Acknowledgements. We would like to thank Mike Jacobson and Darrel Hankerson for answering our questions about the relative speeds of finite field, elliptic

384   A. Menezes, E. Teske, and A. Weng

curve, and hyperelliptic curve operations. Florian Hess kindly provided us with
Theorem 12, which strengthened our results in Section 5.3. Thanks also to Mark
Bauer for reviewing the paper.

References

1. L. ADLEMAN, J. DEMARRAIS AND M. HUANG, "A subexponential algorithm for
   discrete logarithms over the rational subgroup of the jacobians of large genus hyperelliptic curves over finite fields", *Algorithmic Number Theory*, LNCS 877 (1994),
   28–40.
2. P. BARRETT, "Implementing the Rivest Shamir and Adleman public key encryption algorithm on a standard digital signal processor", *Advances in Cryptology—
   CRYPTO '86*, LNCS 263 (1987), 311–323.
3. D. BERNSTEIN, "Circuits for integer factorization: A proposal", preprint, 2001.
4. D. CANTOR, "Computing in the jacobian of a hyperelliptic curve", *Math. Comp.*,
   48 (1987), 95–101.
5. D. CANTOR AND H. ZASSENHAUS, "A new algorithm for factoring polynomials
   over finite fields", *Math. Comp.*, 36 (1981), 587–592.
6. D. COPPERSMITH, A. ODLYZKO AND R. SCHROEPPEL, "Discrete logarithms in
   $GF(p)$", *Algorithmica*, 1 (1986), 1–15.
7. A. ENGE AND P. GAUDRY, "A general framework for subexponential discrete logarithm algorithms", *Acta Arithmetica*, 102 (2002), 83–103.
8. FIPS 186-2, "Digital signature standard (DSS)", Federal Information Processing Standards Publication 186–2, National Institute of Standards and Technology,
   2000.
9. R. FLASSENBERG AND S. PAULUS, "Sieving in function fields", *Experimental Mathematics*, 8 (1999), 339–349.
10. M. FOUQUET, "Anneau d'endomorphismes et cardinalité des courbes elliptiques:
    aspects algorithmiques", PhD thesis, École polytechnique, Palaiseau Cedex, 2001.
11. G. FREY, "Applications of arithmetical geometry to cryptographic constructions",
    *Proceedings of the Fifth International Conference on Finite Fields and Applica-
    tions*, Springer-Verlag, 2001, 128–161.
12. G. FREY AND H. RÜCK, "A remark concerning m-divisibility and the discrete
    logarithm in the divisor class group of curves", *Math. Comp.*, 62 (1994), 865–874.
13. S. GALBRAITH, "Constructing isogenies between elliptic curves over finite fields",
    LMS Journal of Computation and Mathematics, 2 (1999), 118-138.
14. S. GALBRAITH, "Weil descent of jacobians", *Discrete Applied Mathematics*, 12
    (2003), 165–180.
15. S. GALBRAITH, F. HESS AND N. SMART, "Extending the GHS Weil descent at-
    tack", *Advances in Cryptology—EUROCRYPT 2002*, LNCS 2332 (2002), 29–44.
16. R. GALLANT, R. LAMBERT AND S. VANSTONE, "Improving the parallelized Pollard
    lambda search on anomalous binary curves", *Math. Comp.*, 69 (2000), 1699–1705.
17. P. GAUDRY, "An algorithm for solving the discrete log problem in hyperelliptic
    curves", *Advances in Cryptology—EUROCRYPT 2000*, LNCS 1807 (2000), 19–34.
18. P. GAUDRY, F. HESS AND N. SMART, "Constructive and destructive facets of Weil
    descent on elliptic curves", *J. Cryptology*, 15 (2002), 19–46.
19. D. HANKERSON, personal communication, 2003.
20. F. HESS, "The GHS attack revisited", *Advances in Cryptology—EUROCRYPT
    2003*, LNCS 2656 (2003), 374–387.

21. F. HESS, personal communication, 2003.

22. M. JACOBSON, personal communication, 2003.

23. M. JACOBSON, A. MENEZES AND A. STEIN, "Solving elliptic curve discrete logarithm problems using Weil descent", *Journal of the Ramanujan Mathematical Society*, 16 (2001), 231–260.

24. M. JACOBSON AND A. VAN DER POORTEN, "Computational aspects of NUCOMP", *Algorithmic Number Theory—ANTS-IV*, LNCS 2369 (2002), 120–133.

25. D. KOHEL, "Endomorphism rings of elliptic curves over finite fields", PhD thesis, University of California, Berkeley, 1996.

26. F. KUHN AND R. STRUIK, "Random walks revisited: Extensions of Pollard's rho algorithm for computing multiple discrete logarithms", *Selected Areas in Cryptography—SAC 2001*, LNCS 2259 (2001), 212–229.

27. J. LÓPEZ AND R. DAHAB, "High-speed software multiplication in $\mathbb{F}_{2^m}$", *Progress in Cryptology—INDOCRYPT 2000*, LNCS 1977 (2000), 203–212.

28. M. MAURER, A. MENEZES AND E. TESKE, "Analysis of the GHS Weil descent attack on the ECDLP over characteristic two finite fields of composite degree", *LMS Journal of Computation and Mathematics*, 5 (2002), 127–174.

29. A. MENEZES, T. OKAMOTO AND S. VANSTONE, "Reducing elliptic curve logarithms to logarithms in a finite field", *IEEE Transactions on Information Theory*, 39 (1993), 1639–1646.

30. A. MENEZES AND M. QU, "Analysis of the Weil descent attack of Gaudry, Hess and Smart", *Topics in Cryptology—CT-RSA 2001*, LNCS 2020 (2001), 308–318.

31. P. VAN OORSCHOT AND M. WIENER, "Parallel collision search with cryptanalytic applications", *J. Cryptology*, 12 (1999), 1–28.

32. H. ORMAN, "The OAKLEY key determination protocol", RFC 2412, 1998. Available from http://www.ietf.org.

33. S. PAULUS AND A. STEIN, "Comparing real and imaginary arithmetics for divisor class groups of hyperelliptic curves", *Algorithmic Number Theory—ANTS-III*, LNCS 1423 (1998), 576–591.

34. J. POLLARD, "Monte Carlo methods for index computation mod p", *Math. Comp.*, 32 (1978), 918–924.

35. T. SATOH AND K. ARAKI, "Fermat quotients and the polynomial time discrete log algorithm for anomalous elliptic curves", *Commentarii Mathematici Universitatis Sancti Pauli*, 47 (1998), 81–92.

36. I. SEMAEV, "Evaluation of discrete logarithms in a group of p-torsion points of an elliptic curve in characteristic p", *Math. Comp.*, 67 (1998), 353–356.

37. V. SHOUP, *NTL: A library for doing Number Theory*. Available from http://shoup.net/ntl.

38. N. SMART, "The discrete logarithm problem on elliptic curves of trace one", *J. Cryptology*, 12 (1999), 193–196.

39. N. SMART, "How secure are elliptic curves over composite extension fields?", *Advances in Cryptology—Eurocrypt 2001*, LNCS 2045 (2001), 30–39.

40. J. SOLINAS, "Efficient arithmetic on Koblitz curves", *Designs, Codes and Cryptography*, 19 (2000), 195–249.

41. E. TESKE, "Speeding up Pollard's rho method for computing discrete logarithms", *Algorithmic Number Theory*, LNCS 1423 (1998), 541–554.

42. E. TESKE, "On random walks for Pollard's rho method", *Math. Comp.*, 70 (2000), 809–825.

43. E. TESKE, "An elliptic curve trapdoor system", *Cryptology ePrint Archive* Report 2003/058, 2003.

386   A. Menezes, E. Teske, and A. Weng

44. M. WIENER, "The full cost of cryptanalytic attacks", *J. Cryptology*, to appear.
45. M. WIENER AND R. ZUCCHERATO, "Faster attacks on elliptic curve cryptosystems", *Selected Areas in Cryptography—SAC '98*, LNCS 1556 (1999), 190–200.

Author Index

Lecture Notes in Computer Science

For information about Vols. 1–2827

please contact your bookseller or Springer-Verlag

Vol. 2879: R.E. Ellis, T.M. Peters (Eds.), Medical Image Computing and Computer-Assisted Intervention - MICCAI 2003. Proceedings, 2003. XXXIV, 1003 pages. 2003.

Vol. 2878: R.E. Ellis, T.M. Peters (Eds.), Medical Image Computing and Computer-Assisted Intervention - MICCAI 2003. Proceedings, 2003. XXXIII, 819 pages. 2003.

Vol. 2877: T. Böhme, G. Heyer, H. Unger (Eds.), Innovative Internet Community Systems. VIII, 263 pages. 2003.

Vol. 2876: M. Schroeder, G. Wagner (Eds.), Rules and Rule Markup Languages for the Semantic Web. Proceedings, 2003. VII, 173 pages. 2003.

Vol. 2875: E. Aarts, R. Collier, E.v. Loenen, B.d. Ruyter (Eds.), Ambient Intelligence. Proceedings, 2003. XI, 432 pages. 2003.

Vol. 2874: C. Priami (Ed.), Global Computing. XIX, 255 pages. 2003.

Vol. 2871: N. Zhong, Z.W. Raś, S. Tsumoto, E. Suzuki (Eds.), Foundations of Intelligent Systems. Proceedings, 2003. XV, 697 pages. 2003. (Subseries LNAI).

Vol. 2870: D. Fensel, K.P. Sycara, J. Mylopoulos (Eds.), The Semantic Web - ISWC 2003. Proceedings, 2003. XV, 931 pages. 2003.

Vol. 2869: A. Yazici, C. Şener (Eds.), Computer and Information Sciences - ISCIS 2003. Proceedings, 2003. XIX, 1110 pages. 2003.

Vol. 2868: P. Perner, R. Brause, H.-G. Holzhütter (Eds.), Medical Data Analysis. Proceedings, 2003. VIII, 127 pages. 2003.

Vol. 2866: J. Akiyama, M. Kano (Eds.), Discrete and Computational Geometry. VIII, 285 pages. 2003.

Vol. 2865: S. Pierre, M. Barbeau, E. Kranakis (Eds.), Ad-Hoc, Mobile, and Wireless Networks. Proceedings, 2003. X, 293 pages. 2003.

Vol. 2864: A.K. Dey, A. Schmidt, J.F. McCarthy (Eds.), UbiComp 2003: Ubiquitous Computing. Proceedings, 2003. XVII, 368 pages. 2003.

Vol. 2863: P. Stevens, J. Whittle, G. Booch (Eds.), "UML" 2003 - The Unified Modeling Language. Proceedings, 2003. XIV, 415 pages. 2003.

Vol. 2860: D. Geist, E. Tronci (Eds.), Correct Hardware Design and Verification Methods. Proceedings, 2003. XII, 426 pages. 2003.

Vol. 2859: B. Apolloni, M. Marinaro, R. Tagliaferri (Eds.), Neural Nets. X, 376 pages. 2003.

Vol. 2857: M.A. Nascimento, E.S. de Moura, A.L. Oliveira (Eds.), String Processing and Information Retrieval. Proceedings, 2003. XI, 379 pages. 2003.

Vol. 2856: M. Smirnov (Ed.), Quality of Future Internet Services. IX, 293 pages. 2003.

Vol. 2855: R. Alur, I. Lee (Eds.), Embedded Software. Proceedings, 2003. X, 373 pages. 2003.

Vol. 2854: J. Hoffmann, Utilizing Problem Structure in Planing. XIII, 251 pages. 2003. (Subseries LNAI).

Vol. 2853: M. Jeckle, L.-J. Zhang (Eds.), Web Services - ICWS-Europe 2003. VIII, 227 pages. 2003.

Vol. 2852: F.S. de Boer, M.M. Bonsangue, S. Graf, W.-P. de Roever (Eds.), Formal Methods for Components and Objects. VIII, 509 pages. 2003.

Vol. 2851: C. Boyd, W. Mao (Eds.), Information Security. Proceedings, 2003. XI, 453 pages. 2003.

Vol. 2849: N. García, L. Salgado, J.M. Martínez (Eds.), Visual Content Processing and Representation. Proceedings, 2003. XII, 352 pages. 2003.

Vol. 2848: F.E. Fich (Ed.), Distributed Computing. Proceedings, 2003. X, 367 pages. 2003.

Vol. 2847: R.d. Lemos, T.S. Weber, J.B. Camargo Jr. (Eds.), Dependable Computing. Proceedings, 2003. XIV, 371 pages. 2003.

Vol. 2846: J. Zhou, M. Yung, Y. Han (Eds.), Applied Cryptography and Network Security. Proceedings, 2003. XI, 436 pages. 2003.

Vol. 2845: B. Christianson, B. Crispo, J.A. Malcolm, M. Roe (Eds.), Security Protocols. VIII, 243 pages. 2004.

Vol. 2844: J.A. Jorge, N. Jardim Nunes, J. Falcão e Cunha (Eds.), Interactive Systems. Design, Specification, and Verification. XIII, 429 pages. 2003.

Vol. 2843: G. Grieser, Y. Tanaka, A. Yamamoto (Eds.), Discovery Science. Proceedings, 2003. XII, 504 pages. 2003. (Subseries LNAI).

Vol. 2842: R. Gavaldá, K.P. Jantke, E. Takimoto (Eds.), Algorithmic Learning Theory. Proceedings, 2003. XI, 313 pages. 2003. (Subseries LNAI).

Vol. 2841: C. Blundo, C. Laneve (Eds.), Theoretical Computer Science. Proceedings, 2003. XI, 397 pages. 2003.

Vol. 2840: J. Dongarra, D. Laforenza, S. Orlando (Eds.), Recent Advances in Parallel Virtual Machine and Message Passing Interface. Proceedings, 2003. XVIII, 693 pages. 2003.

Vol. 2839: A. Marshall, N. Agoulmine (Eds.), Management of Multimedia Networks and Services. Proceedings, 2003. XIV, 532 pages. 2003.

Vol. 2838: N. Lavrač, D. Gamberger, L. Todorovski, H. Blockeel (Eds.), Knowledge Discovery in Databases: PKDD 2003. Proceedings, 2003. XVI, 508 pages. 2003. (Subseries LNAI).

Vol. 2837: N. Lavrač, D. Gamberger, L. Todorovski, H. Blockeel (Eds.), Machine Learning: ECML 2003. Proceedings, 2003. XVI, 504 pages. 2003. (Subseries LNAI).

Vol. 2836: S. Qing, D. Gollmann, J. Zhou (Eds.), Information and Communications Security. Proceedings, 2003. XI, 416 pages. 2003.

Vol. 2835: T. Horváth, A. Yamamoto (Eds.), Inductive Logic Programming. Proceedings, 2003. X, 401 pages. 2003. (Subseries LNAI).

Vol. 2834: X. Zhou, M. Xu, S. Jähnichen, J. Cao (Eds.), Advanced Parallel Processing Technologies. Proceedings, 2003. XIV, 679 pages. 2003.

Vol. 2833: F. Rossi (Ed.), Principles and Practice of Constraint Programming – CP 2003. Proceedings, 2003. XIX, 1005 pages. 2003.

Vol. 2832: G.D. Battista, U. Zwick (Eds.), Algorithms - ESA 2003. Proceedings, 2003. XIV, 790 pages. 2003.

Vol. 2830: F. Pfenning, Y. Smaragdakis (Eds.), Generative Programming and Component Engineering. Proceedings, 2003. IX, 397 pages. 2003.

Vol. 2828: A. Lioy, D. Mazzocchi (Eds.), Communications and Multimedia Security. Proceedings, 2003. VIII, 265 pages. 2003.